Indian Statistical Institute Series

The **Indian Statistical Institute Series**, a Scopus-indexed series, publishes high-quality content in the domain of mathematical sciences, bio-mathematics, financial mathematics, pure and applied mathematics, operations research, applied statistics and computer science and applications with primary focus on mathematics and statistics. Editorial board comprises of active researchers from major centres of the Indian Statistical Institute.

Launched at the 125th birth Anniversary of P.C. Mahalanobis, the series will publish high-quality content in the form of textbooks, monographs, lecture notes and contributed volumes. Literature in this series are peer-reviewed by global experts in their respective fields, and will appeal to a wide audience of students, researchers, educators, and professionals across mathematics, statistics and computer science disciplines.

More information about this series at https://link.springer.com/bookseries/15910

Surajit Borkotokey · Rajnish Kumar ·
Diganta Mukherjee · K. S. Mallikarjuna Rao ·
Sudipta Sarangi
Editors

Game Theory and Networks

New Perspectives and Directions

Springer

Editors
Surajit Borkotokey
Department of Mathematics
Dibrugarh University
Dibrugarh, Assam, India

Rajnish Kumar
Department of Economics
Queen's University Belfast
Belfast, UK

Diganta Mukherjee
Sampling and Official Statistics Unit
Indian Statistical Institute
Kolkata, West Bengal, India

K. S. Mallikarjuna Rao
IEOR
Indian Institute of Technology Bombay
Mumbai, Maharashtra, India

Sudipta Sarangi
Department of Economics
Virginia Tech
Blacksburg, VA, USA

ISSN 2523-3114 ISSN 2523-3122 (electronic)
Indian Statistical Institute Series
ISBN 978-981-16-4736-9 ISBN 978-981-16-4737-6 (eBook)
https://doi.org/10.1007/978-981-16-4737-6

Mathematics Subject Classification: 91-02, 91-06, 90B10, 90B15, 91A10, 91A12, 91A80, 91D30

This Springer imprint is published by the registered company Springer Nature Singapore Pte Ltd.
The registered company address is: 152 Beach Road, #21-01/04 Gateway East, Singapore 189721, Singapore

Organization

Referees

Saurav Bhattacharya
Pascal Billand
Samarjeet Borah
Anindya Chakrabarty
Subhadip Chakrabarty
Young Sub Chun
Pritha Deb
Anirban Ghatak
Loyimee Gogoi
Anindya Goswami
Chandan Jha
Ruben Juarez
Rajnish Kumar
Ratul Lahkar
Amarjyoti Mahanta
Fenyong Meng
Debasish Mishra
Manipushpak Mitra
Conan Mukherji
Josue Ortega
Hans Peters
Sudipta Sarangi
Indranil Sengupta
A. J. Shaiju
Erika Skrabulakova
Zhenshing Zou

Sponsoring Institutions

Dibrugarh University, Dibrugarh, India
UKIERI Program, India and UK

Preface

Game theory is a formal, mathematical discipline that studies interactive decision-making and is especially relevant for situations of competition and cooperation among a group of agents. The applications of game theory are wide and varied ranging, viz., from addressing strategic negotiations in the international polity to analyzing the economic competitions, from explaining socio-economic issues of bargaining and fair distribution of resources to studying and predicting evolution dynamics of the animal kingdom, and from developing strategic mobile ad hoc networks to building models of voting and social choices, to name just a few.

We live in an interconnected world and an understanding of networks is useful for almost all activities ranging from making a simple decision of buying a product to complex policy decisions by governments. With the incorporation of the strategic interactions through the models of game theory, research in networks became even more critical since it helps in understanding how networks form, which networks are stable, and how people connected via networks make decisions. Social networks and their analysis is an interdisciplinary approach encompassing economics, sociology, political science, epidemiology, psychology, computer science, mathematics, biology, etc.

This compilation contains 14 full-length papers that include both expository and research articles built on networks and game theory's recent developments. It includes chapters on network measures and network formation, application of network theory to contagion, biological data and finance and macroeconomics as expository articles. The volume also contains chapters on fair allocation in the context of queueing, rationing, and cooperative games with transferable utilities for engaged researchers. A few survey chapters on non-cooperative game theory, evolutionary game theory, mechanism design, advances in provision of public goods, and social choice theory are also incorporated to cater to the needs of the beginners in the field. This book brings together new research by scholars in different fields in a manner that is accessible to all. We envisage that this will initiate fruitful interdisciplinary research work and open new directions in game theory with spillover effects on research in networks.

Both the expository and the original research articles were carefully selected from the deliberations made in the International Conference on Game Theory and

Networks held at Dibrugarh University, Dibrugarh, India, during September 6–8, 2019, and the International Seminar on Game Theory and Networks held at Dibrugarh University, Dibrugarh, India, during September 13–14, 2018.

We take this opportunity to thank all our authors whose contributions and timely responses in their efforts starting from the initial submissions to the final revisions made this volume a reality in a trying time that is marked by the arduous challenge the pandemic Covid-19 has brought about. We also thank all the referees whose sincere and involved evaluation of the articles was key to maintaining the volume's scientific quality and standard. The conference is a part of a series of seminars and conferences held under the aegis of a project awarded by the UK-India Education and Research Initiatives (UKIERI) to Surajit Borkotokey, Department of Mathematics, Dibrugarh University, Dibrugarh, India, and Rajnish Kumar, Department of Economics, Queen's Management School, Queen's University, UK. We thank UKIERI for their generous funding and the collaborating institutions for providing the material and technical support. We thank the participants in all these conferences coming from different parts of the world to disseminate their research works. Finally, our deep sense of appreciation goes to Springer for including this volume in their series. We believe that the volume will be useful for the postgraduate and doctoral students, researchers, and practitioners working in various sub-disciplines of game theory and networks.

Dibrugarh, India
Belfast, UK
Kolkata, India
Mumbai, India
Blacksburg, USA
September 2020

Surajit Borkotokey
Rajnish Kumar
Diganta Mukherjee
K. S. Mallikarjuna Rao
Sudipta Sarangi

Introduction

GamesNet: The Genesis

The story of this book is closely tied with GamesNet and the small group of people who have been associated with it. GamesNet (gamesnet.in) is a loosely defined group that brings together researchers across disciplines who have a keen interest in anything and everything that relates to networks, in any tangible or intangible form, and are also enthusiastic about Game Theory. The journey began about 8 years ago, i.e., in 2013 at the famed Indian Statistical Institute, Kolkata. From this humble beginning, and eight conferences later, the GamesNet initiative provides a wonderful example of how a real network can evolve. Over the years, GamesNet has organized multiple programs with visitors from different corners of India as well as from different continents of the world. We proudly enumerate these as follows:

i Research meeting on Strategic Network Formation & Evolution, Indian Statistical Institute, Kolkata, 2013.
ii International Seminar on Game Theory and its applications to Social and Economic Networks, Dibrugarh University, Dibrugarh, Assam 2014.
iii Networks and Games, Indian Institute of Technology, Ropar, 2015.
iv International Conclave on Foundations of Network & Game Theory, IGIDR Mumbai, 2016.
v Mini-Conference on Networks and Games, SOSU, ISI Kolkata, 2017.
vi One-day Workshop on Network and Graphical Statistics, SOSU, ISI Kolkata, 2018.
vii International Workshop on Game Theory and Networks, Dibrugarh University, Dibrugarh, Assam, 2018.
viii International Workshop on Game Theory and Networks, Indian Institute of Technology Bombay, Mumbai, 2019.
ix Workshop on Combinatorial Game Theory, Indian Institute of Technology Bombay, Mumbai, 2019.
x International Conference on Game Theory and Networks, Dibrugarh University, Dibrugarh, Assam, 2019.

GamesNet has been fortunate to receive (financial, technical, and moral) support from various organizations including Department of Science and Technology (DST), National Board for Higher Mathematics (NBHM), IISc Mathematics Initiative, UK-India Education and Research Initiative (UKIERI), and our respective universities and institutes. Several well-known researchers including Tamer Basar, Kaushik Basu, Francis Bloch, Kalyan Chatterjee, Bhaskar Dutta, Sanjeev Goyal, Michel Grabisch, Mathew Jackson, Debasish Mishra, Herve Moulin, Y. Narahari, Indrajit Ray, Agneiszka Rusinowska, Arunava Sen, and Georges Zaccour have participated in our activities and encouraged us. The group has also edited two special issues associated with the journal *Studies in Microeconomics*. During the ongoing Covid pandemic, GamesNet initiated a webinar series on Networks and Games and has reached a far greater audience virtually. The Fall 2020 season even concluded with a paper presentation competition for junior scholars.

The present volume is another way to reach out to researchers in different fields. Several of the papers included in the present volume grew out of the presentations given by the respective authors in some of the conferences organized by GamesNet. Till now, GamesNet has largely been India-centric. We believe this book has become another step that will bring together researchers working in these fields not just in India but also across the globe.

Introduction to the Volume

Game theory studies strategic interactions among agents. The umbrella term "agent" means, for example, individuals in social interactions, attributes in case of economic interactions, or nodes in networks that describe an interaction pattern. It may refer to human beings, computers, or entities like firms and countries. Although notions relating to game theory date back to ancient times and can be found in the works of scholars like Sun Tzu and Chanakya or in medieval times in books like *The Prince*, attempts to formalize it are relatively recent. Scholars like Cournot and Bertrand had started to solve formal models and did Emile Borel who got interested in Colonel Blotto types problems in the early twentieth century. However, formal theory in a systematic way started with the pioneering work of mathematician John von Neumann resulting in a book titled *Theory of Games and Economic Behaviour* jointly written with economist Oskar Morgenstern in 1944. Since then, the subject has attracted researchers from diverse fields and has achieved methodological prominence in multiple disciplines.

Playing games and winning them are fundamental human instincts. The game theory follows these same fundamentals craved by competition, strategies, and monetary gains (loss). It is also interesting to see how cooperation among agents emerges to everyone's benefits under competition! Games may be broadly divided into two categories: non-cooperative games and cooperative games. In non-cooperative games, agents maximize individual utility, do not make any binding agreements, and an equilibrium strategy is the best response of an agent against each of her opponents.

In cooperative games, on the other hand, agents try to maximize their utilities by making binding agreements among themselves to share the total output. The objective is to find what underlies sharing rules that would be attractive and feasible using an axiomatic approach.

The purpose of this book was to provide a compilation of the recent developments in various disciplines of game theory that include primarily the topics on network games, social choice, and distributive justice. Given the vast scope of these topics and the advances, the book is not exhaustive but rather a compendium of recent research and reviews that will introduce new work both to those who are new as well as scholars who have been in the field. With this objective in mind, we have divided the book into three parts: Game Theory and Social Choice, Distributive Justice and, Network Theory and Applications. Before discussing the contributions of the individual chapters, we will briefly introduce our vision of these broad themes to the reader.

Social choice theory is the theoretical study of consensus that brings about social welfare by aggregating individual opinions, preferences, interests, or welfares. Social choice combines welfare economics with voting theory. It builds on the individualistic preferences of the members of a society that lead to a collective decision for the benefits of society. The challenge of social choice theory, therefore, lies in converging to a correct social judgment or acceptable group decision that balances the moral and pragmatic considerations of individuals with possibly varying informational inputs.

Laws, institutions, policies, etc., constitute the framework underlying the very structure of a society. They, in turn, create different distributions of benefits and burdens among the stakeholders. These frameworks are instituted by human beings over the entire spread of political arrangements, their cognition, and functioning. The most interesting part is that such frameworks keep evolving over time both within and across different societies. The structure of these frameworks is an important area of study since the distributions of benefits and burdens it offers to influence the societal behavior at large. Rationalizing the different frameworks and/or their resulting distributions so that they can be morally preferable to the individual agents is a primary concern of the study in distributive justice. Thus, the principles of distributive justice provide moral guidance for the political processes and structures that affect the distribution of benefits and burdens in societies, regardless of the terminology they employ.

Network games are graph restricted games where it is assumed that the interactions among the agents exist only through some network. Roger Myerson in 1977 proposed a communication situation where agents' activities and payoffs are determined by the restrictions imposed on them through their direct or indirect connections. A number of other papers in a similar vein were also published around that time. However, the study of networks took a new turn in the 1990s. These were all aimed at understanding the formation of networks, i.e., how social and economic networks emerged instead of taking them simply as given. In 1996, Jackson and Wolinsky proposed a more sophisticated model of network games, where the network structure is also equally important in defining a game. In 2000, Bala and Goyal proposed the non-cooperative framework of network games. A wide range of theoretical models of network formation has been proposed since then in multiple disciplines to address different types of network situations. In recent years, a number of applications of the theory have also been developed to understand wide-ranging phenomena like homophile, terrorism, and contagion.

Organization of the Book

Part I: Game Theory and Social Choice

This part covers six chapters discussing various aspects of game theory and social choice. The book begins with a chapter on evolutionary game theory by Mahanta titled "Replicator Dynamics and Weak Pay-Off Positive Selection Dynamics: An Overview". In many instances, by engaging in repeated bilateral interactions, they "learn" to play strategies that yield higher benefits and lead to the equilibrium behavior. Among the many dynamic behaviors, replicator dynamics is an important and widely discussed one, which also leads to an evolutionary stable strategy. Another important class of dynamics is the weak-payoff positive selection dynamics. In this chapter, the replicator and weak-payoff dynamics are derived followed by a review of some the most important properties providing the reader with a comprehensive overview.

The second chapter, "Linear Games and Complementarity Problems" by Gokulraj and Chandrashekaran, discusses a generalization of the classical two-person matrix games. A linear game consists of a finite-dimensional inner product space V, a self-dual cone K, and a fixed point e in the interior of K. The players choose from $\Delta = \{x \in K : \langle x, e\rangle = 1$ and the payoff to Player 1 is given by $\langle Lx, y\rangle$, where the first player's choice is $x \in \Delta$ and the second player's choice is $y \in \Delta$. The authors discuss several important properties of the symmetrization of linear games (along the lines of von Neumann symmetrization) and their connections with the complementarity problems using mathematical models involving linear algebra and optimization.

In the chapter "Social Preferences and the Provision of Public Goods", Sarangi and Updadhyay consider the ever-important public goods game. Public goods games are characterized by non-rivalry and non-exclusivity. Non-rivalry implies that multiple players can benefit from the public goods and non-exclusivity makes it difficult to prevent any player to enjoy the benefits. Typically in a Nash equilibrium, players free ride. However, following the more recent advances in behavioral economics, if the utility function considers social preferences, then cooperation may emerge. In this paper, the authors survey the impact of social preferences in the public goods game. Moreover, the authors also consider the situation wherein the players are connected by a network and public good provision happens in the network.

In a society consisting of many agents with conflicting interests, the social choice function attempts to aggregate the individual preferences subject to certain basic axioms. Significant efforts have been made to justify several impossibility results, the chief among them being the Arrow's impossibility theorem. An important desirable property of a social choice function is "strategy-proofness". Strategy-proofness makes it unprofitable for agents to misrepresent their preferences. The Gibbard–Satterthwaite theorem makes it necessary to restrict the class of preferences (else the social choice function will be dictatorial) with some constraints. In the chapter "Recent Results on Strategy-Proofness of Random Social Choice Functions", Roy et al. discuss random social choice functions and their strategy-proofness. This chapter surveys some recent results in this area.

The chapter "Assembly Problems" by Gupta and Sarkar considers an exchange problem in which there is a network of locations and a buyer who wishes to buy a subset of these locations. The objective of the buyer is to buy those locations that constitute a path in the network of the desired length. In this chapter, the authors survey two alternative approaches to this assembly problem, viz., bargaining under complete information and exchange with asymmetric information.

The ranking of different objects is present in almost every situation, especially in these days of wide access to information and the need to identify the relevant information through search. In the chapter "On Different Ranking Methods", Rusinowska considers the situation where the objects are connected by a directed graph and discuss various ranking methods.

Part II: Distributive Justice

This part begins with the chapter "The Efficient, Symmetric and Linear Values for Cooperative Games and Their Characterizations" by Goala and Borkotokey. The main objective of cooperative games with transferable utilities is to provide foundations for distributing the worth of a coalition to the agents in the coalition. In this paper, the authors discuss the Egalitarian Shapley (ESL) value for TU games. The second part of the chapter considers a subclass and provides a characterization for the extended generalized ESL value.

The next chapter "New Characterizations of the Discounted Shapley Values" by Boruah deals with a new class of games that form a basis for the kernel of the discounted Shapley value. The advantage of this model is the minimal requirement of two axioms only, whereas the existing models in the literature require more axioms.

In the next chapter "No-envy in the Queueing Problem with Multiple Identical Machines" by Mitra and Mutuswami, the authors consider queueing problems with multiple machines with the no-envy property, which says that no agent is interested in swapping his allocation with another agent. The paper identifies and analyses the no-envy allocations. The Pareto efficient rule and Lorenz optimal allocation rule are also discussed.

The final chapter in this part "Rationing Rules Under Uncertain Claims: A Survey" by Ertemel and Kumar discusses various rules including the proportional, parametric, equal quantile, and expected-waste constrained uniform gains rules. The paper then discusses the axiomatic characterization of these rules when there is uncertainty in the claims.

Part III: Networks Theory and Applications

Part III of this book consists of four chapters. The first chapter "Building Social Networks Under Consent: A Survey" is an expository chapter by Gilles. The basic problem of network formation is to see how networks of strategic agents form when the decision of the strategic agent is to decide on the links with whom she wishes to form. The typical Nash networks approach, however, does not incorporate consent between agents into the formation problem. The major focus of this paper is to look at the situation where link formation requires mutual consent by both agents. The paper discusses different ways to understand consent in the context of the Nash equilibrium and provides a comparison of the results.

The next chapter "Analysis of Biological Data by Graph Theory Approach Searching of Iron in Biological Cells" by Ždímalová et al. presents the graph-theoretic approach to image processing focusing on the biological data. For obtaining segmentation of biological objects, graph cut algorithms along with some extensions are discussed. As an application, the importance of iron segmentation in Alzheimer's disease is also discussed.

In the real world, the networks (both physical as well as economic) are subject to threats and attacks. Such attacks are caused by nature or by humans. In view of this, one is interested in understanding how robust networks are when they face attacks. In this reprint of a paper "How Do You Defend a Network?" by Dziubiński and Goyal, first published in the Theoretical Economics 12 (2017), 331–376, these aspects are discussed.

The chapter "Macroeconomic and Financial Networks: Review of Some Recent Developments in Parametric and Non-parametric Approaches" by Chakrabarti et al. reviews the recent work in this area using the complex networks approach. The underlying network theory in finance and macroeconomics is surveyed. Topics of discussion include propagation of risk, analyzing social networks, as well as empirical work on financial networks.

The title of the book was carefully chosen to include chapters from both classical game theory and networks and also to highlight the new perspectives and dimensions which, we are sure will prove to be useful for students and researchers (in early as well as senior stages) alike and create exciting (further) interest in various novel perspectives. We believe that the book will span this exciting interdisciplinary ground across economics, mathematics, statistics and also physics, computer science, political science, and sociology and not just be useful to readers but also foster collaborative research.

Contents

Network Theory and Applications

About the Editors

Surajit Borkotokey is Professor at the Department of Mathematics, Dibrugarh University, Assam, India. He completed his Masters from the University of Delhi and Ph.D. from Dibrugarh University. His fields of research are cooperative game theory, networks, aggregation and fuzzy set theory. An Indo-US fellow, Prof. Borkotokey did his post-doctoral research at Louisiana State University, USA. He has been Visiting Professor to the Beijing Institute of Technology, China, and the Slovak University of Technology, Bratislava, besides having collaborations with Virginia Tech, Central Florida University and the University of Paris. He has published over 50 research articles as of now. He has been Guest Editor of a special volume of the *Studies in Microeconomics*, in 2015. Currently he has been working on a UKIERI project awarded by the British Council and the Ministry of HRD, India, with Queen's University, Belfast, UK.

Rajnish Kumar is Senior Lecturer in Economics at Queen's Management School at Queen's University, Belfast, UK. He has earned his Ph.D. in Economics from Rice University, Houston, and the Master's degree from the Indian Statistical Institute, New Delhi, India. His research interests include game theory, network economics, mechanism design and welfare economics. His publications include papers in the fields of fair division, network economics, industrial organization and resource allocation. He has published in journals including the *Economic Theory*, the *European Journal of Operational Research*, the *International Journal of Game Theory*, the *Mathematical Social Sciences*, the *Economics Letters* and the *Manchester School*. Dr. Kumar regularly presents in conferences like the world congress of game theory, social choice and welfare, royal economic society and others. He has been teaching undergraduate and graduate courses on game theory, microeconomics, industrial organization and economics policy. Member of professional societies like the Econometric Society, the Society of Social Choice and Welfare, The Game Theory Society and the Society of Economic Design, Dr. Kumar is recipient of several research grants including from the British Council, the Medical Research Council, the British Academy and the International Growth Centre.

Diganta Mukherjee is Associate Professor at the Sampling and Official Statistics Unit of the Indian Statistical Institute (ISI), Kolkata, India. Earlier, he taught at Jawaharlal Nehru University, New Delhi, India; the University of Essex, UK; and ICFAI Business School, India. He also worked as Consultant for NGOs and other Public Action Groups. Earlier, he has been Technical Advisor to Multi-Commodity Exchange (MCX), Mumbai; the Reserve Bank of India; the Securities and Exchange Board of India; National Sample Survey Office; National Accounts Division (Central Statistical Office), UNDP (through the Ministry of Home Affairs) and Ministry of Electronics and Information Technology. Dr. Mukherjee has been involved in several projects with large corporate houses and various ministries of the Government of India and the Government of West Bengal.

Dr. Mukherjee completed his Ph.D. from the ISI, Kolkata, in 1998, with a thesis titled "Essays on Deprivation, Poverty and Well-being". He also completed his Master of Statistics (1995) and Bachelor of Statistics (1993) degrees from the ISI, Kolkata. His research interests are in welfare and development economics in general and multi-dimensional poverty in particular. He also is active in the area of finance, particularly mathematical models for pricing contracts and analysis and modelling of network data. An author of four books, Dr. Mukherjee has also published over 90 research articles in several journals of repute. He was included in the "2000 Outstanding Intellectuals of the 21st Century" list by the International Biographical Centre, Cambridge, UK, in 2011. A recipient of the National Talent Search Award in 1988 by the National Council of Educational Research and Training, Dr. Mukherjee has delivered invited talks and lectures at universities and institutions around the country.

K. S. Mallikarjuna Rao is Professor at the Department of Industrial Engineering and Operations Research, Indian Institute of Technology Bombay (IIT Bombay), India. He completed his Ph.D. in Mathematics from the Indian Institute of Science (IISc), Bengaluru, India, in 2002, and M.Sc. in Mathematics form Andhra University, Visakhapattanam, India. Earlier, he held postdoc positions at the Universite de Province at Marseille, France; the Indian Statistical Institute, Delhi, India; the University of Texas at Dallas, USA; and the TIFR-CAM, Bengaluru, India. His research interests are in control theory, game theory, networks, probability, optimization and mathematical finance. His research articles have been published in journals of repute.

Sudipta Sarangi is Professor and the Head, Department and of Economics, Virginia Tech, Blacksburg, USA. He did his masters from the Delhi School of Economics, India; and Ph.D. from Virginia Tech, USA, in 2000. He also is Research Associate of DIW Berlin, GATE, University of Lyon-St. Etienne and the Lima School of Economics. Earlier, he served as Program Director at the National Science Foundation during 2013–2015. His research interests range from network theory to development economics. His work on networks focuses on the strategic formation of social and economic networks and how participation in multiple networks affects social outcomes. He also serves on the editorial boards of *Journal of Economic Behavior and*

Organization, *Journal of Public Economic Theory* and *Studies in Microeconomics*. He has been awarded the Merit Scholarship by the Delhi School of Economics (1991–1993) and National Talent Search Scholarship by the National Council for Educational Research and Training, the Government of India (1986–1991).

Game Theory and Social Choice

Replicator Dynamics and Weak Pay-Off Positive Selection Dynamics: An Overview

Amarjyoti Mahanta

1 Introduction

In this review essay, we mainly discuss results related to replicator dynamics and weak pay-off positive selection dynamics. Replicator dynamics[1] is one of the ways of studying the long-term outcome of the interactions of agents who play normal form games and who are not fully rational. By "not fully rational" we mean that the agents playing the game do not have the capacity to form a correct belief about the strategy going to be played by their opponent. Therefore, each agent learns the game by playing repeatedly. In replicator dynamics, each strategy of the game is identified with the type of agents who are playing it. For example, if a game has three strategies and each strategy is played by some agents then we say that there are three types of agents playing the game. The strategies with higher pay-off replicate at a higher rate and become prevalent over time. In other words, the agents playing a strategy with higher pay-off grow at a higher rate than those with lower pay-off. In the long run, the agents playing strategy with lower pay-off get eliminated and only those with higher pay-off prevail. This kind of dynamics belongs to a broader class of learning dynamics called pay-off monotone dynamics. In this kind of dynamics, the ordering of the rate of growth of strategies is the same as the ordering of the pay-off. The strategies with the highest pay-off grow at the highest growth rate and vice versa.

The growth rate of each strategy in replicator dynamics is called the fitness function in the terminology of evolutionary biology. It represents the aggregate behaviour

[1]Taylor and Jonker [13] is the first paper to derive replicator dynamics.

I am grateful to Prof. Surajit Borkotokey, Prof. Sourav Bhattacharya, editors of this volume and an anonymous referee for the important and helpful comments. I thank the participants of International Conference on Game Theory and Network held at Dibrugarh University for helpful comments.

A. Mahanta (✉)
Department of Humanities and Social Sciences, IIT Guwahati, Guwahati, India
e-mail: amarjyoti@iitg.ac.in

S. Borkotokey et al. (eds.), *Game Theory and Networks*, Indian Statistical Institute Series,
https://doi.org/10.1007/978-981-16-4737-6_1

of all the agents playing a particular strategy. The micro-foundation of replicator dynamics can be established in multiple ways and imitation is one of them, because the strategy with higher pay-off is imitated more often, leading to greater replication.

Replication through imitation has been studied using numerous revision protocols, broadly categorized as imitation driven by dissatisfaction and imitation of successful strategies. In these classes of learning dynamics, a type of agent that is following a particular strategy switches to another strategy if the pay-off from the current strategy is less than a threshold level. Here, the agents only need to know the threshold level and the pay-off from its present strategy before switching. In the imitation of a successful strategy, the agents do a pairwise comparison of pay-offs. The agents are randomly matched to play the game. Of the two strategies, the agent switches to one with higher pay-off. The information requirement is not that stringent; the agent has to know about the pay-off of its current strategy and its opponent. Based on this criterion of switching, a number of models have been studied. The consequence of this inflow and outflow of the agents into strategies using the switching criteria leads to the replicator kind of dynamics.

Another class of selection dynamics which is generally compared to replicator dynamics is weak pay-off positive selection dynamics. In such dynamics, at least one of the strategies has a positive rate of growth from the set of strategies having pay-off greater than the expected pay-off. So all the strategies with pay-off greater than the average may not grow at a positive rate. And strategies with lesser pay-off may grow at a higher rate compared to those with a higher pay-off. Sethi [12] provided the micro-foundation of such dynamics based on the imitation of successful strategies, calling it generalized replicator dynamics.

In this review essay[2], we describe the derivation of replicator dynamics and weak pay-off positive selection dynamics based on imitation of successful strategies. We discuss some of the common properties of selection dynamics such as Lipschitz continuity and the fact that the trajectories are bounded within the unit simplex. All the Nash equilibria of a normal form game belong to the set of fixed points of these two dynamics. There are also, however, a few fixed points which are not Nash equilibria. We demonstrate the results showing that those fixed points are not Lyapunov stable. The limitations of the results related to the stability of fixed points are shown. For potential games, the Nash equilibria are asymptotically stable in the case of replicator dynamics. And for the general normal form games, only the evolutionarily stable states which are also fixed points of the replicator dynamics are asymptotically stable. In the case of weak pay-off positive selection dynamics, only monomorphic evolutionarily stable states of any normal form game are asymptotically stable. The presence of cycles in replicator dynamics for the Rock Paper Scissors game has been demonstrated. An example of a stable limit cycle in case of weak pay-off positive selection dynamics is provided.

---

[2] Some of the interesting and expository papers on evolutionary games are Friedman [3], Van Damme [14], Hofbauer and Sigmund [6] and Lahkar [7]. For comprehensive exposition on replicator dynamics and all sorts of imitation dynamics, the reader may refer to Weibull [16], Hofbauer and Sigmund [5], Vega Rodendo [15] and Sandholm [10, 11].

The structure of the paper is as follows: in Sect. 2, we build the model and derive both the selection dynamics from imitation protocols. In Sect. 3, we present the results on the nature of fixed points of these two dynamics and their stability.

2 Model

In this section, we explain how the aggregate imitation dynamics is generated from the simple imitative behaviour of the agents. For simplicity, we consider a symmetric normal form game with three strategies.[3] The pay-off matrix is given below as

$$G = \begin{pmatrix} a_{11} & a_{12} & a_{13} \\ a_{21} & a_{22} & a_{23} \\ a_{31} & a_{32} & a_{33} \end{pmatrix}$$

The social state or population state is given by vector $x = (x_1, x_2, x_3)$ where $x \in [0, 1]^3$ and $x_1 + x_2 + x_3 = 1$. Here, x_i $i \in \{1, 2, 3\}$ where an element of x denotes the fraction of agents playing strategy i. The population state x belongs to the set Δ, $\Delta = \{x : x \in [0, 1]^3 \text{and} \sum_{i=1}^{3} x_i = 1\}$. The strategies are denoted as s_i, $i \in \{1, 2, 3\}$. The corresponding pay-offs are $\pi_1 = a_{11}x_1 + a_{12}x_2 + a_{13}x_3$, $\pi_2 = a_{21}x_1 + a_{22}x_2 + a_{23}x_3$ and $\pi_3 = a_{31}x_1 + a_{32}x_2 + a_{33}x_3$. The expected pay-off of a normal form game G is $\bar{\pi} = \pi_1 x_1 + \pi_2 x_2 + \pi_3 x_3$.

We assume that a continuum of agents is playing the above normal form game. At any time t, agents are randomly matched in pairs to play the game.[4] At time t, x_i, the fraction of agents are playing strategy s_i, $i \in \{1, 2, 3\}$ or x_i fractions of agents are s_i type. An s_1 type of agent can be matched with s_1 type also; the probability of s_1 matched with s_1 is $x_1(t)x_1(t)$ at time t. An s_1 type can be matched with s_2 type; the probability of this event is $x_1(t)x_2(t)$ at time t. An s_1 type can be matched with s_3 type too, the probability of this event being $x_1(t)x_3(t)$ at time t.

When the agents are matched in pairs, each type of agent can compare the pay-off from its own strategy and its opponent. If the pay-off of the strategy of its opponent is higher than its own, then it may switch to the strategy of its opponent, otherwise, it sticks to its own strategy. If similar types of agents are matched, they stick to their own strategy. This imitation rule is called the imitation of successful strategies which is followed by the agents when they are matched in pairs at any time t. When $x_i(t)x_j(t)$ fraction of agents of s_i and s_j types are matched at t, if $\pi_i(t) \geq \pi_j(t)$ then $h(\pi_i - \pi_j)$ fraction of $x_i(t)x_j(t)$ fraction switch to strategy s_i. This $h(\pi_i - \pi_j)$ is a strictly increasing function of $\pi_i - \pi_j$ such that $0 \leq h(\pi_i - \pi_j) \leq 1$ for all values of π_i and π_j, $i, j \in \{1, 2, 3\}$ provided $\pi_i - \pi_j \geq 0$. The $h()$ function is defined for non-negative values only. The probability of switching is taken as a function of the difference of the pay-offs of the matched pair. The information requirement of this

[3] In this paper, we describe all the results by taking symmetric normal form games with three strategies.

[4] This description of imitation protocol is taken from Hofbauer and Sigmund [5].

protocol is very minimal. The agents only have to know about the pay-off of their own strategy and that of their opponents. For a detailed understanding of the informational requirements of different imitation protocols when agents are randomly matched in pairs, see [10, 11].

From the above argument, we get that when $\pi_i(t) \geq \pi_j(t)$, $x_1(t)x_2(t)h(\pi_i(t) - \pi_j(t))$ fraction of agents switch from strategy s_j to strategy s_i. So $x_1(t)x_2(t)h(\pi_i(t) - \pi_j(t))$ are inflows to strategy s_i and at the same time the outflows for the strategy s_j. Taking all these inflows and outflows together at time t, if the pay-offs are $\pi_1(t) \geq \pi_2(t) \geq \pi_3(t)$, we get the following aggregate dynamics of the populated state,

$$\begin{aligned} \dot{x_1}(t) &= x_1(t)x_2(t)h(\pi_1(t) - \pi_2(t)) + x_1(t)x_3(t)h(\pi_1(t) - \pi_3(t)) \\ \dot{x_2}(t) &= -x_1(t)x_2(t)h(\pi_1(t) - \pi_2(t)) + x_2(t)x_3(t)h(\pi_2(t) - \pi_3(t)) \\ \dot{x_3}(t) &= -x_1(t)x_3(t)h(\pi_1(t) - \pi_3(t)) - x_2(t)x_3(t)h(\pi_2(t) - \pi_3(t)) \end{aligned} \tag{1}$$

Here, $x_1(t)x_2(t)h(\pi_1(t) - \pi_2(t))$ term is inflow from strategy s_2 to s_1 and $x_1(t)x_3(t)h(\pi_1(t) - \pi_3(t))$ is the inflow from strategy s_3 to s_1. As at time t the pay-offs are $\pi_1(t) \geq \pi_2(t) \geq \pi_3(t)$, s_1 strategy will have inflows only. The strategy s_2 will have inflows from strategy s_3 and outflows to strategy s_1. But strategy s_3 will have outflows only. Had the pay-offs at time t been $\pi_2(t) \geq \pi_3(t) \geq \pi_1(t)$, the strategy s_1 will have outflows only and strategy s_2 will have inflows only. The strategy s_3 will have inflows from s_1 and outflows to s_2.

In the literature[5] for simplicity, the $h()$ function has been taken as $h(\pi_i(t) - \pi_j(t)) = [\pi_i - \pi_j]_+$. With the incorporation of this change in the equation system (1), we get the general aggregate dynamics for all combinations of pay-offs as

$$\begin{aligned} \dot{x_1}(t) &= x_1(t)x_2(t)[\pi_1(t) - \pi_2(t)] + x_1(t)x_3(t)[\pi_1(t) - \pi_3(t)] \\ \dot{x_2}(t) &= x_1(t)x_2(t)[\pi_2(t) - \pi_1(t)] + x_2(t)x_3(t)[\pi_2(t) - \pi_3(t)] \\ \dot{x_3}(t) &= x_1(t)x_3(t)[\pi_3(t) - \pi_1(t)] + x_2(t)x_3(t)[\pi_3(t) - \pi_2(t)] \end{aligned} \tag{2}$$

With a little manipulation in the equation system (2), we get the replicator dynamics

$$\begin{aligned} \dot{x_1}(t) &= x_1(t)[\pi_1(t) - \bar{\pi}(t)] \\ \dot{x_2}(t) &= x_2(t)[\pi_2(t) - \bar{\pi}(t)] \\ \dot{x_3}(t) &= x_3(t)[\pi_3(t) - \bar{\pi}(t)] \end{aligned} \tag{3}$$

Thus, replicator dynamics can be derived from the aggregate imitation dynamics where the imitation is based on the imitation of successful strategies. It is shown in Weibull [16] and Sandholm [10, 11].

[5] See Sandholm [10, 11], Weibull [16] and Sethi [12].

Next, we discuss the imitation driven by dissatisfaction. In imitation driven by dissatisfaction, an agent following a particular strategy may switch to any other strategy if the pay-off from the present strategy is less than a threshold level. Again in such settings, agents are randomly matched. So s_1 can be matched with s_1 with probability $x_1(t)x_1(t)$ at t. Similarly, s_1 can be matched with s_2 and s_3 at time t with probability $x_1(t)x_2(t)$ and $x_1(t)x_3(t)$, respectively. If the pay-off of s_1 is $\pi_1(t) < k$ where k is the threshold level, then the agent may switch to any other strategy. For example, when s_1 and s_2 are matched and pay-off of strategy s_1 is $\pi_1(t) < k$, then the agent playing s_1 strategy may switch to s_2 or s_3 strategy. The probability of switching to any strategy is $h(k - \pi_1(t))$ for s_1 type of agents at time t. The $h()$ function is same as defined earlier in this section. The consequence of this is that if $\pi_1(t) < k$ at time t, then the outflows from strategy s_1 to strategy s_2 and s_3 are $x_1(t)x_2(t)h(k - \pi_1(t))$ and $x_1(t)x_3(t)h(k - \pi_1(t))$, respectively. These are inflows to s_2 and s_3 strategies. For a sufficiently higher value of k, there is such inflow and outflow to each strategy and the resulting dynamics is as follows:

$$\begin{aligned}
\dot{x_1}(t) =& x_1(t)x_2(t)h(k - \pi_2(t)) + x_1(t)x_3(t)h(k - \pi_3(t)) \\
&- x_1(t)x_2(t)h(k - \pi_1(t)) - x_1(t)x_3(t)h(k - \pi_1(t)) \\
\dot{x_2}(t) =& x_1(t)x_2(t)h(k - \pi_1(t)) + x_2(t)x_3(t)h(k - \pi_3(t)) \\
&- x_1(t)x_2(t)h(k - \pi_2(t)) - x_2(t)x_3(t)h(k - \pi_2(t)) \\
\dot{x_3}(t) =& x_1(t)x_3(t)h(k - \pi_1(t)) + x_2(t)x_3(t)h(k - \pi_2(t)) \\
&- x_1(t)x_3(t)h(k - \pi_3(t)) - x_2(t)x_3(t)h(k - \pi_3(t))
\end{aligned} \tag{4}$$

In the literature, the $h(.)$ function is taken as $h(k - \pi_i(t)) = [k - \pi_i(t)]_+$. By substituting it in Eq. (4) and with a little manipulation, we get the replicator equation:

$$\begin{aligned}
\dot{x_1}(t) &= x_1(t)[\pi_1(t) - \bar{\pi}(t)] \\
\dot{x_2}(t) &= x_2(t)[\pi_2(t) - \bar{\pi}(t)] \\
\dot{x_3}(t) &= x_3(t)[\pi_3(t) - \bar{\pi}(t)]
\end{aligned} \tag{5}$$

Thus, imitation driven by dissatisfaction also provides the micro-foundation of replicator dynamics. The detailed derivation is given in Weibull [16] and Sandholm [10, 11].

One of the generalizations of replicator dynamics is done in the following way. Sethi [12] introduces strategy-specific barriers in the model of imitation of successful strategies. Each strategy in the game G as defined earlier has a barrier. When an agent tries to switch to a strategy, it may not be successful always. A probability is attached to the successful imitation of each strategy. It may be different for each strategy so this is strategy-specific. Each strategy s_i has $\lambda_i,\ \lambda_i \in (0, 1),\ i \in \{1, 2, 3\}$ probability of being successfully imitated.

Drawing from the argument provided in the derivation of the imitation of successful strategies, we obtain that if $\pi_1(t) > \pi_2(t)$ at time t, then $x_1(t)x_2(t)h(\pi_1(t) - \pi_2(t))$ fraction of agents are going to switch from s_2 to s_1 strategy. Now with the introduction of barriers to imitation, we get that only $x_1(t)x_2(t)h(\pi_1(t) - \pi_2(t))\lambda_1$ fraction can successfully imitate strategy s_1 which are switching strategy s_2, so inflow to strategy s_1 is $x_1(t)x_2(t)h(\pi_1(t) - \pi_2(t))\lambda_1$. If $\pi_1(t) < \pi_2(t)$, then the inflow to strategy s_2 from strategy s_1 is $x_1(t)x_2(t)h(\pi_1(t) - \pi_2(t))\lambda_2$. Incorporating this into Eq. (1), we get the aggregate dynamics when $\pi_1(t) \geq \pi_2(t) \geq \pi_3(t)$ as

$$\begin{aligned} \dot{x_1}(t) &= x_1(t)x_2(t)h(\pi_1(t) - \pi_2(t))\lambda_1 + x_1(t)x_3(t)h(\pi_1(t) - \pi_3(t))\lambda_1 \\ \dot{x_2}(t) &= -x_1(t)x_2(t)h(\pi_1(t) - \pi_2(t))\lambda_1 + x_2(t)x_3(t)h(\pi_2(t) - \pi_3(t))\lambda_2 \\ \dot{x_3}(t) &= -x_1(t)x_3(t)h(\pi_1(t) - \pi_3(t))\lambda_1 - x_2(t)x_3(t)h(\pi_2(t) - \pi_3(t))\lambda_2 \end{aligned} \tag{6}$$

As done earlier, we take $h(\pi_i(t) - \pi_j(t)) = [\pi_i(t) - \pi_j(t)]_+$ in the equation system (6). In this scheme of imitation, the switching takes place when agents are matched in these three pairs (s_1, s_2), (s_1, s_3) (s_2, s_3). For notational convenience, we define a set to denote the agent with higher pay-off among a pair of agents when they are matched $B_{ij} = \{i : if\ \pi_i \geq \pi_j, i, j \in \{1, 2, 3\}$ and $i \neq j\}$. Using these two things, the equation system (6) can be written in a more compact form as

$$\begin{aligned} \dot{x_1}(t) &= x_1(t)x_2(t)[\pi_1(t) - \pi_2(t)]\lambda_{i(i\in B_{12})} + x_1(t)x_3(t)[\pi_1(t) - \pi_3(t)]\lambda_{i(i\in B_{13})} \\ \dot{x_2}(t) &= x_1(t)x_2(t)[\pi_2(t) - \pi_1(t)]\lambda_{i(i\in B_{12})} + x_2(t)x_3(t)[\pi_2(t) - \pi_3(t)]\lambda_{i(i\in B_{23})} \\ \dot{x_3}(t) &= x_1(t)x_3(t)[\pi_3(t) - \pi_1(t)]\lambda_{i(i\in B_{13})} + x_2(t)x_3(t)[\pi_3(t) - \pi_2(t)]\lambda_{i(i\in B_{23})} \end{aligned} \tag{7}$$

This is the generalized replicator dynamics which is derived from the imitation of successful strategies with strategy-specific barriers. If all the λ_i are the same, then we get the dynamics of (7) to be exactly the same as replicator dynamics. The important thing to note in the equation system (7) is that the ordering of the rate of growth of strategies may not be the same as the ordering of the pay-offs. However, at least one of the strategies with pay-off greater than the expected value will have a positive growth rate. This is a version of weak pay-off positive selection dynamics. In the preceding section, we discuss the results related to replicator dynamics and weak pay-off positive selection dynamics.

3 Results

At first, we look at the results related to replicator dynamics. Replicator dynamics is part of a larger class of selection dynamics called monotone selection dynamics. The selection dynamics are characterized in the following way. A selection dynamics given by the differential equation $\dot{x}_i = x_i f(x)$ must satisfy the following:

1. Lipschitz continuity;
2. $\sum x_i f_i(x) = 0$;
3. $x_i = 0$ implies $x_i f_i(x) \geq 0$.

Lipschitz continuity ensures that the dynamical system has a unique solution. Since replicator dynamics given in the equation system (3) is continuously differentiable in all the variables (x_1, x_2, x_3), we get that the replicator dynamics has a unique solution.

The second condition says that the unit simplex is the invariant set under the selection dynamics. In the case of replicator dynamics with three strategies, it means $x_1(t)[\pi_1(t) - \bar{\pi}(t)] + x_2(t)[\pi_2(t) - \bar{\pi}(t)] + x_3(t)[\pi_3(t) - \bar{\pi}(t)] = 0$. This is true because $x_1(t) + x_2(t) + x_3(t) = 1$ when $x \in \Delta$. The implication of this condition is that the trajectories of the replicator are bounded within the unit simplex. The third condition means that when population share of a particular strategy is zero, there cannot be a further decrease in the population share of that strategy. These results are shown in Samuelson and Zhang [9], Weibull [16], Cressman [2] and Vega Redondo [15].

The dynamical system given in Eq. (7) also satisfies all the above three conditions. In Sethi [12], it is shown that the dynamical system (7) is Lipschitz continuous. Another way to show Lipschitz continuity in the dynamical system (7) is, each term in each equation of the dynamical system (7) has a max function. We know that a max function is not continuously differentiable but satisfies Lipschitz continuity. Therefore, the weak pay-off positive selection dynamics given by the dynamical system (7) is Lipschitz continuous. Note that the other two conditions are easily met by the weak pay-off positive dynamics given in the dynamical system (7). Based on these conditions, we claim that the dynamical system (7) has a unique solution and the trajectories are bounded within the unit simplex.

Another important characteristic of these two dynamical systems is that the equations are invariant if pay-offs are additively shifted by a common number. This is because the dynamical system is defined based on the difference of pay-offs.

3.1 *Characteristics of Fixed Points*

We now discuss the results related to the fixed points[6] of replicator dynamics and weak pay-off positive selection dynamics. The Nash equilibria of any normal form game are fixed points of the replicator dynamics. All the pure strategy Nash equilibria of a normal form game are monomorphic in nature. A monomorphic population is one in which all the agents follow a single strategy or in other words, only one type of agent are present in the population. For example, in game G given in Sect. 2, $(1, 0, 0)$

[6] It is easy to see that the origin is always a fixed point in these two dynamical systems. The initial point of these two dynamical systems belongs to the unit simplex and as already shown that the trajectories of these two dynamical systems are bounded within the unit simplex. So we are not concerned about the origin as the fixed point of these dynamical systems.

is a monomorphic state which can be a Nash equilibrium and in this pure strategy Nash equilibria all the agents play strategy s_1. Suppose $(1, 0, 0)$ is a Nash equilibrium of game G. In replicator dynamics given by equation system (3), $\dot{x}_2 = 0$ and $\dot{x}_3 = 0$ since $x_2 = 0$ and $x_3 = 0$. And $\dot{x}_1 = 0$ because $\pi_1 = \bar{\pi}$ at $(1, 0, 0)$. In other words, the pay-off of strategy s_1 is the same as the expected pay-off. Thus, pure strategy Nash equilibria of a normal form game are fixed points of the replicator dynamics.

Suppose (x_1^*, x_2^*, x_3^*) is a mixed strategy Nash equilibrium of game G. The mixed strategy Nash equilibria are polymorphic states which means that there is a presence of more than one type of agent. Since (x_1^*, x_2^*, x_3^*) is a mixed strategy Nash equilibrium, the pay-off must be the same from these three strategies, that is $\pi_1 = \pi_2 = \pi_3$ at (x_1^*, x_2^*, x_3^*). It implies that $\pi_1 = \pi_2 = \pi_3 = \bar{\pi}$ at (x_1^*, x_2^*, x_3^*). It is obvious that $\dot{x}_1 = 0,\ \dot{x}_2 = 0,\ \dot{x}_3 = 0$ for the dynamical system (3) at (x_1^*, x_2^*, x_3^*) since $\pi_1 = \pi_2 = \pi_3 = \bar{\pi}$. The mixed strategy Nash equilibria are also the fixed points of the replicator dynamics.

It is obvious that there can be monomorphic states which are fixed points of the replicator dynamics given in (3) but may not be a Nash equilibrium. Is this true for a polymorphic state also? Suppose (x_1^*, x_2^*, x_3^*) is a fixed point of (3) but not a Nash equilibrium. Since it is a fixed point, $\dot{x}_1 = 0,\ \dot{x}_2 = 0,\ \dot{x}_3 = 0$ at (x_1^*, x_2^*, x_3^*). Again $(x_1^*, x_2^*, x_3^*) > 0$, so $\pi_1 = \bar{\pi}$ for $\dot{x}_1 = 0$, $\pi_2 = \bar{\pi}$ for $\dot{x}_2 = 0$ and $\pi_3 = \bar{\pi}$ for $\dot{x}_3 = 0$. This implies $\pi_1 = \pi_2 = \pi_3 = \bar{\pi}$ at (x_1^*, x_2^*, x_3^*). Thus, (x_1^*, x_2^*, x_3^*) is a mixed strategy Nash equilibrium. A polymorphic state fixed point of the replicator dynamics must be a Nash equilibrium. We find these results in Hofbauer and Sigmund [5].

Another type of Nash equilibria are partially mixed strategy Nash equilibria which are also polymorphic states. These Nash equilibria are also fixed points of replicator dynamics. Suppose $x^* = (x_1^*, x_2^*, 0)$ is a partially mixed strategy Nash equilibrium. It implies that $\pi_1 = \pi_2 \geq \pi_3$ at $(x_1^*, x_2^*, 0)$. $x_3 = 0$ implies $\bar{\pi} = \pi_1 = \pi_2$. Therefore, $\dot{x}_1 = 0$, $\dot{x}_2 = 0$ and $x_3^* = 0$ implies $\dot{x}_3 = 0$. Thus, x^* is a fixed point of replicator dynamics. It is easy to see that the converse is always true.

Do we see similar kinds of rest points with regard to weak pay-off positive selection dynamics? It is easy to see that all the monomorphic states are fixed points of the dynamical system (7), the same as replicator dynamics. Thus, pure strategy Nash equilibria are fixed points of the dynamical system (7). Suppose (x_1^*, x_2^*, x_3^*) is a polymorphic state which is a mixed strategy Nash equilibrium. So $\pi_1 = \pi_2 = \pi_3$ at (x_1^*, x_2^*, x_3^*), implying that each term of each equation in the dynamical system (7) is zero. Thus, (x_1^*, x_2^*, x_3^*) is a fixed point of the dynamical system (7). To see whether the converse is true, we take (x_1^*, x_2^*, x_3^*) to be a fixed point of the dynamical system (7). Suppose it is not a mixed strategy Nash equilibrium. In a dynamical system (7), for this to be true, we need $x_2(t)[\pi_1(t) - \pi_2(t)]\lambda_{i(i \in B_{12})} + x_3(t)[\pi_1(t) - \pi_3(t)]\lambda_{i(i \in B_{13})} = 0$ in the first equation of the dynamical system (7). One of the ways the first equation of the dynamical system (7) equals zero is when the pay-offs are $\pi_3 > \pi_1 > \pi_2$ at (x_1^*, x_2^*, x_3^*). Now $\pi_1 > \pi_2$ at (x_1^*, x_2^*, x_3^*) implies that for the second equation of the dynamical system (7) to be equal to zero, $\pi_2 > \pi_3$. So we get the ordering of the pay-offs as $\pi_1 > \pi_2 > \pi_3$. We cannot have two orderings of the pay-offs at the same point

(x_1^*, x_2^*, x_3^*) . Thus, (x_1^*, x_2^*, x_3^*) must be a mixed strategy Nash equilibrium. The polymorphic fixed points of the dynamical system (7) are always Nash equilibria.

Similarly, we obtain that partially mixed strategy Nash equilibria are also the fixed points of weak pay-off positive selection dynamics. A polymorphic fixed point of weak pay-off positive selection dynamics lying in the boundary of the unit simplex is a partially mixed strategy Nash equilibrium. Suppose $(x_1^*, x_2^*, 0)$ is a fixed point of the dynamical system (7). Here $x_3 = 0$ ensures $\dot{x}_3 = 0$ and $\dot{x}_1 = x_1 x_2[\pi_1 - \pi_2]\lambda_{i(i \in B_{12})}$, $\dot{x}_2 = x_1 x_2[\pi_2 - \pi_1]\lambda_{i(i \in B_{12})}$. Since $(x_1^*, x_2^*, 0)$ is a fixed point, $\dot{x}_1 = 0$ and $\dot{x}_2 = 0$. It implies $\pi_1 = \pi_2$ at $(x_1^*, x_2^*, 0)$. Suppose $\pi_1 = \pi_2 < \pi_3$ at $(x_1^*, x_2^*, 0)$. There is a neighbourhood of $(x_1^*, x_2^*, 0)$ such that the pay-offs are $\pi_1 = \pi_2 < \pi_3$ because of the continuity of the pay-off function. In this neighbourhood, $\dot{x}_3 > 0$ since $\pi_1 = \pi_2 < \pi_3$. Again using continuity, we get that $\dot{x}_3 > 0$ at $(x_1^*, x_2^*, 0)$, which leads to a contradiction. Therefore, $\pi_1 = \pi_2 < \pi_3$ is not possible, implying $\pi_1 = \pi_2 \geq \pi_3$. This results in $(x_1^*, x_2^*, 0)$ to be a partially mixed strategy Nash equilibrium when $(x_1^*, x_2^*, 0)$ is a fixed point of the dynamical system (7).

We now analyse the trajectories of the replicator dynamics and weak pay-off positive dynamics. We ask two kinds of questions in this topic. Firstly, if the trajectories are converging to a particular point, what are the characteristics of such a point? The second question is to ascertain whether the trajectories converge to a point or exhibit cyclical behaviour. In Sect. 3.2, we take care of the latter question.

In the case of replicator dynamics, we see that if a fixed point is Lyapunov stable then it must be a Nash equilibrium. This is shown in Hofbauer and Sigmund [5] and Bomze [1]. Suppose $x^* = (x_1^*, x_2^*, x_3^*)$ is Lyapunov stable and not a Nash equilibrium. Since x^* is not a Nash equilibrium, there must be at least one strategy s_i such that $\pi_i(x^*) > \bar{\pi}(x^*)$ at x^*. From the continuity of x, we get that there exist ϵ and a strategy i such that $\pi_i(x^*) - \bar{\pi}(x^*) > \epsilon$ for all x around the ϵ neighbourhood of x^*. This implies that in this ϵ neighbourhood, the growth rate of x_i is positive. Thus, the trajectories have the tendency to move away from the point x^*, therefore it cannot be Lyapunov stable.

A result obtained by Akin (1980) and Samuelson and Zhang [9] shows that a dominated pure strategy will vanish in the long run. The fraction of the population playing a strictly dominated pure strategy or a weakly dominated pure strategy is going to be zero if initially a positive fraction of agents are playing it. An easier proof of this result is given in Cressman [2], Hofbauer and Sigmund [5] and Weibull [16]. Suppose the strategy s_i is dominated by the strategy s_j and s_j is not dominated by any other strategy, then we show the limit of $\dfrac{x_i(t)}{x_j(t)} \longrightarrow 0$ as $t \longrightarrow \infty$. The time derivative of $\dfrac{x_i(t)}{x_j(t)}$ is $\dfrac{x_j(t)x_i(t)[\pi_i(t) - \pi_j(t)]}{x_j(t)^2}$. Since s_i is dominated by s_j, $\pi_i < \pi_j$ for all x, implying $\dfrac{x_j(t)x_i(t)[\pi_i(t) - \pi_j(t)]}{x_j(t)^2} < 0$ for all t . Thus, the limit of $\dfrac{x_i(t)}{x_j(t)} \longrightarrow 0$ as $t \longrightarrow \infty$. It implies that the fraction of agents playing a dominated strategy goes to zero in the long run. Samuelson and Zhang [9] proved further that if a pure strategy is iteratively strictly dominated, then the fraction of agents playing it vanish. The proof is a bit involved so we do not reproduce a version of it here. The idea is more or less similar to the argument related to the dominated strategy. A non-Nash equilibrium cannot be Lyapunov stable under replicator dynamics.

In the case of weak pay-off positive selection dynamics given by the dynamical system (7), a Lyapunov stable state is always a Nash equilibrium. Here again, the argument is similar to replicator dynamics. The proof is obtained using the definition of a Nash equilibrium and continuity in Weibull [16].

3.2 *Stability of Fixed Points*

Next, we present the results related to convergence in these two types of dynamical systems. Hofbaur and Sigmund [5] showed that for a partnership game, the expected pay-off $\bar{\pi}$ is a strict Lyapunov function and the maximum of the expected pay-off is asymptotically stable which is a Nash equilibrium. In a normal game like G, if $a_{ij} = a_{ji}$ then G is a partnership game. A normal form game with a symmetric pay-off matrix is a partnership game. This type of normal form gameis also called a potential game. Partnership games and potential games are similar to normal form games. We demonstrate the convergence result in Appendix. The trajectories converge to the local maximum of the expected pay-off which is a Nash equilibrium. For the trajectories to converge to the local maximum, another requirement is that the initial point should not be a non-Nash equilibrium fixed point. These types of fixed points lie in the boundary of the unit simplex.

A population state x^* is defined[7] as an evolutionarily stable state if the two conditions are satisfied: a) $xGx^* \leq x^*Gx^*$ for all $x \in \Delta$; b) if $x \neq x^*$ and $xGx^* = x^*Gx^*$, then $xGx < x^*Gx$ where G is the pay-off matrix of game G. It is a subset of Nash equilibrium. Strict Nash equilibria are evolutionarily stable states. Weak Nash equilibria are not evolutionarily stable states. The evolutionarily stable states are fixed points of the replicator and weak pay-off positive selection dynamics. The stability of the fixed points of the replicator dynamics of a general normal form game is attained only for evolutionarily stable strategies.

If $x^* \in \Delta$ is an evolutionarily stable state, then it is asymptotically stable. This is shown using the relative entropy function in Taylor and Jonkar [13], and Hofbauer, Schuster and Sigmund [4]. Suppose $x^* \in \Delta$ is an evolutionarily stable state of a game with the pay-off matrix G. The relative entropy function is defined as
$H_{x^*}(x) = \sum_{i=1}^{3} x_i^* log(\frac{x_i^*}{x_i})$ where $x \in \Delta$ and x^* is an evolutionarily stable state. The result is demonstrated in Appendix. We do not have a general convergence result in the case of replicator dynamics. So, weak Nash equilibria which are also fixed points of replicator dynamics may not be asymptotically stable.

[7] This definition is taken from Hofbauer and Sigmund [5].

We next describe the stability results pertaining to the weak pay-off positive selection dynamics. Sethi [12] shows that monomorphic evolutionarily stable states are asymptotically stable in the case of weak pay-off positive selection dynamics given by the dynamical system (7). Suppose $(1, 0, 0) = e_1$ is a monomorphic evolutionarily stable state of game G. As e_1 is an evolutionarily stable state, there exists a neighbourhood $N(e_1)$ of e_1 such that $\pi_1(x) > \pi_i(x)$ for all $x \in N(e_1)$ and $i \in \{2, 3\}$. This implies $\dot{x}_1 > 0$ in the dynamical system (7) when $x \in N(e_1)$. Since the trajectories are always within the unit simplex, $\dot{x}_1 + \dot{x}_2 + \dot{x}_3 = 0$ which implies that $\dot{x}_2 + \dot{x}_3 < 0$. This implies that the trajectories with initial point in the $N(e_1)$ are going to stay in $N(e_1)$. As $\dot{x}_1 > 0$ for all $x \in N(e_1)$, eventually the trajectories are going to hit e_1. Therefore, e_1 is asymptotically stable. We do not have a general result on the stability of all types of evolutionarily stable states in the case of weak pay-off positive selection dynamics. Mahanta [8] shows the convergence to Nash equilibria in case of potential games for weak pay-off positive selection dynamics.

Another interesting character of the trajectories of a dynamical system is the presence of cycles. Next, we look at the possibility of cycles in these two types of selection dynamics. In the Rock Paper Scissors game, replicator dynamics exhibit cycles for some values of pay-offs. Weissing [17] and Weibull[16] show the presence of cycles in the following form of the Rock Paper Scissors game when $a = 0$,

$$A = \begin{pmatrix} 1 & 2+a & 0 \\ 0 & 1 & 2+a \\ 2+a & 0 & 1 \end{pmatrix}$$

In the game with the above pay-off, the only Nash equilibrium is $(\frac{1}{3}, \frac{1}{3}, \frac{1}{3})$. Consider a function $Q(x) = x_1(t)x_2(t)x_3(t)$. The time derivative of $Q(x(t))$ is
$\dot{Q}(x(t)) = x_2(t)x_3(t)\dot{x}_1 + x_1(t)x_3(t)\dot{x}_2(t) + x_2(t)x_1(t)\dot{x}_3(t)$.
Substituting from the replicator dynamics in the above, we get
$\dot{Q}(x(t)) = x_1(t)x_2(t)x_3(t)[\pi_1 + \pi_2 + \pi_3 - 3\bar{\pi}]$.
Again, substituting the pay-off from the pay-off matrix A, we get
$\dot{Q}(x(t)) = x_1(t)x_2(t)x_3(t)a[1 - 3(x_1x_2 + x_3x_2 + x_1x_3)]$.
We know that $(x_1 + x_2 + x_3)^2 = x_1^2 + x_2^2 + x_3^2 + 2(x_1x_2 + x_1x_3 + x_3x_2) = 1$.
Hence, $\dfrac{1 - (x_1^2 + x_2^2 + x_3^2)}{2} = x_1x_2 + x_1x_3 + x_3x_2$.
Using this, we get

$$\dot{Q}(x(t)) = x_1(t)x_2(t)x_3(t)a(3(x_1^2 + x_2^2 + x_3^2) - 1) \tag{8}$$

The term $(3(x_1^2 + x_2^2 + x_3^2) - 1)$ is positive for $x \in \Delta$ and $x \neq (\frac{1}{3}, \frac{1}{3}, \frac{1}{3})$. At $(\frac{1}{3}, \frac{1}{3}, \frac{1}{3})$, $(3(x_1^2 + x_2^2 + x_3^2) - 1) = 0$.

In Eq. (8) when $x \in \Delta$, $a = 0$ implies $\dot{Q}(x(t)) = 0$, $a > 0$ implies $\dot{Q}(x(t)) > 0$ and $a < 0$ implies $\dot{Q}(x(t)) < 0$. Thus, when $a = 0$, for all interior initial points the trajectories of replicator dynamics will generate cycles (closed orbits). Thus, the trajectory from any internal initial point is periodic. When $a > 0$, we get that $(\frac{1}{3}, \frac{1}{3}, \frac{1}{3})$ is not asymptotically stable and when $a < 0$, we get that $(\frac{1}{3}, \frac{1}{3}, \frac{1}{3})$ is asymptotically stable.

In the case of weak pay-off positive selection dynamics, the normal form game with the following pay-off matrix

$B = \begin{pmatrix} 0 & 6 & -4 \\ -3 & 0 & 5 \\ -1 & 3 & 0 \end{pmatrix}$ and $\lambda_1 = 0.8,\ \lambda_2 = 0.1,\ \lambda_3 = 0.4$ exhibit stable limit cycles.

The proof was obtained using the Poincare–Bendixson theorem in Sethi [12]. The main idea is that an invariant set containing no fixed point has been constructed within the unit simplex. So any trajectory originating in this invariant set is going to stay within this set. There is no rest point in this set. Using the Poincare–Bendixson theorem, we get that the trajectories will converge to a stable periodic orbit (stable limit cycle).

These are some of the main results related to replicator dynamics and weak pay-off positive selection dynamics.

4 Conclusion

We conclude by enumerating some possible extensions in these two types of selection dynamics. Some further work can be done related to these two selection dynamics. In the case of replicator dynamics, one can search for games where global convergence of trajectories is possible. As of now, we can show the presence of cycles in the Rock Paper Scissors game in the case of replicator dynamics. We can investigate other games for the presence of cycles. In the case of weak pay-off positive selection dynamics, other forms of strategy-specific barriers can be analysed. For example, the strategy-specific barriers are functions of pay-offs or population states. There are no stability results with respect to imitation driven by dissatisfaction when there are strategy-specific barriers.

Appendix

Convergence of trajectories in partnership games.
The expected pay-off of game G at any x is
$\bar{\pi}(x(t)) = \pi_1(t)x_1(t) + \pi_2(t)x_2(t) + \pi_3(t)x_3(t)$.
The time derivative of $\bar{\pi}(x(t))$ is
$\dot{\bar{\pi}}(x(t)) = \pi_1(t)\dot{x}_1 + \pi_2(t)\dot{x}_2 + \pi_3(t)\dot{x}_3 + \dot{\pi}_1(t)x_1(t) + \dot{\pi}_2(t)x_2(t) + \dot{\pi}_3(t)x_3(t)$.
In partnership games, $\pi_1(t)\dot{x}_1 + \pi_2(t)\dot{x}_2 + \pi_3(t)\dot{x}_3 = \dot{\pi}_1(t)x_1(t) + \dot{\pi}_2(t)x_2(t) + \dot{\pi}_3(t)x_3(t)$.
This implies

$$\dot{\bar{\pi}}(x(t)) = 2(\pi_1(t)\dot{x}_1 + \pi_2(t)\dot{x}_2 + \pi_3(t)\dot{x}_3) \tag{9}$$

Using $\dot{x}_1 + \dot{x}_2 + \dot{x}_3 = 0$ in Eq. (9), we get
$\dot{\bar{\pi}}(x(t)) = 2((\pi_1(t) - \bar{\pi}(t))\dot{x}_1 + (\pi_2(t) - \bar{\pi}(t))\dot{x}_2 + (\pi_3(t) - \bar{\pi}(t))\dot{x}_3)$.
It implies $\dot{\bar{\pi}}(x(t)) = 2((\pi_1(t) - \bar{\pi}(t))^2 x_1 + (\pi_2(t) - \bar{\pi}(t))^2 x_2 + (\pi_3(t) - \bar{\pi}(t))^2 x_3)$.
We obtain $\dot{\bar{\pi}}(x(t)) \geq 0$ for all $x \in \Delta$. For $x > 0$, equality sign holds when $\pi_i = \bar{\pi},\ i \in \{1, 2, 3\}$ which is a mixed strategy Nash equilibrium. Note that the dynamics may not always lead to the global maximum. If the initial point is at some boundary of Δ and the interior mixed strategy Nash equilibrium is the global maximum, the trajectories will never reach it. The important thing to note is that the initial point should not be a non-Nash equilibrium. These types of fixed points are in the boundary of the unit simplex.

Asymptotic stability of evolutionarily stable strategy in Normal form games. In the literature, relative entropy function has been used to show the above result. The relative entropy function is defined as
$H_{x^*}(x) = \sum_{i=1}^{3} x_i^* log(\frac{x_i^*}{x_i})$ where $x \in \Delta$ and x^* is an evolutionarily stable state.

$$H_{x^*}(x) = \sum_{i=1}^{3} x_i^* log(\frac{x_i^*}{x}) = -\sum_{i=1}^{3} x_i^* log(\frac{x_i}{x_i^*}) \geq -log(\sum_{i=1}^{3} \frac{x_i x_i^*}{x_i^*}) = -log(\sum_{i=1}^{3} x_i)$$

$= -log(1) = 0$. The inequality in the above expression is due to the concavity of the log function. Thus, $H_{x^*}(x) \geq 0$ always and holds with equality at $x^* = x$.
The time derivative of $H_{x^*}(x)$ is

$$\begin{aligned}\dot{H}_{x^*}(x) &= -\sum_{i=1}^{3} \frac{x_i^*}{x_i}[\pi_i(x) - \bar{\pi}(x)]x_i. \\ &= -\sum_{i=1}^{3} x_i^*[\pi_i(x) - \bar{\pi}(x)]. \\ &= -\ [\pi_1(x)x_1^* + \pi_2(x)x_2^* + \pi_3(x)x_3^* - \bar{\pi}(x)]\end{aligned}$$

Since x^* is an evolutionarily stable state, there exists a neighbourhood $N(x^*)$ of x^* such that $\pi_1(x)x_1^* + \pi_2(x)x_2^* + \pi_3(x)x_3^* > \bar{\pi}(x)$ for all $x \in N(x^*)$. This implies that for all $x \in N(x^*)$, $\dot{H}_{x^*}(x) < 0$. Thus, x^* is asymptotically stable.

References

1. Bomze, I.M. 1986. Non-cooperative two-person games in biology: A classification. *International Journal of Game Theory* 15 (1): 31–57.
2. Cressman, R. 1997. Local stability of smooth selection dynamics for normal form games. *Mathematical Social Sciences* 34 (1): 1–19.

3. Friedman, D. 1991. Evolutionary games in economics. *Econometrica* 59 (3): 637–666.
4. Hofbauer, J., P. Schuster, and K. Sigmund. 1978. A note on evolutionary stable strategies and game dynamics. *Journal of Theoretical Biology* 81: 609–612.
5. Hofbauer, J., and K. Sigmund. 1998. *Evolutionary games and population dynamics*. Cambridge: Cambridge University Press.
6. Hofbauer, J., and K. Sigmund. 2003. Evolutionary game dynamics. *Bulletin of the American Mathematical Society* 40 (4): 479–519.
7. Lahkar, R. 2012. Evolutionary game theory: An exposition. *Indian Growth and Development Review* 5 (2): 203–213.
8. Mahanta, A. 2019. Learning through imitation with strategy specific barriers, Working Paper.
9. Samuelson, L., and J. Zhang. 1992. Evolutionary stability in asymmetric games. *Journal of Economic Theory* 57 (2): 363–391.
10. Sandholm, W.H. 2010. Population games and evolutionary dynamics. Cambridge: MIT Press.
11. Sandholm, W.H. 2010. Pairwise comparison dynamics and evolutionary foundations for Nash equilibrium. *Games* 1 (1): 3–17.
12. Sethi, R. 1998. Strategy-specific barriers to learning and nonmonotonic selection dynamics. *Games and Economic Behavior* 23 (2): 284–304.
13. Taylor, P.D., and B.L. Jonker. 1978. Evolutionary stable strategies and game dynamics. *Mathematical Biosciences* 40 (1–2): 145–156.
14. Van Damme, E. 1994. Evolutionary game theory. *European Economic Review* 38 (3–4): 847–858.
15. Vega-Redondo, F. 1996. *Evolution, games, and economic behaviour*. Oxford: Oxford University Press.
16. Weibull, J.W. 1997. Evolutionary game theory. Cambridge: MIT Press.
17. Weissing, F.J. 1991. Evolutionary stability and dynamic stability in a class of evolutionary normal form games. In *Game equilibrium models I*, ed. R. Selten. Berlin: Springer.

Linear Games and Complementarity Problems

S. Gokulraj and A. Chandrashekaran

AMS classification(2010): 91A05, 90C33, 15A63, 46N10

1 Introduction

Since Von Neumann and Morgenstern's seminal work in game theory, it has been further developed and used to analyse the conflicts and cooperative situations in biological sciences, social sciences, economics, etc. Depending on the players' antagonistic nature, two-player games are classified as zero-sum games (antagonistic) and non-zero-sum games (not antagonistic). In the following subsections, we briefly survey these games and their generalizations.

1.1 Matrix Games and their Generalizations

A non-cooperative two-person zero-sum game is a game played by two players in which the gain of one player results in a loss for the other. To explain, consider a game G played by the two players say, players I and II. Assume that player I has m possible moves (pure strategies) and player II has n pure strategies. Suppose player I and II play with the ith and jth pure strategy, respectively, then denote the

S. Gokulraj
Industrial Engineering and Operations Research, Indian Institute of Technology Bombay, Powai, Mumbai 400076, India

A. Chandrashekaran (✉)
Department of Mathematics, Central University of Tamil Nadu, Neelakudi campus, Thiruvarur, Tamil Nadu 610005, India

S. Borkotokey et al. (eds.), *Game Theory and Networks*, Indian Statistical Institute Series,
https://doi.org/10.1007/978-981-16-4737-6_2

corresponding pay-off for player I as a_{ij}. If pay-off for player II is $-a_{ij}$, then the game G is said to be a zero-sum game. From this, it could be observed that in any two-person zero-sum finite game (games with a finite number of pure strategies), pay-offs of the players are identified as entries of a matrix. Here, $A = (a_{ij})_{m\times n}$ is the pay-off matrix of player I in the given finite game G. This says that pure strategies of players I and II correspond to the rows and columns of A, respectively, and so we may call player I as a row player and player II as a column player. In case players I and II choose to play with the probability distributions x on rows and y on columns, respectively, then the expected pay-off for player I is $\langle x, Ay\rangle$ and by the definition of a zero-sum game, player II's pay-off is $-\langle x, Ay\rangle$. The set of all possible strategies (probability vectors) for players I and II is denoted as Δ_1 and Δ_2, respectively, and they are defined as follows:

$$\Delta_1 = \{x \in \mathbb{R}^m | x \geq 0 \text{ and } \sum_{i=1}^{m} x_i = 1\},\ \Delta_2 = \{y \in \mathbb{R}^n | y \geq 0 \text{ and } \sum_{j=1}^{n} y_j = 1\}.$$

The solution concept of zero-sum games is the existence of an optimal strategy pair (equilibrium strategy pair). A strategy pair (x^*, y^*) is said to be an optimal strategy pair if the following inequality holds:

$$\langle x, Ay^*\rangle \leq \langle x^*, Ay^*\rangle \leq \langle x^*, Ay\rangle \ \ \forall x \in \Delta_1 \text{ and} y \in \Delta_2.$$

This says, both the players cannot improve their expected pay-off by unilaterally changing their strategy from the optimal strategy (strategy in the optimal strategy pair). The expected pay-off $\langle x^*, Ay^*\rangle$ at the optimal strategy pair (x^*, y^*) is said to be the value of the game G, or the value of A and is denoted by $v(A)$. When the context is clear, we simply write v in place of $v(A)$. The existence of an optimal strategy pair is known from the celebrated minimax theorem for finite games. For more details on minimax theorem, we refer to [25] and the references therein. A strategy is said to be completely mixed if it is in the interior of the strategy set. A game is called a completely mixed game if all the optimal strategies are completely mixed.

Non-cooperative two-person zero-sum games, also known as matrix games, are well-studied in the game theory literature and have a wide variety of applications in economic theory, social sciences, matrix theory, etc. For instance, in matrix theory the Perron-Frobenius theorem and its extension to the positive operators have been proved by using the minimax theorem [3, 26]. In [28], game-theoretic proofs for the well-known characterizations of non-singular **M**-matrix have been given by Raghavan. In particular, he shows that the game corresponding to the **Z**-matrix (matrix with non-positive off-diagonal entries) is completely mixed if the value is positive. By exploiting the value being positive, he proves a number of equivalent conditions for the **Z**-matrix such as the positive stable property, P-property and semi-positive property.

We see from [28] that for a **Z**-matrix A, the value of A is positive if and only if there exists an $x^* \in \mathbb{R}^n$ such that

$$x^* > 0 \text{ and } Ax^* > 0. \tag{1}$$

Here, the inequality $x > 0$ means entrywise positive vector in $\mathbb{R}^n$. From the study of a linear continuous dynamical system, we know that for a given real matrix $A_{n\times n}$, the continuous dynamical system $\frac{dy}{dt} + Ay = 0$ is asymptotically stable in $\mathbb{R}^n$ (that is, any trajectory starting from any point in $\mathbb{R}^n$ converges to the origin) if and only if there exists a real symmetric matrix X in $\mathcal{S}^n$ (space of all $n \times n$ real symmetric matrices) such that

$$X > 0 \text{ and } L_A(X) > 0. \tag{2}$$

Here $X > 0$ means that X is a positive definite matrix and L_A denotes the well-known Lyapunov transformation defined on $\mathcal{S}^n$. That is, for $X \in \mathcal{S}^n$, $L_A(X) := AX + XA^t$. A similar result holds for the discrete dynamical system also. Motivated by the similarity between the inequalities (1) and (2), Gowda and Ravindran extended the concept of value to the general linear transformations defined on the finite-dimensional real inner product spaces [12]. In fact, they defined the concept of linear games over the self-dual cone as the generalization of the non-cooperative two-person zero-sum game and obtained the results of dynamical systems in terms of the value. Also, they extended the classical matrix game results of Kaplansky [17] and Raghavan [28] to the general linear game setting. Further studies on the linear games over the proper cone appeared in [24].

In [11], Gowda, studied the completely mixed linear games (linear games with all its optimal strategies in the interior of the strategy set). He classified those linear transformations for which the corresponding linear game is completely mixed even if we change the strategy set in the ambient space. In addition to that, he generalized the result of Kaplansky in [17] which describes the necessary and sufficient conditions for the linear game to be completely mixed when the value is zero.

In [9], Gokulraj and Chandrashekaran defined symmetric linear games which are linear games corresponding to skew-symmetric linear transformations. And they discussed the symmetrization procedure for the general linear games. In particular, they show that for a given linear game, there exists a symmetric linear game whose solution yields the solution to the underlying linear game. Also, they discussed the results on the nature of the optimal strategies of the linear game and its symmetrized game. In fact, they proved the following theorem.

Theorem 1.1 (Theorem 3.6 in [9]) *For a given linear game $G = (L, K, e)$, consider its generalized Von Neumann symmetrization $\widehat{G} = (\widehat{L}, \widehat{K}, \widehat{e})$. If $\widehat{G}$ has a pure (or mixed or completely mixed) strategy equilibrium, then G also has a pure (or mixed or completely mixed) strategy equilibrium.*

In addition to that, they consider the gRPS symmetric linear game which is the generalization of the well-known symmetric zero-sum game "Rock Paper Scissor" and proved that a symmetric linear game has a pure strategy equilibrium if and only if it is not gRPS. Also, it is shown that a completely mixed symmetric linear game is a gRPS game.

1.2 Two-Person Non-zero Sum Games

Though some situations are modelled as zero-sum games, most situations are non-zero-sum. For instance, models in social sciences like prisoner's dilemma, models of interaction between landlord and tenant, husband and wife, etc., are not always antagonistic.

Two-person non-zero-sum game, also known as a bimatrix game, is a game played by two players say player I (row player) and player II (column player) whose pay-off matrices are A and B, respectively, in $M_{m \times n}(\mathbb{R})$. Suppose player I plays with the probability vector x and player II with y, then their expected pay-offs are $\langle x, Ay \rangle$ and $\langle x, By \rangle$, respectively.

In non-zero-sum games, we use the existence of equilibrium as the solution of the game, whereas in a zero-sum game we use the optimality concept as the solution. In the early 1950s, John Nash, in his seminal papers [21, 22], proved the existence of equilibrium pairs for non-zero-sum games in mixed strategies. That is, every two-person non-zero-sum game has a mixed strategy equilibrium. In fact, he proved this result for N-person games. In the literature, this equilibrium concept is called Nash equilibrium. In this article, by the word equilibrium we mean Nash equilibrium. A pair of probability vectors (strategies) (x^*, y^*) is said to be an equilibrium pair if the following inequalities hold for all the probability vectors $x \in \mathbb{R}^m$ and $y \in \mathbb{R}^n$.

$$v_1 := \langle x^*, Ay^* \rangle \geq \langle x, Ay^* \rangle,$$

$$v_2 := \langle x^*, By^* \rangle \geq \langle x^*, By \rangle.$$

That is, x^* is the best response for player I against player II's y^*, likewise player II's best response against x^* is y^*. The expected pay-offs v_1 and v_2 at the equilibrium pair (x^*, y^*) are called as the equilibrium value of player I and player II, respectively.

Consider a two-player game with the pay-off matrix A being **Z**-matrix. That is, $A = sI - B$ where s is a scalar and B is a non-negative matrix. It is known that properties of A can be obtained by analysing the matrix game $\Gamma(A)$. Here, we emphasize that in this matrix game, diagonal and off-diagonal entries of the pay-off matrix might have different signs. So the process of iterative elimination of dominated strategy (IEDS) is not beneficial in this matrix game consideration. In contrast, if we consider the bimatrix game $\Gamma(A, B)$, then player II's pay-off matrix has non-negative entries and applying the process of IEDS possibly reduces the dimension of the game, which eventually helps in finding the equilibria. This situation is illustrated in the following example.

Example 1.2 Let $B = \begin{bmatrix} 5 & 2 \\ 2 & 1 \end{bmatrix}$ and $A = 6I_2 - B = \begin{bmatrix} 1 & -4 \\ -4 & 5 \end{bmatrix}$ where I_2 denotes the identity matrix of order 2. Consider the bimatrix game $\Gamma(A, B)$. Here, we note that player II's pay-off matrix have non-negative entries and applying the IEDS process eliminates the dominated strategy (2nd column) of player II; by that, the dimension of the game is reduced to 2×1 from 2×2. In the next iteration, player I's dominated 2nd row is eliminated. By this, the dimension of the game gets reduced to 1×1 from 2×1. This does not happen if we consider as a matrix game $\Gamma(A)$.

Motivated by these possible benefits of the bimatrix game consideration for a **Z**-matrix, bi-linear games have been introduced as a generalization of the bimatrix game in [10]. And results related to **Z**-transformations were proved through this bi-linear game concept.

Motivated by the study on linear and bi-linear games and their connections to dynamical systems and matrix theory, since zero-sum games are a special case of non-zero-sum games, we ask if the results of the two-person non-zero-sum game can be extended to the general linear maps. In what follows, we extend some of the results on the bimatrix game to the general linear maps defined on a finite-dimensional real inner product space. In particular, we extend the results on bimatrix games due to Parthasarathy and Raghavan [25, 27] to this general bi-linear game setting. We show that the existence of an equilibrium pair for a given bi-linear game is equivalent to the existence of a solution to a corresponding conic optimization problem. Further, we observe some relationships between bi-linear and linear games under special circumstances. We also characterize the equilibrium pairs of completely mixed bi-linear games corresponding to **Z**-transformations. Using this for a special type of **Z**-transformation, we relate the bi-linear game results and some known classical results in matrix theory and the theory of complementarity problems. In addition to that, we discuss the converse of Theorem 1.1.

The organization of the article is as follows. In the immediate section, we provide some preliminary definitions and results on linear complementarity problems, **Z**-transformations, linear games and tensor product of finite-dimensional spaces. In Sect. 3, we recall generalized Von Neumann symmetrization of a linear game and prove the converse of Theorem 1.1. In Sect. 4, we recall bi-linear games and show that the existence of equilibrium pairs for bi-linear game is equivalent to the existence of solutions for an associated conic optimization problem. In addition to this, we consolidate some results about completely mixed bi-linear games. In Sect. 5, we discuss the relationship between the game-theoretic value of linear and bi-linear games corresponding to a given **Z**-transformation. Also, we characterize the equilibrium pairs of the completely mixed bi-linear games corresponding to the given **Z**-transformations.

2 Preliminaries

In this article, $(V, \langle\cdot,\cdot\rangle)$ and $(W, \langle\cdot,\cdot\rangle)$ are finite-dimensional real inner product spaces where $\langle\cdot,\cdot\rangle$ denotes the inner product on the respective spaces. For a linear map $L : V \to W$, we denote its transpose as L^t. For $x \in V$, let $x^\perp := \{y \in V | \langle y, x\rangle = 0\}$.

A non-empty subset K in V is said to be a convex cone if $px + qy \in K$ for all $x, y \in K$ and $p, q \geq 0$. A convex cone K is closed if it is topologically closed and pointed if $K \cap -K = \{0\}$. A closed convex pointed cone is said to be a proper cone if it has a non-empty interior. A convex cone K is said to be self-dual if its dual $K^* := \{x \in V | \langle x, y\rangle \geq 0 \;\forall\; y \in K\}$ is K itself. For a set K, we use K° and ∂K to denote the interior and boundary of K, respectively. We consider the partial ordering induced by K as $x \succeq_K y$ when $x - y \in K$ and $x \succ_K y$ when $x - y \in K^\circ$.

Let K be a self-dual cone in V. For a fixed $e \in K^\circ$, consider the subset $\Delta := \{x \in K | \langle x, e\rangle = 1\}$. It is clear that Δ is a compact convex set and forms a base for K. That is, every element in K is a positive scalar multiple of an element in Δ.

Lemma 2.1 ([12], Lemma 1) *Let $x \in \Delta$ and $z \in V$.*

1. *If $z \neq 0$ and $\langle z, e\rangle = 0$, then there exists $t > 0$ such that $x - tz \in \partial K$.*
2. *If $z \neq x$ and $\langle z, e\rangle = 1$, then there exists $t > 0$ such that $(1 + t)x - tz \in \partial K$.*

2.1 Linear Complementarity Problems

For a given matrix $A \in M_n(\mathbb{R})$ and a vector $q \in \mathbb{R}^n$, the standard linear complementarity problem $\big(LCP(A, q)\big)$ is to find an $x \in \mathbb{R}^n$ such that

$$x \geq 0,\;\; y := Ax + q \geq 0 \text{ and } x^t y = 0.$$

Here, $x \geq 0$ means entrywise non-negative vector in $\mathbb{R}^n$ $(x \in \mathbb{R}^n_+)$. If such an x exists, we call it a solution of the problem $LCP(A, q)$. For the detailed literature on LCP theory and its applications, we refer to [4, 5, 14, 20] and references therein.

Cone linear complementarity problem is a generalization of the above standard linear complementarity problem. That is, for a given proper cone K and a vector q in V and a linear transformation $L : V \to V$, the cone linear complementarity problem $(LCP(L, K, q))$ is to find an $x \in K$ such that $y := L(x) + q \in K^*$ and $\langle x, y\rangle = 0$. Here, K and its dual cone K^* play the role of $\mathbb{R}^n_+$ in the standard LCP setting.

A linear transformation L is said to have the **Q**-property if for all $q \in V$, $LCP(L, K, q)$ has a solution. A sufficient condition for a linear transformation to have the **Q**-property is given by Karamardian in [18]. Similarly, L has the **GUS**-property if for all $q \in V$, $LCP(L, K, q)$ has a unique solution. In the standard LCP setting, these transformations are known as the **Q**-matrix and **P**-matrix, respectively, and they play a prominent role in matrix theory and economic theory. For further details, see [6, 15].

2.2 Z-Transformations and their Properties

An $n \times n$ real matrix A is said to be a **Z**-matrix, if all its off-diagonal entries are non-positive. In [6], Fiedler and Ptak study these types of matrices in detail. Matrices of these types are studied not just in matrix theory but also in numerical analysis and economic theory. In numerical analysis, the study of matrix splitting and asymptotic rate of convergence of various iterative methods involves such kinds of matrices [30]. In Economics, these matrices are called "matrices of the Leontief type", and they play an important role in the study of Leontief's input-output system and factor-price equalization; for details, see [7, 23] and references therein. Properties of a **Z**-matrix related to the LCP theory are found in [2, 16] and references therein.

From [2], we see that for a **Z**-matrix A, the following are equivalent:

(1) A is a **P**-matrix.
(2) A is a **Q**-matrix.
(3) A is semi-positive (that is, there exists an $x > 0$ such that $Ax > 0$).
(4) A^{-1} exists and is non-negative.
(5) A is positive stable (that is, the real part of every eigenvalue is positive).

It is easy to observe that for non-negative vectors x, y with $x^t y = 0$, A being **Z**-matrix, we get $y^t Ax \leq 0$. This observation is used to extend the **Z**-matrix property to the general linear transformations in the following way:

Definition 1 Given a closed convex cone K in V and a linear transformation L defined on V, we say L is a **Z**-transformation on K (or has the **Z**-property) if

$$[x \in K, y \in K^* \text{ and } \langle x, y\rangle = 0] \implies \langle L(x), y\rangle \leq 0.$$

Linear transformations of these types are introduced in the form of cross-positive matrices in [29]. In [13], Gowda and Tao introduced the above formal definition of **Z**-transformation on proper cones and extended the **Z**-matrix results to the **Z**-transformations defined on a proper cone. We have the following equivalent conditions for a **Z**-transformation defined on any proper cone K [13, 29].

(a) $e^{-tL}(K) \subseteq K$ for all $t \geq 0$.
(b) $L = \lim_{n\to\infty} (\alpha_n I - S_n)$ where $\alpha_n \in \mathbb{R}$ and S_n is a linear transformation on V with $S_n(K) \subseteq K$ for all n.

The last item helps to generate **Z**-transformations. In this article, we consider transformations of the form $\alpha I - S$ with $S(K) \subseteq K$ which is a natural extension of the **Z**-matrix. Such transformations are important in matrix theory; see [1, 2].

2.3 Linear Games and Related Results on Z-Transformations

Given a linear transformation $L : V \to V$ and a self-dual cone K, a fixed $e \in K^\circ$, we consider the linear game denoted by $\Gamma(L) := (L, K, e)$ as a game played by two players, say, players I and II in the following way: If players I and II choose $x \in \Delta$ and $y \in \Delta$, respectively, as their strategy, then the pay-off for player I is $\langle L(x), y\rangle$ and the pay-off for player II is $-\langle L(x), y\rangle$. Since Δ is a compact convex set and L is linear, by the min-max Theorem of Von Neumann ([19], Theorems 1.5.1 and 1.3.1), there exist optimal strategies (equilibriums) x' for player I and y' for player II such that

$$\langle L(x), y'\rangle \leq \langle L(x'), y'\rangle \leq \langle L(x'), y\rangle \ \forall\, x, y \in \Delta.$$

The pair (x', y') is called an optimal strategy pair and the expected pay-off at the optimal strategy pair $\langle L(x'), y'\rangle$ is called the value of the game $\Gamma(L)$ and is denoted by $v(L)$. The strategies in the interior of Δ are called completely mixed. The linear game $\Gamma(L)$ is completely mixed if the strategies in all the optimal pairs are completely mixed. The following theorem due to Gowda and Ravindran appeared in [12].

Theorem 2.2 ([12], Theorem 6) *Suppose L is a* **Z**-*transformation. Then the following are equivalent:*

1. *L is positive stable.*
2. *For every $q \in V$, $LCP(L, K, q)$ has a solution.*
3. *L is invertible with $L^{-1}(K) \subseteq K$.*
4. *There exists $x \in K^\circ$ such that $L(x) \in K^\circ$.*
5. *There exists $x \in K^\circ$ such that $L^t(x) \in K^\circ$.*
6. *A dynamical system $\dot{x} + L(x) = 0$ is asymptotically stable.*
7. *$v(L) > 0$.*

Moreover, when $v(L) > 0$ the game $\Gamma(L)$ is completely mixed.

We recall that the linear transformation L is said to be Lyapunov-like if both L and $-L$ are **Z**-transformations on K; it is Stein-like if $L = I - T$ where I is the identity transformation and $T \in \overline{Aut(K)}$ (closure of the set of all invertible linear transformation on V which maps K onto K). In addition to the above result, it is shown that the game with a non-zero value is completely mixed when L is a Lyapunov/Stein-like transformation. Also, the positive stable and Schur stable (all the eigenvalues lie inside the unit disc) properties were related to the value being positive for Lyapunov and Stein-like transformations, respectively.

2.4 Tensor Product of Vector Spaces

The concept of tensor product of V and W can be defined in terms of elementary tensors which in turn can be defined as the bi-linear functional on the Cartesian product $V \times W$. For $x \in V$ and $y \in W$, the elementary tensor $x \otimes y$ is defined to be a bi-linear functional from $V \times W$ to $\mathbb{R}$ such that $x \otimes y(v, w) := \langle x, v\rangle \cdot \langle y, w\rangle$ for all (v, w) in $V \times W$. The tensor product of V and W, denoted by $V \otimes W$, is an inner product space consisting of all finite sums of elementary tensors $x_i \otimes y_i$ where x_i and y_i are elements of V and W, respectively. The inner product on the space $V \otimes W$ is defined as follows:

$$\left\langle \sum_{i=1}^{m_1} x_i \otimes y_i \, , \sum_{j=1}^{m_2} v_j \otimes w_j \right\rangle := \sum_{i=1}^{m_1} \sum_{j=1}^{m_2} \langle x_i, v_j\rangle \langle y_i, w_j\rangle,$$

for every $\sum_{i=1}^{m_1} x_i \otimes y_i$ and $\sum_{j=1}^{m_2} v_j \otimes w_j$ in $V \otimes W$ where $m_1, m_2 \in \mathbb{N}$. The norm on this tensor product space is the one induced by the inner product defined as above. For a fixed basis $\{v_1, v_2 \cdots v_{\dim(V)}\}$ of V and $\{w_1, w_2 \cdots w_{\dim(W)}\}$ of W, the set of all elementary tensors $v_i \otimes w_j$ forms a basis for $V \otimes W$. By this, we can see that the dimension of the space $V \otimes W$ equals the product of dimensions of the spaces V and W.

The tensor product of two operators $T \in B[V]$ and $S \in B[W]$ is the operator $T \otimes S : V \otimes W \to V \otimes W$ in $B[V \otimes W]$ defined as follows:

$$(T \otimes S) \sum_{i=1}^{m_1} x_i \otimes y_i := \sum_{i=1}^{m_1} T(x_i) \otimes S(y_i) \quad \text{for all} \sum_{i=1}^{m_1} x_i \otimes y_i \in V \otimes W.$$

It is to be noted that in the matrix case, tensor product of operators is called the Kronecker product of matrices. In the following proposition, we summarize some of the properties of tensor product spaces which will be used in the later chapter.

Proposition 2.3 *For $\alpha \in \mathbb{R}$, $x_1, x_2 \in V$, $y_1, y_2 \in W$, $T \in B[V]$ and $S \in B[W]$, the following are true:*

1. $\alpha(x \otimes y) = (\alpha x \otimes y) = (x \otimes \alpha y)$,
2. $(x_1 + x_2) \otimes y = x_1 \otimes y + x_2 \otimes y$,
3. $x \otimes (y_1 + y_2) = x \otimes y_1 + x \otimes y_2$,
4. $||x \otimes y|| = ||x|| \, ||y||$,
5. $(T \otimes S)^t = T^t \otimes S^t$.

3 Symmetrization of Linear Games

Let (L, K, e) be the given linear game. We recall that Δ is the strategy set for both the players in (L, K, e) and denote the set of all extreme points of Δ in K by $\text{ext}(\Delta)$.

Let us denote the inner product space $V \otimes V$ as $\widehat{V}$. Consider the skew-symmetric linear map $\widehat{L}$ defined as follows:

$$\widehat{L} := P \circ (E \otimes L) - (E \otimes L^t) \circ P, \tag{3}$$

where P and E, respectively, are symmetric linear maps on $\widehat{V}$ and V such that $P(x \otimes y) := y \otimes x$ and $E(x) := \langle e, x \rangle e$ for all $x, y \in V$.

Consider the set $\text{ext}(\Delta) \otimes \text{ext}(\Delta) := \{x_i \otimes y_i | x_i, y_i \in \text{ext}(\Delta)\} \subseteq \widehat{V}$. Let $\widehat{K}$ be the proper cone defined as $\text{Pos}\left(\text{ext}(\Delta) \otimes \text{ext}(\Delta)\right)$. That is,

$$\widehat{K} = \left\{ \sum_{i=1}^{m} \lambda_i (x_i \otimes y_i) \middle| \lambda_i \geq 0,\ x_i, y_i \in \text{ext}(\Delta),\ m \in \mathbb{N} \right\}. \tag{4}$$

For a fixed $\widehat{e} := e \otimes e \in \widehat{K}$, $\widehat{\Delta} := \{Z \in \widehat{K} | \langle Z, \widehat{e} \rangle = 1\}$ is a base of $\widehat{K}$. It is known that $\Delta \otimes \Delta := \{x \otimes y | x, y \in \Delta\} \subset \widehat{\Delta}$. Here, the set inclusion is strict since for any λ_1 and λ_2 with $\lambda_1 + \lambda_2 = 1$, $\lambda_1 x \otimes x + \lambda_2 y \otimes y \in \widehat{\Delta}$ for all $x, y \in \Delta$ but does not belong to $\Delta \otimes \Delta$ whenever $x \neq y$. Also, we can observe that the set of all extreme points of $\widehat{\Delta}$ consists only of elementary tensors $x \otimes y$ where $x, y \in \text{ext}(\Delta)$. For more details, see [9].

Theorem 3.1 ([9], Theorem 3.4) *Given a linear game (L, K, e) there exists a symmetric linear game $(\widehat{L}, \widehat{K}, \widehat{e})$ whose solution yields a solution to the given linear game.*

The symmetric linear game $\Gamma(\widehat{L})$ derived in the above theorem is called the generalized Von Neumann symmetrization of the given linear game $\Gamma(L)$. Linear games which may have different strategy sets for two players are studied in [24] as linear games over proper cones. Linear games over a proper cone (need not be self-dual) can also be symmetrized in the same procedures given above.

Definition 2 (*Pure strategies*) For the given linear game (L, K, e), an element in the strategy set Δ is said to be a pure strategy if and only if it is an extreme point of the strategy set Δ. Any other strategy is called a mixed strategy. In particular, strategies in the interior are called completely mixed strategies.

The following theorem deals with the converse of Theorem 1.1.

Theorem 3.2 *Let $\Gamma(L)$ be the given linear game and $\Gamma(\widehat{L})$ be its generalized Von Neumann symmetrization. $\Gamma(\widehat{L})$ has a pure strategy equilibrium if and only if $\Gamma(L)$ has a pure strategy equilibrium. A similar result holds for mixed and completely mixed strategy equilibrium.*

Proof Assume that $\Gamma(L)$ has a pure strategy equilibrium pair (x^*, y^*). We claim that $(y^* \otimes x^*, y^* \otimes x^*)$ is a pure strategy equilibrium pair for the symmetric linear game $\Gamma(\widehat{L}) = (\widehat{L}, \widehat{K}, \widehat{e})$. That is, for all z in $\widehat{\Delta}$,

$$\langle y^* \otimes x^*, \widehat{L}(z) \rangle \leq \langle y^* \otimes x^*, \widehat{L}(y^* \otimes x^*) \rangle = 0 \leq \langle z, \widehat{L}(y^* \otimes x^*) \rangle.$$

Now, consider $z := \sum_{i=1}^{n} \alpha_i (x_i \otimes y_i)$ in $\widehat{\Delta}$ where $\alpha_i \geq 0$ and $x_i, y_i \in \Delta$ for all i. Then,

$$\begin{aligned}\langle z, \widehat{L}(y^* \otimes x^*)\rangle &= \langle z, [P \circ (E \otimes L) - (E \otimes L^t) \circ P](y^* \otimes x^*)\rangle \\ &= \langle z, L(x^*) \otimes E(y^*)\rangle - \langle z, E(x^*) \otimes L^t(y^*)\rangle \\ &= \langle \sum_{i=1}^{n} \alpha_i (x_i \otimes y_i), L(x^*) \otimes e\rangle - \langle \sum_{i=1}^{n} \alpha_i (x_i \otimes y_i), e \otimes L^t(y^*)\rangle \\ &= \langle \sum_{i=1}^{n} \alpha_i x_i, L(x^*)\rangle - \langle \sum_{i=1}^{n} \alpha_i y_i, L^t(y^*)\rangle.\end{aligned}$$

Since z is arbitrary and (x^*, y^*) is an equilibrium pair for the linear game (L, K, e), the above equality becomes the following inequality:

$$\langle z, \widehat{L}(y^* \otimes x^*)\rangle = \langle y, L(x^*)\rangle - \langle x, L^t(y^*)\rangle \geq 0, \tag{5}$$

where $y = \sum_{i=1}^{n} \alpha_i x_i$ and $x = \sum_{i=1}^{n} \alpha_i y_i$ belongs to Δ. Since $\widehat{L}$ is skew-symmetric, we have

$$\langle y^* \otimes x^*, \widehat{L}(y^* \otimes x^*)\rangle = 0 \text{ and } \langle y^* \otimes x^*, \widehat{L}(z)\rangle \leq 0. \tag{6}$$

From (5) and (6), it is clear that $(y^* \otimes x^*, y^* \otimes x^*)$ is an equilibrium pair for the symmetric linear game $(\widehat{L}, \widehat{K}, \widehat{e})$. Since y^*, x^* is in $\text{ext}(\Delta)$, $y^* \otimes x^*$ is in $\text{ext}(\widehat{\Delta})$. Thus, $(y^* \otimes x^*, y^* \otimes x^*)$ is a pure strategy equilibrium. The converse part is clear from Theorem 1.1. Similar proof can be given for mixed and completely mixed strategies. □

4 Results on Bi-linear Games

Let us recall the definition of the bi-linear game defined as in [10]. Let K_1 and K_2 be self-dual cones in V and W, respectively. For a fixed $e_1 \in K_1^\circ$ and $e_2 \in K_2^\circ$, consider the sets Δ_1 and Δ_2 defined as follows:

$$\Delta_1 := \{x \in K_1 | \langle x, e_1\rangle = 1\}, \;\; \Delta_2 := \{y \in K_2 | \langle y, e_2\rangle = 1\}.$$

Let $L_1, L_2 : V \to W$ be two linear transformations. The bi-linear game corresponding to L_1 and L_2 denoted as $\Gamma(L_1, L_2)$ is defined to be the game played by two players, say players I and II. If the players I and II choose $x \in \Delta_1$ and $y \in \Delta_2$, respectively, as their strategies then their pay-offs $P_I(x, y)$ and $P_{II}(x, y)$ are defined as follows:

$$P_I(x, y) := \langle y, L_1(x)\rangle, \;\; P_{II}(x, y) := \langle y, L_2(x)\rangle.$$

A strategy x of player I (II) is said to be completely mixed if $x \in \Delta_1^\circ$ (Δ_2°).

Remark 1 It is to be noted that "bi-linear games" considered in this article are different from the one that appeared in [8] where bi-linear games are defined as a very general class of games, for which bimatrix games, two-person Bayesian games, polymatrix games, etc., are a special case. One technical difference between these two definitions is the ambient space where the game is defined. The game considered here is named as "bi-linear" since it involves two linear maps which define the pay-off functions.

Definition 3 (*Equilibrium pair*) A pair of strategies $(x^*, y^*) \in \Delta_1 \times \Delta_2$ is said to be an equilibrium pair for $\Gamma(L_1, L_2)$ if it satisfies the following inequalities:

$$\begin{aligned} \langle y^*, L_1(x^*)\rangle &\geq \langle y^*, L_1(x)\rangle \ \forall\, x \in \Delta_1, \\ \langle y^*, L_2(x^*)\rangle &\geq \langle y, L_2(x^*)\rangle \ \forall\, y \in \Delta_2. \end{aligned} \tag{7}$$

The pay-offs $v_1 := \langle y^*, L_1(x^*)\rangle$ and $v_2 := \langle y^*, L_2(x^*)\rangle$ at the equilibrium pair is called the equilibrium value of player I and II, respectively. That is, v_1 and v_2 are the value of the respective player corresponding to the equilibrium pair (x^*, y^*).

Let $\mathcal{E}$ denote the set of all equilibrium pairs of the game $\Gamma(L_1, L_2)$, and we say $\mathcal{E}$ is completely mixed if the strategies in all the equilibrium pairs are completely mixed. In this case, we say that the bi-linear game $\Gamma(L_1, L_2)$ is completely mixed. We define the following subsets:

$$S(y^*) := \{x \in \Delta_1 | (x, y^*) \in \mathcal{E}\},$$

$$T(x^*) := \{y \in \Delta_2 | (x^*, y) \in \mathcal{E}\}.$$

Remark 2 In a bi-linear game $\Gamma(L_1, L_2)$, if $L_2 = -L_1$ then the bi-linear game definition is consistent with the definition of the linear game $\Gamma(L_1)$ defined as in [12, 24].

Theorem 4.1 ([10], Theorem 3.2) *For any bi-linear game $\Gamma(L_1, L_2)$, there exists a equilibrium pair (x^*, y^*).*

The above theorem ensures the existence of Nash equilibrium for this general bi-linear game setting. The following are the results characterizing equilibrium pairs of the given bi-linear game.

Theorem 4.2 *A pair (x^*, y^*) is an equilibrium for the bi-linear game $\Gamma(L_1, L_2)$ if and only if there exists $p, q \in \mathbb{R}$ such that the following holds:*

$$\begin{aligned} pe_1 - L_1^t(y^*) &\in K_1, \\ qe_2 - L_2(x^*) &\in K_2, \end{aligned} \tag{8}$$

$$\langle x^*, L_1^t(y^*)\rangle + \langle y^*, L_2(x^*)\rangle = p + q. \tag{9}$$

Proof Let (x^*, y^*) be an equilibrium pair of $\Gamma(L_1, L_2)$. Now, consider $p := \langle L_1(x^*), y^*\rangle$ and $q := \langle L_2(x^*), y^*\rangle$. Now, (8) follows from the fact that (x^*, y^*) is an equilibrium pair and by the assumption on p and q, Eq. (9) is also true. Conversely, assume that (x^*, y^*) satisfies (8) and (9) for some $p, q \in \mathbb{R}$. From (8), we see that $p \geq \langle L_1(y^*), x\rangle \ \forall\ x \in \Delta_1$ and $q \geq \langle L_2(x^*), y\rangle \ \forall\ y \in \Delta_2$. This together with (9) implies that $p = \langle L_1(y^*), x^*\rangle$ and $q = \langle L_2(x^*), y^*\rangle$. Thus, (x^*, y^*) is the equilibrium pair for $\Gamma(L_1, L_2)$. □

Theorem 4.3 *A pair (x^*, y^*) is an equilibrium for the bi-linear game $\Gamma(L_1, L_2)$ if and only if for some $p, q \in \mathbb{R}$, (x^*, y^*) solves the following conic optimization problem:*

$$\begin{aligned} \max\ & \langle y, L_1(x)\rangle + \langle y, L_2(x)\rangle - p - q, \\ s.t\ & qe_2 - L_2(x) \succeq_{K_2} 0, \\ & pe_1 - L_1^t(y) \succeq_{K_1} 0, \\ & x \in \Delta_1, y \in \Delta_2. \end{aligned} \tag{10}$$

Proof Let (x^*, y^*) be an equilibrium pair of $\Gamma(L_1, L_2)$. Since the objective function of the conic optimization problem (COP) (10) is non-positive, by Theorem 4.2 we can see that (x^*, y^*) satisfies the COP (10). Conversely, assume that (x^*, y^*) solves the COP (10) for some $p, q \in \mathbb{R}$. Since the objective function is non-positive and attains its maximum, $\langle y^*, L_1(x^*)\rangle + \langle y^*, L_2(x^*)\rangle - p - q = 0$. Thus, by Theorem 4.2, (x^*, y^*) is the equilibrium pair of $\Gamma(L_1, L_2)$. □

Consider a linear map $E : V \to W$ where $E(y) := \langle y, e_1\rangle e_2$. We observe and state the following propositions without proof.

Proposition 4.4 *If $\Gamma(L_1, L_2)$ has equilibrium values v_1 and v_2, then for any $a, b \in \mathbb{R}$ the bi-linear game $\Gamma(L_1 + aE, L_2 + bE)$ has equilibrium values $v_1 + a$ and $v_2 + b$.*

Proposition 4.5 *If (x^0, y^0) is an equilibrium pair for $\Gamma(L_1, L_2)$, then (x^0, y^0) is also an equilibrium pair for $\Gamma(L_1 + aE, L_2 + bE)$ where $a, b \in \mathbb{R}$. That is, for any $a, b \in \mathbb{R}$, $\Gamma(L_1, L_2)$ and $\Gamma(L_1 + aE, L_2 + bE)$ have same set of equilibrium pairs.*

Theorem 4.6 *Let $\Gamma(L_1, L_2)$ be a bi-linear game with $v_2 = 0$ and for some $(x^0, y^0) \in \mathcal{E}$, $S(y^0)$ be completely mixed (i.e.) $S(y^0) \subseteq \Delta_1^\circ$. Then,*

(i) $\dim(\ker L_2) = 0$ *or* 1.
(ii) *If* $\dim(\ker L_2) = 1$ *then* $S(y^0) = \{x^0\}$.

Proof Consider a bi-linear game $\Gamma(L_1, L_2)$ with $v_2 = 0$. Let $(x^0, y^0) \in \mathcal{E}$ with $S(y^0) \subseteq \Delta_1^\circ$. From the inequalities in (7), we can observe that $v_1 e_1 - L_1^t(y^0) \in K_1$ and $L_2(x^0) \in -K_2$. Since $x^0 \in K_1^\circ$, $v_1 e_1 - L_1^t(y^0) = 0$. Clearly, $\dim(\ker L_2) \geq 0$. Suppose it is non-zero, we claim that $\dim(\ker L_2) = 1$. On the contrary, let us assume $\dim(\ker L_2) \geq 2$. Now, $\ker L_2$ has at least two linearly independent vectors, say z^1, z^2 in V and either one of them has to be linearly independent with x^0. Without loss of generality, we consider z^1 is linearly independent with x^0. In case $\langle z^1, e_1\rangle = 0$, then by Lemma 2.1 there exists a $t > 0$ such that $x^* := x^0 - tz^1 \in \partial(K_1)$. In case $\langle z^1, e_1\rangle \neq 0$, without loss of generality we can take $\langle z^1, e_1\rangle = 1$. Then again by Lemma 2.1, there exists a $t > 0$ such that $x^* := (1 + t)x^0 - tz^1 \in \partial(K_1)$. In both cases, x^* is in $\partial\Delta_1$ and we claim that $x^* \in S(y^0)$. In case $\langle z^1, e_1\rangle = 1$, since $v_1 e_1 = L_1^t(y^0)$, $v_1 = \langle y^0, L_1(x^*)\rangle$ and so

$$\forall\, x \in \Delta_1 \;\; \langle v_1 e_1 - L_1^t(y^0), x\rangle \geq 0 \implies v_1 \geq \langle y^0, L_1(x)\rangle.$$

Now $\forall\, y \in \Delta_2$,

$$\begin{aligned}
\langle y^0, L_2(x^*)\rangle &= \langle y^0, L_2\big((1+t)x^0 - tz^1\big)\rangle \\
&= \langle y^0, (1+t)L_2(x^0)\rangle \\
&\geq \langle y, (1+t)L_2(x^0)\rangle \\
&= \langle y, (1+t)L_2(x^0)\rangle - \langle y, tL_2(z^1)\rangle.
\end{aligned}$$

Hence, $\langle y^0, L_2(x^*)\rangle \geq \langle y, L_2(x^*)\rangle$ for all y in Δ_2. This implies $x^* \in \partial\Delta_1 \cap S(y^0)$ which is a contradiction to the assumption that $S(y^0)$ is completely mixed. Similarly, in $\langle z^1, e_1\rangle = 0$ case also we arrive at a contradiction. Thus, $\dim(\ker L_2) = 1$. To prove the second part of the theorem, let $\dim(\ker L_2) = 1$. Suppose there exists $x \in S(y^0)$ different from x^0. If x is linearly independent with x^0, then as in the above proof, we can find a $x^* \in S(y^0)$ which is not completely mixed. Thus, x has to be linearly dependent with x^0. Since $x \in \Delta_1$, $x = x^0$. Hence $S(y^0)$ has unique element. □

Remark 3 Similar results hold for L_1^t if $v_1 = 0$ and $T(x^0)$ is completely mixed.

Theorem 4.7 *Let $\Gamma(L_1, L_2)$ be a bi-linear game. If* $\dim W < \dim V$*, then there exist x^* in $S(y^0)$ such that x^* is not completely mixed $\big((i.e.)\ x^* \in \partial\Delta_1\big)$.*

Proof Let $\dim W := m < \dim V := n$ and $(x^0, y^0) \in \mathcal{E}$. Without loss of generality, we assume that the equilibrium value of the player II (v_2) is zero. We claim $S(y^0)$ is not completely mixed. On the contrary, let us assume that $S(y^0)$ is completely mixed. From Theorem 4.6 and given hypothesis, it is clear that $\text{rank}(L_2) = n - 1 = m$. This implies $\text{null}(L_2) = 1$ and so by Theorem 4.6, $L_2(x^0) = 0$. Consider a subspace decomposition $V = \{x^0\} \oplus \{x^0\}^\perp$. Now consider a linear map $L := L_2|_{\{x^0\}^\perp}$. Here, $\text{rank}(L) = \dim\{x^0\}^\perp = m = \dim W$. Thus, there exists a vector, say $\pi \in \{x^0\}^\perp$, such that $L(\pi) = e_2$. Clearly, π and x^0 are linearly independent and so as in the proof of Theorem 4.6, there exists $x^* \in \partial\Delta_1$. Here, $L_2(x^*) = \alpha e_2$ for some non-zero $\alpha \in \mathbb{R}$ and further $v_1 e_1 - L_1^t(y^0) = 0$. It is easy to verify that $x^* \in S(y^0)$ which is a contradiction to our assumption that $S(y^0)$ is completely mixed. Hence $S(y^0)$ is not completely mixed. □

Remark 4 Similarly, we can prove if $\dim W > \dim V$ then $T(x^0)$ is not completely mixed.

Theorem 4.8 *If $\mathcal{E}$ is completely mixed, then* $\dim V = \dim W$ *and $\mathcal{E}$ has a unique pair.*

Proof Let $\mathcal{E}$ be completely mixed. From Theorem 4.7, it is clear that $\dim V = \dim W$. To show $\mathcal{E}$ has a unique equilibrium pair, let $(x^*, y^*) \in \mathcal{E}$ and without loss of generality we assume $v_2 = 0$. Since $S(y^*)$ is completely mixed and $v_2 = 0$, by Theorem 4.6, $\text{rank}(L_2) = n - 1$ and there exists unique x^* in $S(y^*)$. Now suppose $(x^1, y^1) \in \mathcal{E}$. Since $x^1 \in K_1^\circ$, $y^1 \in K_2^\circ$, it is easy to observe the following:

$$v_1' e_1 = L_1^t(y^1),$$

$$v_2' e_2 = L_2(x^1),$$

where $v_1' = \langle y^1, L_1(x^1)\rangle$ and $v_2' = \langle y^1, L_2(x^1)\rangle$. By using the above equations, we can observe that $(x^1, y^*) \in \mathcal{E}$. Since $S(y^*)$ has a unique element, $x^1 = x^*$. In a similar way, we can show that $y^1 = y^*$. Hence $\mathcal{E}$ has a unique equilibrium pair. □

5 Bi-linear Games and Complementarity Problems

In this section, we consider only the bi-linear games over the self-dual cones. Let K be a self-dual cone and $\widehat{L}$ be a linear map defined on V such that $\widehat{L}(K) \subseteq K$. Consider a linear map L defined as $L := sI - \widehat{L}$ where s is a fixed scalar and I is the identity map on V. Clearly, this is a **Z**-transformation. Any such linear map is called **M**-transformation if $\rho(\widehat{L}) \leq s$ and a non-singular **M**-transformation if $\rho(\widehat{L}) < s$. In this section, for a linear transformation of the form $L = sI - \widehat{L}$, we fix $e \in K^\circ \cap \widehat{L}(K)$ whenever $\widehat{L}$ is invertible or simply $e \in K^\circ$ otherwise. We consider the set Δ as the strategy set for both the players in $\Gamma(L)$ as well as $\Gamma(L, \widehat{L})$.

Since $\widehat{L}(K) \subseteq K$, we can observe that the equilibrium value of the player II(v_2) in the bi-linear game $\Gamma(L, \widehat{L})$ is always non-negative. So, we define the value of $\Gamma(L, \widehat{L})$ with respect to the value of the player I(v_1) as in the following definition:

Definition 4 Let L be a **Z**-transformation such that $L = sI - \widehat{L}$ with $\widehat{L}(K) \subseteq K$. Then we say that the value of the bi-linear game $\Gamma(L, \widehat{L})$ is positive if there exists an equilibrium pair (x^*, y^*) such that $\langle y^*, L(x^*)\rangle$ is positive.

Theorem 5.1 (Theorem 4.1 in [10]) *If L is semi-positive, then $\Gamma(L, \widehat{L})$ has value positive.*

From the above theorem, we can observe that the bi-linear game $\Gamma(L, \widehat{L})$ has a value positive whenever the linear game $\mathbf{\Gamma(L)}$ has a value positive. It is evident from the following example that the converse of Theorem 5.1 need not be true.

Example 5.2 Let $\widehat{L} : \mathbb{R}^2 \to \mathbb{R}^2$ be the linear transformation defined by the matrix $[\widehat{L}] = \begin{bmatrix} 4 & 3 \\ 2 & 2 \end{bmatrix}$ and $L : \mathbb{R}^2 \to \mathbb{R}^2$ be defined by $[L] = 5I_2 - [\widehat{L}] = \begin{bmatrix} 1 & -3 \\ -2 & 3 \end{bmatrix}$ with respect to the standard basis where I_2 denotes the identity matrix of order 2.

In the linear game settings, Gowda and Ravindran showed that for a **Z**-transformation, semi-positivity is equivalent to the value being positive but it is not true in the bi-linear game setting. However, the converse of Theorem 5.1 holds with the assumption of some weaker conditions that are shown in the following theorem.

Theorem 5.3 (Theorem 4.3 in [10]) *For some $(x^*, y^*) \in \mathcal{E}$, if $\Gamma(L, \widehat{L})$ has value positive and $S(y^*)$ has at least one completely mixed strategy, then L is semi-positive.*

The following example illustrates that the converse of Theorem 5.3 need not be true.

Example 5.4 Let $\widehat{L} : \mathbb{R}^2 \to \mathbb{R}^2$ be the linear transformation defined by the matrix $[\widehat{L}] = \begin{bmatrix} 4 & 3 \\ 2 & 2 \end{bmatrix}$ and $L : \mathbb{R}^2 \to \mathbb{R}^2$ be defined by $[L] = 6I_2 - [\widehat{L}] = \begin{bmatrix} 2 & -3 \\ -2 & 4 \end{bmatrix}$ with respect to the standard basis.

Lemma 4.5 in [10] says that for a **Z**-transformation $L = sI - \widehat{L}, s > 0$ whenever the value of $\Gamma(L, \widehat{L})$ is positive. From Example 5.2, we can observe that it is not necessary to hold $s \geq \rho(\widehat{L})$. However, it does hold with some condition on $\Gamma(L, \widehat{L})$ that is shown in Theorem 4.6 in [10]. In particular, they proved the following theorem.

Theorem 5.5 *Consider a* **Z**-*transformation* $L := sI - \widehat{L}$. *Suppose there exists* $(x^*, y^*) \in \mathcal{E}$ *such that the value of* $\Gamma(L, \widehat{L})$ *is positive and* $x^* \in K^\circ$. *Then the following statements hold:*

(1) Linear complementarity problem (L, K, q) *has global solvable property.*
(2) The dynamical system $\dot{x} + L(x) = 0$ *is asymptotically stable.*
(3) L *is invertible with* $L^{-1}(K) \subseteq K$.
(4) L *is a non-singular* **M**-*transformation.*

The following examples illustrate the connection between the bi-linear games and complementarity problems given in the above theorem.

Example 5.6 Consider the following **Z**-transformations defined on $\mathbb{R}^2$ and $\mathbb{R}^3$.

$$(i)\ [L_1] := \begin{bmatrix} 4 & -4 \\ -3 & 5 \end{bmatrix};\ (ii)\ [L_2] := \begin{bmatrix} 7 & -3 & -5 \\ -3 & 4 & -2 \\ -5 & 0 & 7 \end{bmatrix}$$

.

(i) By considering $[\widehat{L_1}] = \begin{bmatrix} 2 & 4 \\ 3 & 1 \end{bmatrix}$, we can see that the bi-linear game $\Gamma(L_1, \widehat{L_1})$ has $(\frac{3}{4}, \frac{1}{4})$ as an equilibrium strategy and $\frac{1}{2}$ as an equilibrium value for player I.

(ii) By considering $[\widehat{L_2}] = \begin{bmatrix} 1 & 3 & 5 \\ 3 & 4 & 2 \\ 5 & 0 & 1 \end{bmatrix}$, we can see that for the bi-linear game $\Gamma(L_2, \widehat{L_2})$, $(\frac{13}{27}, \frac{4}{27}, \frac{10}{27})$ and $\frac{1}{71}$, respectively, are equilibrium strategy and equilibrium value of player I. Thus, we can conclude from the above theorem that both L_1 and L_2 have **Q**-property.

In [10], a theorem which is analogous to Theorem 6 in [12] is proved. Here, we state the theorem without the proof which will be used to characterize the equilibrium pairs of the completely mixed bi-linear games corresponding to the **Z**-transformations.

Theorem 5.7 (Theorem 4.8 in [10]) *Suppose both* L *and* $\widehat{L}$ *are invertible with* $L^{-1}(K) \subseteq K$. *Then,*

$$v_1 = \frac{1}{\langle (L^t)^{-1}(e), e\rangle} \text{ and } v_2 = \frac{1}{\langle (\widehat{L})^{-1}(e), e\rangle},$$

respectively, are equilibrium values of players I and II corresponding to the equilibrium pair (x^*, y^*) *where* $x^* := v_2(\widehat{L})^{-1}(e)$ *and* $y^* := v_1(L^t)^{-1}(e)$.

It is to be noted that strategies x^* and y^* defined in the above theorem are in fact completely mixed strategies. In the following theorem, we discuss the converse of Theorem 5.3, equivalence of the statements in Theorem 5.5 and the equilibrium characterization of the completely mixed bi-linear games. Though part of the theorem has appeared in [10], for completeness we provide a proof for all the statements.

Theorem 5.8 *Consider a* **Z***-transformation* $L := sI - \widehat{L}$. *Suppose* $\widehat{L}$ *is invertible. Then the following are equivalent:*

1. L *is invertible with* $L^{-1}(K) \subseteq K$.
2. *There exist an equilibrium pair* (x^*, y^*) *such that the value of* $\Gamma(L, \widehat{L})$ *is positive and* x^* *is completely mixed.*
3. L *is semi-positive.*

Moreover, when $\Gamma(L, \widehat{L})$ *is completely mixed, the following holds.*

(a) $\Gamma(L, \widehat{L})$ *has the unique optimal strategy pair* (x^*, y^*) *where* $x^* := \frac{1}{\langle (\widehat{L})^{-1}(e), e\rangle}(\widehat{L})^{-1}(e)$ *and* $y^* := \frac{1}{\langle (L^t)^{-1}(e), e\rangle}(L^t)^{-1}(e)$.

(b) *The values of the players I and II are* $v_1 = \frac{1}{\langle (L^t)^{-1}(e), e\rangle}$ *and* $v_2 = \frac{1}{\langle (\widehat{L})^{-1}(e), e\rangle}$, *respectively.*

Proof (1) $\Rightarrow$ (2) : Assume L^{-1} exists and $L^{-1}(K) \subseteq K$. Now, consider x^*, y^* defined as in Theorem 5.7. It is clear that (x^*, y^*) is an equilibrium pair with x^* completely mixed and the equilibrium value corresponding to (x^*, y^*) is positive. (2) $\Rightarrow$ (1) follows from Theorem 5.5. We know the equivalence of (1) and (3) from Theorem 2.2. Now assume $\Gamma(L, \widehat{L})$ is completely mixed.

(a) By Theorem 5.7, we see that the pair (x^*, y^*) where $x^* := \frac{1}{\langle (\widehat{L})^{-1}(e), e\rangle}(\widehat{L})^{-1}(e)$ and $y^* := \frac{1}{\langle (L^t)^{-1}(e), e\rangle}(L^t)^{-1}(e)$ is an equilibrium pair. And from Theorem 4.8, we can say that (x^*, y^*) is unique.

(b) Again by Theorem 5.7, we see that the values of players I and II are $\frac{1}{\langle (L^t)^{-1}(e), e\rangle}$ and $\frac{1}{\langle (\widehat{L})^{-1}(e), e\rangle}$, respectively. □

Concluding Remarks

In this article, we have given a brief survey on the linear and bi-linear games and proved the converse of Theorem 1.1. Then we have consolidated some results on bi-linear games. In particular, we proved the uniqueness of equilibrium pairs for completely mixed bi-linear games. Using this, we characterized the equilibrium pairs of completely mixed bi-linear games corresponding to a special type of **Z**-transformations. In addition to that, we have related the bi-linear game results to the well-known results on **Z**-transformations in matrix theory and linear complementarity problems.

Acknowledgements The authors thank the anonymous referees for their valuable comments and suggestions on the earlier version of this article. The first author thanks the University Grants Commission, India, for the financial support via UGC Senior Research Fellowship (SRF-S.ID: 421898).

References

1. Berman, A., M. Neumann, and R..J. Stern. 1989. *Nonnegative matrices in dynamic systems*. Pure and Applied Mathematics (New York). New York: Wiley Inc. A Wiley-Interscience Publication.
2. Berman, A., and R..J.. Plemmons. 1994. *Nonnegative matrices in the mathematical sciences*, vol. 9. Classics in Applied Mathematics. Philadelphia, PA: Society for Industrial and Applied Mathematics (SIAM). https://doi.org/10.1137/1.9781611971262. Revised reprint of the 1979 original.
3. Blackwell, D. 1961. Minimax and irreducible matrices. *Journal of Mathematical Analysis and Applications* 3: 37–39. https://doi.org/10.1016/0022-247X(61)90005-1.
4. Cottle, R.W., J.S. Pang, and R.E. Stone. 1992. *The linear complementarity problem*. Computer Science and Scientific Computing. Boston, MA: Academic Press Inc.
5. Ferris, M.C., and J.S. Pang. 1997. Engineering and economic applications of complementarity problems. *SIAM Review* 39 (4): 669–713. https://doi.org/10.1137/S0036144595285963.
6. Fiedler, M., and V. Pták. 1962. On matrices with non-positive off-diagonal elements and positive principal minors. *Czechoslovak Mathematical Journal* 12 (87): 382–400.
7. Gale, D., and H. Nikaidô. 1965. The Jacobian matrix and global univalence of mappings. *Mathematische Annalen* 159: 81–93. https://doi.org/10.1007/BF01360282.
8. Garg, J., A.X. Jiang, and R. Mehta. 2011. Bilinear games: Polynomial time algorithms for rank based subclasses. In *International Workshop on Internet and Network Economics*, 399–407. Berlin, Heidelberg: Springer.
9. Gokulraj, S., and A. Chandrashekaran. 2019. On symmetric linear games. *Linear Algebra and its Applications* 562: 44–54. https://doi.org/10.1016/j.laa.2018.10.004.
10. Gokulraj, S., and A. Chandrashekaran. 2020. Linear complementarity problems and bi-linear games. *Applications of Mathematics* 65: 665–675. https://doi.org/10.21136/AM.2020.0371-19
11. Gowda, M.S. 2016. Completely mixed linear games on a self-dual cone. *Linear Algebra and its Applications* 498: 219–230. https://doi.org/10.1016/j.laa.2015.05.025.
12. Gowda, M.S., and G. Ravindran. 2015. On the game-theoretic value of a linear transformation relative to a self-dual cone. *Linear Algebra and its Applications* 469: 440–463. https://doi.org/10.1016/j.laa.2014.11.032.
13. Gowda, M.S., and J. Tao. 2009. **Z**-transformations on proper and symmetric cones: **Z**-transformations. *Mathematical Programming* 117 (1–2, Ser. B): 195–221. https://doi.org/10.1007/s10107-007-0159-8.
14. Isac, G. 1992. *Complementarity problems*, vol. 1528. Lecture Notes in Mathematics. Berlin: Springer.
15. Kaneko, I. 1978. A linear complementarity problem with an $n \times 2n$ "P"-matrix. complementarity and fixed point problems. *Mathematical Programming Studies* 7: 120–141.
16. Kaneko, I. 1978. Linear complementarity problems and characterizations of Minkowski matrices. *Linear Algebra and its Applications* 20 (2): 111–129. https://doi.org/10.1016/0024-3795(78)90045-9.
17. Kaplansky, I. 1945. A contribution to von Neumann's theory of games. *Annals of Mathematics* 46 (2): 474–479. https://doi.org/10.2307/1969164.
18. Karamardian, S. 1972. The complementarity problem. *Mathematical Programming* 2 (1): 107–129. https://doi.org/10.1007/BF01584538.

19. Karlin, S. 1959. *Mathematical methods and theory in games, programming and economics. Vol. I: Matrix games, programming, and mathematical economics*. Reading, Mass.-London: Addison-Wesley Publishing Co., Inc.
20. Murty, K.G. 1988. *Linear complementarity, linear and nonlinear programming*, vol. 3. Sigma Series in Applied Mathematics. Berlin: Heldermann.
21. Nash, J. 1950. Equilibrium points in n-person games. *Proceedings of the National Academy of Sciences of the United States of America* 36: 48–49. https://doi.org/10.1073/pnas.36.1.48.
22. Nash, J. 1951. Non-cooperative games. *Annals of Mathematics* 2 (54): 286–295. https://doi.org/10.2307/1969529.
23. Nikaidô, H. 1968. *Convex structures and economic theory*, vol. 51. Mathematics in Science and Engineering. New York-London: Academic Press.
24. Orlitzky, M.J. 2017. *Positive Operators, Z-Operators, Lyapunov Rank, and Linear Games on Closed Convex Cones*. Ann Arbor, MI: ProQuest LLC. http://gateway.proquest.com/openurl?url_ver=Z39.88-2004&rft_val_fmt=info:ofi/fmt:kev:mtx:dissertation&res_dat=xri:pqm&rft_dat=xri:pqdiss:10268358. Thesis (Ph.D.)–University of Maryland, Baltimore County
25. Parthasarathy, T., and T.E.S. Raghavan. 1971. *Some topics in two-person games*. Modern Analytic and Computational Methods in Science and Mathematics, No. 22. New York: American Elsevier Publishing Co., Inc.
26. Raghavan, T.E.S. 1965. On linear operators leaving a convex set invariant in Banach spaces. *Sankhyā Series A* 27: 293–302.
27. Raghavan, T.E.S. 1970. Completely mixed strategies in bimatrix games. *Journal of the London Mathematical Society* 2 (2): 709–712. https://doi.org/10.1112/jlms/2.Part_4.709.
28. Raghavan, T.E.S. 1978. Completely mixed games and M-matrices. *Linear Algebra and Its Applications* 21 (1): 35–45. https://doi.org/10.1016/0024-3795(87)90198-4.
29. Schneider, H., and M. Vidyasagar. 1970. Cross-positive matrices. *SIAM Journal on Numerical Analysis* 7: 508–519. https://doi.org/10.1137/0707041.
30. Varga, R.S. 2000. *Matrix iterative analysis*, vol. 27. Expanded. Springer Series in Computational Mathematics. Berlin: Springer. https://doi.org/10.1007/978-3-642-05156-2.

Social Preferences and the Provision of Public Goods

Sudipta Sarangi and Sakshi Upadhyay

How selfish soever man may be supposed, there are evidently some principles in his nature, which interest him in the fortune of others, and render their happiness necessary to him, though he derives nothing from it, except the pleasure of seeing it [45, p. 9]

JEL Classification: D91 · H41 · D85 · C72

1 Introduction

Public goods are characterized by *non-rivalry*, meaning that more than one person can simultaneously benefit from them, and *non-exclusivity*, meaning that it is difficult to prevent any individual from enjoying their benefits. They simultaneously benefit many people and their creation requires the coordinated actions of people who will subsequently enjoy its benefits. Environmental protection, research and innovation, vaccination, health care services, highways, and public parks are just a few important examples.

Despite receiving benefits from public goods, individuals tend to free ride on the contributions of others in a group. Given that these goods are non-rival and non-excludable, it is evident that once the goods have been produced, every agent can consume them regardless of their contribution. Unless there exist mechanisms to make individuals act in their common interests, rational or self interested individuals will not act to achieve their common or group interests [37]. There are various

S. Sarangi · S. Upadhyay (✉)
Department of Economics, 880 West Campus Drive Virginia Tech, Blacksburg, VA 24061, USA
e-mail: usakshi@vt.edu

S. Sarangi
e-mail: ssarangi@vt.edu

S. Borkotokey et al. (eds.), *Game Theory and Networks*, Indian Statistical Institute Series,
https://doi.org/10.1007/978-981-16-4737-6_3

mechanisms to ensure cooperation or reduce free riding. We discuss punishment, commitment, and communication as some of the mechanisms which help increase cooperation in a society.

An alternative explanation for the cooperation seen can be provided by social preferences. Social norms and preferences also matter in the provision of public goods. For example, local resources can be managed well when users care about others and can organize and enforce their own rules, instead of following externally imposed norms [20].

The importance of social norms and preferences can also be seen in our daily lives. Consider a family with grandparents/parents and their children. Parents or grandparents might invest in infrastructure, environment, technology to mitigate climate change so that it can benefit their future generations. They have an added incentive to invest in highways, public schools, public parks, environment-friendly vehicles or practices because of the concern for their children. Another way to understand this idea is that your family member might contribute more to public goods that will benefit you in the future as compared to a stranger.

In the following subsection, we start with a simple public goods model and show how free riding is an equilibrium in this game. Our goal in this chapter is to examine the reasons for the absence of free riding, in particular, by focusing on the role of social preferences. In Sect. 2, we provide evidence on the absence of free riding and also discuss mechanisms available in the literature to ensure cooperation. Section 3 describes how social preferences can explain cooperation. The section also entails a model on social preferences and the equilibrium after incorporating social preferences. In Sect. 4, we will discuss models of various types of social preferences available in the literature. These theoretical models are also supplemented with experimental evidence. Section 5 discusses how social preferences influence public goods provisioning in a coalition or network framework. Section 6 concludes the chapter.

1.1 A Simple Model of Public Goods

To fix ideas formally, we now present the public goods model in Fehr and Schmidt [18]. We will use this model to arrive at a fundamental result in public goods which will also be the first Proposition of this chapter. Let there be $n \geq 2$ individuals in a society who simultaneously decide on their contribution levels $g_i \in [0, y]$, $i \in [1, 2, \ldots, n]$ to the public good. Each player has an endowment of y. The monetary payoff of player i is given by [18, p. 836, Eq. 11]

$$x_i(g_1, g_2, \ldots, g_n) = y - g_i + a \sum_{j=1}^{n} g_j, \qquad 1/n < a < 1 \tag{1}$$

Here a denotes the constant marginal return to public good $G = \sum_{i=1}^{n} g_i$. Since $a < 1$, contributing to G leads to loss of $1 - a$. The dominant strategy of an individual i is to choose $g_i = 0$.

Definition 1 A strategy g_i^* is a Nash equilibrium of this game if for all $i \in [1, 2, \ldots n]$, $x_i(g_i^*, g_{-i}) \geq x_i(g_i^{'}, g_{-i})$ for all $g_i' \in (0, y]$.

In this game, we have $g_i^* = 0$ as the Nash equilibrium strategy for any player $i \in [1, \ldots\ldots, n]$. Strategy other than $g_i* = 0$ is denoted by $g_i^{'}$. Strategy of players other than i is given by g_{-i}.

Thus, the standard model predicts $g_i = 0$ for all $i \in [1, 2, \ldots, n]$. However, since $a > 1/n$, aggregate monetary payoff is maximized at $g_i = y$. This observation leads us to fundamental result about public goods in economics summarized in our first Proposition.

Proposition 1 *Suppose the payoff function is given by Eq. 1 and satisfies* $1/n < a < 1$, *then in Nash equilibrium* $g_i = 0$ *for all* $i \in [1, 2, \ldots, n]$.

Not contributing to the public goods is termed as 'free riding'. Kim and Walker [30] summarize the 'free-rider' problem in their theoretical model: "If the method of voluntary contributions is used to determine the level at which public goods will be provided, then the resulting provision level will be far below the optimal level, and many individuals will contribute nothing at all." Free riding is also evident in their experimental results.

In practice, however, we may not always see free riding. In the next section, we provide experimental and empirical evidence on lack of free riding. Various mechanisms to ensure cooperation or avoid free riding will also be discussed in the next section. In a later section, we examine how social preferences can be used to explain cooperation or absence of free riding.

2 Evidence on Free Riding Behavior

Free riding has been a widely accepted notion in the literature of public goods games. Previous theory suggests that players try to get the benefit from public goods without contributing towards it. However, those results are in sharp contrast to the existence of cooperative behavior among individuals in real-life public goods games. This behavior has been substantiated by data from national surveys as shown in Andreoni [1], who states: "Around 85% of households make donations to charity, 50% of tax returns include charitable deductions". Another related evidence of cooperative behavior can be found in voting in elections. Individuals tend to vote in elections, even though economic theory predicts that free riding will be higher as the decisive power of one vote is low. Countries joining International Environment Agreements (IEA) to solve environmental issues is also an example of cooperation. Group of

77 (G77), United Nations Framework Convention on Climate Change (UNFCCC), Kyoto Protocol are some of the existing IEA's.

The contrast between the theoretical predictions and real-life evidence motivated the testing of 'free-rider' hypothesis in the lab. Ledyard [34] surveys the experimental literature on public goods before 1995. Some of the prominent papers included in the survey are Marwell and Ames [35], Isaac et al. [29], Isaac and Walker [28], and Andreoni [2]. One of the findings from these experiments suggests that individuals contribute more than the Nash equilibrium prediction in a public goods game. As we saw in the previous section, the Nash equilibrium in a public good game is to free ride, however, it is optimal to contribute the full amount. On an average, contributions were about 40–60% of the optimal level in these experiments. However, the contributions varied over individuals. The other common observation is that contributions start at 40–60% of the optimal level but over the periods decline to 'free riding' outcome.

This decay in contribution levels was further analyzed in Andreoni [2] through 'learning' and 'strategies' hypothesis. According to the learning hypothesis, repeated periods allow individuals to learn the incentives from the game which can explain the fall in contribution levels. At the same time learning also allow players to signal future moves to each other. This leads to the strategy hypothesis, where in a repeated games a rational player will develop multi period strategies that can lead to cooperative behavior. However, Andreoni finds no significant support for either of these hypotheses which could have explained the decay experienced in these games. We now state our first observation from findings in this section.

Observation 1: *Empirical and experimental evidence show that individuals cooperate and contribute to public goods as opposed to the theoretical predictions.*

These experimental results motivated research on the importance of institutional environment which can help in achieving the optimal outcome or reduce free riding. In next subsection, we discuss these mechanisms which can help further increase cooperation in a public goods game.

2.1 Mechanisms that Avoid Free Riding

This section summarizes the institutional environments which have been used in the literature to reduce the incidence of 'free riding'. Institutional environment refers to the context or setting in which individuals would make their decisions. The payoff function remains the same as in Eq. 1, however, we look into different settings under which the public goods game is played. We discuss three such institutional environments: communication, commitment, and punishment.

Communication

Communication between the participants regarding their strategies or intentions can help in increasing contributions to a public goods game. Isaac and Walker [28] were the first to test face-to-face communication as a means to reduce 'free riding'. According to the authors, "The role of communication is to a) help the group understand the group profit implications for different allocations and b) build credibility to the expected decisions of group members". Communication thus helps in learning the optimal strategy (contributing to public goods). Ostrom [38] also finds that face-to-face communication can sustain cooperation even through the last period. Communication enforces no verbal agreement and hence can be thought of as 'cheap talk' [39]. This paper summarizes the findings on collective action and one of the findings suggests that when communication is implemented by allowing subjects to signal promises through their computer terminals, much less cooperation is observed as compared to the case when subjects are allowed face-to-face communication.

Communication enhances cooperation, however, the effectiveness of communication depends on its structure and the level of private information among players. Palfrey et al. [41] provide an answer to this problem both theoretically and empirically. They find theoretical bounds on efficiency gains that can be attained through different modes of communication by using the Bayesian-mechanism design. The bounds depend upon the distribution of private information(value of endowed unit of output) and on the richness of the message space (communication structure). The authors choose three forms of pre-play communication: binary message (intention to contribute or not), practice game (announce their contribution against different contribution costs), and natural language communication (exchange of chat messages) in order to test their theoretical bounds. The results from their experiment find efficiency and public goods provisioning to be significantly higher in case of natural language communication as efficiency bounds predicted by the theoretical model were only achieved in this treatment. This might be because "unrestricted chats give subjects an opportunity to understand each other's intentions and messages". Natural language communication can be thought of as a more personal form of communication which gives more scope to convey an individual's message and intentions than a restricted message or any other form.

Commitment

Commitment can also be used as a strategy to enhance cooperation. "Commitment is a means by which players can assure one another that they are not going to free ride on others' contributions, so that group members can contribute without fearing that they will be free ridden" [32]. Chen [14] was one of the first paper to use 'pledge to contribute' as a commitment. The authors find that group-based pledge (subjects make a pledge before making a contribution, are given feedback and have to then contribute a proportion of the mean pledge) and face-to-face communication have similar results in enhancing cooperation. Through commitment, individuals can eliminate free riding. However, once an individual makes a commitment, he/she is more vulnerable for being free ridden [32]. This is because individuals might use commitment by others as an opportunity to 'free ride' on their contribution. To respond to this issue authors design a mechanism, where players can commit to cooperating to a small degree and then observe other player's reciprocal contributions. The mechanism allows participants to signal their commitment without exposing them to be 'free-ridden'. They test for the efficiency of different 'pledges for contribution'. The study finds that 'increase only' pledge is effective in increasing cooperation. This mechanism works as a commitment strategy by not letting the players reverse their contributions and allows players to reduce their extent of free riding by limiting their commitments.

Punishment

People who cooperate might be willing to 'punish' the free riders. Ostrom et al. [40] was one of the first papers to test the impact of punishment in a public goods framework. The authors allow for costly punishments in a repeated common pool resource game and find that participants punish free riders in their experiment. However, in their paper, the same subjects interacted for multiple periods, thus giving them an incentive to cooperate and punish free riders. To rule out these incentives, Fehr and Gachter [19] in their experiment have a punishment and non-punishment treatment crossed with a stranger (group composition changes every period) and partner treatment (group composition is fixed). The authors find that in both the treatments, the punishment is heavier if the more negatively individual deviates from the contributions of group members. The average contribution goes up in both the stranger and partner treatment when punishment is allowed and approaches to full cooperation in partner treatment.

Previous experiments which studied the role of punishments could not elicit much about the robustness of punishment schemes. Nikiforakis and Normann [36] in their paper provide a comparative statistics of punishment in public goods games. They find that contributions to public goods increase monotonically in the effectiveness of punishment (factor by which the punishment reduces the punished player's income). Higher effectiveness leads us near to social optimal outcome.

Individuals do not contribute in a public goods game, due to the chance of being 'free ridden' by others. All the mechanisms discussed above change the environmental setting of a game in a manner which increases the incentive to cooperate. The success of the mechanism depends upon how effective it is in reducing chances of being 'free ridden'.

3 Social Preferences: An Alternative Explanation

While the experimental literature has provided us with examples of several mechanisms that can lead to free riding and reduce cooperation, we now focus on an alternative approach to explain these findings: the presence of *social preferences*. Theories of others regarding preferences/social preferences are based on the assumption (and observation) that people care about the well-being of others. In his paper, Andreoni [5] shows that, on an average, about half of all cooperation is due to subjects who understand free riding but cooperate due to kindness. The author also suggests that the decline in cooperation observed in multiple trials of public goods experiment might not be due to learning, but maybe a result of frustrated attempts at kindness.

According to Fehr and Fischbacher [20]: 'An individual exhibits social preference if the person cares about material resources allocated to relevant reference agents'. The relative reference agent can vary according to different domains, thus resulting in various types of social preferences. The authors empirically also show that it is difficult to understand concepts of competition on market outcomes, laws governing cooperation and collective action, optimal contracts and property rights, social norms and market failures without incorporation of social preferences.

Nash equilibrium strategy of players in a public goods game is to contribute nothing. However, past literature suggests clear evidence of cooperation among players. Players' incentive to contribute positively can be predicted theoretically by including social preferences in their payoff functions. Examples of such social preferences include the responsibility of the older generation (grandparents/parents) towards their future generation. Such responsibility drives elders to contribute positively towards any public goods or service which will guarantee a secure future for their children. We illustrate such cooperative behavior using a model of social preferences from Fehr and Schmidt [18]. In the later subsections, we introduce different models of social preferences.

3.1 A Simple Model of Public Goods with Social Preferences

In order to show how the results in a public goods model (Proposition 1) change after incorporation of social preferences, we use the inequity aversion model of Fehr and Schmidt [18]. In this model, in addition to purely selfish individuals, the authors assume the presence of subjects who dislike inequity both when they are worse off than other players and also when they are better off than other players.

Consider a set of n players indexed by $i \in [1, 2, \ldots, n]$ and let $x = x_1, x_2, \ldots, x_n$ denote vector of monetary payoffs. The utility function of $i \in [1, 2, \ldots, n]$ is given by

$$U_i(x) = x_i - \alpha_i\left(\frac{1}{n-1}\sum_{j\neq i} max|x_j - x_i, 0|\right) - \beta_i\left(\frac{1}{n-1}\sum_{j\neq i} max|x_i - x_j, 0|\right) \tag{2}$$

The second term in Eq. 2 measures loss from disadvantageous inequality, the third term measures loss from advantageous inequality. The two parameters α_i and β_i measure player i's utility loss from disadvantageous inequality and from advantageous inequality. The authors assume that $\beta_i \leq \alpha_i$ and $0 \leq \beta_i < 1$. $\beta_i \leq \alpha_i$ implies, players suffer more from inequality that is to their disadvantage, i.e., the subject is loss averse in social comparisons. $\beta_i \geq 0$, rules out the subjects who like to be better than others.

We now substitute Eq. 1 in Eq. 2 to see how public goods provisioning changes due to presence of inequity aversion. For this result, we focus on Proposition 4c of Fehr and Schmidt [18] which discusses positive contribution levels of individuals.[1] Player i who does not contribute ($g_i = 0$) is a 'free rider'. Let number of free riders be represented by k. Recall from Eq. 1, g_i and a denote the contribution levels and marginal return to public good, respectively.

Proposition 2 ([18, p. 839 Proposition 4(c)]) *If $k/(n-1) < (a + \beta_j - 1)/(\alpha_j + \beta_j)$ for all players $j \in [1, 2, \ldots, n]$ with $a + \beta_j > 1$, then other equilibria with positive contribution levels does exist. In this equilibria, all k players with $a + \beta_i < 1$ must choose $g_i = 0$, while all other players contribute $g_i = g \in [0, y]$. Note further that $(a + \beta_j - 1)(\alpha_j + \beta_j) < a/2$.*

We first discuss the author's intuition behind the proof and then move towards the sketch of the proof. If there are sufficiently many players with $a + \beta_j > 1$, they can sustain cooperation among themselves even when other players are free riding. This only holds when contributors are not affected much by the disadvantageous inequality. This is because if α_j increases, it is less likely to be the case that: $k/(n - 1) < (a + \beta_j - 1)/(\alpha_j + \beta_j)$

Sketch of the proof

- Following from the author's Proposition 4a, the dominant strategy of k free riders, with $a + \beta_i < 1$, is $g_i = 0$ (not contribute). This is because free rider's return from public good (a) and non-pecuniary benefit from reducing inequality(β_i) is less than 1.
- The remaining $n - k$ or j players with $a + \beta_j > 1$ contribute positively with $g_i = g \in [0, y]$. j's payoff is given by

$$U_j(g) = y - g + (n-k)ag - \alpha_j\left(\frac{1}{n-1}kg\right) \tag{3}$$

Any individual who contributes is deprived of the advantageous utility which reduces the third term in Eq. 2 to zero, thereby forming Eq. 3.

[1] The case of free rider ($g_i = 0$) is studied in part a and b of the Proposition 4 of the original paper.

Suppose player deviates from contributing g to $g - \Delta$, such that $\Delta > 0$. The deviation strategy towards contributing less than g will not payoff if and only if $U(g - \Delta) \leq U(g)$. Simplifying this inequality leads us to the following condition: $k/(n-1) \leq (a + \beta_j - 1)/(\alpha_j + \beta_j)$.

- Following from author's Proposition 4b, if there are only a few players with $a + \beta_i > 1$, they would suffer too much loss from the disadvantageous inequality caused by the free riders. The proof given by the authors shows that if a potential contributor knows that the number of free riders, k, is larger than $a(n-1)/2$, then he will not contribute either.

4 Types of Social Preferences

As seen in the previous section, incorporation of inequity aversion in the standard utility functions, predicts cooperation in a public goods game. Depending on the assumptions of the model and the type of social preference, the model can look different. For instance, we can have altruism as a social preference, incorporated into the standard utility function. However, the mechanism to arrive at the equilibrium will be similar and will lead to positive contributions being made to the public goods game. In the next subsection, we will discuss other papers on fairness and inequity aversion. In the later subsections, we explain models with different social preferences and their outcomes.

4.1 Fairness and Inequity Aversion

Inequity aversion implies that individuals care for equitable distribution of resources or equal outcomes. These models consider an individual 'fair' if the individual is willing to give up their payoff to help others. A model of fairness is represented in Eq. 2. The second and third term which measures the individual loss from disadvantageous and advantageous inequality, respectively, are a measure of fairness in their model.

Rabin [42] was one of the first to develop game-theoretic solution concept "fairness equilibria". An outcome is considered to be fair if the intention behind the action is kind, whereas if the intention is hostile, the action is considered to be unfair. The model is applicable to all finite-strategy games involving two players. Each player's expected subjective utility depends on: his strategy, his beliefs about other player's strategy choices, and his beliefs about other player's beliefs about his strategy.

Let $a_1 \in S_1$ and $a_2 \in S_2$ represent strategies chosen by two players; $b_1 \in S_1$ and $b_2 \in S_2$ represent player 2's belief about strategy player 1 is choosing, and player 1's belief about what strategy player 2 is choosing. $c_1 \in S_1$ and $c_2 \in S_2$ represent player 1's belief about what player 2 believes player 1's strategy is, and player 2's beliefs about what player 1 believes player 2's strategy is.

Each player i chooses a_i to maximize expected utility

$$U_i(a_i, b_j, c_i) = \pi_i(a_i, b_j) + \bar{f}_j(b_j, c_i).[1 + f_i(a_i, b_j)] \tag{4}$$

- $\pi_i(a_i, b_j)$ is individual i's material payoff.
- Player i's kindness to player j is measured by $f_i(a_i, b_j)$. The function measures how much more than or less than player j's equitable payoff[2] player i believes he is giving to player j. When $f_i = 0$, player i is giving j her equitable payoff. If $f_i > 0$, player i is giving j more than her equitable payoff. When $f_i < 0$, player i is giving j less than her equitable payoff.
- $\bar{f}_i(b_j, C_i)$ measures player i's belief about how kind player j is being to him. If player i believes that player j is treating him badly ($\bar{f}_i(b_j, C_i) < 0$), then i chooses a_i such that $f_i(a_i, b_j)$ is low or negative. The opposite situation occurs when $\bar{f}_i(b_j, C_i) > 0$.

The above game by Rabin is a psychological game of the type described by Geanakoplos et al. [25]. The equilibrium concept in these games is called psychological Nash equilibrium which is an analog of Nash equilibrium. The psychological Nash equilibrium concept imposes an additional condition that all higher order beliefs match actual behavior. Rabin uses psychological Nash equilibrium to arrive at the *fairness equilibrium*, which we describe in the next definition.

Definition 2 ([42, p. 1288, Definition 3]) The pair of strategies $(a_1, a_2) \in (S_1, S_2)$ is fairness equilibrium if for $i = 1, 2, j \neq i$

- $a_i \in \arg\max_{a \in S_i} U_i(a, b_j, c_i)$
- $c_i = b_i = a_i$

According to the above definition, an individual i's strategy (a_i) should maximize her payoff. The strategy should also be equal to player j's belief about player i's strategy (b_i) and player i's belief about what player j believes player i's strategy is (c_i). Thus, individuals actions and their higher order beliefs both match their actual behavior.

A mutual-max(min) outcome is the one where the player's mutually maximize (minimize) each other's payoffs. We now discuss one of the Propositions which talks about two types of Nash equilibrium being 'fairness equilibrium'.

Proposition 3 ([42, p. 1290, Proposition 1]) *Suppose that (a_1, a_2) is a Nash equilibrium, and either a mutual-max outcome or a mutual-min outcome. Then (a_1, a_2) is a fairness equilibrium.*

The proof is intuitive. First, suppose (a_1, a_2) is mutual-max outcome, then both f_1 and f_2 are non-negative. This implies players have positive regard for each other. Since both players are choosing a strategy that maximizes their payoff and payoff of

[2] Equitable payoff is the average of highest and lowest payoff of player j.

other players, this must maximize their own utility. Now suppose (a_1, a_2) is mutual-min outcome, then f_1 and f_2 will be non-positive, both players would like to decrease the well-being of others. Simultaneously, player also maximizes his own utility by maximizing his material well-being.

We can draw predictions by applying a prisoner dilemma game into a public goods framework with only 2 players. The Nash equilibrium in a prisoner's dilemma game is to defect, which can be interpreted as no cooperation in a public goods game. Incorporation of reciprocal motives, in a public goods game can lead to full cooperation as one of the equilibrium. The implications will be difficult if there are more than two people (which is usually the case). The payoff function incorporates the stylized facts evident in many experiments: people are willing to sacrifice their own well-being to help those who are kind, people are willing to sacrifice their own well-being to punish those who are unkind. However, Rabin [42] model can only be applied to two persons game.

Fehr and Schmidt [18] models fairness as a self-entered inequity aversion, i.e., individuals are willing to give up some payoff to move in the direction of an equitable outcome. Individuals in these models are concerned about their relative utility or payoff as compared to others. Unlike Rabin [42], Fehr and Schmidt [18] do not model intentions explicitly and use standard game theory in order to analyze n-person public goods game. The authors assume that subjects suffer more from inequity due to their material disadvantage than from inequity due to their material advantage (see Sect. 3.1). In the presence of inequity-averse people, the authors can explain "fair" and "cooperative" as well as "competitive" and "non-cooperative" behavioral patterns. The model also accounts for the interaction between distribution of preferences in a given society. For instance, the presence of 'free riders' in the society induces many inequity-averse individuals to behave in a selfish manner. This happens because if there are only a few individuals who have $\alpha + \beta_i > 1$, they suffer too much loss from disadvantageous inequality caused by free riders. This is Proposition 4(b) in their paper, discussed briefly in our Sect. 3.1.

The experimental evidence on fairness and inequity aversion is not obvious. Dannenberg et al. [16] test for inequity aversion using model from Fehr and Schmidt [18]. The experiment is a two-step procedure using within-subject design. In the first step, subjects played selected games to estimate their individual other regarding preferences. In the second step, subjects with preferences (fair and selfish) according to Fehr and Schmidt [18] were matched into pairs and interacted with the possibility of punishment. They find a significant effect of advantageous inequity aversion (third term in Eq. 2) on an individual's contribution to public goods. Another paper, Blanco et al. [8] also uses within-subject design to assess predictive power of Fehr and Schmidt [18] model. They find that inequity aversion can explain an individual's behavior in a public goods game at an aggregate level, however, not at the individual level. Aggregate level tests compare the distribution of outcomes across different experiments that were run with different samples and thereby check for consistency. Individual level analysis on the other hand uses within-subject design to test for decisions in different experiments with the same sample. The model of inequity aversion was based on the relative payoff of individuals. In our next section, we discuss models of altruism focusing on absolute payoff of individuals.

4.2 Altruism

Standard utility function as defined in Eq. 1 focus on individual's monetary payoff. Models of giving or donating to a charity have been based on 'altruism', where an individual is assumed to contribute to the public goods because they simply demand more of public goods. However, these models have low predictive power and were not able to incorporate the empirical findings, this lead to the development of models with 'impure altruism'. In the models with 'impure altruism', individuals are assumed to contribute to public goods because of two reasons: (1) altruism: people demand more of public goods, (2) people get some private goods benefits from the gift per se which is called 'warm-glow'. The second motive is also termed as 'egoistic motive'.

Andreoni [3, 4] presents the model of giving that incorporates a warm glow in a public goods game. Suppose there is one private good and one public good. Individuals are endowed with wealth w_i, which they can allocate between consumption of private good x_i and their gift to the public good g_i. Let n be total number of individuals and $G = \sum_{i=1}^{n} g_i$. In order to explain how a utility function transforms in case of impure altruism, we use the utility function from [4, p. 465, Eq. 1] as stated below

$$U_i = U_i(x_i, G, g_i), \qquad i = 1, 2, \ldots, n \tag{5}$$

Here, U_i is assumed to be strictly quasi concave. Notice that g_i enters twice in the utility function, once as part of public good G, and as private good g_i. This captures the fact that an individual's contribution/gift (g_i) has properties of a private good that are independent of its properties as a public good.

If the utility function is of the form $U_i = U_i(x_i, G)$ then preferences are purely altruistic. This is because individual does not get any private goods to benefit from the contribution. In contrast, if the utility function is of the form $U_i = U_i(x_i, g_i)$, then the preferences are purely egoistic and the individual is only motivated to give because of the warm glow. Individual only derives private goods benefit from contributing to the public goods.

Let gift/contribution of all the other players except i be denoted by $G_{-i} = \sum_{j \neq i} g_j$, individual donations/contributions can be found by solving

$$\begin{aligned} \max_{x_i, g_i, G} \quad & U_i(x_i, G, g_i) \\ \text{s.t.} \quad & x_i + g_i = w_i \\ & G_{-i} + g_i = G \end{aligned}$$

Under the Nash equilibrium, G_{-i} is treated exogenously, thus we can rewrite $g_i = G - G_{-i}$. Substituting the budget constraints given above into utility function (Eq. 5), we get

$$\max_G \quad U_i(w_i + G_{-i} - G, G, G - G_{-i}) \tag{6}$$

Differentiating Eq. 6 w.r.t G and solving leads a donation function that is given by the following:

$$G = f_i(w_i + G_{-i}, G_{-i}) \tag{7}$$

$$g_i = f_i(w_i + G_{-i}, G_{-i}) - G_{-i} \tag{8}$$

The first argument in Eq. 7 is from the public dimension of the utility function. The second argument is from the private goods dimension of the utility function. The partial derivative of f_i with respect to first argument is denoted by $f_{i\alpha}$. This is i's marginal propensity to donate for altruistic reasons. f_{ie} represents the partial derivative of f_i with respect to second argument. This is i's marginal propensity to donate due to egoistic reasons. Thus, the model incorporates both altruistic and egoistic reasons for contributing to public goods. In the model, $0 < f_{i\alpha} < 1$ and $f_{ie} > 0$. From the equations above, we can say that individual's contribution g_i is increasing in both egoistic and altruistic motives.

Incorporation of these motives can lead to positive contribution ($g_i > 0$) unlike the standard model, which will predict no contribution ($g_i = 0$). The predictions from their model are also consistent with various empirical findings mentioned by Andreoni. Including private provisioning of public goods or impure altruism also increase the predictive power of the models. For instance, the pure altruism model predicts that an increase in the amount of public good provided (G), implies a dollar-for-dollar decrease in an individual's own contribution (g_i). If there is a dollar-for-dollar decrease in g_i for any increase in G, we call such crowding out of g_i as complete. However, empirically, the magnitude of such crowding out is found to be incomplete or proportionately less than the magnitude of change in G. Such findings are consistent with the theoretical predictions from impure altruism models.

Andreoni et al. [6] provide evidence of altruistic preferences in various games: prisoner's dilemma, public goods game, dictator game, trust games, and gift exchange games. The survey also suggests the formation of altruistic preferences can be due to cultural norms, psychological development, socialization, and neural foundations.

The model of impure altruism also predicts that individuals will reduce their contribution to the public goods when other individuals increase their contributions. However, this observation is in contrast to various other outcomes in a public goods game. For instance, conditional cooperation (discussed in the next section) is observed in a public goods game, where individuals cooperate if they see others contributing. Reciprocity observed in many games also contradicts the assumption of altruism. "An altruistic person's kindness does not depend on behavior of others, whereas the kindness of a strong reciprocator is conditional on the perceived kindness of other players" Fehr et al. [21]. We next discuss reciprocity and conditional cooperation.

4.3 Reciprocity and Conditional Cooperation

Theories of reciprocity and conditional cooperation incorporate an individual's willingness to cooperate if others are cooperating as well. In these models, individuals are also concerned about the intentions behind other's decisions. We can also apply the concept of 'conditional cooperation' to the section of inequity aversion, wherein individuals contribute if they believe others will contribute due to his/her concern for equity in payoffs [13]. However, models of reciprocity or conditional cooperation capture the intentions or beliefs of individuals as compared to models of inequity aversion.

Reciprocal motivation is modeled in Rabin [42], however, the model does not apply to sequential games. Falk and Fischbacher [17] extend the notion of reciprocity in a sequential game. The authors present a formal theory of reciprocity (Eq. 9), where the players utility now depends upon an individual's payoff and also on the kindness (how kind a person perceives action by another player) and the reciprocation term (response to the experienced kindness). We now represent their utility function in a 'reciprocity game' [17, p. 301, Definition 3]

$$U_i(f, s_i{''}, s_i') = \pi_i(f) + \rho_i \sum_{n \to f} \psi_j(n, s_i{''}, s_i')\sigma_i(n, s_i{''}, s_i') \qquad n \in N_i \tag{9}$$

The game is a two-player extensive form game with finite number of stages. Let $i \in \{1, 2\}$ be a player in the game and let player j be the other player. N denotes the set of nodes and N_i is the set of nodes where player i has the first move. $n \in N$ is one of the node in the game. S_i and S_j is behavioral strategy space of player i and j respectively. $s_i \in S_i$, $s_j \in S_j$ are behavior strategy of player i and j, respectively. s_i' denotes *first order belief* of player i and captures i's belief about the behavior strategy player j will choose. $s_i{''}$ denotes the *second order belief* of player i and captures i's belief about j's belief about which strategy player i will choose. F denotes the set of end nodes of the game. The models fixes, f as an end node that follows (directly or indirectly) node n.

The first term in Eq. 9: $\pi_i(f)$ is individual i's material payoff. The second term is the reciprocity utility and comprises of

- Reciprocity parameter: ρ_i is a positive constant and common knowledge. It captures the strength of player i's reciprocal preferences. A high ρ_i implies reciprocal utility is more important as compared to the other utility. If $\rho_i = 0$, then utility equals material payoff $\pi_i(f)$.
- Kindness term: $\psi_j(n, s_i{''}, s_i')$ which measures how kind i perceives action by another player j. It depends upon the consequence or outcome of that action and underlying intention. ψ_j is product of outcome term ($\triangle_j$) and intention factor (υ_j). Outcome term measures the output, $\triangle_j > 0$ expresses advantageous outcome for i, $\triangle_j < 0$ expresses disadvantageous outcome for i. The intention factor measures the intention behind the outcome. $\upsilon_j = 1$ captures a situation where $\triangle_j$ is the result

of an action which j completed intentionally and $\upsilon_j < 1$ implies j's action was not fully intentional.

- Reciprocation term: $\sigma_i(n, s_i", s_i')$ expresses response to experienced kindness, how much i alters payoff of j with his move in node n. A rewarding action implies a positive reciprocation term, whereas a punishment implies a negative reciprocation term.
- Product of kindness term and reciprocation term measures the reciprocal utility in a particular node. If the kindness term in a particular node n is positive, then individual i's utility increases if he/she chooses an action in that node which increases j's payoff. The opposite holds when the kindness term is negative and i has an incentive to reduce j's payoff. The model measures kindness in each node where i has the move, hence the overall reciprocity utility is the sum of the reciprocity utility in all nodes (before the considered end node), weighted with the reciprocity parameter (ρ_i).

The authors discuss the intuition of their theoretical prediction for public goods game. According to the authors: 'the strategic structure of a prisoner's dilemma is very similar to a public goods game'. In a sequential prisoner's dilemma, player 1 either cooperates or defects, and after observing player 1's outcome, player 2 also chooses to cooperate or defect. The subgame perfect solution is for both the players to defect. The above model predicts that if player 2 is sufficiently reciprocally motivated there is a positive probability that player 2 rewards player 1's cooperation with cooperation. Also, player 2 will defect if player 1 defects. Conditional cooperation between subjects can also be seen through the payoff function (Eq. 9), subjects contribute more if the contributions of other group members are also higher. In terms of model parameters: if player j is cooperative then kindness term ψ_j will be positive and if player i expresses response to experienced kindness then, σ_i is also positive. Since $\rho_i > 0$, i's contribution increases in response to higher contribution by j.

The authors also substantiate their theoretical predictions with experimental results from Fischbacher et al. [23]. In their experiment, subjects could conditionally indicate how many tokens they wanted to contribute to public goods. The best strategy is to contribute nothing irrespective of others contributions, however, subjects' average contribution was increasing in the mean contributions of others. Using this conditional-cooperation strategy, more than half of the subjects were classified as "conditional cooperators" and the rest were classified as free riders.

Fehr and Gachter [19] also find evidence of reciprocity in their public goods experiment, the more a subject free rides relative to others the more he/she is punished. In order to test multiple preferences, Croson [15] conducts an experiment to test theories of commitment, theories of altruism, and theories of reciprocity in a public goods game. Almost all subjects demonstrate a positive correlation between their own contribution and belief of others' contributions, consistent with the theory of reciprocity.

4.4 Heterogeneous Social Preferences

Positive contribution towards public goods game can be explained and sustained through incorporation of others regarding preferences. We have discussed theoretical models of three types of social preferences along with their experimental evidence. These models incorporate heterogeneity, thereby allowing for the presence of different equilibria.

Experimental evidence in the public goods game, further add to the above observation. Chaudhuri [13] in his survey summarizes the advances made in the literature since Ledyard [34] by agreeing upon the presence of distinct types of players. These players differ in social preferences and/or their beliefs about others, which can explain their behavior being contrary to the standard theoretical prediction of free riding. Gunnthorsdottir et al. [27] in their voluntary contribution mechanism (VCM) public goods experiment classify the subjects into 'free riders' (contributes 30% or less of his/her endowment) and 'cooperators' (contributes more than 30%) based on their first round contribution.

Heterogeneous preferences can also explain the decline in cooperation in these experiments due to the presence of free riders [22]. The decline in cooperation over periods is suggested due to the"presence of imperfect conditional cooperators", those who match others contributions but only partially. Interaction of "imperfect conditional cooperators" with free riders leads to an increase in free-rider behavior.

5 Extensions

In this section, we discuss extensions of the model of social preferences in public goods game. We explain how the effectiveness of 'Coalition' and 'Networks' increases through the incorporation of social preferences.

5.1 Coalition Formation

Coalitions, subgroups of individuals who agree to act collectively to produce public goods, represent a possible solution to the public goods problem. Coalitions such as International Environmental Agreements (IEA) where countries cooperate for an environmental cause are also observed in practice. Agents in a coalition first decide whether or not to join a coalition, then members decide how much to contribute. Social preferences also influence coalition size and their inclusion can lower the threshold for contributing to the public goods.

Kolstad [31] assumes homogeneous Charness and Rabin [12] preferences. Let there be $i = 1, 2, \ldots, N$ countries, each with potential to emit w_i. Each country chooses level of abatement (a public good) given by g_i or level of emissions $x_i = w_i - g_i$.

Welfare is positively affected by

1. Direct benefits of emitting: x_i. If the country has more emissions, the production cost reduces.
2. Aggregate level of abatement: G. Higher abatement leads to lower pollution levels or reduces the environmental damage.

National welfare (u_i) depends upon egoistic component/self centered (π_i) and pro-social or altruistic component (α_i). Altruistic component (α_i) depends on the vector of egoistic payoffs of other countries. The payoff is given as follows:

$$u_i(x_i, G) = \lambda_i \pi_i(x_i, G) + (1 - \lambda_i)\alpha_i(\pi) \quad (10)$$

Here, $\lambda_i \in [0, 1]$ reflects extent to which country is selfish or altruistic. Egoistic component can be described as following in a public goods game framework:

$$\pi_i = x_i + aG, \quad \text{where} \quad x_i + g_i = w_i, G = \sum g_i \quad (11)$$

$$= w_i - g_i + aG, \quad \text{where} \quad G = \sum g_i; 0 \le g_i \le w_i \quad (12)$$

Here, w_i is the maximum possible emissions for country i, g_i is the level of abatement for country i and G is aggregate abatement over all the countries. a represents the marginal per capita return (MPCR) and indicates how much an investment in abatement returns privately. The authors assume $a \in (1/(N-1), 1)$. This is because, for $a = 1$, the individual will be indifferent between abating and non-abating (welfare from emitting and abatement have the same return). Small values of a are also excluded because coordination might not be enough for abatement.

Now we talk about the altruistic payoff in the utility function which is taken from Charness and Rabin [12].

$$\alpha_i(\pi) = [\delta_i(min_{j \neq i}\pi_j) + \epsilon_i \sum_j \pi_j]/(1 - \lambda_i) \quad \text{where} \quad \delta_i, \epsilon_i \ge 0; \delta_i + \epsilon_l + \lambda_i = 1 \quad (13)$$

Herc, δ_i reflects relative importance of agent i of distribution/equity and ϵ_i reflects importance of efficiency. Equity is represented in the model by a Rawlsian preference which is the minimum monetary payoff over the rest of the population. Efficiency is represented by total monetary payoffs over the population. The inclusion of social preferences in the model reduces the threshold for contributing to the public goods. This result is given by Proposition 1 in the paper

Proposition 4 ([31, p. 15, Proposition 1]) *Assuming the N homogeneous player public goods game with Charness and Rabin social preferences, then*

1. *Efficient (Pareto Optimal) outcomes involve all countries undertaking maximal abatement; and*
2. *The Non-cooperative Nash equilibrium involves each agent either not abating ($gi = 0$) or fully abating ($gi = wi$) according to*

$$g_i = 0 \quad \textit{if} \quad a < \bar{a}_i \tag{14}$$

$$g_i = w_i \quad \textit{if} \quad a > \bar{a}_i \tag{15}$$

$$\textit{where} \quad \bar{a}_i = (\lambda_i + \epsilon_i)/[1 + \epsilon_i(N - 1)]. \tag{16}$$

In case of standard preferences, the cutoff for abating and not abating is $a = 1$, with social preferences cutoffs are lower, by construction $\bar{a}_i < 1$. $\bar{a}_i$ can also be interpreted as MPCR between cooperation and non-cooperation. Concerns for efficiency ($\epsilon_i > 0$), keeping δ_i constant also lowers $\bar{a}_i$. Thus, inclusion of social preferences reduces the cutoff for abating or not abating. With the presence of social preferences, countries find it individually rational to abate (provide public goods).

Ringius et al. [43] identify 'fairness' as a motivation for countries in environmental negotiation. The study also analyzes various IEA's with negotiations leading to the Kyoto protocol and find considerations of fairness and equity to be building characteristics of these negotiations. In their empirical analysis, Lange et al. [33] show that equity issues are considered highly important in international climate negotiations by using a worldwide survey of people involved in international climate policy. Polluter pays rule (rule of equal ratio between abatement costs and emissions) and the accompanying poor losers rule (exempting due to GDP) are the most widely accepted equity principles according to this study.

Grüning et al. [26] in their paper incorporate fairness and justice in countries' preferences. We now illustrate their utility function to understand how coalitions/IEA incorporate social preferences. The public goods problem arises because each country can choose their level of abatement (say reducing carbon emissions) and benefits from the reduced emissions by all the other countries as well. Country j's payoff can be represented by the following quasi-linear logarithmic function (consisting of benefit minus abatement cost) minus a term which measures heterogeneity by means of variance in all abatement strategies [26, p. 141, Eq. 1].

$$P_j = ln\left(\sum_i a_i\right) - a_j - \theta.\sigma(a_1, a_2, \ldots, a_N) \tag{17}$$

In the above payoff, $ln(\sum_i a_i)$, measure the benefit from abatement of all the countries. (a_i, a_j) measures the abatement levels of country i and j. $\sigma(a_1, a_2, \ldots, a_N)$ measures variance in the environmental policies of all the countries. Variance is a measure of fairness and justice in their model since countries prefer a more egalitarian cost sharing. Variance in their model is defined as $\sum_i \frac{(a_i - \bar{a})}{N}$, where $\bar{a}$ is the global average of all countries environmental policies. A country's payoff is also assumed to be concave in its own strategy and continuous in that of the opponents. $\theta \geq 0$

represents preference intensity for welfare loss due to cost dispersion. For instance, $\theta = 0$ corresponds to the case of pure selfishness, increasing θ corresponds with stronger concern for 'fair or just' cost sharing. Equity concerns are homogeneous in this symmetric payoff function, however, their model is also robust to heterogeneous countries. The authors extend their results to asymmetry in measuring equity (σ is modified by incorporating countries self interest), heterogeneous countries (countries have different θ or different abatement costs).

The authors find that stronger fairness attitudes lead to homogeneous results as countries both inside and outside IEA adjust abatement levels to each other. This is explained by Proposition 1 from their paper

Proposition 5 ([26, p. 143, Proposition 1]) *Abatements inside and outside the coalition.*

1. *For signatories, stronger fairness preferences result in smaller abatement activities. If θ exceeds the threshold level $\tilde{\theta}$, even an outsider becomes active. The stronger θ, the more abatements an outsider carries out. In the limit (for $\theta \to \infty$) there is no difference between an insider and an outsider.*
2. *The aggregate does not significantly change in θ. For $\theta < \tilde{\theta}$; fairness has a negative impact on global abatements, while $A(S; \theta)$ remains constant for all θ exceeding the threshold $\tilde{\theta}$.*

Here $A(S; \theta)$, is the aggregate abatement activity and is given by

$$A(S; \theta) = Sa_S^*(S, \theta) + (N - S)a_o^*(S, 0) \tag{18}$$

Here, $a_S^*(S, \theta)$ and $a_o^*(S, 0)$ are the abatement activities of countries inside and outside IEA, respectively. Countries in IEA are signatories and are represented by S. According to the above Proposition, if $\theta < \tilde{\theta}$[3]; outsiders are free riders and fairness concern leads to signatories reducing their abatements. This leads to a lower $A(S; \theta)$ or loss in environmental quality. For $\theta > \tilde{\theta}$, $A(S; \theta)$ does not change and for stronger θ countries abatement becomes similar, leading to not much difference between insider and outsider. In other words, stronger fairness preferences lead to more abatement by non-signatories. Fairness concerns imply that countries should not deviate too much from other countries' environmental policies. This deviation is measured by the variance in the payoff function (Eq. 17). Fairness concern thus leads to similar abatements by signatories as well.

Thus, either all or almost none of the countries form an IEA. Internalization of the global environmental externality stabilizes IEA's, whereas free riding hinders larger coalitions. Thus, stronger fairness preferences are needed to overcome the instability of grand coalition as these preferences favor similar behavior with respect to abatement.

Sarangi and Upadhyay [44] study the role of social preference in a two-stage public goods game where, in the first stage, heterogeneous agents first choose whether or not

[3] $a_o^*(S, 0) = 0$ for $\theta < \tilde{\theta}$, see Grüning and Peters [26, p. 142, Eq. 5].

to join a coalition then, in the next stage, the coalition votes on whether its members will contribute. The preferences are assumed to be Rawlsian, wherein the individuals care about the least well-off person in the society.

Let there be $i = 1, 2, \ldots, n$ players. The individuals payoff depend on their own payoff and the payoff of the least well-off person. λ_i is the weight on their won payoff. The utility function in their model is as follows:

$$\pi_i = \lambda_i(P_i) + (1 - \lambda_i)(min(P_j)) \tag{19}$$

Here P_i is the monetary payoff of i and min (P_j) is the lowest monetary payoff of any player j.[4]

They find that individuals with stronger social preferences are more likely to join the coalition and vote for the coalition to contribute to the public goods. This can be summarized from Proposition 2 in their paper given below. Let the decision to join be given by j_i, $j_i = 1$ means individual joins the coalition and $j_i = 0$ implies individual does not join the coalition.

Proposition 6 ([44, p. 10, Proposition 2]) *In the subgame perfect Nash equilibrium, if $\lambda_i \leq \gamma$ then $j_i = 1$. If $\lambda_i > \gamma$ then $j_i = 0$.*

The threshold for joining the coalition ($\lambda_i \leq \gamma$), also satisfies the threshold for contributing to the public goods in their paper. Thus, incorporation of social preferences can result in a larger coalition. The result is intuitive since individuals with stronger social preferences are more likely to join the coalition and contribute to public goods.

The inclusion of social preferences lowers the thresholds for contribution and increases the likelihood of a larger coalition/grand coalition. Accounting for social preferences in the coalition framework helps in learning about the successful development of coalitions.

5.2 Network Formation

Bramoullé et al. [10] provide the first network model of public goods and answer how social or geographical structure affects the level and pattern of public goods. The study finds that individuals who have active social neighbors usually gain more from the contribution of others (due to more links) but contribute less to public goods. This is similar to the concept of free riding observed in a general public goods game. For similar reasons, an addition of a new link increases access to public goods, however, reduces an individual's incentive to contribute. Galeotti et al. [24] suggest that the effect of adding links to a network depends upon where is the link added.

Galeotti et al. [24] in their paper also examine patterns of social communication in a network. In the game, individuals choose to personally acquire information

[4] Player i can also have the least payoff, in that case $\pi_i = P_i$.

and form connections with others. Main findings of the paper suggest that strict equilibrium of the game exhibits the "law of the few". According to the "law of the few", in a social group, small subset of individuals personally acquire information (called hub), while rest of the population forms connections with this small set of information acquirers. Individual information acquisition is a local public good game and implies that an equilibrium which entails links can lead to under provision of information acquisition. The socially optimal output is when central player in a star network acquires information and the others form links with hub. This happens when cost for forming an additional link is less than cost of acquiring information. If this is not the case, then in the social optimal outcome, all players acquire information and no one forms links.

Caria and Fafchamps [11] conduct a public good experiment on a star network (one central player and seven spokes). The design of the experiment is based on the theoretical work of Bramoullé et al. [10]. By design, the contribution of the center player benefits all individuals located at the spokes, while the contributions of the spokes only benefits the center. Following prediction from equity and efficiency, center player should be motivated to contribute more than the spokes. Also, the central player experiences 'social pressure' as other players also expect central player to contribute more than others. This is captured using the 'guilt aversion' model from Battigalli and Dufwenberg [7], where subjects experience guilt if their actions determine a payoff for other players that is lower than what these players expect.

Guilt aversion that star center i feel towards player j can be captured by [11, p. 397]

$$G_{ij}(c_i, \alpha_j, z) = max\{E_j[\pi_j] - \pi_j, 0\} = max\{r(\alpha_j^z - c_i^z), 0\} \tag{20}$$

Here, $E_j[\pi_j]$ is the expected payoff of spoke player j and π_j is the actual payoff of player j. c_i is the contribution profile of the star player, z indicates the average contribution of all spoke players. Thus, c_i^z indicates the contribution of player i when seven spokes have contributed on average z. α_j^z is the expectation profile of player j from player i when spoke members contribute on average z. r indicates the rate of return to public good contributions. Thus, guilt is a measure of the difference between player i's contribution and what spoke members expect i to contribute.

Each player was also asked to predict the average value of contribution among the other 7 players for each level of z. α_i^z records how much player i expects other 7 players to contribute when they play as center of the star and spokes on average contribute z. $\bar{\alpha}_z$ is the contribution that individuals in the network, on an average, expect from a player at the center of the star. This is arrived at by taking the group expectations: average of α_i^z over all the eight players.

We now use their utility function to illustrate incorporation of social preference (guilt aversion here) in a Network [11, p. 397, Eq. 2].

$$u_i(c_i, \bar{\alpha}, z) = \pi_i - \frac{1}{7}\sum_{j \neq i} 7g_i * G_{ij}(c_i, \alpha_j, z) \tag{21}$$

Utility is monetary payoff minus cost of guilt central player experiences for each of the 7 spoke members. The authors assume that player i believes that each spoke has the same expectations, so that individual expectation coincide with group expectation ($\alpha_j = \bar{\alpha}$). Hence, the central player experiences same guilt towards each of the 7 spoke players. The utility function simplifies to [11, p. 397, Eq. 3]

$$u_i(c_i, \bar{\alpha}, z) = \pi_i - g_i * G_{ij}(c_i, \bar{\alpha}, z) \tag{22}$$

Here $G_{ij}(c_i, \bar{\alpha}, z) = max\{r(\bar{\alpha}^z - c_i^z), 0\}$. The first term in the utility function, Eq. 22 reflects concern for monetary payoff and second term is cost of guilt. If player i is sufficiently averse to guilt, he/she will align his/her contributions to expectations of other players to minimize guilt. For instance, if player i contributes an amount lower than what other players expected, he/she will be guilty. Suppose player i increases contribution by one unit, this will decrease guilt of player i by $g_i r$.

The Contribution decisions in the game were made before assigning positions in the network. This was done in order to ask subjects how much they would like to contribute: (i) if they are assigned the spoke position and (ii) if they are assigned the center position. Contribution in case (i) is denoted by denoted by s_i and in case (ii) is denoted by c_i. Each player had three notes worth 50 INR and had to decide how many notes to contribute, thus $z \in \{0, 1, 2, 3\}$. For each value of z, central player has to decide how much he would like to contribute. Vector $c_i = (c_i^0, c_i^1, c_i^2, c_i^3)$ collects four conditional decision of player i. Subjects were also asked to predict the average value of contribution (c_j^z) among the other seven players for each value of z. This helps to get an estimate of α_j^z, and we thereby arrive at $\bar{\alpha}^z$. The results from the experiment suggest that subjects in the center contribute as much as the average contribution, thus suggesting evidence of 'conditional cooperation'. Subjects play 'conditional cooperation' even when efficiency and equity concerns would require star player to contribute more than others. Disclosing group expectations significantly increases the contribution made by the star central player, thus confirming evidence of guilt aversion.

Altruism has been studied in a network framework [9], however, in the context of transfers. The structure of the network again plays a role in determining how income shocks lead to change in inequality. The consequence of change in altruism network is uncertain and also depends upon where the expansion takes place.

Zhang [47] investigate social preferences in a networks game. Their models incorporate inequity aversion [18] and welfare preferences [12]. The experiment manipulated the network structure: star or circle and the return from public good. Subjects at the core/center of a network contribute more than others in a star network. Subjects in a circular network, who earned less than others also contribute more than the predicted outcome in the subsequent period. This behavior suggests individuals exhibit welfare preferences rather than inequity aversion.

Cooperation is reinforced when conditional cooperators are more likely to interact. Thus, cooperation should fare better in highly clustered networks. Suri and Watts [46] conducted a series of web experiments in which individuals play local public goods game with network topology varying across the sessions. In contrast to the earlier results, they find that network topology had no significant effect on an average contribution. Players were as likely to decrease their contributions for low contributing neighbors as they were to increase their contributions in response to high contributing neighbors, thereby suggesting evidence of conditional cooperation. Positive effects of cooperation were contagious only to direct neighbors in the network.

6 Conclusion

Public goods simultaneously benefit many people and are vital to individuals and societies which further fosters economic growth. A key theme in public goods research is deciding how much of public goods to produce and how to pay for it. While public goods theory predicts free riding and inefficient outcomes, experimental results suggest the existence of cooperation, with contribution rates at 40–60 percent of the efficient level. Donations to charity, payment of taxes, voting in elections, and countries participating in IEA's are some of the other examples which support the claim that cooperation does exist. There are various mechanisms in the literature to reduce 'free riding'. Face-to-face communication, pledging the contribution, punishing the free riders are some of the effective tools to increase and sustain cooperation over the periods.

Studies on public goods highlight that human behavior is not entirely motivated by pure self interest. This has led to the formalization of others regarding preferences in the standard utility function. We have classified social preferences as impure altruism, fairness and inequity aversion, reciprocity and conditional cooperation. These models with social preferences are able to generate predictions for positive contribution and cooperation in a public goods game. However, there are variations in predictions of these models which arise from the heterogeneity of preferences of individuals. Further, the preference of one individual might vary, contingent on the situation. Thereby in some situations, an individual's behavior can be driven by fairness, while in other scenarios, he might be influenced by reciprocity.

Another possible solution to the public goods problem can be carried out through coalitions among individuals who agree to act collectively. Incorporation of social preferences in a public goods framework with coalition can then explain the existence of groups like IEA. Further incorporation of social preferences in a network framework can lead to interesting insights into a public goods game.

References

1. Andreoni, James. 1988. Privately provided public goods in a large economy: The limits of altruism. *Journal of Public Economics* 35 (1): 57–73.
2. Andreoni, James. 1988. Why free ride?: Strategies and learning in public goods experiments. *Journal of Public Economics* 37 (3): 291–304.
3. Andreoni, James. 1989. Giving with impure altruism: Applications to charity and Ricardian equivalence. *Journal of Political Economy* 97 (6): 1447–1458.
4. Andreoni, James. 1990. Impure altruism and donations to public goods: A theory of warm-glow giving. *The Economic Journal* 100 (401): 464–477.
5. Andreoni, James. 1995. Cooperation in public-goods experiments: Kindness or confusion? *American Economic Review* 891–904.
6. Andreoni, James, William T Harbaugh, and Lise Vesterlund. 2010. Altruism in experiments. *Behavioural and Experimental Economics*, 6–13.
7. Battigalli, Pierpaolo, and Martin Dufwenberg. 2007. Guilt in games. *American Economic Review* 97 (2): 170–176.
8. Blanco, Mariana, Dirk Engelmann, and Hans-Theo Normann. 2010. A within-subject analysis of other-regarding preferences. *Available at SSRN 934700*.
9. Bourlès, Renaud, Yann Bramoullé, and Eduardo Perez-Richet. 2017. Altruism in networks. *Econometrica* 85 (2): 675–689.
10. Bramoullé, Yann, Rachel Kranton, et al. 2007. Public goods in networks. *Journal of Economic Theory* 135 (1): 478–494.
11. Caria, A Stefano, and Marcel Fafchamps. 2019. Expectations, network centrality, and public good contributions: Experimental evidence from India. *Journal of Economic Behavior & Organization* 167: 391–408.
12. Charness, Gary, and Matthew Rabin. 2002. Understanding social preferences with simple tests. *The Quarterly Journal of Economics* 117 (3): 817–869.
13. Chaudhuri, Ananish. 2011. Sustaining cooperation in laboratory public goods experiments: A selective survey of the literature. *Experimental Economics* 14 (1): 47–83.
14. Chen, Xiao-Ping. 1996. The group-based binding pledge as a solution to public goods problems. *Organizational Behavior and Human Decision Processes* 66 (2): 192–202.
15. Croson, Rachel T.A. 2007. Theories of commitment, altruism and reciprocity: Evidence from linear public goods games. *Economic Inquiry* 45 (2): 199–216.
16. Dannenberg, Astrid, Thomas Riechmann, Bodo Sturm, and Carsten Vogt. 2007. Inequity aversion and individual behavior in public good games: An experimental investigation. *ZEW-Centre for European Economic Research Discussion Paper*, (07-034).
17. Falk, Armin, and Urs Fischbacher. 2006. A theory of reciprocity. *Games and Economic Behavior* 54: 293–315.
18. Fehr, Ernst, and Klaus M. Schmidt. 1999. A theory of fairness, competition, and cooperation. *The Quarterly Journal of Economics* 114 (3): 817–868.
19. Fehr, Ernst, and Simon Gachter. 2000. Cooperation and punishment in public goods experiments. *American Economic Review* 90 (4): 980–994.
20. Fehr, Ernst, and Urs Fischbacher. 2002. Why social preferences matter-the impact of non-selfish motives on competition, cooperation and incentives. *The Economic Journal* 112 (478): C1–C33.
21. Fehr, Ernst, Urs Fischbacher, and Simon Gächter. 2002. Strong reciprocity, human cooperation, and the enforcement of social norms. *Human Nature* 13 (1): 1–25.
22. Fischbacher, Urs, and Simon Gachter. 2010. Social preferences, beliefs, and the dynamics of free riding in public goods experiments. *American Economic Review* 100 (1): 541–56.
23. Fischbacher, Urs, Simon Gächter, and Ernst Fehr. 2001. Are people conditionally cooperative? Evidence from a public goods experiment. *Economics Letters* 71 (3): 397–404.
24. Galeotti, Andrea, Sanjeev Goyal, Matthew O. Jackson, Fernando Vega-Redondo, and Leeat Yariv. 2010. Network games. *The Review of Economic Studies* 77 (1): 218–244.

25. Geanakoplos, John, David Pearce, and Ennio Stacchetti. 1989. Psychological games and sequential rationality. *Games and Economic Behavior* 1 (1): 60–79.
26. Grüning, Christine, and Wolfgang Peters. 2010. Can justice and fairness enlarge international environmental agreements? *Games* 1 (2): 137–158.
27. Gunnthorsdottir, Anna, Daniel Houser, and Kevin McCabe. 2007. Disposition, history and contributions in public goods experiments. *Journal of Economic Behavior & Organization* 62 (2): 304–315.
28. Isaac, R Mark, and James M. Walker. 1988. Communication and free-riding behavior: The voluntary contribution mechanism. *Economic Inquiry* 26 (4): 585–608.
29. Isaac, R Mark, James M. Walker, and Susan H. Thomas. 1984. Divergent evidence on free riding: An experimental examination of possible explanations. *Public Choice* 43 (2): 113–149.
30. Kim, Oliver, and Mark Walker. 1984. The free rider problem: Experimental evidence. *Public Choice* 43 (1): 3–24.
31. Kolstad, Charles D. 2014. International environmental agreements among heterogeneous countries with social preferences. *National Bureau of Economic Research.*
32. Kurzban, Robert, Kevin McCabe, Vernon L. Smith, and Bart J. Wilson. 2001. Incremental commitment and reciprocity in a real-time public goods game. *Personality and Social Psychology Bulletin* 27 (12): 1662–1673.
33. Lange, Andreas, Carsten Vogt, and Andreas Ziegler. 2007. On the importance of equity in international climate policy: An empirical analysis. *Energy Economics* 29 (3): 545–562.
34. Ledyard, O. 1995. Public goods: Some experimental results. In *Handbook of Experimental Economics*, vol. 1.
35. Marwell, Gerald, and Ruth E. Ames. 1981. Economists free ride, does anyone else. *Journal of Public Economics* 15 (3): 295–310.
36. Nikiforakis, Nikos, and Hans-Theo. Normann. 2008. A comparative statics analysis of punishment in public-good experiments. *Experimental Economics* 11 (4): 358–369.
37. Olson, Mancur. 2012. The logic of collective action [1965]. In *Contemporary Sociological Theory*, vol. 124.
38. Ostrom, Elinor. 1997. *Neither markets nor states: Linking transformation process in collective action arenas*. Perspectives on Public Choice: A handbook.
39. Ostrom, Elinor. 2000. Collective action and the evolution of social norms. *Journal of Economic Perspectives* 14 (3): 137–158.
40. Ostrom, Elinor, James Walker, and Roy Gardner. 1992. Covenants with and without a sword: Self-governance is possible. *American Political Science Review* 86 (2): 404–417.
41. Palfrey, Thomas, Howard Rosenthal, and Nilanjan Roy. 2017. How cheap talk enhances efficiency in threshold public goods games. *Games and Economic Behavior* 101: 234–259.
42. Rabin, Matthew. 1993. Incorporating fairness into game theory and economics. *American Economic Review*, 1281–1302.
43. Ringius, Lasse, Asbjørn Torvanger, and Arild Underdal. 2002. Burden sharing and fairness principles in international climate policy. *International Environmental Agreements* 2 (1): 1–22.
44. Sarangi, Sudipta, and Sakshi Upadhyay. 2019. To Join or not to Join: Coalition Formation in Public Good Games.
45. Smith, Adam. 1976. The Theory of Moral Sentiments. *Liberty Classics*.
46. Suri, Siddharth, and Duncan J. Watts. 2011. Cooperation and contagion in web-based, networked public goods experiments. *ACM SIGecom Exchanges* 10 (2): 3–8.
47. Zhang, Yang. 2018. Social Preferences in Network Games: Theory and Laboratory Evidence. *Available at SSRN 3059710*.

[illegible]

Recent Results on Strategy-Proofness of Random Social Choice Functions

Souvik Roy, Soumyarup Sadhukhan, and Arunava Sen

1 Introduction

Randomization has long been recognized as a useful device to achieve fairness in allocation problems. For example, in a cricket match, the first use of the pitch for batting may be important for the determination of the result, and a flip of coin is the device used to decide this issue. In addition to resolving fairness, randomization is also useful for incentivizing people to reveal their private information truthfully in mechanism design problems. In this essay, we will briefly survey some of the main results in randomized mechanism design problem in the context of voting models.

A voting model is one where individuals/agents/voters have to choose one among a number of alternatives or candidates. Each individual has a ranking or preference over all alternatives and a *(deterministic) social choice function* picks an alternative for every tuple of individual ranking of alternatives. An important feature of the voting model is that monetary payments or transfers are not permitted—this assumption is entirely in keeping with our understanding of voting. Individual preferences are private information and are known only to the individuals themselves. A social choice function is *strategy-proof* if no individual can gain by misrepresenting her preference. A fundamental question in mechanism design theory is the following: what is the set of strategy-proof social choice functions? If a social choice function is not strategy-proof, there are strong grounds to conclude that the social goals represented by the social choice functions are unattainable in the presence of private information.

S. Roy · S. Sadhukhan
Economic Research Unit, Indian Statistical Institute, Kolkata, India
e-mail: souvik.2004@gmail.com

S. Sadhukhan
e-mail: soumyarup.sadhukhan@gmail.com

A. Sen (✉)
Economics and Planning Unit, Indian Statistical Institute, Delhi, India
e-mail: asen@isid.ac.in

S. Borkotokey et al. (eds.), *Game Theory and Networks*, Indian Statistical Institute Series,
https://doi.org/10.1007/978-981-16-4737-6_4

The classical result on strategy-proofness is the Gibbard-Satterthwaite Theorem [25, 45]. According to the theorem, a social choice function is strategy-proof and satisfies a mild range condition only if it is *dictatorial*. Thus, there exists an agent whose most preferred alternative is always chosen. This result relies on the assumption of an *unrestricted domain*, i.e., it is assumed that every individual can have *any* preference over the alternatives. However, in several situations, it is natural to assume that individuals can never have certain preferences. In an extreme case, all individuals have a single ranking; the problem is then rendered trivial because individuals do not have any opportunity for misrepresentation. In general, considering plausible restrictions on possible preferences, called the *restricted domain* approach to the mechanism design problem, has produced important insights. For instance, the domain of *single-peaked* preferences admits a variety of well-behaved strategy-proof social choice functions (see Sect. 4.2 for further details).

There has been a great deal of research since the 1970s on the structure of strategy-proof (deterministic) social choice functions on both restricted and unrestricted domains. In contrast, there has been much less work on strategy-proof random social choice functions. There has, however, been some recent literature addressing these issues, and our goal in this paper is to survey some of these results. We focus mainly on two questions. Does randomization help in escaping the well-known negative results such as the Gibbard-Satterthwaite Theorem? Secondly, in restricted domain environments, does randomization further enrich the class of well-behaved deterministic social choice functions?

The paper is organized as follows. Section 2 introduces the problem and discusses various properties of random social choice functions. Sections 3 and 4 present results on unrestricted domains and various restricted domains, respectively. Section 5 discusses the deterministic extreme point property while Sect. 6 concludes.

2 Preliminaries

We consider a society consisting of a (finite) set of individuals $N = \{1, \dots, n\}$ with at least two individuals. Except in Sect. 4.5, the set A is assumed to be finite. The set of alternatives or candidates is A with $|A| \geq 2$. Society faces the problem of choosing a probability distribution over alternatives based on the "preferences" of individuals in the society.

For notational convenience, we do not use braces for singleton sets whenever it is clear from the context; for instance, we denote the set $\{i\}$ by i.

2.1 Preferences

A complete, reflexive, asymmetric, and transitive binary relation over A (also called a linear order) is called a *preference*. A preference can be viewed as a strict ranking over the alternatives. We denote by $\mathbb{L}(A)$ the set of all preferences over A. For $P \in \mathbb{L}(A)$ and $a, b \in A$, aPb is interpreted as "a is strictly preferred to b according to P". For $P \in \mathbb{L}(A)$ and $1 \leq k \leq m$, by $r_k(P)$ we denote the k-th ranked alternative in P, i.e., $r_k(P) = a$ if and only if $|\{b \in A \mid bPa\}| = k$. We denote the top-ranked alternative of a preference P by $\tau(P)$ (instead of $r_1(P)$). For $P \in \mathscr{D}$ and $a \in A$, the *upper contour set* of a at P, denoted by $U(a, P)$, is defined as the set of alternatives that are as good as a in P, that is, $U(a, P) = \{b \in A \mid bPa\}$. We call a set U an upper contour set at a preference P if it is the upper contour set of some alternative at P.

A set of admissible preferences (henceforth referred to as a *domain*) is denoted by $\mathscr{D} \subseteq \mathbb{L}(A)$. For $a \in A$, we denote by $\mathscr{D}^a$ the preferences in $\mathscr{D}$ that have a as the top-ranked alternative. For a domain $\mathscr{D}$, the *top-set* of $\mathscr{D}$, denoted by $\tau(\mathscr{D})$, is the set of alternatives that appear as a top-ranked alternative in some preference in $\mathscr{D}$, that is, $\tau(\mathscr{D}) = \cup_{P \in \mathscr{D}} \tau(P)$.

A *preference profile* (or simply a profile), denoted by $P_N = (P_1, \ldots, P_n)$, is an element of $\mathscr{D}^n = \mathscr{D} \times \cdots \times \mathscr{D}$ that represents a collection of preferences one for each individual.

2.2 Random Social Choice Functions

In this section, we define social choice functions and discuss their properties. We denote the set of probability distributions over A by ΔA. A **random social choice function (RSCF)** is a function $\varphi : \mathscr{D}^n \rightarrow \Delta A$ that assigns a probability distribution or lottery over A at every profile. For $a \in A$ and $P_N \in \mathscr{D}^n$, the probability of a at the outcome $\varphi(P_N)$ is denoted by $\varphi_a(P_N)$, and for $B \subseteq A$, the total probability of the alternatives in B at $\varphi(P_N)$ is denoted by $\varphi_B(P_N) = \sum_{a \in B} \varphi_a(P_N)$. Some examples of RSCFs are provided below.

Example 1 (*RSCFs based on scoring rules*) A score vector s is an m-dimensional vector $(s_1, s_2, \ldots, s_m)$ such that $s_1 \geq s_2 \geq \cdots \geq s_m$ with $s_1 > s_m$.[1] For any individual i, any preference P_i, and any alternative a, the score assigned by i to a in P_i is $s(a, P_i) = s_k$ where k is the rank of a in P_i, i.e., $r_k(P_i) = a$. The score of a at profile P_N is $\mathbf{s}(a, P_N) = \sum_{i \in N} s_i(a, P_i)$. We now define two RSCFs based on score vectors (for other such RSCFs see [7])

The Proportional Scoring Rule φ^{PS}: for all $a \in A$ and profiles P_N,

$$\varphi_a^{PS}(P_N) = \frac{\mathbf{s}(a, P_N)}{\sum_{a \in A} \mathbf{s}(a, P_N)}.$$

Let $M(P_N)$ denote the set of alternatives that attain the maximum score at profile P_N, i.e., $M(P_N) = \arg\max_{a \in A} \mathbf{s}(a, P_N)$.

---

[1] Well-known score vectors are the *Plurality vector* $(1, 0, \ldots, 0)$ and the *Borda vector* $(m-1, m-2, \ldots, 0)$.

Table 1 The proportional scoring rule

1\2	*abc*	*acb*	*bac*	*bca*	*cab*	*cba*
abc	$(\frac{6}{12}, \frac{4}{12}, \frac{2}{12})$	$(\frac{6}{12}, \frac{3}{12}, \frac{3}{12})$	$(\frac{5}{12}, \frac{5}{12}, \frac{2}{12})$	$(\frac{4}{12}, \frac{5}{12}, \frac{3}{12})$	$(\frac{5}{12}, \frac{3}{12}, \frac{4}{12})$	$(\frac{4}{12}, \frac{4}{12}, \frac{4}{12})$
acb	$(\frac{6}{12}, \frac{3}{12}, \frac{3}{12})$	$(\frac{6}{12}, \frac{2}{12}, \frac{4}{12})$	$(\frac{5}{12}, \frac{4}{12}, \frac{3}{12})$	$(\frac{4}{12}, \frac{4}{12}, \frac{4}{12})$	$(\frac{5}{12}, \frac{2}{12}, \frac{5}{12})$	$(\frac{4}{12}, \frac{3}{12}, \frac{5}{12})$
bac	$(\frac{5}{12}, \frac{5}{12}, \frac{2}{12})$	$(\frac{5}{12}, \frac{4}{12}, \frac{3}{12})$	$(\frac{4}{12}, \frac{6}{12}, \frac{2}{12})$	$(\frac{3}{12}, \frac{6}{12}, \frac{3}{12})$	$(\frac{4}{12}, \frac{4}{12}, \frac{4}{12})$	$(\frac{3}{12}, \frac{5}{12}, \frac{4}{12})$
bca	$(\frac{4}{12}, \frac{5}{12}, \frac{3}{12})$	$(\frac{4}{12}, \frac{4}{12}, \frac{4}{12})$	$(\frac{3}{12}, \frac{6}{12}, \frac{3}{12})$	$(\frac{2}{12}, \frac{6}{12}, \frac{4}{12})$	$(\frac{3}{12}, \frac{4}{12}, \frac{5}{12})$	$(\frac{2}{12}, \frac{5}{12}, \frac{5}{12})$
cab	$(\frac{5}{12}, \frac{3}{12}, \frac{4}{12})$	$(\frac{5}{12}, \frac{2}{12}, \frac{5}{12})$	$(\frac{4}{12}, \frac{4}{12}, \frac{4}{12})$	$(\frac{3}{12}, \frac{4}{12}, \frac{5}{12})$	$(\frac{4}{12}, \frac{2}{12}, \frac{6}{12})$	$(\frac{3}{12}, \frac{3}{12}, \frac{6}{12})$
cba	$(\frac{4}{12}, \frac{4}{12}, \frac{4}{12})$	$(\frac{4}{12}, \frac{3}{12}, \frac{5}{12})$	$(\frac{3}{12}, \frac{5}{12}, \frac{4}{12})$	$(\frac{2}{12}, \frac{5}{12}, \frac{5}{12})$	$(\frac{3}{12}, \frac{3}{12}, \frac{6}{12})$	$(\frac{2}{12}, \frac{4}{12}, \frac{6}{12})$

Table 2 The uniform maximal scoring rule

1\2	*abc*	*acb*	*bac*	*bca*	*cab*	*cba*
abc	$(1, 0, 0)$	$(1, 0, 0)$	$(\frac{1}{2}, \frac{1}{2}, 0)$	$(0, 1, 0)$	$(1, 0, 0)$	$(\frac{1}{3}, \frac{1}{3}, \frac{1}{3})$
acb	$(1, 0, 0)$	$(1, 0, 0)$	$(1, 0, 0)$	$(\frac{1}{3}, \frac{1}{3}, \frac{1}{3})$	$(\frac{1}{2}, 0, \frac{1}{2})$	$(0, 0, 1)$
bac	$(\frac{1}{2}, \frac{1}{2}, 0)$	$(1, 0, 0)$	$(0, 1, 0)$	$(0, 1, 0)$	$(\frac{1}{3}, \frac{1}{3}, \frac{1}{3})$	$(0, 1, 0)$
bca	$(0, 1, 0)$	$(\frac{1}{3}, \frac{1}{3}, \frac{1}{3})$	$(0, 1, 0)$	$(0, 1, 0)$	$(0, 0, 1)$	$(0, \frac{1}{2}, \frac{1}{2})$
cab	$(1, 0, 0)$	$(\frac{1}{2}, 0, \frac{1}{2})$	$(\frac{1}{3}, \frac{1}{3}, \frac{1}{3})$	$(0, 0, 1)$	$(0, 0, 1)$	$(0, 0, 1)$
cba	$(\frac{1}{3}, \frac{1}{3}, \frac{1}{3})$	$(0, 0, 1)$	$(0, 1, 0)$	$(0, \frac{1}{2}, \frac{1}{2})$	$(0, 0, 1)$	$(0, 0, 1)$

The Uniform Maximal Scoring Rule φ^{UMS}: for all $a \in A$ and profiles P_N,

$$\varphi_a^{UMS}(P_N) = \begin{cases} \frac{1}{|M(P_N)|} & \text{if } a \in M(P_N), \\ 0 & \text{otherwise.} \end{cases}$$

Tables 1 and 2 illustrate the Proportional Scoring Rule and the Uniform Maximal Scoring Rule, respectively, in the case where $N = \{1, 2\}$, $A = \{a, b, c\}$, and s is the Borda score vector.

A RSCF is a **deterministic social choice function (DSCF)** if it selects a degenerate probability distribution at every profile. Formally, an RSCF $\varphi : \mathscr{D}^n \to \Delta A$ is a DSCF if $\varphi_a(P_N) \in \{0, 1\}$ for all $a \in A$ and all $P_N \in \mathscr{D}^n$. For convenience, we shall sometimes write a DSCF as a function $f : \mathscr{D}^n \to A$.

In our model, as in the standard model of mechanism design, an individual's preference is her *private information*, i.e., known only to herself. Since the mechanism designer goals depend on this information, it must be elicited from the individuals. The property of *strategy-proofness* ensures that individuals have the correct incentives to report their true preferences. Specifically, she cannot obtain a strictly better alternative according to her true preference irrespective of her beliefs about the reports of other individuals. In game-theoretic terms, truth-telling constitutes a dominant strategy for every individual in the direct revelation game.

Strategy-proofness for a DSCF can be defined straightforwardly along the lines of the discussion in the preceding paragraph: a DSCF $f : \mathscr{D}^n \rightarrow A$ is strategy-proof if either $f(P_i, P_{-i}) = f(P_i', P_{-i})$ or $f(P_i, P_i) P_i f(P_i', P_{-i})$ for all $P_i, P_i' \in \mathscr{D}$ for all $P_{-i} \in \mathscr{D}^{n-1}$ and all individuals i. Consider an individual i whose true preference is P_i and "believes" that all other individuals will announce $P_{-i} \in \mathscr{D}^{n-1}$. If she is truthful, the outcome is $f(P_i, P_{-i})$. On the other hand, suppose she considers manipulating or misrepresenting her preference as P_i', the new outcome is $f(P_i', P_{-i})$. If f is strategy-proof, the misrepresentation will either keep the outcome unchanged or lead to a worse outcome according to her true preference P_i. Importantly, i cannot gain by the misrepresentation no matter what she believes about the preferences of others.

There are some conceptual difficulties in extending the same idea to RSCFs. The strategy-proofness property involves the comparison of the outputs of a DSCF or RSCF at two profiles—one where the individual is truthful and the other, where she misrepresents her preference. In the case of a DSCF, these two outputs are alternatives and can be compared using the individual's (true) preference. However, in the case of a RSCF, the relevant outputs are lotteries and it is not obvious how preferences over alternatives can be extended to rankings over lotteries.

In some cases, there is a natural way to evaluate lotteries given an individual's preferences. Suppose $A = \{a, b, c\}$, and an individual has the preference $P = abc$.[2] Consider the lotteries $p = (0.5, 0.3, 0.2)$ and $q = (0.6, 0.35, 0.05)$.[3] Observe that q can be obtained from p by transferring probabilities from lower to higher ranked alternatives. Therefore, requiring the individual to prefer q to p would appear entirely reasonable. However, this argument cannot be applied while comparing p with $r = (0.4, 0.5, 0.1)$. Here, probabilities are simultaneously shifted from lower to higher ranked alternatives *and* from higher to lower ranked alternatives.

In this essay, we focus on the *stochastic dominance* approach introduced in Gibbard [26]. Following Von Neumann and Morgenstern [46], the standard approach to lottery comparisons is via expected utility comparisons: thus, lottery q is preferred to lottery p if the expected utility from q is greater than the expected utility from p. The difficulty in following this approach is that inputs to the RSCF are preferences (ordinal rankings) rather than utility functions. A natural way to deal with this issue is to consider *utility representations* of preferences. For example, a utility representation of the preference $P = abc$, consists of real numbers $u(a)$, $u(b)$, and $u(c)$ with $u(a) \succ u(b) \succ u(c)$. Observe that for any such representation, the expected utility from q is greater than the expected utility from p. However, the expected utility from p can be greater or less than that of r depending on the utility representation chosen.[4] According to the stochastic dominance criterion, the expected utility of the

[2] By $P = abc$, we mean the preference where a is the top-ranked, b is the second-ranked, and c is the third-ranked alternative.

[3] By (p_1, p_2, p_3), we denote the lottery where a, b, and c receive probabilities p_1, p_2, and p_3, respectively.

[4] To see this, choose $u(a) = 1$ and $u(c) = 0$. If $u(b)$ is close to one, r will have a higher expected utility than p. The opposite will be true if $u(b)$ is chosen sufficiently close to zero.

lottery obtained from truth-telling must not be lower than the expected utility of any lottery arising from misrepresentation of preferences for *any* representation of true preferences. This is stated formally below.[5]

Let P be a preference ordering. The function $u : A \rightarrow \Re$ is a *utility representation* of P if $u(a) > u(b)$ whenever aPb. The RSCF φ is **stochastic dominance strategy-proof** if $\sum_{a\in A} u(a)\varphi_a(P_i, P_{-i}) \geq \sum_{a\in A} u(a)\varphi_a(P_i', P_{-i})$ for all $P_i, P_i' \in \mathscr{D}$, for all $P_{-i} \in \mathscr{D}^{n-1}$, and for all utility representations u of P_i. This notion of strategy-proofness places a heavy burden on the truth-telling lottery. In the example discussed previously, φ will fail to be strategy-proof if p and r arise from truth-telling and misrepresentation, respectively, because *there exists* a utility representation of abc according to which r has a higher expected utility than p. Thus, we may be confident that a RSCF that is strategy-proof in this sense will induce individuals to be truthful. However, we may be excessively cautious in eliminating from consideration RSCFs that fail to satisfy this property. A weaker notion of strategy-proofness would only require the expected utility from the lottery from truth-telling not be smaller than that from misrepresentation for all utility representation of the true preference. In the previous example, a RSCF which produced p and q from truth-telling and misrepresentation, respectively, would fail strategy-proofness. However, it would not violate the condition if misrepresentation yielded r instead of q.[6]

We now present an alternative formulation of stochastic dominance strategy-proofness. The lottery p **stochastically dominates** lottery q at a preference P if $p(U) \geq q(U)$ for all upper contour sets U of P. Another equivalent way to define stochastic dominance is as follows. A RSCF $\varphi : \mathscr{D}^n \rightarrow \Delta A$ is **stochastic dominance strategy-proof** if $\varphi(P_i, P_{-i})$ stochastically dominates $\varphi(P_i', P_{-i})$ for all $P_i, P_i' \in \mathscr{D}$, for all $P_{-i} \in \mathscr{D}^{n-1}$ and all individuals i. It is straightforward to verify that the two notions of stochastic dominance strategy-proofness are equivalent and reduce to the notion of strategy-proofness for DSCFs. Henceforth, we shall refer to stochastic dominance strategy-proofness simply as strategy-proofness. If a RSCF is not strategy-proof, we shall say it is *manipulable*.

The proportional scoring rule is strategy-proof, while the uniform maximal scoring rule is not. For instance, individual 2 can manipulate φ^{UMS} at the profile (abc, cba) via the preference cab as $\varphi_U(abc, cba) > \varphi_U(abc, cab)$ for the upper contour set $U = c$ of the preference cba (see Table 2).

The next property of a RSCF ensures that it is minimally responsive to the preferences of individuals. This property requires an alternative to be chosen with probability one if this alternative is top-ranked by all individuals. Formally, a RSCF $\varphi : \mathscr{D}^n \rightarrow \Delta A$ satisfies **unanimity** if for all $a \in A$ and all $P_N \in \mathscr{D}^n$, $[\tau(P_i) =$

---

[5] It is important to emphasize that cardinal mechanisms are not considered here as the private information of agents is not cardinal (i.e., utility functions). There is a small literature on strategy-proof RSCFs on cardinal preferences (see [19, 20, 27] for details).

[6] For further discussion of alternate notions of strategy-proofness, see [3–6, 9].

a for all $i \in N] \Rightarrow [\varphi_a(P_N) = 1]$.[7] Note that the proportional scoring rule is not unanimous, whereas the uniform maximal scoring rule is unanimous.

There is a natural way to generate "new" RSCFs from any given collection of RSCFs. Let φ^j, $j = 1, \ldots, K$ be a collection of RSCFs and let λ^j, $j = 1, \ldots, K$ be non-negative real numbers such that $\sum_{j=1}^{K} \lambda_j = 1$. Define $\varphi = \sum_{j=1}^{K} \lambda^j \varphi^j$ where $\varphi_a(P_N) = \sum_{j=1}^{K} \lambda^j \varphi_a^j(P_N)$ for all $P_N \in \mathscr{D}^n$ and all $a \in A$. We shall refer to φ as a *convex combination* of the RSCFs φ^j. Since the convex combination of a collection of lotteries is a lottery, φ is a RSCF. We record some important properties of convex combinations of RSCFs below. They can be easily verified and are stated without proof.

Remark 2.1 Let φ be a convex combination of φ^j, $j = 1, \ldots, K$. If each φ^j is strategy-proof and satisfies unanimity, then φ is strategy-proof and satisfies unanimity.

The set of strategy-proof RSCFs satisfying unanimity is, therefore, a convex set. This set can, therefore, be characterized by its extreme points. Note that the RSCFs φ^j could be deterministic. Since DSCFs cannot be written as convex combinations of other RSCFs, it follows that strategy-proof DSCFs satisfying unanimity are extreme points of the set of strategy-proof RSCFs satisfying unanimity. A question of considerable theoretical and conceptual interest is whether they are the *only* extreme points. We shall discuss this issue in greater detail in Sect. 5.

3 Results on the Unrestricted Domain

In this section, we present characterization results for unanimous and strategy-proof RSCFs on the unrestricted domain. A domain $\mathscr{D}$ is **unrestricted** if it contains *all* preferences over A, i.e., $\mathscr{D} = \mathbb{L}(A)$. We distinguish two cases based on the number of alternatives in A.

3.1 The Case of Two Alternatives

An important class of social choice problems is concerned with the case of two alternatives. Among such problems are those where individuals have to vote Yes or No to a proposal, to Approve or Disapprove a resolution or if there are two candidates in an election.

We introduce a class of DSCFs on the unrestricted domain with two alternatives. A *committee* $\mathscr{W}$ is a set of subsets of N such that:

[7] It is worth mentioning that under strategy-proofness, unanimity can be weakened in the following way: a RSCF $\varphi : \mathscr{D}^n \to \Delta A$ satisfies unanimity if for all $a \in A$ and all $P_N \in \mathscr{D}^n$, $[P_i = P_j$ for all $i, j \in N] \Rightarrow [\varphi_a(P_N) = 1]$ where $a = \tau(P_i)$ for some (and hence, all) $i \in N$.

(i) $\emptyset \notin \mathscr{W}$ and $N \in \mathscr{W}$,
(ii) for all $S, T \subseteq N$, if $S \subseteq T$ and $S \in \mathscr{W}$, then $T \in \mathscr{W}$.

The elements of $\mathscr{W}$ are called *winning coalitions*, and other subsets of N are called *losing coalitions*.

Let us assume $A = \{a, b\}$. For $P_N \in \mathscr{D}^n$, by $N^a(P_N)$, we denote the set of individuals $i \in N$ who have a as their top-ranked alternative, that is, $\tau(P_i) = a$. For a committee $\mathscr{W}$, a DSCF $f_{\mathscr{W}}$ is called a **voting by committees** rule with respect to a and b if at any profile P_N, a is chosen as the outcome if and only all members of *some* winning coalition vote for a, that is, if for every $P_N \in \mathscr{D}^n$,

$$f_{\mathscr{W}}(P_N) = \begin{cases} a \text{ if } N^a(P_N) \in \mathscr{W} \\ b \text{ if } N^a(P_N) \notin \mathscr{W}. \end{cases}$$

Voting by Committees is a rich class of rules. It includes *majority voting* where a coalition is winning only if it contains at least half the members of the society, the *unanimity rule* where only the coalition of all individuals is winning, and *dictatorship* where a coalition is winning if and only if it contains a specific individual called the *dictator*.

A RSCF is called a *random voting by committees* rule with respect to a and b if it is a convex combination of voting by committees rules with respect to the same alternatives.

Theorem 1 ([33, 35]) *A RSCF on a domain over two alternatives is unanimous and strategy-proof if and only if it is a random voting by committees rule.*

3.2 *The Case of More Than Two Alternatives*

It is well-known in social choice theory that the set of strategy-proof DSCFs shrinks dramatically if the set of alternatives increases beyond two. According to the celebrated **Gibbard-Satterthwaite** Theorem, every strategy-proof DSCF satisfying unanimity must be dictatorial. Formally, a DSCF $f : \mathscr{D}^n \to A$ is **dictatorial** or is a **dictatorship** if there is an individual $i \in N$ called the dictator such that f selects the top-ranked alternative of i at every profile P_N, i.e., $f(P_N) = \tau(P_i)$ for all $P_N \in \mathscr{D}^n$. Thus, all the well-behaved rules such as majority rule are no longer strategy-proof once there are at least three alternatives. Gibbard [26] provides a complete answer to the following question: does the negative result for DSCFs extend to RSCFs as well?

A RSCF $\varphi : \mathscr{D}^n \to \Delta A$ is **unilateral** if (i) φ is strategy-proof and (ii) there exists an individual $i \in N$ such that $\varphi(P_N) = \varphi(P'_N)$ for all profiles P_N, P'_N such that $P_i = P'_i$. In other words, a RSCF is a unilateral if it is strategy-proof and depends only on the preference of a single individual. An example of a unilateral is a RSCF that gives probability 0.5, 0.4, and 0.1 to individual i's first-ranked, second-ranked, and third-ranked alternatives, respectively. A **duple** is a RSCF φ if (i) φ is strategy-proof and (ii) there exist alternatives $a, b \in A$ such that for all profiles P_N, $\varphi_c(P_N) > 0$ only if $c \in \{a, b\}$. A duple is a strategy-proof RSCF that gives positive probability to at most two fixed alternatives at any profile.

Remark 2.1 implies that convex combinations of unilaterals and duples are strategy-proof. Gibbard [26] shows that the converse is also true: every strategy-proof RSCF (on the unrestricted domain) is a convex combination of unilaterals and duples.

Theorem 2 ([26]) *Assume $m \geq 3$. A RSCF on the unrestricted domain is strategy-proof if and only if it is a convex combination of unilateral and duple rules.*

The class of strategy-proof RSCFs on the unrestricted domain is not as restricted as the class of strategy-proof DSCFs. Although a unilateral only considers the preference of a single individual while assigning probabilities, unilaterals for different individuals can be combined (using convex combinations) to generate a more acceptable RSCF which is also strategy-proof. Similarly, duples over different pairs of alternatives can be combined to produce RSCFs that have full support at every profile. Unilaterals and duples can be combined as well. Consequently, the class of strategy-proof RSCFs is "large" and includes many RSCFs that have attractive features from an ethical point of view (unlike dictatorship, for example). One such RSCF is the Proportional Scoring Rule in Example 1 which can be expressed as a convex combination of unilaterals and duples. Further examples and results can be found in Barbera [7].

The discussion in the previous paragraph is subject to an important caveat. A duple does not satisfy unanimity since it assigns zero probability to all except two alternatives. Nor can duples be convexified in a manner that the resulting RSCF satisfies unanimity. A unilateral satisfies unanimity only if the first-ranked alternative of an individual gets probability one. Recall that a RSCF is a random dictatorship if it is a convex combination of dictatorial DSCFs. Combining these observations, we obtain the following result.

Theorem 3 ([26]) *Assume $m \geq 3$. A RSCF on the unrestricted domain is strategy-proof and satisfies unanimity if only if it is a random dictatorship.*

We provide a proof of this result in the case where there are two individuals. An induction argument can be used to extend the argument to an arbitrary number of individuals.[8]

Proof It is left to the reader to verify that every random dictatorship is unanimous and strategy-proof. We prove the converse. Let $N = \{1, 2\}$. Assume that $\varphi : [\mathbb{L}(A)]^2 \rightarrow \Delta A$ satisfies unanimity and strategy-proofness.

Lemma 1 *Let $(P_1, P_2) \in [\mathbb{L}(A)]^2$ be such that $\tau(P_1) \neq \tau(P_2)$. Then $[\varphi_a(P_1, P_2) > 0] \implies [a \in \{\tau(P_1), \tau(P_2)\}]$.*

---

[8] Duggan [17] provides a geometric proof of the result.

Proof Suppose not, i.e., suppose that there exists $P_1, P_2 \in \mathbb{L}(A)$ and $a, b \in A$ such that $\tau(P_1) = a \neq b = \tau(P_2)$ and $\varphi_a(P_1, P_2) + \varphi_b(P_1, P_2) < 1$. Let $\alpha = \varphi_a(P_1, P_2)$ and $\beta = \varphi_b(P_1, P_2)$. Let $P_1' = ab\cdots$ and $P_2' = ba\cdots$. Then strategy-proofness implies $\varphi_a(P_1', P_2) = \alpha$. Furthermore, it must be that $\varphi_a(P_1', P_2) + \varphi_b(P_1', P_2) = 1$ as otherwise voter 1 will manipulate via P_2 and thereby obtaining probability one on b by unanimity. Hence, $\varphi_b(P_1', P_2) = 1 - \alpha$. Note that strategy-proofness also implies $\varphi_b(P_1', P_2') = \varphi_b(P_1', P_2) = 1 - \alpha$ and $\varphi_a(P_1', P_2') = \alpha$.

By a symmetric argument, $\varphi_b(P_1', P_2') = \varphi_b(P_1, P_2') = \beta$ and $\varphi_a(P_1', P_2') = 1 - \beta$. Comparing the probabilities on a and b given by φ at the profile (P_1', P_2'), it follows that $\alpha + \beta = 1$ contradicting the earlier conclusion. ■

Lemma 2 *Let* (P_1, P_2), $(\bar{P}_1, \bar{P}_2) \in [\mathbb{L}(A)]^2$ *be such that* $\tau(P_1) = a \neq b = \tau(P_2)$ *and* $\tau(\bar{P}_1) = c \neq d = \tau(\bar{P}_2)$. *Then* $[\varphi_a(P_1, P_2) = \varphi_c(\bar{P}_1, \bar{P}_2)]$ *and* $[\varphi_b(P_1, P_2) = \varphi_d(\bar{P}_1, \bar{P}_2)]$.

Proof Let $P_1 = a\cdots, P_2 = b\cdots$. Let $(\hat{P}_1, \hat{P}_2)$ be an arbitrary profile where $\tau(\hat{P}_1) = a$ and $\tau(\hat{P}_2) = b$. Then strategy-proofness implies that $\varphi_a(\hat{P}_1, P_2) = \varphi_a(P_1, P_2)$. Lemma 1 implies $\varphi_b(\hat{P}_1, P_2) = \varphi_b(P_1, P_2)$. Now changing voter 2's ordering from P_2 to $\hat{P}_2$ and applying the same arguments, it follows that $\varphi_a(\hat{P}_1, \hat{P}_2) = \varphi_a(P_1, P_2)$ and $\varphi_b(\hat{P}_1, \hat{P}_2) = \varphi_b(P_1, P_2)$.

Assume that $c \neq b$. The argument in the previous paragraph implies that it can be assumed without loss of generality that c is the second-ranked outcome at P_1 (if a and c are distinct), i.e., it can be assumed that $P_1 = ac\cdots$. Let $\bar{P}_1 = ca\cdots$. Then strategy-proofness implies $\varphi_a(\bar{P}_1, P_2) + \varphi_c(\bar{P}_1, P_2) = \varphi_a(P_1, P_2) + \varphi_c(P_1, P_2)$. By Lemma 1, $\varphi_c(P_1, P_2) = \varphi_a(\bar{P}_1, P_2) = 0$. Hence, $\varphi_a(P_1, P_2) = \varphi_c(\bar{P}_1, P_2)$ while $\varphi_b(P_1, P_2) = \varphi_b(\bar{P}_1, P_2)$. Assume $b \neq d$. Switching voter 2's preferences from P_2 to $\bar{P}_2$ and applying the same argument as above, it follows that $\varphi_c(\bar{P}_1, P_2) = \varphi_c(\bar{P}_1, \bar{P}_2)$ while $\varphi_b(\bar{P}_1, P_2) = \varphi_d(\bar{P}_1, \bar{P}_2)$.

The arguments above can deal with all cases except the case where $c = b$ and $d = a$. Since $m \geq 3$, there exists $x \in A$ distinct from a and b. Let $\tilde{P}_1$ be such that $\tau(\tilde{P}_1) = x$. From earlier arguments, $\varphi_a(P_1, P_2) = \varphi_x(\tilde{P}_1, \bar{P}_2)$ and $\varphi_b(P_1, P_2) = \varphi_a(\tilde{P}_1, \bar{P}_2)$.

Applying these arguments again, it can be inferred that $\varphi_x(\tilde{P}_1, \bar{P}_2) = \varphi_b(\bar{P}_1, \bar{P}_2)$ and $\varphi_a(\tilde{P}_1, \bar{P}_2) = \varphi_a(\bar{P}_1, \bar{P}_2)$ establishing the Lemma. ■

Lemmas 1 and 2 above establish that φ is a random dictatorship. ■

We now proceed to examine the structure of strategy-proof RSCFs on restricted domains.

4 Results on Restricted Domains

In many mechanism design problems, the mechanism designer has a-priori information about the preferences of individuals. For instance, a and c may represent candidates with "extreme" positions while b is a "moderate" candidate. The designer may know (without preference revelation) that b always lies between a and c in the preferences of all individuals. As a consequence, RSCFs need to be defined only over a subset of the set of all preferences. The designer also has to consider a narrower class of preferences while checking for possible deviations from truth-telling. Of course, various types of restricted domains can be considered. In this section, we review results on several well-known restricted domains.

4.1 Dictatorial Domains

A domain $\mathscr{D}$ is a **dictatorial domain** if *every* unanimous and strategy-proof DSCF $f : \mathscr{D}^n \to A$ is dictatorial. Similarly, a domain $\mathscr{D}$ is a **random dictatorial domain** if *every* unanimous and strategy-proof RSCF $\varphi : \mathscr{D}^n \to \Delta A$ is a random dictatorship. A random dictatorial domain is clearly a dictatorial domain. The unrestricted domain is both a dictatorial domain as well as a random dictatorship domain by virtue of the Gibbard-Satterthwaite Theorem and Theorem 3, respectively. This observation motivates the following general observation: is every dictatorial domain a random dictatorial domain?

Aswal et al. [1] provide sufficient conditions for a domain to be dictatorial in terms of a graph induced by the domain.[9] Two alternatives a and b are defined to be *linked* if there exist two preferences in the domain, one where a is ranked first and b is ranked second and another preference where the reverse is true. Consider a graph where the nodes are alternatives. There is an edge between two nodes if they are linked. Aswal et al. [1] show that a condition on this graph is sufficient for the underlying domain to be dictatorial. They refer to this as the *linked domain* condition. It can be described as follows: there is an ordering of the nodes such that the first two in the ordering are linked; in addition, every subsequent node is linked to at least two others that are predecessors of the node in the ordering. An implication of this result is that there are several domains other than the unrestricted domain that are dictatorial. These domains can be much smaller than the unrestricted domain (which has $m!$ orderings). There are, in fact, linked domains whose size is a linear function of m.

Chatterji et al. [14] investigate the relationship between linked domains and random dictatorial domains. They provide examples of linked domains that are not random dictatorial. In other words, there are domains where every DSCF that is strategy-proof and satisfies unanimity is dictatorial but admit RSCFs that are strategy-proof, satisfy unanimity but are not random dictatorships. In these domains, the randomization allows the mechanism designer to escape the straightjacket of (random) dictatorship. Chatterji et al. [14] also provide additional conditions on linked domains that make them random dictatorial domains. One such condition is the *hub condition* according to which there is a node that is linked to every other node in the graph. Examples suggest that strong conditions are required to make linked domains, random dictatorial domains.

We now consider several domains that are not random dictatorial domains.

[9] See also Sato [44] and Pramanik [36].

4.2 Single-Peaked Domains

Single-peaked preferences are the bedrock of the theory of political economy (see [2] for example). There is an underlying structure on alternatives with respect to which preferences are described. We proceed to details.

We let $A = \{a_1, \ldots, a_m\}$. There is a prior ordering $\prec$ on the elements of A given by $a_1 \prec \cdots \prec a_m$. We write $x \preceq y$ to mean that either $x \prec y$ or $x = y$. For all $a, b \in A$, we define $[a, b] = \{c \mid \text{ either } a \preceq c \preceq b \text{ or } b \preceq c \preceq a\}$ as the set of alternatives that lie "between" a and b. For any $B \subseteq A$, $[a, b]_B = [a, b] \cap B$ denotes the set of alternatives in B that lie in the interval $[a, b]$. Whenever we refer to the maximum or minimum of a subset of alternatives, we are referring to the maximum and minimum with respect to the ordering $\prec$. Whenever we write $\tau(\mathscr{D}) = \{b_1, \ldots, b_k\}$, we assume without loss of generality that $b_1 \prec \cdots \prec b_k$.

A preference P is **single-peaked** if for all $a, b \in A$, $[\tau(P) \preceq a \prec b$ or $b \prec a \preceq \tau(P)]$ implies aPb. A domain is called *single-peaked* if each preference in the domain is single-peaked and is called *maximal single-peaked* if it contains all single-peaked preferences.

A preference is single-peaked if there exists a unique alternative that is first-ranked (sometimes referred to as the *peak*). Moving farther away from the peak in any direction leads to a decline in preferences. Consider the problem of finding a location on a street to build a public facility such as a hospital or school. Every individual has a unique location on the street which is her peak. While comparing two possible locations for the public good on the same side of her peak, she strictly prefers the location closer to her peak. The street can also be interpreted as the political spectrum. If a and b are two political candidates with $a \prec b$, then a is more "left-wing" than b. If a voter's preferences are single-peaked and her peak (or ideal candidate) c is more left-wing than a, i.e., $c \prec a$, then she will prefer candidate a to b. If on the other hand, the voter's peak is b, she will prefer a to c. Figure 1 is a diagrammatic representation of a single-peaked preference.

An important class of DSCFs on single-peaked domains is *min-max rules*. These rules were introduced in Moulin [30] and constitute the set of all unanimous and strategy-proof DSCFs on the maximal single-peaked domain. Min-max rules are based on a class of parameters, one for each subset of individuals, which we denote

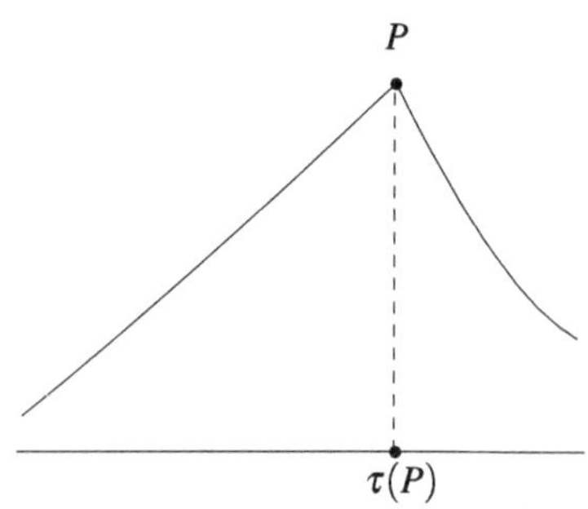

Fig. 1 A graphic illustration of a single-peaked preference

Table 3 Parameters of the min-max rule f

β	β_1	β_2	β_3	$\beta_{\{1,2\}}$	$\beta_{\{1,3\}}$	$\beta_{\{2,3\}}$
	a_8	a_9	a_7	a_4	a_5	a_2

by $(\beta_S)_{S\subseteq N}$. These parameters are required to satisfy some boundary conditions and some monotonicity properties. As the name suggests, the outcome at any profile is calculated by taking suitable minima and maxima of the top-ranked alternatives at the profile and the parameters.

Definition 4.1 A DSCF f on $\mathscr{D}^n$ is called a **min-max** rule if for all $S \subseteq N$, there exists $\beta_S \in A$ satisfying

$$\beta_\emptyset = a_m,\ \ \beta_N = a_1,\ \text{ and } \beta_T \preceq \beta_S \text{ for all } S \subseteq T$$

such that

$$f(P_N) = \min_{S\subseteq N}\left[\max_{i\in S}\{\tau(P_i), \beta_S\}\right].$$

A property that occurs frequently in social choice theory is *tops-onlyness*. A RSCF is tops-only if its outcome at a profile depends only on the top-ranked alternatives in that profile. Two profiles $P_N, P'_N \in \mathscr{D}^n$ are *tops-equivalent* if each individual has the same top-ranked alternative in the two profiles, i.e., $\tau(P_i) = \tau(P'_i)$ for all $i \in N$. A RSCF $\varphi : \mathscr{D}^n \to \Delta A$ is **tops-only** if $\varphi(P_N) = \varphi(P'_N)$ for all tops-equivalent $P_N, P'_N \in \mathscr{D}^n$. Note that min-max rules are tops-only by definition. In what follows, we provide an example of a min-max rule.

Example 2 Let $A = \{a_1, \ldots, a_{10}\}$ and $N = \{1, 2, 3\}$. Consider the min-max rule, say f, with parameters as given in Table 3.

The outcome of the min-max rule at the profile (a_5, a_3, a_8), where a_5, a_3, and a_8 are the top-ranked alternatives of individuals 1, 2, and 3, respectively, is determined as follows.

$$\begin{aligned}
f(P_N) &= \min_{S\subseteq\{1,2,3\}}\left[\max_{i\in S}\{P_i(1), \beta_S\}\right]\\
&= \min\big[\max\{\beta_\emptyset\}, \max\{\tau(P_1), \beta_1\}, \max\{\tau(P_2), \beta_2\}, \max\{\tau(P_3), \beta_3\},\\
&\qquad \max\{\tau(P_1), \tau(P_2), \beta_{\{1,2\}}\}, \max\{\tau(P_1), \tau(P_3), \beta_{\{1,3\}}\}, \max\{\tau(P_2), \tau(P_3), \beta_{\{2,3\}}\},\\
&\qquad \max\{\tau(P_1), \tau(P_2), \tau(P_3), \beta_{\{1,2,3\}}\}\big]\\
&= \min\left[a_{10}, a_8, a_9, a_8, a_5, a_8, a_8, a_8\right]\\
&= a_5.\square
\end{aligned}$$

It is shown in Moulin [30] and Weymark [48] that a DSCF on the *maximal* single-peaked domain is unanimous and strategy-proof if and only if it is a min-max rule. In this section, we present results for RSCFs for a large class of single-peaked domains, which we call *top-connected* single-peaked domains.

For a domain $\mathscr{D}$, the *top-interval* $I(\mathscr{D})$ is the set of alternatives $[\min(\tau(\mathscr{D})), \max(\tau(\mathscr{D}))]$.

Definition 4.2 A single-peaked domain $\mathscr{D}$ is **top-connected** if for every two consecutive alternatives a_r and a_s in $\tau(\mathscr{D})$ with $\min(\tau(\mathscr{D})) \preceq a_r \prec a_s \preceq \max(\tau(\mathscr{D}))$, there exist $P \in \mathscr{D}^{a_r}$ and $P' \in \mathscr{D}^{a_s}$ such that $a_s P a_{r-1}$ if $a_{r-1} \in I(\mathscr{D})$ and $a_r P' a_{s+1}$ if $a_{s+1} \in I(\mathscr{D})$.

Observe that some alternative may not appear as a top-ranked alternative in any preference in a top-connected single-peaked domain, in other words, the top-set of such a domain does not necessarily contain all alternatives.

Remark 4.1 Note that top-connectedness does not impose any restriction (except from single-peakedness) on any preference with the top-ranked alternative as $\min(\tau(\mathscr{D}))$ or $\max(\tau(\mathscr{D}))$. To see this, take, for instance, $\min(\tau(\mathscr{D})) = a_r \prec a_s \preceq \max(\tau(\mathscr{D}))$. Definition 4.2 says that there must exist a single-peaked preference $P \in \mathscr{D}^{a_r}$ such that $a_s P a_{r-1}$ if $a_{r-1} \in I(\mathscr{D})$. However, since $a_r = \min(\tau(\mathscr{D}))$, it must be that $a_{r-1} \notin I(\mathscr{D})$. Therefore, this condition does not apply to P. Similar logic applies to any preference with the top-ranked alternative as $\max(\tau(\mathscr{D}))$.

For a sequence of alternatives $b_1, \ldots, b_k$, denote by $\langle b_1, \ldots, b_k \rangle \ldots$ a preference where $P(l) = b_l$ for all $l = 1, \ldots, k$. Then, the top-connectedness property of a domain $\mathscr{D}$ assures that for every two consecutive alternatives a_r and a_s in $\tau(\mathscr{D})$ with $\min(\tau(\mathscr{D})) \preceq a_r \prec a_s \preceq \max(\tau(\mathscr{D}))$, there are two single-peaked preferences P and P' such that $P = \langle a_r, a_{r+1}, \ldots, a_{s-1}, a_s \rangle \ldots$ if $a_{r-1} \in I(\mathscr{D})$ and $P' = \langle a_s, a_{s-1}, \ldots, a_{r+1}, a_r \rangle \ldots$ if $a_{s+1} \in I(\mathscr{D})$. For example, if $A = \{a_1, \ldots, a_{15}\}$ and $\tau(\mathscr{D}) = \{a_3, a_4, a_5, a_8, a_{10}\}$, then top-connectedness ensures, for instance, that preferences such as $\langle a_5, a_6, a_7, a_8 \rangle \ldots$ and $\langle a_8, a_7, a_6, a_5 \rangle \ldots$ are present in the domain. Note that as we mention in Remark 4.1, top-connectedness does not impose any restriction (except from single-peakedness) on the preferences with top-ranked alternatives a_3 or a_{10}. Thus, the top-connectedness property of a domain $\mathscr{D}$ guarantees that for every two consecutive alternatives a_r and a_s in $\tau(\mathscr{D})$ with $\min(\tau(\mathscr{D})) \preceq a_r \prec a_s \preceq \max(\tau(\mathscr{D}))$, there are two single-peaked preferences P and P' such that $P|_{I(\mathscr{D})} = \langle a_r, a_{r+1}, \ldots, a_{s-1}, a_s \rangle \ldots$ and $P'|_{I(\mathscr{D})} = \langle a_s, a_{s-1}, \ldots, a_{r+1}, a_r \rangle \ldots$.

We provide an example of a top-connected single-peaked domain in Example 3.

Example 3 Let $A = \{a_1, \ldots, a_{10}\}$ be the set of alternatives. Consider the top-connected single-peaked domain $\mathscr{D} = \{P_1, \ldots, P_9\}$ given in Table 4. Here, $\tau(\mathscr{D}) = \{a_3, a_4, a_7, a_9\}$.

It is worth noting that the number of preferences in a top-connected single-peaked domain can range from $2|\tau(\mathscr{D})| - 1$ to 2^{m-1}. Thus, the class of such domains is quite large. It should be further noted that any single-peaked domain $\mathscr{D}$ with $|\tau(\mathscr{D})| = 2$ is a top-connected single-peaked domain. This is because top-connectedness does not impose any condition on the preferences with top-ranked alternatives $\min(\tau(\mathscr{D}))$ or $\max(\tau(\mathscr{D}))$.

Table 4 Preference domain for Example 3

P_1	P_2	P_3	P_4	P_5	P_6	P_7	P_8	P_9
a_3	a_3	a_4	a_4	a_4	a_7	a_7	a_9	a_9
a_4	a_2	a_3	a_5	a_5	a_6	a_8	a_{10}	a_8
a_5	a_4	a_2	a_6	a_6	a_5	a_9	a_8	a_7
a_2	a_1	a_5	a_3	a_7	a_4	a_6	a_7	a_6
a_1	a_5	a_6	a_7	a_3	a_3	a_5	a_6	a_{10}
a_6	a_6	a_1	a_8	a_2	a_2	a_4	a_5	a_5
a_7	a_7	a_7	a_9	a_8	a_8	a_3	a_4	a_4
a_8	a_8	a_8	a_{10}	a_9	a_9	a_{10}	a_3	a_3
a_9	a_9	a_9	a_2	a_1	a_1	a_2	a_2	a_2
a_{10}	a_{10}	a_{10}	a_1	a_{10}	a_{10}	a_1	a_1	a_1

Our next theorem provides a characterization of unanimous and strategy-proof RSCFs on top-connected single-peaked domains. A *random min-max* rule is a convex combination of min-max rules.

Theorem 4 ([40]) *A RSCF on a top-connected single-peaked domain is unanimous and strategy-proof if and only if it is a random min-max rule.*

Ehlers et al. [21] consider the case where the set of alternatives is continuous, say the interval $[0, 1]$. They provide a different characterization of unanimous and strategy-proof RSCFs on the maximal single-peaked domain by means of a class of RSCFs called *probabilistic fixed ballot rule* (PFBR). Below, we define these rules for the case of finitely many alternatives.

A PFBR φ is based on a collection of parameters $(\beta_S)_{S \subseteq N}$, called **probabilistic ballots**. Each probabilistic ballot β_S, which is associated to the coalition $S \subseteq N$, is a probability distribution on A satisfying the following two properties.

- **Ballot unanimity**: β_N assigns probability 1 to a_1, and $\beta_\emptyset$ assigns probability 1 to a_m.
- **Monotonicity**: probabilities according to β_S move toward left as S gets bigger, i.e., $\beta_S([a_1, a_k]) \leq \beta_T([a_1, a_k])$ for all $S \subset T$ and all $a_k \in A$.[10]

For example, suppose there are two individuals $\{1, 2\}$ and four alternatives $\{a_1, a_2, a_3, a_4\}$. A choice of probabilistic ballots could be $\beta_\emptyset = (0, 0, 0, 1)$, $\beta_{\{1\}} = (0.5, 0.2, 0.1, 0.2)$, $\beta_{\{2\}} = (0.4, 0.3, 0.2, 0.1)$, and $\beta_N = (1, 0, 0, 0)$.[11]

A PFBR φ w.r.t. a collection of probabilistic ballots $(\beta_S)_{S \subseteq N}$ works as follows. For each $1 \leq k \leq m$, let $S(k, P_N) = \{i \in N : \tau(R_i) \preceq a_k\}$ be the set of individuals whose peaks are not to the right of a_k. Consider an arbitrary profile P_N and an arbitrary alternative a_k. We induce the probabilities $\beta_{S(k,P_N)}([a_1, a_k])$ and $\beta_{S(k-1,P_N)}([a_1, a_{k-1}])$. If

[10] For a subset B of A, we denote the probability of B according to β_S by $\beta_S(B)$.

[11] Here (x, y, w, z) is the probability distribution where a_1, a_2, a_3, and a_4 receive probabilities x, y, w, and z, respectively.

$a_k = a_1$, then set $\beta_{S(0,P_N)}([a_1, a_0]) = 0$. The probability of the alternative a_k selected at the profile P_N is defined as the difference between these two probabilities, i.e., $\varphi_{a_k}(P_N) = \beta_{S(k,P_N)}([a_1, a_k]) - \beta_{S(k-1,P_N)}([a_1, a_{k-1}])$.[12] Consider, for example, the PFBR φ w.r.t. the parameters described in the previous paragraph. Let $P_N = (P_1, P_2)$ be a profile where $\tau(P_1) = a_2$ and $\tau(P_2) = a_4$. Then,

$$\begin{aligned}
\varphi_{a_1}(P_N) &= \beta_{S(1,P_N)}([a_1, a_1]) - 0 = 0,\\
\varphi_{a_2}(P_N) &= \beta_{S(2,P_N)}([a_1, a_2]) - \beta_{S(1,P_N)}([a_1, a_1])\\
&= \beta_{\{1\}}([a_1, a_2]) - \beta_{\emptyset}([a_1, a_1]) = 0.7 - 0 = 0.7,\\
\varphi_{a_3}(P_N) &= \beta_{S(3,P_N)}([a_1, a_3]) - \beta_{S(2,P_N)}([a_1, a_2])\\
&= \beta_{\{1\}}([a_1, a_3]) - \beta_{\{1\}}([a_1, a_2]) = 0.8 - 0.7 = 0.1, \text{ and}\\
\varphi_{a_4}(P_N) &= \beta_{S(4,P_N)}([a_1, a_4]) - \beta_{S(3,P_N)}([a_1, a_3])\\
&= \beta_N([a_1, a_4])\beta_{\{1\}}([a_1, a_3]) = 1 - 0.8 = 0.2.
\end{aligned}$$

Clearly, the PFBR satisfies the tops-only property.

It is important to note that the probabilistic ballot β_S for a coalition $S \subseteq N$ represents the outcome of φ at the "boundary profile" where individuals in S have the preference $\underline{P}_i = (a_1 \cdots a_{k-1}\, a_k \cdots a_m)$, while the others have the preference $\overline{P}_i = (a_m \cdots a_k\, a_{k-1} \cdots a_1)$. We call such a profile a *S-boundary profile*.[13] If a PFBR φ is unanimous, then it follows that $\beta_\emptyset$ assigns probability 1 to a_m and β_N assigns probability 1 to a_1, which in turn implies ballot unanimity. We now argue that $(\beta_S)_{S\subseteq N}$ is monotonic if φ is strategy-proof. Consider a proper subset $S \subset N$ and $i \in N \setminus S$. Let P_N and P'_N be the S-boundary and $S \cup \{i\}$-boundary profiles, respectively. In other words, only individual i changes her preference $\underline{P}_i$ in the $S \cup \{i\}$-boundary profile to $\overline{P}_i$. Strategy-proofness of φ implies that the probability of each upper contour set of $\underline{P}_i$ is weakly increased from $\varphi(P_N)$ to $\varphi(P'_N)$. Since the interval $[a_1, a_k]$ coincides with the upper contour set of a_k at $\underline{P}_i$, it follows that $\beta_S([a_1, a_k]) \leq \beta_{S\cup\{i\}}([a_1, a_k])$. Monotonicity of $(\beta_S)_{S\subseteq N}$ follows from the repeated application of this argument.

The outcome of a PFBR at any profile is uniquely determined by its outcomes at boundary profiles. It is shown in Ehlers et al. [21] that every PFBR is unanimous and strategy-proof on the single-peaked domain. Thus, unanimity and strategy-proofness of a PFBR at every profile can be ensured by imposing those conditions only on boundary profiles.

The deterministic versions of PFBRs can be obtained by additionally requiring the probabilistic ballots be degenerate, i.e., $\beta_S(a_k) \in \{0, 1\}$ for all $S \subseteq N$ and $a_k \in A$. These DSCFs were introduced by Moulin [30]; we refer to these as *Fixed Ballot*

---

[12] Since $S(k-1, P) \subseteq S(k, P_N)$ and $[a_1, a_{k-1}] \subset [a_1, a_k]$, monotonicity ensures $\varphi_{a_k}(P_N) = \beta_{S(k,P_N)}([a_1, a_k]) - \beta_{S(k-1,P_N)}([a_1, a_{k-1}]) \geq 0$. Moreover, note that $\sum_{k=1}^{m} \varphi_{a_k}(P_N) = \sum_{k=1}^{m} \beta_{S(k,P_N)}([a_1, a_k]) - \beta_{S(k-1,P_N)}([a_1, a_{k-1}]) = \beta_{S(m,P_N)}([a_1, a_m]) = 1$. Therefore, $\varphi(P_N) \in \Delta(A)$ and φ is a well-defined RSCF.

[13] Note that for every $S \subseteq N$, there is a unique S-boundary profile.

Rules (or FBRs).[14] Moulin [30] showed that a DSCF is unanimous, tops-only, and strategy-proof on the single-peaked domain if and only if it is an FBR. It can be easily verified that an arbitrary mixture of FBRs is unanimous and strategy-proof on the single-peaked domain and is a PFBR. Theorem 3 of [34] and Theorem 5 of [37] prove that the converse is also true.

Below, we present the formal definition of PFBRs.

Definition 4.3 A RSCF $\varphi : \mathscr{D}^n \to \Delta A$ is called a **Probabilistic Fixed Ballot Rule** (or **PFBR**) if there exists a collection of probabilistic ballots $(\beta_S)_{S \subseteq N}$ satisfying ballot unanimity and monotonicity such that for all $P_N \in \mathscr{D}^n$ and $a_k \in A$, we have

$$\varphi_{a_k}(P_N) = \beta_{S(k,P_N)}([a_1, a_k]) - \beta_{S(k-1,P_N)}([a_1, a_{k+1}]),$$

where $\beta_{S(0,P_N)}([a_1, a_0]) = 0$.

Theorem 5 ([21]) *A RSCF on the maximal single-peaked domain is unanimous and strategy-proof if and only if it is a PFBR.*

It follows from Theorem 5 and Theorem 4 that every PFBR is a random min-max rule and vice versa.[15]

4.3 *Single-Dipped Domains*

Single-dipped preferences are the reverse of single-peaked preferences. In the latter, preferences decline as one moves farther away from its peak. On the other hand, preferences increase in single-dipped preferences as one moves farther away from its "dip". These preferences are appropriate for the location of "public bads" such as nuclear plants and garbage dumps. All individuals want such facilities to be located as far away as possible from their location.

A preference P is **single-dipped** if it has a unique minimal element $d(P)$, the *dip* of P, such that for all $a, b \in A$, $[d(P) \preceq a \prec b$ or $b \prec a \preceq d(P)] \Rightarrow bPa$. A domain is *single-dipped* if each preference in it is single-dipped. A single-dipped preference is illustrated in Fig. 2.

Peters et al. [33] introduce the notion of binary restricted domains and show that every unanimous and strategy-proof RSCF on a binary restricted domain is a convex combination of unanimous and strategy-proof DSCFs on the same domain. It is easy to verify that every top-connected single-dipped domain is a binary restricted domain.

[14] Moulin [30] called these Augmented Median Voter Rules, while [8] called these Generalized Median Voter Schemes. For an FBR φ, the subtraction form in Definition 4.3 can be simplified to a min-max form (see Definition 10.3 in [31]).

[15] In a related paper, Dutta et al. [18] analyzed the structure of strategy-proof and unanimous RSCFs on domains containing strictly convex continuous single-peaked preferences on a convex subset of Euclidean space. They show that every strategy-proof and unanimous RSCF on any such domain must be a random dictatorship.

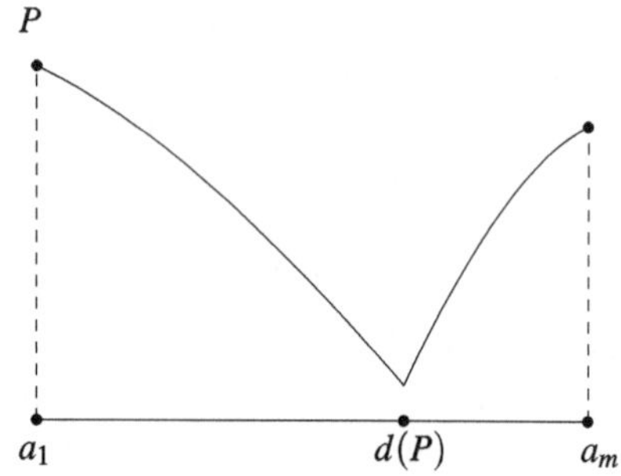

Fig. 2 An illustration of a single-dipped preference

It follows that every strategy-proof RSCF satisfying unanimity on a top-connected single-dipped domain is a random voting by committees rules (recall the definition of such rules in Sect. 3.1) with respect to the boundary alternatives a_1 and a_m.

Theorem 6 ([33]) *A RSCF on a top-connected single-dipped domain is strategy-proof and satisfies unanimity if and only if it is a random voting by committees rule with respect to* a_1 *and* a_m.

It follows from Theorem 6 that any strategy-proof RSCF that satisfies unanimity on a top-connected single-dipped domain can assign positive probability to only the "boundary" alternatives a_1 and a_m.

4.4 Single-Crossing Domains

The *single-crossing* property is a familiar one in economic theory.[16] It appears frequently in models of income taxation and redistribution [29, 38], local public goods and stratification [22, 23, 47], and coalition formation [16, 28]. A more detailed discussion of applications and other issues can be found in Saporiti [42].

A domain $\mathscr{D}$ is a **single-crossing** domain if there exists an ordering $\prec$ over A and an ordering $\lhd$ over $\mathscr{D}$ such that for all $a, b \in A$ and all $P, P' \in \mathscr{D}$, $[a \prec b, P \lhd P', \text{ and } bPa] \implies bP'a$. Preferences in a single-crossing domain can be ordered in such a way that every pair of alternatives switch their relative ranking at most once along the ordering. A single-crossing domain $\bar{\mathscr{D}}$ is *maximal* if there does not exist another single-crossing domain that is a strict superset of $\bar{\mathscr{D}}$. Note that a maximal single-crossing domain with m alternatives contains $m(m-1)/2+1$ preferences.[17] A domain $\mathscr{D}$ is *successive single-crossing* if there is a maximal single-crossing domain $\bar{\mathscr{D}}$ with respect to some ordering $\lhd$ and two preferences $P', P'' \in \bar{\mathscr{D}}$ with $P' \unlhd P''$ such that $\mathscr{D} = \{P \in \bar{\mathscr{D}} \mid P' \unlhd P \unlhd P''\}$.[18]

---

[16] See, for example, Romer [39], p. 181, and Austen-Smith and Banks [2], pp. 114–115.

[17] For details, see Saporiti [42].

[18] By $P \unlhd P'$, we mean either $P = P'$ or $P \lhd P'$.

Examples of a maximal single-crossing domain and a successive single-crossing domain with five alternatives are shown below.

Example 4 Let the set of alternatives be $A = \{a_1, a_2, a_3, a_4, a_5\}$ with the prior order $a_1 \prec \cdots \prec a_5$. The domain $\bar{\mathscr{D}} = \{a_1a_2a_3a_4a_5,\ a_2a_1a_3a_4a_5,\ a_2a_3a_1a_4a_5,\ a_2a_3a_4a_1a_5,\ a_2a_4a_3a_1a_5,\ a_4a_2a_3a_1a_5,\ a_4a_2a_3a_5a_1,\ a_4a_3a_2a_5a_1, a_4a_3a_5a_2a_1,\ a_4a_5a_3a_2a_1,\ a_5a_4a_3a_2a_1\}$ is a maximal single-crossing domain with respect to the ordering $\lhd$ given by $a_1a_2a_3a_4a_5 \lhd a_2a_1a_3a_4a_5 \lhd a_2a_3a_1a_4a_5 \lhd a_2a_3a_4a_1a_5 \lhd a_2a_4a_3a_1a_5 \lhd a_4a_2a_3a_1a_5 \lhd a_4a_2a_3a_5a_1 \lhd a_4a_3a_2a_5a_1 \lhd a_4a_3a_5a_2a_1 \lhd a_4a_5a_3a_2a_1 \lhd a_5a_4a_3a_2a_1$ since every pair of alternatives change their relative ordering at most once along this ordering. Note that the cardinality of A is 5 and that of $\bar{\mathscr{D}}$ is $5(5-1)/2+1 = 11$. The domain $\mathscr{D} = \{a_1a_2a_3a_4a_5,\ a_2a_1a_3a_4a_5,\ a_2a_3a_1a_4a_5,\ a_2a_3a_4a_1a_5,\ a_2a_4a_3a_1a_5,\ a_4a_2a_3a_1a_5\}$ is a successive single-crossing domain since it contains all the preferences between $a_1a_2a_3a_4a_5$ and $a_4a_2a_3a_1a_5$ in the maximal single-crossing domain $\bar{\mathscr{D}}$. □

In what follows, we introduce a restricted version of min-max rules called tops-restricted min-max rule. For such a min-max rule, all the parameters are required to come from the top-set of the domain. Formally, a DSCF $f : \mathscr{D}^n \to A$ is a **tops-restricted min-max (TM)** rule if for all $S \subseteq N$, there exists $\beta_S \in \tau(\mathscr{D})$ satisfying the conditions that $\beta_\emptyset = \max(\tau(\mathscr{D}))$, $\beta_N = \min(\tau(\mathscr{D}))$, and $\beta_T \preceq \beta_S$ for all $S \subseteq T$ such that

$$f(P_N) = \min_{S \subseteq N}\left[\max_{i \in S}\{\tau(P_i), \beta_S\}\right].$$

Note that if $\tau(\mathscr{D}) = A$, then a TM rule becomes a min-max rule. For an example of a TM rule, consider the DSCF f in Example 2 and a domain $\mathscr{D}$ with $\tau(\mathscr{D}) = \{a_2, a_3, a_4, a_5, a_7, a_8, a_9\}$. Since all parameters of f take values in $\tau(\mathscr{D})$, f becomes a TM rule on $\mathscr{D}$.

It is worth noting that the outcome of a min-max rule at a profile is either some top-ranked alternative at that profile or some parameter value (that is, β_S for some $S \subseteq N$). Since for a TM rule f, all these alternatives must be in the top-set of the corresponding domain, its outcome also lies in the same set, that is, $f(P_N) \in \tau(\mathscr{D})$ for all $P_N \in \mathscr{D}^n$.

A crucial property of a single-crossing domain is that the outcome of a unanimous and strategy-proof DSCF always lies in the top-set of the domain. This implies that one can restrict a single-crossing domain to its top-set for the purpose of analyzing unanimous and strategy-proof DSCFs on it. It can be verified that a single-crossing domain restricted to its top-set is a top-connected single-peaked domain. Therefore, by Theorem 4, it follows that a DSCF on a single-crossing domain is unanimous and strategy-proof if and only if it is a TM rule. These results are formally proved in Saporiti [43].[19] Subsequently, [41] have shown that these properties hold for RSCFs on single-crossing domains as well and provide a characterization of unanimous and strategy-proof RSCFs on these domains.

[19] Saporiti [43] uses the term *augmented representative voter schemes* for TM rules.

A RSCF $\varphi : \mathscr{D}^n \to \Delta A$ is a **tops-restricted random min-max (TRM)** rule if φ can be written as a convex combination of some TM rules on $\mathscr{D}^n$. As we have explained earlier, if $\tau(\mathscr{D}) = A$, then a TRM rule $\varphi : \mathscr{D}^n \to \Delta A$ becomes a random min-max rule.

Theorem 7 ([41]) *A RSCF on a successive single-crossing domain is unanimous and strategy-proof if and only if it is a tops-restricted random min-max rule.*

4.5 Euclidean Domains

Euclidean preferences are a special case of single-peaked preferences. Alternatives are located on the real line (or the unit interval without loss of generality) and $\prec$ is the natural order $<$ on the real numbers. We let the set of alternatives A be the interval $[0, 1]$. A preference P is **Euclidean** if there is $x \in [0, 1]$, such that $\tau(P) = x$ and for all alternatives $a, b \in A$, $|x - a| < |x - b|$ implies aPb. A domain is *Euclidean* if it contains *all* Euclidean preferences.

A preference is Euclidean if an alternative a is preferred to another alternative b only if the distance from a to the peak is smaller than the distance of b to the peak. If both a and b lie on the same side of the peak, then single-peakedness would imply that the alternative closer to the peak would be preferred. However, Euclidean preferences also compare alternatives on different sides of the peak unlike single-peakedness. Euclidean preferences are determined completely by the peak of an individual's peak. Consequently, the domain of Euclidean preferences is a strict subset of the set of the maximal single-peaked domain.

Since the Euclidean domain is a strict subset of the maximal single-peaked domain, the possibility that there are unanimous and strategy-proof DSCFs on the domain apart from the min-max rules cannot be excluded. However, [11] show that this case does not arise: a DSCF on the Euclidean domain is unanimous and strategy-proof if and only if it is a min-max rule. Furthermore, [40] show that the same holds even for RSCFs on the Euclidean domain.

Theorem 8 ([40]) *A RSCF on the Euclidean domain is unanimous and strategy-proof if and only if it is a random min-max rule.*

4.6 Dichotomous Domains

Dichotomous preferences are a generalization of binary preferences. There are an arbitrary number of alternatives but each alternative can belong to exactly one of two indifference classes—a "good" set and a "bad" set. An important point of departure from our earlier discussion is that an individual can be indifferent between alternatives. A *dichotomous* domain is the set of all dichotomous preferences. Dichotomous domains have been studied extensively in Bogomolnaia et al. [10].

A dichotomous preference for individual i can be represented by a subset X_i of A with the interpretation that X_i is the good set of i. A profile is n-tuple $(X_1, \ldots, X_n)$. Let $\mathscr{X}^n$ denote the set of all profiles.

A characterization of strategy-proof DSCFs satisfying unanimity remains an open and difficult problem. However, [24] provide a necessary condition, called the pair-triple property, for a strategy-proof RSCF to be representable as a convex combination of strategy-proof DSCFs.

A RSCF $\varphi : \mathscr{X}^n \to \Delta A$ satisfies the **Pair-Triple (PT) Property** if for all $i, j \in N$, all $a, b, c \in A$, and all $X_{-\{i,j\}} \in \mathscr{X}^{n-2}$, we have

$$\varphi_a(\{b\}, \{c\}, X_{-\{i,j\}}) + \varphi_b(\{c\}, \{a\}, X_{-\{i,j\}}) + \varphi_c(\{a\}, \{b\}, X_{-\{i,j\}}) \leq 1.$$

In the notation above, the first component of a profile denotes the preference of individual i and the second one denotes that of individual j.

Theorem 9 ([24]) *A strategy-proof RSCF on the dichotomous domain satisfying unanimity can be represented as a convex combination of strategy-proof DSCF satisfying unanimity only if it satisfies the PT property. In the case of three alternatives, the converse also holds.*

A more complete result on the structure of strategy-proof RSCFs satisfying unanimity on dichotomous domains is not yet available.

4.7 Additional Literature

In this subsection, we briefly review some related results in the literature.

Chatterji and Sen [13] provide conditions on a domain which ensure that every unanimous and strategy-proof DSCF on it has the tops-only property. Subsequently, [15] consider the same problem for RSCFs. They identify two conditions, the *interior property* and the *exterior property*, and show that on every domain satisfying these two properties, a strategy-proof RSCF satisfying unanimity also satisfies the tops-only property. This result is particularly useful in characterizing strategy-proof RSCFS on various domains.

Chatterji et al. [12] investigate *hybrid domains*. Given an ordering $\prec$ over the alternatives, a preference is *hybrid* if there exist *threshold* alternatives $a_{\underline{k}}$ and $a_{\overline{k}}$ with $a_{\underline{k}} \prec a_{\overline{k}}$ such that preferences over the alternatives in the interval between $a_{\underline{k}}$ and $a_{\overline{k}}$ are "unrestricted" relative to each other, while preferences over other alternatives retain features of single-peakedness. Thus, the set A can be decomposed into three parts: left interval $L = \{a_1, \ldots, a_{\underline{k}}\}$, right interval $R = \{a_{\overline{k}}, \ldots, a_m\}$, and middle interval $M = \{a_{\underline{k}}, \ldots, a_{\overline{k}}\}$. Formally, a preference is $(\underline{k}, \overline{k})$*-hybrid* if the following holds: (i) for a voter whose best alternative lies in L (respectively in R), preferences over alternatives in the set $L \cup R$ are conventionally single-peaked, while preferences over alternatives in M are arbitrary subject to the restriction that the best alternative in

M is the left threshold $a_{\underline{k}}$ (respectively, right threshold $a_{\overline{k}}$), and (ii) for a voter whose peak lies in M, preferences restricted to $L \cup R$ are single-peaked but arbitrary over M. Observe that if $\underline{k} = 1$ and $\overline{k} = m$, then preferences are unrestricted, while the case where $\overline{k} - \underline{k} = 1$ coincides with the case of single-peaked preferences. They characterize all strategy-proof RSCFs satisfying unanimity on these domains.

Peters et al. [32] consider domains on graphs. In such domains, there is a graph with the alternatives as nodes with preferences declining as one moves away from the top-ranked alternative along any spanning tree of the graph. Note that if the underlying graph is a line graph, then the resulting domain becomes single-peaked. They characterize all strategy-proof RSCFs satisfying unanimity.

5 The Deterministic Extreme Point Property

In this subsection, we discuss the following issue: in what sense does randomization enlarge the possibilities for a mechanism designer? As we have noted earlier, an implication of Remark 2.1 is that a convex combination of strategy-proof DSCFs satisfying unanimity is a strategy-proof RSCF that satisfies unanimity. A domain $\mathscr{D}$ satisfies the **deterministic extreme point property (DEP)** if the converse is true: i.e., if every unanimous and strategy-proof RSCF can be written as a convex combination of unanimous and strategy-proof DSCFs. If a domain satisfies DEP, the only additional possibility afforded by randomization is that before the elicitation of preferences from individuals, the designer can pick a strategy-proof DSCF satisfying unanimity according to a fixed probability distribution. Thereafter, the designer simply follows the DSCF chosen. Such a procedure does not exhaust all possibilities if the domain does not satisfy DEP. In particular, there will exist strategy-proof RSCFs satisfying unanimity, where the designer will have to randomize over alternatives *after* the elicitation of preferences. For this reason, we regard DEP as a benchmark property for domains. Randomization expands the possibilities available to the designer only if the domain under consideration violates the DEP property.

The DEP property of a domain can be utilized in finding optimal mechanisms on it. Consider an optimization problem with incentive constraints and unanimity constraints. Since these are linear constraints, the constraint space is a polytope and the results identify its extreme points. If the objective function is linear, the DEP property implies that an optimal solution is a deterministic mechanism. This fact may help in finding optimal random mechanisms using the knowledge of the same for deterministic mechanisms as optimizing over an infinite set of random mechanisms may be harder than optimizing over a finite set of deterministic mechanisms. Further, the incentive constraints may simplify with deterministic mechanisms.

It follows from Theorems 1, 2, 3, 4, 6, 7, and 8 that several well-known domains of strict preferences, namely the unrestricted, single-peaked, single-dipped, single-crossing, and Euclidean domains, satisfy the DEP property. However, as we have seen, there are dictatorial domains [14] and hybrid domains [12] that are not random dictatorial. Peters et al. [32] show that DEP is satisfied for a domain on graph only when the underlying graph is a line, i.e., only when the domain is single-peaked. The dichotomous domain (a domain where indifference is permitted) also does not satisfy DEP. This conclusion follows from Theorem 9 since the TP property is not vacuous.

In spite of its theoretical significance, there is as yet, no general analysis of domains satisfying DEP. A more challenging open question is to characterize the extreme points of strategy-proof RSCFs satisfying unanimity in domains that do not satisfy DEP.

6 Conclusion

We have attempted to provide a brief survey of recent results pertaining to the structure of strategy-proof RSCFs on various preference domains. Although considerable progress has been made, some key issues, such as the precise relationship between strategy-proof DSCFs and strategy-proof RSCFs on a given domain, require further investigation.

References

1. Aswal, N., S. Chatterji, and A. Sen. 2003. Dictatorial domains. *Economic Theory* 22 (1): 45–62.
2. Austen-Smith, D., and J.S. Banks. 2000. *Positive political theory I: collective preference*, vol. 1. University of Michigan Press.
3. Aziz, H., F. Brandl, F. Brandt, and M. Brill. 2018. On the tradeoff between efficiency and strategyproofness. *Games and Economic Behavior* 110: 1–18.
4. Aziz, H., F. Brandt, and M. Brill. 2013. On the tradeoff between economic efficiency and strategy proofness in randomized social choice. In *Proceedings of the 2013 international conference on Autonomous agents and multi-agent systems*, 455–462.
5. Aziz, H., and P. Stursberg. 2014. A generalization of probabilistic serial to randomized social choice. In *Twenty-Eighth AAAI Conference on Artificial Intelligence*.
6. Balbuzanov, I. 2016. Convex strategyproofness with an application to the probabilistic serial mechanism. *Social Choice and Welfare* 46 (3): 511–520.
7. Barbera, S. 1979. Majority and positional voting in a probabilistic framework. *The Review of Economic Studies* 46 (2): 379–389.
8. Barberà, S., F. Gul, and E. Stacchetti. 1993. Generalized median voter schemes and committees. *Journal of Economic Theory* 61 (2): 262–289.
9. Bogomolnaia, A., and H. Moulin. 2001. A new solution to the random assignment problem. *Journal of Economic theory* 100 (2): 295–328.
10. Bogomolnaia, A., H. Moulin, and R. Stong. 2005. Collective choice under dichotomous preferences. *Journal of Economic Theory* 122 (2): 165–184.
11. Border, K.C., and J.S. Jordan. 1983. Straightforward elections, unanimity and phantom voters. *The Review of Economic Studies* 50 (1): 153–170.
12. Chatterji, S., S. Roy, S. Sadhukhan, A. Sen, and H. Zeng. 2020. Restricted probabilistic fixed ballot rules and hybrid domains.
13. Chatterji, S., and A. Sen. 2011. Tops-only domains. *Economic Theory* 46 (2): 255–282.
14. Chatterji, S., A. Sen, and H. Zeng. 2014. Random dictatorship domains. *Games and Economic Behavior* 86: 212–236.

15. Chatterji, S., and H. Zeng. 2018. On random social choice functions with the tops-only property. *Games and Economic Behavior* 109: 413–435.
16. Demange, G. 1994. Intermediate preferences and stable coalition structures. *Journal of Mathematical Economics* 23 (1): 45–58.
17. Duggan, J. 1996. A geometric proof of gibbard's random dictatorship theorem. *Economic Theory* 7 (2): 365–369.
18. Dutta, B., H. Peters, and A. Sen. 2002. Strategy-proof probabilistic mechanisms in economies with pure public goods. *Journal of Economic Theory* 106 (2): 392–416.
19. Dutta, B., H. Peters, and A. Sen. 2007. Strategy-proof cardinal decision schemes. *Social Choice and Welfare* 28 (1): 163–179.
20. Ehlers, L., D. Majumdar, D. Mishra, and A. Sen. 2020. Continuity and incentive compatibility in cardinal mechanisms. *Journal of Mathematical Economics.*
21. Ehlers, L., H. Peters, and T. Storcken. 2002. Strategy-proof probabilistic decision schemes for one-dimensional single-peaked preferences. *Journal of Economic Theory* 105 (2): 408–434.
22. Epple, D., and G.J. Platt. 1998. Equilibrium and local redistribution in an urban economy when households differ in both preferences and incomes. *Journal of Urban Economics* 43 (1): 23–51.
23. Epple, D., T. Romer, and H. Sieg. 2001. Interjurisdictional sorting and majority rule: an empirical analysis. *Econometrica* 69 (6): 1437–1465.
24. Gaurav, A., J. Picot, and A. Sen. 2017. The decomposition of strategy-proof random social choice functions on dichotomous domains. *Mathematical Social Sciences* 90: 28–34.
25. Gibbard, A. 1973. Manipulation of voting schemes: a general result. *Econometrica* 41 (4): 587–601.
26. Gibbard, A. 1977. Manipulation of schemes that mix voting with chance. *Econometrica* 45 (3): 665–681.
27. Hylland, A. 1980. Aggregation procedure for cardinal preferences: a comment. *Econometrica: Journal of the Econometric Society*, 539–542.
28. Kung, F.C. 2006. An algorithm for stable and equitable coalition structures with public goods. *Journal of Public Economic Theory* 8 (3): 345–355.
29. Meltzer, A.H., and S.F. Richard. 1981. A rational theory of the size of government. *Journal of political Economy* 89 (5): 914–927.
30. Moulin, H. 1980. On strategy-proofness and single peakedness. *Public Choice* 35 (4): 437–455.
31. Nisan, N., T. Roughgarden, E. Trados, and V.V. Vazirani. 2007. *Algorithmic Game Theory.* Cambridge University Press. https://doi.org/10.1017/CBO9780511800481.
32. Peters, H., S. Roy, and S. Sadhukhan. 2019. Unanimous and strategy-proof probabilistic rules for single-peaked preference profiles on graphs.
33. Peters, H., S. Roy, S. Sadhukhan, and T. Storcken. 2017. An extreme point characterization of strategy-proof and unanimous probabilistic rules over binary restricted domains. *Journal of Mathematical Economics* 69: 84–90.
34. Peters, H., S. Roy, A. Sen, and T. Storcken. 2014. Probabilistic strategy-proof rules over single-peaked domains. *Journal of Mathematical Economics* 52: 123–127.
35. Picot, J., and A. Sen. 2012. An extreme point characterization of random strategy-proof social choice functions: The two alternative case. *Economics Letters* 115 (1): 49–52.
36. Pramanik, A. 2015. Further results on dictatorial domains. *Social Choice and Welfare* 45 (2): 379–398.
37. Pycia, M., and M.U. Ünver. 2015. Decomposing random mechanisms. *Journal of Mathematical Economics* 61: 21–33.
38. Roberts, K.W. 1977. Voting over income tax schedules. *Journal of Public Economics* 8 (3): 329–340.
39. Romer, T. 1975. Individual welfare, majority voting, and the properties of a linear income tax. *Journal of Public Economics* 4 (2): 163–185.
40. Roy, S., and S. Sadhukhan. 2019. A characterization of random min-max domains and its applications. *Economic Theory* 68 (4): 887–906.
41. Roy, S., and S. Sadhukhan. 2019. A unified characterization of randomized strategy-proof rules. ISI working paper.

42. Saporiti, A. 2009. Strategy-proofness and single-crossing. *Theoretical Economics* 4 (2): 127–163.
43. Saporiti, A. 2014. Securely implementable social choice rules with partially honest agents. *Journal of Economic Theory* 154: 216–228.
44. Sato, S. 2010. Circular domains. *Review of Economic Design* 14 (3–4): 331–342.
45. Satterthwaite, M.A. 1975. Strategy-proofness and arrow's conditions: Existence and correspondence theorems for voting procedures and social welfare functions. *Journal of Economic Theory* 10 (2): 187–217.
46. Von Neumann, J., and O. Morgenstern. 1944. *Theory of games and economic behavior*. Princeton: Princeton University Prress.
47. Westhoff, F. 1977. Existence of equilibria in economies with a local public good. *Journal of Economic Theory* 14 (1): 84–112.
48. Weymark, J.A. 2011. A unified approach to strategy-proofness for single-peaked preferences. *SERIEs* 2 (4): 529–550.

Assembly Problems

Dhritiman Gupta and Soumendu Sarkar

JEL Classifications: C78 and D82

1 Introduction

A set of sellers own one unit each of an indivisible good. A buyer wants to purchase a subset of these units. Additionally, the units in the subset are required to constitute a path of a feasible length in a graph. The nodes of this graph represent units of the good, and edges between pair of nodes represent the complementarity of the pair in the production process used by the buyer. The sellers have non-negative valuations for the units they own. The buyer has a non-negative valuation for every subset of units on a feasible path. These valuations may be common knowledge or private information. An Assembly Problem is the exchange problem described by the graph, the minimal size of a feasible subset, and the valuations of the agents.

Efficient assembly is obtained easily in static models with complete information. It is the prospect of strategic delays or private information that makes the assembly problem interesting. Games of complete information multi-period bargaining are used to model the former, while static games of incomplete information are used to model the latter. This chapter provides a brief survey of the literature and discusses some recent results using these approaches.

We are grateful to Professors Prabal Roy Chowdhury and Arunava Sen and participants of the International Conference on Game Theory and Networks 2019, Dibrugarh University, for their comments and suggestions.

D. Gupta
O. P. Jindal Global University, Sonipat, India

S. Sarkar (✉)
University of Delhi, New Delhi, India

S. Borkotokey et al. (eds.), *Game Theory and Networks*, Indian Statistical Institute Series,
https://doi.org/10.1007/978-981-16-4737-6_5

Practical examples of assembly problems include assembling of patent rights for manufacturing life-saving drugs, copyright of musical pieces for composing scores for movies or concerts, and land rights for industrial development among others (see [9] for a survey). In what follows, the problem of assembling land for industrial development is treated as a leading example. For detailed discussions of these applications, the interested reader may refer to Sarkar [28] and Gupta and Sarkar [8].

Assembly is a problem of exchange between a buyer and a set of sellers. Therefore, it is a general version of the bilateral trade problem investigated extensively in the literature on strategic bargaining as well as mechanism design. While the buyer cannot extract full surplus in the unique equilibrium of an alternate offer bilateral bargaining problem [27], private information prevents mechanisms for bilateral trade to be successful [16]. Till recently, the analysis of assembly problems was restricted to the case of fully complementary items. Consequently, the results available mimicked the negative results in the bilateral trade problem. This chapter provides a general model that accommodates various degrees of complementarity and substitutability—thus providing a set of results the nature of which ranges from negative to positive.

The next section provides a brief survey of the literature on assembly problems. A general model is described in the subsequent section which is exploited further to drive some of the results on the efficiency of equilibria. The final section discusses and compares the results under alternative approaches and indicates some directions for further research.

2 Literature

We explore two alternative assumptions about the information structure: (a) agents have complete information about the valuations of other agents and (b) agents have private information about their respective valuations. The natural way to model an assembly problem under the assumption (a) is strategic bargaining among the buyer and sellers, while the approach taken for characterizing satisfactory equilibrium outcomes under the assumption (b) is mechanism design.

In strategic bargaining games, agents on one side of the market propose prices, and those on the other side accept or reject. The range of price offers, sequencing of the offers, and length of the negotiation process are given by the bargaining protocol which is common knowledge (see [21]). Consider the one-period bilateral trade problem where the buyer makes the first offer which the seller may accept or reject. This game has a unique subgame perfect Nash equilibrium outcome if the seller accepts any offer that does not make him strictly worse-off: the buyer offers the seller his exact valuation, the seller accepts, and thus the buyer extracts the entire surplus. In contrast, in the infinite horizon alternate offer bargaining model due to Rubinstein [27], the buyer has to offer a strictly positive share of the surplus to the seller to avoid strategic delay. This share of surplus can be viewed as a cost of the holdout.

The holdout problem has been studied in the land assembly context [1–3, 6, 14, 15, 20, 25]. Secret offers [11, 19, 25] and the choice of bargaining order over sellers [12, 32] are two other topics of interest.

Roy Chowdhury and Sengupta [25] use a protocol which is a natural extension of the protocol by Rubinstein [27] to the assembly problem where all items are complementary. If the offers are public, the buyer who has an outside option extracts a higher share surplus relative to the buyer without an outside option. Holdout may be unavoidable when offers are less transparent even if the buyer has an outside option.

The general model introduced in the next section potentially accommodates more number of sellers than the number of items required by the buyer. The buyer may also require the purchased items to form a path on a given graph. This model allows for various degrees of complementarity. Secret offers or outside options are not explored here, and instead, the focus is on competition among sellers. It uses the public offer protocol due to Roy Chowdhury and Sengupta [25].

The holdout problem has been modeled using the Coalitional Bargaining approach. In the first of such models, Chatterjee et al. [4] studied sequential offers of n-person coalitional bargaining with transferable utility and time discounting. They showed that the efficient coalition may not form for a certain order of proposers. Ray and Vohra [24] study the same problem where externalities across coalitions are a possibility. Myerson [17] provides a complementary approach to coalitional bargaining, analyzing bargaining on networks, where edges between agents are used to model some specific relationship.

Mechanism Design theory lays down rules for "satisfactory" allocation in the presence of private information [26]. Myerson and Satterthwaite [16] have provided such a set of desirable properties: maximum welfare or gains at every allocation (ex-post efficiency), truthful reporting in expectation (interim incentive compatibility), participation in expectation (interim individual rationality), and balanced payments (budget balance).

Consider bilateral trade under private information. The double auction mechanism due to Chatterjee and Samuelson [5] is described as follows: trade takes place if the buyer's reported valuation exceeds that of the seller's, at a price equal to the average of these two reports. When all valuations are distributed uniformly over [0, 1], the double auction mechanism maximizes expected welfare subject to interim incentive compatibility and individual rationality [16]. But it is not efficient in the ex-post sense: the double auction mechanism forgoes some efficient trade opportunities.

Early literature on mechanism design for land assembly primarily look for *second-best* mechanisms in exchange models without any contiguity restrictions (e.g., see [7, 10, 23]). The question of existence of satisfactory mechanisms for general environments remained unresolved till recently.

Williams [31] finds that a satisfactory mechanism can be constructed if and only if there is a Groves mechanism for the problem that results in an expected budget surplus. In a closely related paper, Krishna and Perry [13] show that a *successful* mechanism can be constructed if and only if the VCG mechanism for the problem results in a positive expected budget surplus. The second half of the next section shows how the results due to Williams [31] and Krishna and Perry [13] can resolve the question of satisfactory mechanism design in the assembly problem. We primarily confine the discussion to the existence of first-best mechanisms in the independent private value settings following Sarkar [28–30].

3 The Models

For a given production process, the nature of complementarity among items held by the sellers can be modeled through a graph, say Γ. In any such graph, items, or equivalently, corresponding sellers, are represented by nodes. An edge connects a pair of nodes in this graph if the corresponding inputs are complementary in the buyer's production process. A path is a sequence of connected nodes. The buyer wants to purchase a path of the desired length, say k. This implies that the buyer can combine any k complementary inputs to produce output. We will denote a path by $\mathcal{P}$ and the corresponding sum of seller valuations by $\mathcal{S}$.

A seller is critical if he lies on every feasible path (see Fig. 1). This implies that the corresponding input is complementary with respect to every feasible production plan. If there is only one feasible path in Γ, all sellers in that path are critical. But if there are multiple feasible paths, a seller must belong to their intersection in order to qualify as critical. If there are multiple feasible paths, the number of critical sellers cannot exceed $k-1$: not all sellers on a single path can be critical. Paths of length less than k that do not have an intersection with any feasible path can be excluded from the analysis, because the buyer's valuation over such paths is zero.

A classification of graphs with at least two feasible paths is useful in this context.

In cycles of order $k+1$, referred to as $\Gamma^{\triangle}$ (see Fig. 2), every input on a feasible path can be substituted by another input on the graph.

Fig. 1 A feasible path in the star graph when $k = 3$; seller 1 is critical

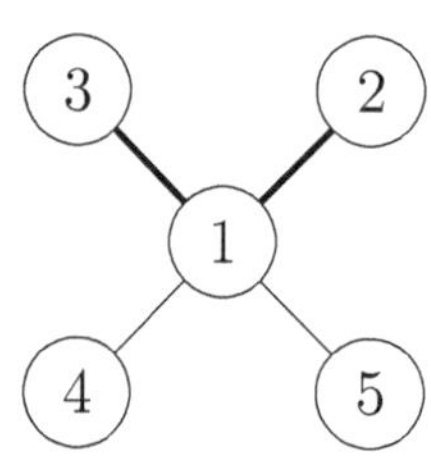

Fig. 2 A cycle of length 4

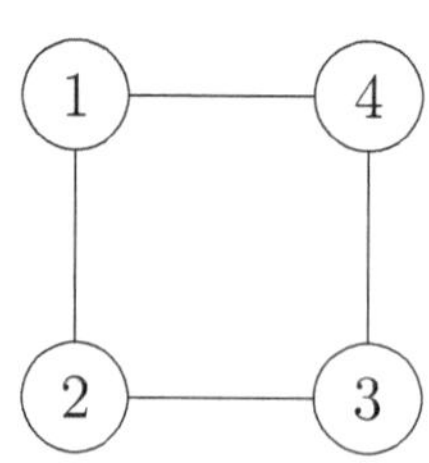

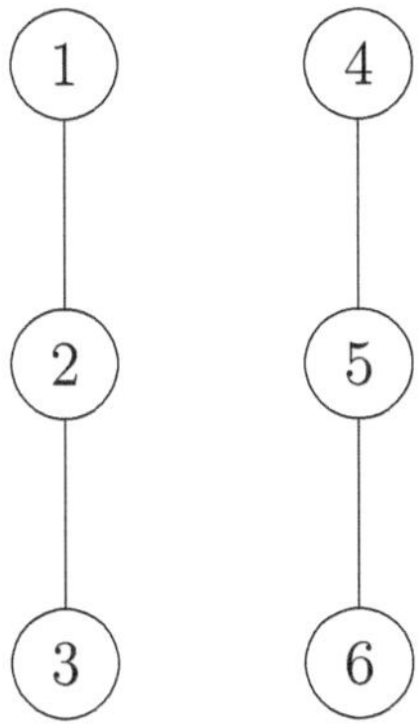

Fig. 3 Graph with disjoint feasible paths; k = 3

Fig. 4 A line graph with two critical sellers marked red; k = 3

In graphs with two disjoint paths, referred to as Γ^D (see Fig. 3), no input is completely substitutable, but a feasible path can be substituted by another feasible path.

In graphs with critical sellers, referred to as Γ^* (see Figs. 4 and 1), inputs corresponding to critical sellers are not substitutable but those corresponding to non-critical sellers are substitutable in a limited sense.

Finally, consider graphs where (i) there is no cycle of length $k + 1$, (ii) no two paths are disjoint, and (iii) the intersection of all feasible paths is empty, referred to as Γ^O (see Fig. 5), referred to as *oddball*. In such graphs, inputs in the intersection of two or more feasible paths cannot be substituted with respect to these feasible paths, but they are substitutable with respect to inputs on other feasible paths.

Facts 1–5 below imply that single component graphs with (a) critical sellers, (b) $k + 1$-cycle, (c) disjoint paths, and (d) oddball are four mutually exclusive and exhaustive categories. A graph may have multiple components from different classes.

Fig. 5 An oddball graph, $n = 5, k = 3$

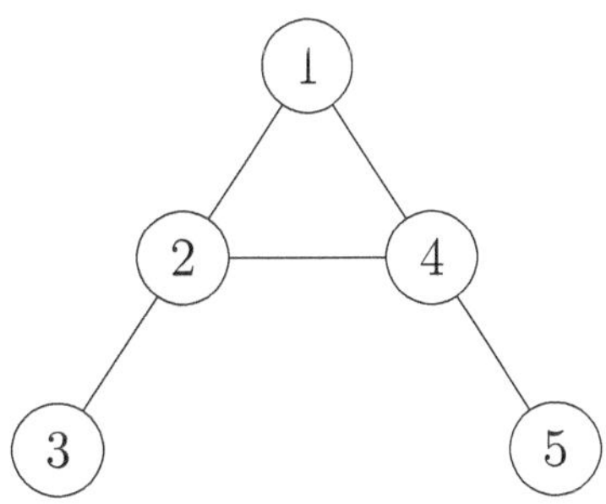

- **Fact 1:** All sellers on a single path of length k are critical, regardless of whether this path is a cycle.
- **Fact 2:** The number of critical sellers on a single path reduces with its length.
- **Fact 3:** No cycle of length more than k has a critical seller.
- **Fact 4:** Cycles of length $2k$ or more have at least two disjoint feasible paths and hence, no critical seller.
- **Fact 5:** The oddball class covers all cycles of length larger than $k + 1$ but smaller than $2k$. Further, since every pair of feasible paths on an oddball graph intersect at least once, it also covers graphs containing cycles of length less than or equal to k.

In a complete information environment, the valuation of the buyer and sellers is given by a vector $\mathbf{v} \equiv (v_0, v_1, \ldots, v_n)$. The first component of this vector denotes the valuation of the buyer for a path of length k or more, and other components denote the valuation of the respective sellers for their items. In a private information environment, agents only know their own valuation and the support of the valuations of other agents; a commonly known prior μ describes their beliefs over possible valuation profiles.

We assume that there exists a path $\mathcal{P} \in \Gamma$, such that it results in a positive surplus: $v_0 - \sum_{i \in \mathcal{P}} v_i > 0$. Given such a graph Γ, the expression $\max_{\mathcal{P} \in \Gamma} \left(v_0 - \sum_{i \in \mathcal{P}} v_i\right)$ is referred to as "full surplus" or "efficient surplus".

In complete information strategic bargaining games, a discount factor is applied to compare payoffs that arise in different time periods. We assume that the agents use the same discounting factor $\delta \in [0, 1]$.

An assembly problem with complete information is a tuple: $\langle \Gamma, k, \mathbf{v}, \delta \rangle$. When Γ is a complete graph of order n, an assembly problem is referred to by the tuple $\langle n, k, \mathbf{v}, \delta \rangle$. An assembly problem with private information is a triple: $\langle \Gamma, k, \mu \rangle$.

The results on complete information bargaining and mechanism design for the assembly problem are discussed in the next two subsections. Only a brief sketch of the argument is provided below each result. The interested reader is referred to the original papers for detailed proofs.

3.1 Bargaining with Complete Information

The bargaining protocol due to Rubinstein [27] and its different extended versions have been used in many contexts. A slightly general version of this protocol due to Roy Chowdhury and Sengupta [25] can be described as follows. In each period, active agents on one side of the market make offers of surplus shares to the other side—this gives rise to two alternative cases, where buyers make offers in odd periods and sellers in even periods and vice versa. The offers made are either accepted or rejected. If accepted, the deal is implemented, i.e., the seller sells his item at the agreed offer and leaves the market with his payment immediately. The game proceeds with the reduced set of agents. The ones making offers in the previous period now take on the role of responders. Offers are made and are either accepted or rejected. And so the game proceeds till the buyer is able to pick up at least one feasible path.

There are usually multiple equilibria in multiagent bargaining problems like assembly, some of which may be non-stationary. The nature of the equilibria also depends on which side of the market proposes first. In the discussion below, our focus will be to characterize bounds on equilibrium surplus shares under this protocol. Consequently, we are able to avoid details like stationarity or sequencing of the offers.

It is a standard practice in bargaining theory to express payoffs in terms of surplus shares instead of net payoffs. For instance, in the bilateral trade game when the buyer has valuation v_0 and seller v_1, the surplus realized on trade is $v_0 - v_1$. If the surplus shares in an equilibrium are α and $1 - \alpha$, the net payoffs are $\alpha(v_0 - v_1)$ and $(1 - \alpha)(v_0 - v_1)$—indicating that trade takes place at the price of $v_1 + (1 - \alpha)(v_0 - v_1)$, which the buyer pays and the seller receives.

The buyer can utilize negative surplus offers to exclude some sellers from the bargaining process, i.e., choose the sellers to bargain with in each period. Notice that a seller will not possibly make a negative offer to the buyer in our setting, since it delays trade with the buyer or eliminates the prospect of a trade. Bilateral bargaining models like that by Rubinstein [27] do not use this feature, while multilateral models like Roy Chowdhury and Sengupta [25] do.

The bilateral game studied by Rubinstein [27] is a special assembly problem with $n = k = 1$. Here, the only seller present is critical. The Subgame Perfect Nash Equilibrium of this game, which is now a standard result, is presented below.

Theorem 1 ([27]) *Consider the model where the buyer bargains with one seller for one input: $\langle n = 1, k = 1, v_0 > v_1, \delta\rangle$. There is a unique SPNE of the model described as follows:*

> *Agent i proposes a share $\frac{\delta}{1+\delta}$ of the surplus to j whenever she has to propose, and accept any share at least equal to $\frac{\delta}{1+\delta}$ whenever j has to propose.*

The game ends in the first period itself, with agent i proposing $\frac{\delta}{1+\delta}$ to the seller and the seller accepting it.

To see that the strategies proposed above constitute an equilibrium, apply the "one-shot deviation principle": no agent can gain by deviating from these strategies in any period for one period and conforming in the preceding and succeeding periods. If agent i proposes a higher share, it will be rejected and the play in the succeeding periods can only guarantee a lower payoff; if she proposes a lower share, it will be accepted immediately. Accepting lower shares is not profitable. Proving the uniqueness of this equilibrium is a rather involved exercise (see [22]).

The model studied by Roy Chowdhury and Sengupta [25] is a special assembly problem with $n = k \geq 2$ and all seller valuations are identical. Since the buyer wants all n plots, all sellers are critical here.

Theorem 2 ([25]) *Consider the model $\langle n \geq 2, k = n, v_1 \leq \cdots \leq v_n, v_0 > \sum_{i=1}^{n} v_i, \delta\rangle$. The buyer's equilibrium payoff cannot be more than $\frac{1-\delta}{1+\delta}$ of the full surplus for any $\delta > 0$.*

The proof of this result for $n = k = 2$ shows profitable deviations for one of the agents when the bound $\frac{1-\delta}{1+\delta}$ is crossed. An induction argument is used for the general case.

Both of these results correspond to the situation where all sellers are critical. The result below, in contrast, shows the possibility of full surplus extraction when there is no critical seller on the underlying graph, and seller valuations are identical.

Theorem 3 ([8]) *Consider $\langle \Gamma, k, \mathbf{v}, \delta \rangle$ where Γ has at least two different feasible paths and $\mathbf{v}$ is any arbitrary valuation profile.There exists $\bar{\delta} \in [0, 1)$ such that for all $\delta > \bar{\delta}$ the buyer extracts full surplus in at most two periods in an equilibrium if and only if*

- $\Gamma \neq \Gamma^*$, *i.e., there does not exist a critical seller in the underlying graph, and*
- $\mathcal{S}_1 = \mathcal{S}_2$, *i.e., there exist at least two paths with the minimum sum of valuations.*

This result characterizes equilibrium outcomes when the valuations of sellers are equal and the underlying graph does not contain a critical seller, i.e., either the graph has a $k + 1$-cycle, or it has at least a pair of disjoint paths, or it is an oddball graph. These three graphs have special properties—each node on a feasible path is substitutable by another node in a $k + 1$ cycle, every path is substitutable by another path in a graph with a pair of disjoint paths, and each node in a feasible path is substitutable by a set of nodes in an oddball graph. The first class of graphs exhibits full substitutability, while the other two exhibit limited substitutability. Consider the first class of graphs. If the buyer is the first to make offers, she can make offers of zero surplus shares to all sellers on a chosen feasible path in an equilibrium: any seller rejecting such offers must compete with corresponding substitute sellers in the next period. If the sellers are making first offers, competition ensures that sellers make no positive claims. Consequently, full surplus extraction takes place in the first period itself. In the other two classes, the buyer may be required to exclude all sellers in the first period, to achieve full surplus extraction in the second period.

The buyer cannot extract full surplus when the underlying graph contains at least one critical seller.

Theorem 4 ([8]) *Suppose $\Gamma = \Gamma^*$. The buyer cannot extract full surplus in an equilibrium.*

This result is obtained since at least one of the critical sellers can keep rejecting the offers of the buyer till all other sellers have accepted. He can then claim a positive surplus share in the ensuing subgame, by Theorem 1.

When seller valuations are not equal, the sum of seller valuations may differ over paths. The path corresponding to the least sum of seller valuations is efficient in the sense that it corresponds to the highest potential surplus. It follows that if possible, the buyer would prefer to purchase the efficient path.

Let $\mathcal{P}_i$ denote the path corresponding to the i-th smallest sum of valuations on a path in Γ. We will refer to a set of assembly problems as *rich* if there does not exist two disjoint paths $\mathcal{P}_1$ and $\mathcal{P}_2$ such that $\mathcal{S}_1 = \mathcal{S}_2$. Suppose the richness condition is not satisfied. The buyer, if offering first, can offer negative surplus shares to all sellers who reject such offers. In the next period, sellers on $\mathcal{P}_1$ and $\mathcal{P}_2$ cannot claim any surplus: the buyer extracts full surplus in the second period. If the sellers are making offers first, sellers on these two paths cannot claim any surplus share.

Theorem 5 ([8]) *Consider the rich class of assembly problems* $\langle\Gamma, k, \mathbf{v}, \delta\rangle$. *There does not exist any equilibrium where the buyer extracts full surplus.*

The proof of this result shows that at least one seller getting zero surplus share has a profitable deviation. Thus, full surplus extraction is not an equilibrium outcome.

By Theorem 4, the buyer cannot extract full surplus when the underlying graph contains critical sellers. The final results of this section provide bounds on the buyer's surplus share in such a problem.

Theorem 6 ([8]) *Consider an assembly problem* $\langle\Gamma, k, \mathbf{v}, \delta\rangle$ *with exactly one critical seller. In any equilibrium buyer's share of surplus cannot exceed* $\frac{1}{1+\delta}$.

Theorem 7 ([8]) *Consider an assembly problem* $\langle\Gamma, k, \mathbf{v}, \delta\rangle$ *with* m *critical sellers, where* $2 \le m \le k$. *In any equilibrium buyer's share of surplus cannot exceed* $\frac{1-\delta}{1+\delta}$.

The proof of Theorem 6 closely follows that of Theorem 4, while the proof of Theorem 7 follows that of Theorem 2. There exist assembly problems where these bounds are exactly achieved: for example, when $n = k = 1$, the corresponding bound is exactly achieved if the buyer is making the first offer (recall Theorem 1). It is also exactly achieved when Γ is a single line graph with three nodes, $k = 2$, and the buyer is making the first offer. When $n = k = 2$, the corresponding bound is exactly achieved if the buyer is making the first offer (recall Theorem 2). It is also exactly achieved when Γ is a single line graph with four nodes, $k = 3$, and the buyer is making the first offer.

3.2 Mechanism Design

Due to the well-known Revelation Principle (see [18]), it suffices to assume that the buyer and the sellers directly report their individual valuations to a central planner, who then decides allocations and payments according to a declared rule. A set of essential definitions is provided below.

A deterministic allocation $x \in \mathbb{R}^{n+1}$ is described as follows: for components $i = 1, \ldots, n$, x_i is -1 if seller i sells and 0 otherwise; $x_0 = 1$ if $\sum_{i=1}^{n} |x_i| \ge k$ and 0 otherwise. Let $\mathbb{X}$ be the set of all deterministic allocations.

Definition 1 (*Allocation Rule*) A deterministic allocation rule $P : [\underline{v_0}, \bar{v_0}] \times [\underline{v}, \bar{v}]^n \rightarrow \mathbb{X}$ maps each profile of reported values to a deterministic allocation.

For any agent j, $P_j(\mathbf{v})$ is the j-th component of $P(\mathbf{v})$.

Definition 2 (*Transfer Rule*) A transfer rule t is a map $t : [\underline{v_0}, \bar{v_0}] \times [\underline{v}, \bar{v}]^n \rightarrow \mathbb{R}^{n+1}$.

If $t_j(\mathbf{v}) > 0$ (resp. $t_j(\mathbf{v}) < 0$), then agent j pays (resp. receives) the amount $t_j(\mathbf{v})$.

Definition 3 (*Payoffs*) Given a mechanism (P, t). The (ex-post) utility of agent j with valuation v_j reporting $\hat{v}_j$ in mechanism (P, t) is

$$U_j^{(P,t)}(\hat{v}_j, v_{-j}|v_j) = v_j P_j(\hat{v}_j, v_{-j}) - t_j(\hat{v}_j, v_{-j}).$$

For convenience, the superscript (P, t) in the notation will be henceforth dropped.

Bayesian incentive compatibility requires that truthful reporting is optimal for each agent and for each valuation in *expectation.* This expectation is computed with respect to the prior distribution of valuations of other agents.

Definition 4 (*Bayesian Incentive Compatibility*) A mechanism is Bayesian incentive compatible (BIC) if for all j,

$$E_{-j}U_j(v_j, v_{-j}|v_j) \geq E_{-j}U_j(\hat{v}_j, v_{-j}|v_j) \text{ for all } v_j \text{ and } \hat{v}_j,$$

where $E_{-j}(\cdot)$ denotes expectation taken over v_{-j}.

Definition 5 (*Interim Individual Rationality*) A mechanism is interim individually rational (IIR) if for all j,

$$E_{-j}U_j(v_j, v_{-j}|v_j) \geq 0 \quad \text{for all } v_j.$$

The rest of the chapter uses the notation $U_j(\mathbf{v})$ and $U_j(v_j)$ for the ex-post and interim utilities in an equilibrium, respectively. Also, E is used to denote expectation operator over profile $\mathbf{v}$.

Definition 6 (*Efficiency*) An allocation rule P is ex-post efficient if for all $\mathbf{v}$,

$$\sum_j v_j P_j(\mathbf{v}) \geq \sum_j v_j P'_j(\mathbf{v}) \text{ for any allocation rule } P'.$$

To define ex-post efficient allocations in this setting, denote the feasible paths in Γ by $\mathcal{P}_1, \ldots, \mathcal{P}_q$ where $q \geq 1$. Consider a valuation profile $\mathbf{v}$. The sum of valuations in path $\mathcal{P}_i$ will be denoted by $S_i(\mathbf{v})$, $i = 1, \ldots, q$. These sums are ordered as follows: $S_{[1]}(\mathbf{v}) \leq \cdots \leq S_{[q]}(\mathbf{v})$. The paths corresponding to these sums are denoted by $\mathcal{P}_{[1]}(\mathbf{v}), \ldots, P_{[q]}(\mathbf{v})$, respectively. Efficiency requires trade to take place with sellers in $\mathcal{P}_{[1]}(\mathbf{v})$ if $v_0 > S_{[1]}(\mathbf{v})$; if $v_0 \leq S_{[1]}(\mathbf{v})$, then trade does not occur. For example, in the graph in Fig. 6, there are two feasible paths $\{1 - 2 - 3\}$ and $\{2 - 3 - 4\}$ when $k = 3$. If the valuations of the sellers are as indicated in the diagram, efficiency requires trade with sellers 1, 2, and 3 if $v_0 > 19$. Since the subsequent analysis will not require any special treatment of tie-breaking, any rule satisfying the condition above is called an efficient rule, denoted by P^*.

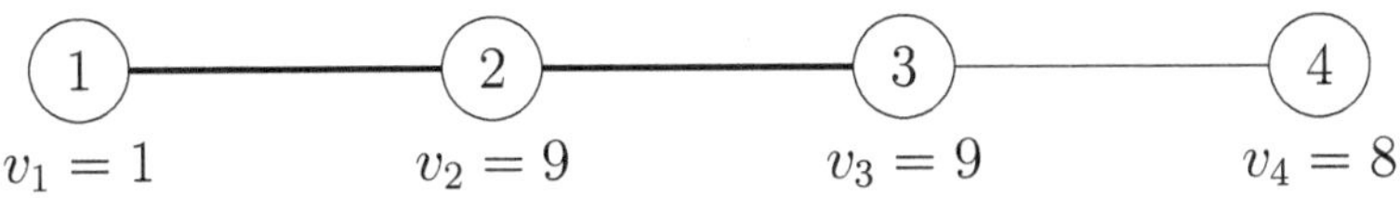

Fig. 6 $\mathcal{P}_{[1]}(\mathbf{v})$

Definition 7 (*Budget Balance*) A mechanism (P, t) satisfies ex-post budget balance if, for all $\mathbf{v}$,

$$\sum_{j=0}^{n} t_j(\mathbf{v}) = 0. \tag{1}$$

A mechanism achieves the *first-best* if it satisfies efficiency, IIR, and BB. A mechanism is *successful* if (a) it is BIC with respect to some prior μ and (b) it achieves the first-best.

Part A of the following result provides a sufficient condition and a weaker necessary condition for the existence of a successful mechanism when the number of feasible paths in the underlying graph is more than one. Part B states that no successful mechanism exists when there is only one feasible path.

Theorem 8 ([29])

A. *Let $\langle \Gamma, k, \mu \rangle$ be an assembly problem with $q > 1$.*

I. Suppose μ satisfies the following condition:

$$\underline{v}_0 \geq E\left(\sum_{i \in \mathcal{P}_{[1]}(\mathbf{v})} S_{[1]}(\bar{v}, v_{-i}) \right) - (k-1) E\left(S_{[1]}(\mathbf{v}) \right). \tag{2}$$

Then there exists a successful mechanism with respect to μ.

II. Suppose there exists a successful mechanism with respect to μ. Then the following holds:

$$\underline{v}_0 > E\left(\sum_{i \in \mathcal{P}_{[1]}(\mathbf{v})} S_{[1]}(\bar{v}, v_{-i}) - (k-1) S_{[1]}(\mathbf{v}) \middle| \mathbf{v} \in \tilde{V} \right), \tag{3}$$

where

$$\tilde{V} = \{\mathbf{v} \in [\underline{v}_0, \bar{v}_0] \times [\underline{v}, \bar{v}]^n : \underline{v}_0 > S_{[1]}(v) \text{ and } v_0 > S_{[1]}(\bar{v}, v_{-i}) \text{ for all } i \in \mathcal{P}_{[1]}(\mathbf{v})\}.$$

B. *Let $\langle \Gamma, k, \mu \rangle$ be an assembly problem with $q = 1$. The Myerson-Satterthwaite negative result applies, i.e., there does not exist any successful mechanism.*

This result is proved using the WKP condition due to Williams [31] and Krishna and Perry [13]: there exists a successful mechanism if and only if the well-known

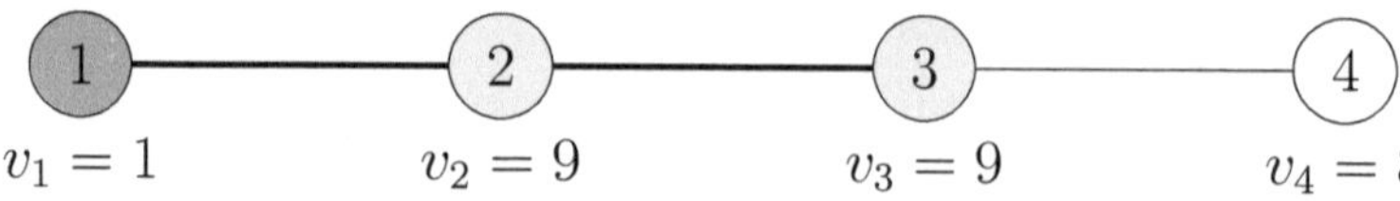

Fig. 7 Pivotal seller in darker shade and non-pivotal sellers in lighter shade at $\mathbf{v}$

VCG mechanism runs an expected budget surplus. The discussion below provides further interpretation.

Suppose $q > 1$ and efficiency requires trade to take place at a profile $\mathbf{v}$. A seller is called successful at this profile if he trades under the efficient rule. A successful seller $i \in \mathcal{P}_{[1]}(\mathbf{v})$ is trade-pivotal at $\mathbf{v}$ if trade is not efficient at $(\bar{v}, v_{-i})$, i.e., when seller i reports his highest possible valuation. Trade-pivotality is illustrated in the examples below. Let $n = 4, k = 3$, and $q = 2$. The supports of the prior distributions are as follows: $\underline{v}_0 = 25$, $\bar{v}_0 = 35$, $\underline{v} = 0$, $\bar{v} = 10$. Let $v_0 = 26$.

Recall from the example corresponding to Fig. 6 that sellers 1, 2, and 3 trade at $\mathbf{v}$. If seller 1's valuation is 10 instead of 1, the sum of the valuations on paths $\{1 - 2 - 3\}$ and $\{2 - 3 - 4\}$ are 28 and 26, respectively. Hence, trade does not take place at $(10, v_{-1})$, i.e., seller 1 is trade-pivotal at $\mathbf{v}$. But sellers 2 and 3 are not trade-pivotal at $\mathbf{v}$: if seller 2 has a valuation of 10, the sum of valuations on $\{1 - 2 - 3\}$ is 20 and trade can take place at $(10, v_{-2})$; same follows for seller 3. Pivotal and non-pivotal sellers and the efficient feasible path are shown in Fig. 7.

Now consider the profile $v'_0 = 28$, , $v'_1 = 1$, $v'_2 = 2$, $v'_3 = 3$, and $v'_4 = 2$. Trade takes place at $\mathbf{v}'$ with sellers 1, 2, and 3. Note that trade also takes place when the buyer's valuation is the lowest possible, i.e., $\underline{v}_0 = 25$. Furthermore, no successful seller at $\mathbf{v}'$ is trade-pivotal: if seller 1 reports a valuation of 10, efficiency requires trade with sellers 2, 3, and 4; if sellers 2 or 3 report a valuation of 10, efficiency requires trade with sellers 1, 2, and 3. This is illustrated in Fig. 8.

In the statement of Theorem 8, the set $\tilde{V}$ is the set of profiles v such that (i) it is efficient to trade at $(\underline{v}_0, v_{-0})$ and, therefore, also at $\mathbf{v}$, and (ii) all successful sellers are non-pivotal at $\mathbf{v}$. Hence, $\mathbf{v}' \in \tilde{V}$ but $\mathbf{v} \notin \tilde{V}$.

Pick $\mathbf{v} \in \tilde{V}$ and a successful seller i. Suppose i's valuation changes to $\bar{v}$. Since i is not trade-pivotal, trade still takes place and the sum of valuations of the successful sellers in the profile $(\bar{v}, v_{-i})$ is $S_{[1]}(\bar{v}, v_{-i})$. The sum of valuations of all other successful sellers at v is $S_{[1]}(\mathbf{v}) - v_i$. The difference of these two terms, summed over all successful sellers, is $\sum_{i \in \mathcal{P}_{[1]}(\mathbf{v})} S_{[1]}(\bar{v}, v_{-i}) - (k-1)S_{[1]}(\mathbf{v})$. Part A-I of Theorem 8 states that there exists a successful mechanism if the expectation of this term

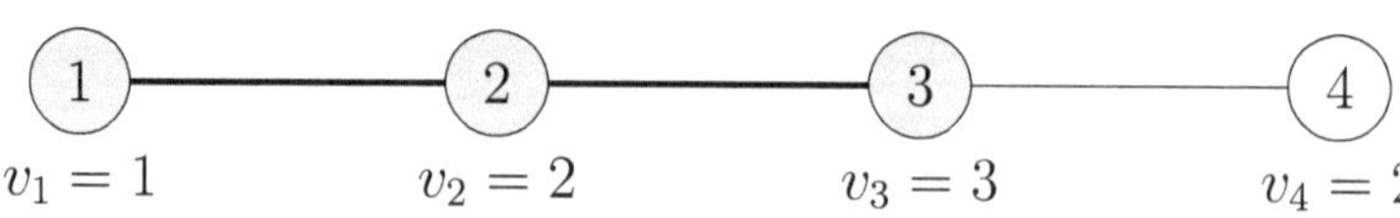

Fig. 8 Shaded nodes representing non-pivotal sellers at $\mathbf{v}'$

is at most $\underline{v}_0$. Part A-II states that if there exists a successful mechanism, then the expectation of this term, conditional on the profile belonging to $\tilde{V}$, is less than $\underline{v}_0$.

For illustration, consider the simple example: sellers are located on a complete graph and seller valuations are distributed uniformly in [0, 1]. If $\underline{v}_0 \geq \frac{k(k+1)}{n+1}$, then the existence of BIC mechanisms that achieve the first-best is guaranteed. For instance, if $n = 2$ and $k = 1$, $\underline{v}_0 \geq \frac{2}{3}$ is the required condition. Since $\frac{k(k+1)}{n+1} \to 0$ as $n \to \infty$, it becomes easier to satisfy the sufficient condition as the number of sellers increase.

To examine the role of critical sellers with respect to existence of a successful mechanism, let $c(\Gamma)$ denote the set of critical sellers in $\langle \Gamma, k, \mu \rangle$. If $q > 1$, then $|c(\Gamma)| \leq k - 1$. Conditions (2) and (3) can be reformulated to account for critical nodes.

Theorem 9 ([29]) *Let $\langle \Gamma, k, \mu \rangle$ be an assembly problem with $q > 1$.*

I. Suppose μ satisfies the following condition:

$$\underline{v}_0 \geq |c(\Gamma)|\bar{v} + E\left(\sum_{i \in \mathcal{P}_{[1]}(\mathbf{v}) \setminus c(\Gamma)} \left(S_{[1]}(\bar{v}, v_{-i}) + v_i \right) - (k - |c(\Gamma)|) S_{[1]}(\mathbf{v}) \right). \tag{4}$$

Then there exists a successful mechanism with respect to μ.

II. Suppose there exists a successful mechanism with respect to μ. Then the following holds:

$$\underline{v}_0 > |c(\Gamma)|\bar{v} + E\left(\sum_{i \in \mathcal{P}_{[1]}(\mathbf{v}) \setminus c(\Gamma)} \left(S_{[1]}(\bar{v}, v_{-i}) + v_i \right) - (k - |c(\Gamma)|) S_{[1]}(\mathbf{v}) \middle| \mathbf{v} \in \tilde{V} \right), \tag{5}$$

where

$$\tilde{V} = \{\mathbf{v} \in [\underline{v}_0, \bar{v}_0] \times [\underline{v}, \bar{v}]^n : \underline{v}_0 > S_{[1]}(\mathbf{v}) \text{ and } v_0 > S_{[1]}(\bar{v}, v_{-i}) \text{ for all } i \in \mathcal{P}_{[1]}(\mathbf{v})\}.$$

Corollary 1 *Suppose there exists a successful mechanism with respect to μ. Then*

$$\underline{v}_0 > |c(\Gamma)|\bar{v}. \tag{6}$$

Theorem 9 and Corollary 1 state that the count of critical nodes puts a lower bound on the support of the buyer's valuation essential for the existence of a successful mechanism.

4 Discussion

The results presented in Sect. 3.1 show the bearing of the degree of complementarity among inputs and asymmetry of valuations on full surplus extraction in the assembly problem. While the presence of critical sellers or sufficient asymmetry in seller valuations prevents full surplus extraction, even limited substitutability enables the buyer to extract full surplus in two periods, provided she is sufficiently patient. The number of critical sellers present in the problem also imposes an upper bound on the volume of surplus that can be extracted by the buyer.

In the light of these results, the formation of seller coalitions becomes one of the possible explanations of a holdout in the assembly problem. Consider the following example for an illustration. In the problem where one item is required and two sellers are present, by making alternate offers to one of the sellers according to the equilibrium strategy specified in Theorem 1 and by excluding the other seller using negative offers, the buyer can assure herself $\frac{\delta}{1+\delta}$ share of the full surplus. If sellers are allowed to use trigger strategies, there exists an equilibrium where both sellers collude to claim $\frac{1}{1+\delta}$ of the full surplus and the buyer picks one of them with equal probability provided $\delta > \frac{1}{\sqrt{2}}$. This equilibrium is sustained by the following trigger strategy: if any seller deviates by charging less than $\frac{1}{1+\delta}$, the other seller charges zero surplus share in the subsequent period. The buyer then rejects the deviating seller's offer and chooses to purchase from the other seller. The collusive payoff $\frac{1}{2(1+\delta)}$ is greater than the non-collusive payoff $1 - \delta$ if $\delta > \frac{1}{\sqrt{2}}$. In this equilibrium, both sellers get a positive expected payoff. If $\delta < \frac{1}{\sqrt{2}}$, sellers compete and earn zero surplus shares in the equilibrium. A complete characterization of possible coalitions and corresponding surplus shares in assembly problems is an open agenda for future work.

The results in Sect. 3.2 are not strategically informative, since the discussion here involves direct mechanisms. But the role of critical sellers turns out to be prominent here as well.

The nature of the first-best mechanism is not described in these results. As shown by Krishna and Perry [13], it is essentially a projection of the well-known VCG payments in the space of balanced transfers. Implementing the first-best requires knowledge of the prior, and hence turn out to be costly in the informational requirement in many real-life applications. This remark is also applicable to the optimal mechanism (see [28]).

The VCG mechanism itself exhibits several good properties like ex-post efficiency, dominant strategy incentive compatibility, and ex-post individual rationality. Further, it is also ex-post budget balanced in the limit as the number of sellers becomes large (see [30]), making it attractive for many applications with independent private values.

References

1. Asami, Y. 1985. A game-theoretic approach to the division of profits from economic land development. *Regional Science and Urban Economics* 18: 233–246.
2. Cai, H. 2000. Delay in multilateral bargaining under complete information. *Journal of Economic Theory* 93: 260–276.
3. Cai, H. 2003. Inefficient markov perfect equilibria in multilateral bargaining. *Economic Theory*.
4. Chatterjee, K., B. Dutta, D. Ray, and K. Sengupta. 1993. A noncooperative theory of coalitional bargaining. *The Review of Economic Studies* 60: 463–477.
5. Chatterjee, K., and W. Samuelson. 1983. Bargaining under Incomplete Information. *Operations Research* 31: 835–851.
6. Göller, D., and M. Hewer. 2015. Breakdown in multilateral negotiations. *Journal of Economic Theory* 157: 478–484.
7. Grossman, Z., J. Pincus, and P. Shapiro. 2010. A second-best mechanism for land assembly. Working paper.
8. Gupta, D., and S. Sarkar. 2018. Bargaining for Assembly. Working Paper.
9. Heller, M. 2008. *The Gridlock Economy: How Too Much Ownership Wrecks Markets Stops Innovation, and Costs Lives*. Basic Books.
10. Kominers, S.D., and E.G. Weyl. 2011. Concordance among holdouts. In *Proceedings of the 12th ACM conference on Electronic commerce*, ACM, 219–220.
11. Krasteva, S., and H. Yildirim. 2012. On the role of confidentiality and deadlines in bilateral negotiations. *Games and Economic Behavior* 75: 714–730.
12. Krasteva, S., and H. Yildirim. 2012. Payoff uncertainty, bargaining power, and the strategic sequencing of bilateral negotiations. *The RAND Journal of Economics* 43: 514–536.
13. Krishna, V., and M. Perry. 2000. Efficient mechanism design. Working Paper.
14. Menezes, F., and R. Pitchford. 2004. A Model of Seller Holdout. *Economic Theory*.
15. Miceli, T.J., and K. Segerson. 2012. Land assembly and the holdout problem under sequential bargaining. *American Law and Economics Review* 14: 372–390.
16. Myerson, R., and M. Satterthwaite. 1983. Efficient Mechanisms for Bilateral Trading. *Journal of Economic Theory* 29: 265–281.
17. Myerson, R.B. 1977. Graphs and cooperation in games. *Mathematics of Operations Research* 2: 225–229.
18. Myerson, R.B. 1979. Incentive compatibility and the bargaining problem. *Econometrica*, 61–73.
19. Noe, T.H., and J. Wang. 2004. Fooling all of the people some of the time: a theory of endogenous sequencing in confidential negotiations. *The Review of Economic Studies* 71: 855–881.
20. O'Flaherty, B. 1994. Land assembly and urban renewal. *Regional Science and Urban Economics* 24: 287–300.
21. Osborne, M.J., and A. Rubinstein. 1990. *Bargaining and markets*, Academic press.
22. Osborne, M.J., and A. Rubinstein. 1994. *A course in game theory*, MIT press.
23. Plassmann, F., and T. Tideman. 2010. Providing Incentives for Efficient Land Assembly. Tech. rep., working paper.
24. Ray, D., and R. Vohra. 1999. A theory of endogenous coalition structures. *Games and economic behavior* 26: 286–336.
25. Roy Chowdhury, P., and K. Sengupta. 2012. Transparency, complementarity and holdout. *Games and Economic Behavior* 75: 598–612.
26. Royal Swedish Academy of Sciences. 2007. Scientific Bacground on Mechanism Design Theory. http://www.nobelprize.org/nobel_prizes/economic-sciences/laureates/2007/advanced-economicsciences2007.pdf.
27. Rubinstein, A. 1982. Perfect equilibrium in a bargaining model. *Econometrica*, 97–109.
28. Sarkar, S. 2014. Mechanism Design for Land Acquisition. Unpublished PhD thesis submitted to Indian Statistical Institute.
29. Sarkar, S. 2017. Mechanism design for land acquisition. *International Journal of Game Theory* 46: 783–812.

30. Sarkar, S. 2018. Convergence of VCG mechanism to ex-post budget balance in a model of land acquisition. *Mathematical Social Sciences* 93: 37–46.
31. Williams, S.R. 1999. A characterization of efficient, Bayesian incentive compatible mechanisms. *Economic Theory* 14: 155–180.
32. Xiao, J. 2018. Bargaining orders in a multi-person bargaining game. *Games and Economic Behavior* 108: 364–379.

On Different Ranking Methods

Agnieszka Rusinowska

JEL Classification: D71 · D85 · C02

1 Related Literature on Ranking Methods

Ranking accompanies our everyday activities and is crucial in various situations, in particular, when facing competitive issues and having to choose from a set of alternatives. As a consequence, the investigation of appropriate ranking methods is particularly important. Which method should be used when one needs to rank, for instance, political candidates or parties in election, teams in sport competition, universities or institutes in excellence competition, scientific candidates for academic positions?

The literature on ranking methods and their applications is very rich and gets a lot of interest for many years; for some examples see, e.g., [16, 57, 67], for ranking scientific journals, web pages on the internet, and alternatives in social choice, respectively. There exists a vast literature on the classical problem of ranking objects, based on a binary relation between the objects (e.g., [8, 30, 49, 62]).

In this short overview, we will briefly recall some selected ranking methods. We will focus on ranking methods for directed graphs, where nodes have different interpretations, depending on the ranking subject and environment. A ranking method is then formally defined as a mapping which assigns to every directed graph a complete preorder on the set of nodes. Every node gets its ranking score, and a node is ranked higher, the higher is its score. In this stream of literature, usually axiomatic characterizations to ranking methods and ranking scores are provided.

A. Rusinowska (✉)
CNRS – Paris School of Economics, Centre d'Economie de la Sorbonne, 106-112 Bd de l'Hôpital, 75647 Paris Cedex 13, France
e-mail: agnieszka.rusinowska@univ-paris1.fr

S. Borkotokey et al. (eds.), *Game Theory and Networks*, Indian Statistical Institute Series,
https://doi.org/10.1007/978-981-16-4737-6_6

One of the well-known ranking methods is based on Copeland score ([22]). When defining an outdegree (respectively, an indegree) of a node in a directed graph as the number of its outgoing (respectively, ingoing) arcs, the Copeland score of the node simply counts the difference between its outdegree and its indegree. In the literature, there exist several axiomatizations of the ranking by Copeland score; see e.g., [7, 41] (also [15] for a related method). Reference [62] provides an axiomatic characterization of the ranking by Copeland score on the class of tournaments, where the ranking coincides with the one by outdegree.

In order to measure domination in directed graphs, [18] characterize two relational power measures: the score measure and the β-measure; see also [42, 70], as well as [17] for the case of undirected graphs. Reference [19] characterize the ranking induced by the score measure (that they call the ranking by outdegree) for arbitrary directed graphs. The ranking induced by the β measure (the β-ranking) is axiomatically characterized in [14]. A related idea underlies the ranking for chess players investigated earlier in [34], where defeating a strong opponent gives more points than defeating a weak one.

[21] introduce a ranking method based on the degree ratio of a node, which is its outdegree divided by its indegre, and a ranking method based on a modified degree ratio. The authors provide axiomatic characterizations of these two ranking methods as well as an alternative axiomatization of the Copeland score.

Some ranking methods have been also introduced for weighted directed graphs. The outflow as a relational power measure for weighted (and also non-weighted) directed graphs is axiomatically characterized by [18]. Also [20] deliver an axiomatic characterization of the outflow ranking method for weighted directed graphs.

There is a variety of other methods that are based on hyperlinks for ranking web pages or citations for ranking academic journals. Reference [58] presents an economic analysis of many ranking methods and the use of citations in the law. Ranking methods based on evaluations or citations consider a one-sided setting in which experts evaluate some items for ranking, and a peers' setting when the experts coincide with the items.

The first citation index for articles published in journals is the Science Citation Index (SCI), which uses the counting method, based on counting the total number of citations received by a journal (see [35]). The Impact Factor ([36]) of an academic journal counts the average number of citations received by articles published on it; see also [37]. Reference [53] use an iteration (impact adjusted) method to examine the impact factors of economic journals.

The Markov-chain approach comes originally from [75] and [45]. Reference [57] introduce the influence measure which counts both direct and indirect citations. Google's Page Rank ([16]) uses a similar recursive approach and is based on the invariant method. The axiomatic approach to the invariant method and several axiomatizations of eigencentrality (used in the eigenvalue centrality method) is presented, e.g., in [1, 47, 56, 69]; see also [2, 68]. [28] propose a "market" approach to ranking items in a network, e.g., ranking web pages connected by links or papers connected by citations. Their set of methods includes the eigencentrality method. Also the so-called mutual centrality method characterized in [27] is related to the eigenvalue centrality. Reference [26] introduces and axiomatically characterizes the handicap-based method, which assigns both scores to the items and weights to the experts. References [24, 25] investigates rankings in a dynamic setting.

The related contributions come also from the extremely vast literature on bibliometrics. We mention just few of them. Although the Impact Factors of journals are among the oldest bibliometric indices used for evaluating journals (see, e.g., [5, 38, 39] for surveys), many others have been introduced. The well-known h-index (the Hirsch index, [43]) widely examined from an axiomatic point of view (see e.g., [12, 54, 60, 77]) induces a ranking method that supports evaluations of researchers. [33] introduces another bibliometric index, the so-called g-index, axiomatically characterized, e.g., in [76]; see also [32, 55, 61, 78], as well as [4, 31, 44, 48] for some other bibliometric indices. Also [10] provide an axiomatic foundation of the ranking of journals based on Impact Factors and suggest alternative rankings that use some generalizations of Impact Factors.

Within this bibliometric literature, numerous works discuss in detail the importance of some properties (e.g., independence and consistency) for bibliometric rankings of authors and journals; see e.g., [9–11, 54, 55, 71, 74]. Some other properties might be subject to discussion, for instance, totality for ranking departments, saying that when two equal-size departments have the same citation distribution, they must be equivalent. For various works on ranking departments we also refer to, e.g., [6, 23, 29, 46, 59, 64].

An important issue in ranking is related to the fact that in some situations we are faced to compare authors, journals, departments belonging to different fields of research (e.g., [3, 63, 65, 66]). There exist several research directions on how to solve this normalization problem between different fields. One of the ideas lies on the fractional counting of citations, meaning that the value of a citation given by an article is inversely proportional to the total number of articles that it cites. Fractional counting of citations is proposed in [40, 50, 51]. [13] axiomatically characterize the ranking authors by using the fractional counting of citations. There exist also some empirical studies on this concept; see e.g., [52, 72, 73].

In the following two sections, we briefly present preliminaries and then recall several ranking methods for directed graphs that use outdegree (and indegree) of a node.

2 Notation and Basic Definitions

We introduce some basic notation and definitions, as in [19].

Directed graphs A *directed graph* (or *digraph*) is a pair (N, D), where $N = \{1, 2, \ldots, n\}$ is a finite set of nodes and $D \subset N \times N$ is a set of *arcs* on N. We only consider digraphs (N, D) that are irreflexive, i.e., $(i, i) \notin D$ for every $i \in N$. Since the set of nodes N is fixed, a digraph (N, D) can be represented by its binary relation D. The collection of all digraphs on N is denoted by $\mathcal{D}$. For $i \in N$ and

$D \in \mathcal{D}$, we define the set of *successors* of node $i \in N$ in digraph D by

$$S_D(i) = \{j \in N \mid (i, j) \in D\}$$

and the set of *predecessors* of i in D by

$$P_D(i) = \{j \in N \mid (j, i) \in D\}.$$

The cardinalities of $S_D(i)$ and $P_D(i)$ are called the *outdegree* $out_i(D)$ and the *indegree* $in_i(D)$ of node i in D, i.e.,

$$out_i(D) = \#S_D(i) \text{ and } in_i(D) = \#P_D(i).$$

Preorder A *preorder* on N is a binary relation $\mathcal{R} \subset N \times N$ that is reflexive (i.e., $(i, i) \in \mathcal{R}$ for all $i \in N$) and transitive (i.e., if $(i, j) \in \mathcal{R}$ and $(j, h) \in \mathcal{R}$, then $(i, h) \in \mathcal{R}$ for every $i, j, h \in N$). A preorder $\mathcal{R}$ on N is complete if $(i, j) \in \mathcal{R}$ or $(j, i) \in \mathcal{R}$ or both for every pair $i, j \in N, i \neq j$. We use the standard notation, i.e.,

$$i \succeq j \text{ if and only if } (i, j) \in \mathcal{R} \text{ (} i \text{ is ranked at least as high as } j\text{)},$$

$$i \succ j \text{ if and only if } [i \succeq j \text{ and not } j \succeq i] \text{ (} i \text{ is ranked higher than } j\text{)},$$

$$i \sim j \text{ if and only if } [i \succeq j \text{ as well as } j \succeq i] \text{(} i \text{ and } j \text{ are ranked equally)}.$$

We denote the collection of all complete preorders by $\mathcal{W}$.

Ranking methods A *ranking method* is a mapping $R \colon \mathcal{D} \to \mathcal{W}$ which assigns to every digraph $D \in \mathcal{D}$ on N a complete preorder $R(D) \in \mathcal{W}$. We use the notation

$$i \succeq_D j \text{ if and only if } (i, j) \in R(D).$$

A digraph $D \in \mathcal{D}$ is a *tournament* on N if

$$\#[\{(i, j), (j, i)\} \cap D] = 1 \text{ for all } i, j \in N, \ i \neq j.$$

Note that every tournament is a complete digraph, where by a complete digraph we mean $D \in \mathcal{D}$ such that $(i, j) \in D$ or $(j, i) \in D$ or both for every pair $i, j \in N, i \neq j$. Let $\mathcal{CD} \subset \mathcal{D}$ be the collection of all complete digraphs on N, and let $\mathcal{T} \subset \mathcal{CD} \subset \mathcal{D}$ denote the class of all tournaments on N.

The *ranking method by outdegree* is the ranking method $R^{out} : \mathcal{D} \to \mathcal{W}$ which assigns to every digraph $D \in \mathcal{D}$ on N a complete preorder $R^{out}(D) \in \mathcal{W}$ given by

$$(i, j) \in R^{out}(D) \text{ if and only if } out_i(D) \geq out_j(D).$$

We use the notation

$$i \succeq_D^{out} j \text{ if and only if } (i, j) \in R^{out}(D).$$

The *Copeland score* $cop_i(D)$ of node $i \in N$ in digraph D is defined by

$$cop_i(D) = 2\#\left(S_D(i) \setminus P_D(i)\right) + \#\left(S_D(i) \cap P_D(i)\right).$$

For $D \in \mathcal{CD}$, $\#S_D(i) + \#P_D(i) - \#\left(S_D(i) \cap P_D(i)\right) = n - 1$. Hence, $2\#\left(S_D(i) \setminus P_D(i)\right) + \#\left(S_D(i) \cap P_D(i)\right) = 2\#S_D(i) - \#\left(S_D(i) \cap P_D(i)\right) = \#S_D(i) - \#P_D(i) + n - 1$, and therefore

$$cop_i(D) = 2\#\left(S_D(i) \setminus P_D(i)\right) + \#\left(S_D(i) \cap P_D(i)\right) = out_i(D) - in_i(D) + n - 1.$$

The *ranking method by Copeland score* is the ranking method given by

$$i \succeq_D^{cop} j \text{ if and only if } cop_i(D) \geq cop_j(D) \text{ for all } i, j \in N.$$

Note that for tournaments the ranking by outdegree and the ranking by Copeland score are the same, since $S_D(i) \cap P_D(i) = \emptyset$ for all $i \in N$ and $D \in \mathcal{T}$.

However, these two ranking methods are different on $\mathcal{D}$.

For a digraph $D \in \mathcal{D}$ and a permutation $\pi \colon N \to N$, the permuted digraph $\pi D \in \mathcal{D}$ is given by $(\pi(i), \pi(j)) \in \pi D$ if and only if $(i, j) \in D$.

The *β-measure* on N (introduced in [18]) is the function $\beta \colon \mathcal{D} \to \mathbb{R}^N$ defined by

$$\beta_i(D) = \sum_{j \in S_D(i)} \frac{1}{in_j(D)} \quad \text{for all } i \in N \text{ and } D \in \mathcal{D}.$$

The β-measure equally distributes the domination power over a node $j \in N$ in a digraph D over all its predecessors.

The *ranking method by the β-measure* or *the β-ranking* is the ranking method given by

$$i \succeq_D^{\beta} j \text{ if and only if } \beta_i(D) \geq \beta_j(D) \text{ for all } i, j \in N.$$

3 Axiomatizations of the Ranking Methods

Rubinstein's result on the ranking in a tournament On the class of tournaments $\mathcal{T}$, [62] provides an axiomatic characterization of the ranking by Copeland score (i.e., by outdegree, since for tournaments the rankings by outdegree and by Copeland score are the same).

The following three axioms (as formulated in [19]) are used for Rubinstein's characterization:

(i) *Anonymity*:
Permuting the nodes in a digraph permutes accordingly the ranking, i.e.,

$$\text{For every } D \in \mathcal{D} \text{ and permutation } \pi : N \to N \text{ it holds that}$$
$$i \succeq_D j \text{ if any only if } \pi(i) \succeq_{\pi D} \pi(j).$$

(ii) *Positive responsiveness*:
If i is ranked at least as high as j, then increasing the outdegree of i makes i being ranked higher than j, i.e.,

$$\text{Let } D \in \mathcal{D} \text{ and } i, j, h \in N, i \neq j \text{ be such that } (i, h) \notin D, \text{ and let } D' = D \cup \{(i, h)\}.$$
$$\text{Then } i \succeq_D j \text{ implies that } i \succ_{D'} j.$$

(iii) *Independence of irrelevant arcs*:
The order between two nodes does not change if changes only take place with respect to arcs on which they are neither the predecessor nor the successor, i.e.,

$$\text{Let } D, D' \in \mathcal{D} \text{ and } i, j \in N \text{ be such that } S_D(i) = S_{D'}(i), S_D(j) = S_{D'}(j),$$
$$P_D(i) = P_{D'}(i), \text{ and } P_D(j) = P_{D'}(j). \text{ Then } i \succeq_D j \text{ if and only if } i \succeq_{D'} j.$$

Ranking by outdegree [19] generalize Rubinstein's result by characterizing the ranking by outdegree for arbitrary digraphs. The first two axioms introduced in [62], i.e., anonymity and positive responsiveness are the same, while independence of irrelevant arcs is generalized in a straightforward way to independence of non-dominated arcs.

Formally, for a ranking method represented by $\{\succeq_D \mid D \in \mathcal{D}\} \subset \mathcal{W}$, we consider the following three axioms ([19]):

(i) *Anonymity*:
Permuting the nodes in a digraph permutes accordingly the ranking, i.e.,

$$\text{For every } D \in \mathcal{D} \text{ and permutation } \pi : N \to N \text{ it holds that}$$
$$i \succeq_D j \text{ if any only if } \pi(i) \succeq_{\pi D} \pi(j).$$

(ii) *Positive responsiveness*:
If i is ranked at least as high as j, then increasing the outdegree of i makes i being ranked higher than j, i.e.,

$$\text{Let } D \in \mathcal{D} \text{ and } i, j, h \in N, i \neq j \text{ be such that } (i, h) \notin D, \text{ and let } D' = D \cup \{(i, h)\}.$$
$$\text{Then } i \succeq_D j \text{ implies that } i \succ_{D'} j$$

(iii) *Independence of non-dominated arcs*:
The order between two nodes does not change if changes only take place in arcs on which they are not the predecessors, i.e.,

$$\text{Let } D, D' \in \mathcal{D} \text{ and } i, j \in N \text{ be such that } S_D(i) = S_{D'}(i) \text{ and } S_D(j) = S_{D'}(j).$$
$$\text{Then } i \succeq_D j \text{ if and only if } i \succeq_{D'} j.$$

Reference [19] prove (their Theorem 2.4) that a ranking method is equal to the ranking method by outdegree if and only if it satisfies anonymity, positive responsiveness, and independence of non-dominated arcs.

Ranking by Copeland score [7] presents an alternative generalization of Rubinstein's result by providing an axiomatic characterization of the ranking by Copeland score for arbitrary digraphs.

More precisely, [7] characterizes the Copeland score by the following axioms (that we state by using the same notation borrowed from [19], the first two being the same as in [19]):

(i) *Anonymity*:
Permuting the nodes in a digraph permutes accordingly the ranking, i.e.,

$$\text{For every } D \in \mathcal{D} \text{ and permutation } \pi : N \to N \text{ it holds that}$$
$$i \succeq_D j \text{ if any only if } \pi(i) \succeq_{\pi D} \pi(j).$$

(ii) *Positive responsiveness*:
If i is ranked at least as high as j, then increasing the outdegree of i makes i being ranked higher than j, i.e.,

$$\text{Let } D \in \mathcal{D} \text{ and } i, j, h \in N, i \neq j \text{ be such that } (i, h) \notin D, \text{ and let } D' = D \cup \{(i, h)\}.$$
$$\text{Then } i \succeq_D j \text{ implies that } i \succ_{D'} j.$$

(iii) *Independence of 2- or 3-cycles*:
Deleting or adding a cycle of length 2 or 3 to a digraph does not change the ranking of the nodes, i.e.,

$$\text{Let } D, D' \in \mathcal{D} \text{ be such that } D' = D \cup \{(h, g), (g, h)\} \text{ for some } h, g \in N \text{ with } \{(h, g), (g, h)\} \cap D = \emptyset, \text{ or } D' = D \cup \{(h, g), (g, f), (f, h)\} \text{ for some } h, g, f \in N \text{ with } \{(h, g), (g, f), (f, h)\} \cap D = \emptyset. \text{ Then } i \succeq_D j \text{ if and only if } i \succeq_{D'} j \text{ for all } i, j \in N.$$

(iv) *Negative responsiveness*:
If i is ranked at least as high as j, then increasing the indegree of j makes i being ranked higher than j, i.e.,

$$\text{Let } D \in \mathcal{D} \text{ and } i, j, h \in N, i \neq j \text{ be such that } (h, j) \notin D, \text{ and let } D' = D \cup \{(h, j)\}.$$
$$\text{Then } i \succeq_D j \text{ implies that } i \succ_{D'} j.$$

As mentioned in [19], the ranking by Copeland score does not satisfy independence of non-dominated arcs on $\mathcal{D}$. Moreover, the ranking by outdegree does not satisfy independence of 2- or 3-cycles nor negative responsiveness for arbitrary digraphs. Furthermore, note that while independence of non-dominated arcs generalizes independence of irrelevant arcs, independence of 2- or 3-cycles does not.

More precisely, [19] prove the following results (their Proposition 3.4) for a ranking method R on $\mathcal{D}$:

- If R satisfies independence of non-dominated arcs, then R satisfies independence of irrelevant arcs.
- R satisfies independence of non-dominated arcs on $\mathcal{T}$ if and only if R satisfies independence of irrelevant arcs on $\mathcal{T}$.
- On $\mathcal{D}$, independence of 2- or 3-cycles and independence of irrelevant arcs are two independent properties.

Reference [41] provides an axiomatic characterization of the ranking by Copeland score restricted to the *class of complete 2-digraphs*, which are modified digraphs such that there exist exactly two (possibly the same) arcs between every pair of nodes $i, j \in N, i \neq j$.

As emphasized in [19], the notions of 2-digraphs and "standard" digraphs recalled in Sect. 2 are different.

Reference [41] shows that for complete 2-digraphs the ranking by Copeland score is characterized by the following three properties:

(i) *Anonymity* (stated for complete 2-digraphs);

(ii) *Positive responsiveness* (stated for complete 2-digraphs);

(iii) *Independence of reversing cycles*:
Reversing a cycle in a complete 2-digraph does not change the ranking of the nodes.

Reference [19] point out that for complete 2-digraphs the ranking by Copeland score is the same as the ranking by outdegree with the outdegree defined for such graphs by $out_i(D) = \#\{(h, j) \in D \mid h = i\}$. Both ranking methods also satisfy independence of reversing cycles on $\mathcal{CD}$.

Ranking by the β-measure [14] characterize the β-ranking by using the following axioms:

(i) *Anonymity*;

(ii) *Positive responsiveness*;

(iii) *Independence of irrelevant arcs*:
Some arcs are irrelevant for comparing two nodes, i.e., arcs which do not "involve" the two nodes.

(iv) *Node addition*:
Adding nodes that are not linked to any other node has no influence on the ranking.

(v) *Independence of local density*:
Increasing the number of successors of a node and simultaneously increasing their number of predecessors, in the same proportion, does not change (improve or worsen) the position of that node.

When comparing the above conditions with the axioms stated in [19], the first two properties (anonymity, positive responsiveness) are the same, while independence of irrelevant arcs is strictly weaker than the independence of non-dominated arcs (as pointed out before). The last two properties (node addition, independence of local density) are not related to any of the conditions in [19].

References

1. Altman, A., and M. Tennenholtz, 2005. Ranking systems: The PageRank axioms. In: EC'05 proceedings of the 6th ACM conference on electronic commerce, 1–8. New York: ACM.
2. Altman, A., and M. Tennenholtz. 2008. Axiomatic foundations for ranking systems. *Journal of Artificial Intelligence Research* 31: 473–495.
3. Amin, M., and M.A. Mabe. 2000. Impact factors: Use and abuse. *Perspectives in Publishing* 1: 1–6.
4. Anderson, T.R., R.K.S. Hankin, and P.D. Killworth. 2008. Beyond the Durfee square: Enhancing the h-index to score total publication output. *Scientometrics* 76 (3): 577–588.
5. Archambault, E., and V. Larivière. 2009. History of the journal impact factor: Contingencies and consequences. *Scientometrics* 79 (3): 635–649.
6. Arencibia-Jorge, R., I. Barrios-Almaguer, S. Fernández-Hernández, and R. Carvajal-Espino. 2008. Applying successive h-indices in the institutional evaluation: A case study. *Journal of the American Society for Information Science and Technology* 59: 155–157.
7. Bouyssou, D. 1992. Ranking methods based on valued preference relations: A characterization of the net flow method. *European Journal of Operational Research* 60: 61–67.
8. Bouyssou, D. 2004. Monotonicity of 'ranking by choosing' procedures: A progress report. *Social Choice and Welfare* 23 (2): 249–273.
9. Bouyssou, D., and T. Marchant. 2010. Consistent bibliometric rankings of authors and journals. *Journal of Informetrics* 4: 365–378.
10. Bouyssou, D., and T. Marchant. 2011. Bibliometric rankings of journals based on Impact Factors: An axiomatic approach. *Journal of Informetrics* 5 (1): 75–86.
11. Bouyssou, D., and T. Marchant. 2011. Ranking scientists and departments in a consistent manner. *Journal of the American Society for Information Science and Technology* 62 (9): 1761–1769.
12. Bouyssou, D., and T. Marchant. 2014. An axiomatic approach to bibliometric rankings and indices. *Journal of Informetrics* 8: 449–477.
13. Bouyssou, D., and T. Marchant. 2016. Ranking authors using fractional counting of citations: An axiomatic approach. *Journal of Informetrics* 10 (1): 183–199.
14. Bouyssou, D., and T. Marchant. 2018. The β-ranking and the β-measure for directed networks: Axiomatic characterizations. *Social Networks* 52 (1): 145–153.
15. Bouyssou, D., and P. Perny. 1992. Ranking methods for valued preference relations: A characterization of a method based on leaving and entering flows. *European Journal of Operational Research* 61: 186–194.

16. Brin, S., and L. Page. 1998. The anatomy of large-scale hypertextual Web search engine. *Computer Networks and ISDN Systems* 30: 107–117.
17. van den Brink, R., P. Borm, R. Hendrickx, and G. Owen. 2008. Characterizations of the β- and the degree network power measure. *Theory and Decision* 64: 519–536.
18. van den Brink, R., and R.P. Gilles. 2000. Measuring domination in directed networks. *Social Networks* 22: 141–157.
19. van den Brink, R., and R.P. Gilles. 2003. Ranking by outdegree for directed graphs. *Discrete Mathematics* 271: 261–270.
20. van den Brink, R., and R.P. Gilles. 2009. The outflow ranking method for weighted directed graphs. *European Journal of Operational Research* 193: 484–491.
21. van den Brink, R., and A. Rusinowska. 2021. The degree ratio ranking method for directed graphs. *European Journal of Operational Research* 288(2): 563–575.
22. Copeland, A.H. 1951. A Reasonable Social Welfare Function. Mimeographed Notes, University of Michigan Seminar on Applications of Mathematics to the Social Sciences.
23. Coupé, T. 2003. Revealed performances: Worldwide rankings of economists and economics departments, 1990–2000. *Journal of the European Economic Association* 1 (6): 1309–1345.
24. Demange, G. 2012. On the influence of a ranking system. *Social Choice and Welfare* 39: 431–455.
25. Demange, G. 2014. Collective attention and ranking methods. *Journal of Dynamics and Games* 1: 17–43.
26. Demange, G. 2014. A ranking method based on handicaps. *Theoretical Economics* 9 (3): 915–942.
27. Demange, G. 2017. Mutual rankings. *Mathematical Social Sciences* 90: 35–42.
28. Du, Y., E. Lehrer, and A. Pauzner. 2015. Competitive economy as a ranking device over networks. *Games and Economic Behavior* 91: 1–13.
29. Dusansky, R., and C.J. Vernon. 1998. Rankings of U.S. economics departments. *Journal of Economic Perspectives* 12:157–170
30. Dutta, B., and J.F. Laslier. 1999. Comparison functions and choice correspondences. *Social Choice and Welfare* 16: 513–532.
31. van Eck, N.J., and L. Waltman. 2008. Generalizing the h- and g-indices. *Journal of Informetrics* 2 (4): 263–271.
32. Egghe, L. 2006. An improvement of the h-index: The g-index. *ISSI Newsletter* 2 (1): 8–9.
33. Egghe, L. 2006. Theory and practice of the g-index. *Scientometrics* 69 (1): 131–152.
34. Elo, A.E. 1978. *The rating of chess players, past and present*. New York: Arco Pub.
35. Garfield, E. 1955. Citation indexes to science: A new dimension in documentation through the association of ideas. *Science* 122: 108–111.
36. Garfield, E. 1972. Citation analysis as a tool in journal evaluation. *Science* 178 (4060): 471–479.
37. Garfield, E. 1979. Is citation analysis a legitimate evaluation tool? *Scientometrics* 1: 359–375.
38. Garfield, E. 2006. The history and meaning of the journal impact factor. *Journal of the American Medical Association* 295: 90–93.
39. Glänzel, W., and H.F. Moed. 2002. Journal impact measures in bibliometric research. *Scientometrics* 53 (2): 171–193.
40. Glänzel, W., Schubert, A., Thijs, B., and K. Debackere. 2011. A priori vs. a posteriori normalisation of citation indicators. The case of journal rankig. *Scientometrics* 87(2):415–424.
41. Henriet, D. 1985. The Copeland choice function - an axiomatic characterization. *Social Choice and Welfare* 2: 49–63.
42. Herings, P.J.J., G. van der Laan, and D. Talman. 2005. The positional power of nodes in digraphs. *Social Choice and Welfare* 24: 439–454.
43. Hirsch, J.E. 2005. An index to quantify an individual's scientific research output. *Proceedings of the National Academy of Sciences* 102: 16569–16572.
44. Jin, B., L. Liang, R. Rousseau, and L. Egghe. 2007. The R- and AR-indices: Complementing the h-index. *Chinese Science Bulletin* 52 (6): 855–863.
45. Kendall, M.G. 1955. Further contributions to the theory of paired comparisons. *Biometrics* 11: 43.

46. Kinnucan, H.W., and G. Traxler. 1994. Ranking agricultural economics departments by AJAE page counts: A reappraisal. *Agricultural and Resource Economics Review* 23: 194–199.
47. Kitti, M. 2016. Axioms for centrality scoring with principal eigenvectors. *Social Choice and Welfare* 46 (3): 639–653.
48. Kosmulski, M. 2006. A new Hirsch-type index saves time and works equally well as the original h-ndex. *ISSI Newsletter* 2 (3): 4–6.
49. Laslier, J.F. 1997. *Tournament solutions and majority voting*. Berlin: Springer.
50. Leydesdorff, L., and T. Opthof. 2010. Normalization at the field level: Fractional counting of citations. *Journal of Informetrics* 4 (4): 644–646.
51. Leydesdorff, L., and T. Opthof. 2010. Scopus's Source Normalized Impact per Paper (SNIP) versus a journal impact factor based on fractional counting of citations. *Journal of the American Society for Information Science and Technology* 61 (11): 2365–2369.
52. Leydesdorff, L., and J.C. Shin. 2011. How to evaluate universities in terms of their relative citation impacts: Fractional counting of citations and the normalization of differences among disciplines. *Journal of the American Society for Information Science and Technology* 62 (6): 1146–1155.
53. Liebowitz, S.J., and J.P. Palmer. 1984. Assessing the relative impacts of economics journals. *Journal of Economic Literature* 22: 77–88.
54. Marchant, T. 2009. An axiomatic characterization of the ranking based on the h-index and some other bibliometric rankings of authors. *Scientometrics* 80: 325–342.
55. Marchant, T. 2009. Score-based bibliometric rankings of authors. *Journal of the American Society for Information Science and Technology* 60: 1132–1137.
56. Palacios-Huerta, I., and O. Volij. 2004. The measurement of intellectual influence. *Econometrica* 72: 963–977.
57. Pinski, G., and F. Narin. 1976. Citation influence for journal aggregates of scientific publications: Theory, with application to the literature of physics. *Information Processing and Management* 12: 297–312.
58. Posner, R.A. 2000. An economic analysis of the use of citations in the law. *American Law and Economics Review* 2: 381–406.
59. Prathap, G. 2006. Hirsch-type indices for ranking institutions' scientific research output. *Current Science* 91 (11): 1439.
60. Quesada, A. 2009. Monotonicity and the Hirsch index. *Journal of Informetrics* 3 (2): 158–160.
61. Rousseau, R. 2008. Woeginger's axiomatisation of the h-index and its relation to the g-index, the $h^{(2)}$-index and the r^2-index. *Journal of Informetrics* 2 (4): 335–340.
62. Rubinstein, A. 1980. Ranking the participants in a tournament. *SIAM Journal of Applied Mathematics* 38 (1): 108–111.
63. Ruiz-Castillo, J., and R. Costas. 2014. The skewness of scientific productivity. *Journal of Informetrics* 8 (4): 917–934.
64. Schubert, A. 2007. Successive h-indices. *Scientometrics* 70: 201–205.
65. Schubert, A., and T. Braun. 1986. Relative indicators and relational charts for comparative assessment of publication output and citation impact. *Scientometrics* 9 (5–6): 281–291.
66. Schubert, A., and T. Braun. 1996. Cross-field normalization of scientometric indicators. *Scientometrics* 36 (3): 311–324.
67. Sen, A.K. 1979. *Collective Choice and Social Welfare*. Amsterdam, The Netherlands: North Holland Publishing Company.
68. Slutzki, G., and O. Volij. 2005. Ranking participants in generalized tournaments. *International Journal of Game Theory* 33: 255–270.
69. Slutzki, G., and O. Volij. 2006. Scoring of web pages and tournaments - axiomatizations. *Social Choice and Welfare* 26: 75–92.
70. van den Brink, R., and P. Borm. 2002. Digraph competitions and cooperative games. *Theory and Decision* 53: 327–342.
71. Waltman, L.R., and N.J.P. van Eck. 2009. A taxonomy of bibliometric performance indicators based on the property of consistency. In: Proceedings of the 12th international conference on scientometrics and informetrics, international society of scientometrics and informetrics (ISSI), 1002–1003.

72. Waltman, L.R., and N.J.P. van Eck. 2013. Source normalized indicators of citation impact: An overview of different approaches and an empirical comparison. *Scientometrics* 96 (3): 699–716.
73. Waltman, L.R., and N.J.P. van Eck. 2013. A systematic empirical comparison of different approaches for normalizing citation impact indicators. *Journal of Informetrics* 7 (4): 833–849.
74. Waltman, L.R., N.J.P. van Eck, T.N. van Leeuwen, M.S. Visser, and A.F.J. van Raan. 2011. Towards a new crown indicator: Some theoretical considerations. *Journal of Informetrics* 5 (1): 37–47.
75. Wei, T.H. 1952. *The algebraic foundations of ranking theory*. London: Cambridge University Press.
76. Woeginger, G.J. 2008. An axiomatic analysis of the Egghe's *g* index. *Journal of Informetrics* 2 (4): 364–368.
77. Woeginger, G.J. 2008. An axiomatic characterization of the Hirsch-index. *Mathematical Social Sciences* 56: 224–232.
78. Woeginger, G.J. 2008. A symmetry axiom for scientific impact indices. *Journal of Informetrics* 2 (4): 298–303.

Distributive Justice

The Efficient, Symmetric and Linear Values for Cooperative Games and Their Characterizations

Sujata Goala and Surajit Borkotokey

Mathematics Subject Classification: 91A10 · 91A12

1 Introduction

A cooperative game with transferable utilities or a TU game, in short, describes situations where players make binding agreements to generate some worth or profit together. The problem is then how to share the profit among the players in a rational manner. The value is a function that prescribes a scheme of sharing the profit among the players. The most popular value in TU games till date is the Shapley value [20] which gives every player the average of her marginal contributions stemming out from all possible coalitions she can make with her peers under the given binding agreements. The Shapley value is the unique value that satisfies four properties, namely, efficiency, symmetry, linearity, and dummy axiom or the null player property. Another very popular value found in the literature is the equal division rule (ED) that splits the profits equally among the players irrespective of their productivities. The ED also satisfies efficiency, symmetry,[1] and linearity. There is a large class of values that satisfies these three properties, we call them ESL values.

In this paper, we survey the recent developments in the ESL values and their characterizations. We also make a brief discourse of some of the subclasses of the ESL values that build on these characterizations. We show some interesting results

[1]In many occasions, the symmetry we are considering here is called equal treatment to equals and symmetry is another axiom where the permutation of a player does not effect her payoff till she generates the same worth under different permutations in a coalition, however, for linear and efficient values, the two axioms are equivalent ([13], Theorem 2).

S. Goala (✉) · S. Borkotokey
Department of Mathematics, Dibrugarh University, Dibrugarh 786004, India

S. Borkotokey et al. (eds.), *Game Theory and Networks*, Indian Statistical Institute Series,
https://doi.org/10.1007/978-981-16-4737-6_7

that give us these subclasses after adding extra axioms with efficiency, symmetry and linearity.

The first-ever instance of obtaining a unique and independent representation of the ESL values without any additional axiom is seen in the work of Ruiz et al. [19]. They [19] define the least square pre-nucleolus as a solution to a minimization problem that satisfies efficiency, symmetry and linearity. Driessen and Radzik [9] obtain an alternative representation of the ESL values given in [19]. Hernandez-Lemoneda et al. [10] and Chameni-Nembua and Andjiga[4] also propose two similar representations of the ESL values, however, in [10], it is discussed under the name of semi-values without dummy axiom. A more formal representation of the ESL values described by Chameni-Nembua and Andjiga [4] can be seen in the work of Chameni-Nembua [5] with an alternative and realistic interpretation. It is worth mentioning at this point that the Shapley value builds on the extreme case of marginalism and the ED builds on the extreme case of egalitarianism. There is a large class of values that consolidate egalitarianism and marginalism. The Egalitarian Shapley value due to Joosten [11], the discounted Shapley value again by Joosten [11], the solidarity value by Nowak and Radzik [15], the consensus value by Ju et al. [12] are also members of this class. Radzik and Driessen [17] introduce additional axioms to efficiency, symmetry and linearity and obtain the Shapley value [20], the solidarity value [15], the least square pre-nucleolus [19] and the consensus value [12] from the ESL representation described in [9]. They further provide two types of acceptability, namely, social and general acceptability to characterize these four values. Meanwhile, Casajus and Huettner [2] use the representation of the ESL values by Chameni-Nembua [5] to characterize the Egalitarian Shapley value. In [14], Malawski introduces the procedural value based on the characterization of the ESL values by Ruiz et al. [19]. Following [5], Casajus and Huettner [3] obtain a new subclass of ESL values which they call the class of generalized solidarity values. In a recent work by Béal et al. [1], an important class of solidarity values is proposed that combines marginalism and egalitarianism based on the size of coalitions. Their characterization of the value uses the ESL representation of Ruiz et al. [19]. Choudhury et al. [6] propose a generalization of the egalitarian Shapley value based on the coalition size. The characterization of this class also uses the representation of the ESL values by Ruiz et al. [19].

In this survey, we make an account of the developments of the ESL values, their representations and how these representations help in characterizing the specific values mentioned above. The rest of the paper proceeds as follows. In Sect. 2, we present the preliminary notions. Section 3 includes the different representations and characterizations of the ESL values. Section 4 deals with some special subclasses of ESL values and their characterizations. Finally, in Sect. 5, we introduce a new class of ESL values that includes most of the above mentioned classes followed by the concluding remarks in Sect. 6.

2 Preliminary

Let $N = \{1, 2, ..., n\}$ denote the player set. The subsets of N are called coalitions. A TU game is the pair (N, v) where the function $v : 2^N \to \mathbb{R}$ is such that $v(\emptyset) = 0$. We call v the characteristic function of the TU game (N, v). For each coalition, $S \subseteq N$, the real number $v(S)$ denotes the worth or profit generated by the players in S. If no ambiguity arises over the player set N, a TU game is also denoted by its characteristic function v. The size of the coalitions S, T, etc., are denoted by their respective small letters, viz., s, t, etc. Let $G(N)$ denote the family of all TU games defined over N. Under standard addition and scalar multiplication of functions, $G(N)$ is a linear space of dimension $2^n - 1$ over $\mathbb{R}$. Two special games, namely, the unanimity and the standard or identity games are of particular interest to us as both these classes make two standard bases for $G(N)$. In the following, we define these two classes of games.

Let $T \subseteq N$ be fixed, then for each coalition $S \subseteq N$ define,

(i) **The unanimity game**: $u_T(S) = \begin{cases} 1 & T \subseteq S \\ 0 & \text{otherwise.} \end{cases}$

(ii) **The standard game** : $e_T(S) = \begin{cases} 1 & T = S \\ 0 & \text{otherwise.} \end{cases}$

The null game $v_0 \in G(N)$ is defined as $v_0(S) = 0$ for all $S \subseteq N$. The marginal contribution of a player $i \in N$ from the coalition $S \subseteq N$ with regard to the TU game v is given by $v(S \cup i) - v(S)$. Given $v \in G(N)$, player $i \in N$ is called a null player in v if her marginal contributions from all the coalitions with regard to v are zero, i.e., $v(S \cup i) - v(S) = 0$ for each $S \subseteq N$. A value is a mapping $\Phi : G(N) \mapsto \mathbb{R}^n$ that uniquely determines for each $v \in G(N)$, the distribution of total worths produced by the cooperation of all the players in N. Two players $i, j \in N$ are called *symmetric* with respect to the game v if for all $S \subseteq N \setminus \{i, j\}$, $v(S \cup i) = v(S \cup j)$. The three axioms, namely efficiency, symmetry, and linearity for a value $\Phi : G(N) \to \mathbb{R}^n$ are listed below.

Axiom 1. Efficiency: For any $v \in G(N)$, we have $\sum_{i \in N} \Phi_i(v) = v(N)$.
Axiom 2. Symmetry: For every pair of symmetric players $i, j \in N$ with respect to the game $v \in G(N)$, we have $\Phi_i(v) = \Phi_j(v)$.
Axiom 3. Linearity: For all $u, w \in G(N)$, every pair $\gamma, \eta \in \mathbb{R}$, and every player $i \in N$, we have

$$\Phi_i(\gamma u + \eta w) = \gamma \Phi_i(u) + \eta \Phi_i(w); \tag{2.1}$$

Φ is additive if in particular (2.1) holds for $\gamma = \eta = 1$.

Some important values for TU games mentioned in Sect. 1 are listed below.

(a) The equal division rule (ED) denoted by Φ^{ED} is given by

$$\Phi_i^{ED}(v) = \frac{v(N)}{n}. \tag{2.2}$$

(b) The Shapley value due to [20], $\Phi^{Sh}(v)$ is given by

$$\Phi_i^{Sh} = \sum_{S \subseteq N \setminus i} \frac{s!(n-s-1)!}{n!}[v(S \cup i) - v(S)]. \tag{2.3}$$

(c) For $\alpha \in [0, 1]$, the α-Egalitarian Shapley value $\Phi^{\alpha-ES}$ due to [11] is given by

$$\Phi^{\alpha-ES}(v) = \alpha\Phi^{ED}(v) + (1-\alpha)\Phi^{Sh}(v) \text{ for } v \in G(N). \tag{2.4}$$

(d) For $\delta \in [0, 1]$, the δ-discounted Shapley value Φ^{δ} again due to [11] is given by

$$\Phi_i^{\delta}(v) = \sum_{\substack{S \subseteq N \\ i \in S}} \frac{(s-1)!(n-s)!}{n!}\delta^{n-s}[v(S) - \delta v(S \setminus i)], \quad \forall i \in N. \tag{2.5}$$

(e) The solidarity value Φ^{sol} due to Nowak and Radzik [15], which was later simplified by Radzik and Driessen [17] is given by

$$\Phi_i^{sol}(v) = \frac{v(N)}{n} - \frac{v(N \setminus i)}{n^2} + \sum_{S \subsetneq N \setminus i} \frac{s!(n-s-1)!}{n!}[\frac{v(S \cup i)}{s+2} - \frac{v(S)}{s+1}], \quad \forall i \in N. \tag{2.6}$$

(f) The least square pre-nucleolus due to Ruiz et al. [19] is define as the vector $\Phi^L(v) = (\Phi_1^L, ..., \Phi_n^L) = (x_1^*, ..., x_n^*)$ that minimizes the function

$$f(x_1, ..., x_n) = \sum_{\emptyset \neq S \subseteq N} \{v(S) - \sum_{j \in S} x_j\}^2,$$

subject to $\sum_{j \in N} x_j = v(N)$ such that $(x_1, ..., x_n) \in \mathbb{R}^n$.

(g) The equal surplus division $\Phi^E(v)$ due to Driessen and Funaki [7] is given by

$$\Phi_i^E(v) = v(i) + \frac{v(N) - \sum_{j \in N} v(j)}{n}, \quad \forall i \in N. \tag{2.7}$$

(h) The consensus value Φ^{co} due to Ju et al. [12] is given by

$$\Phi_i^{co}(v) = \frac{v(i)}{2} + \frac{1}{2}\{v(i) + \frac{v(N) - \sum_{j \in N} v(j)}{n}\}, \quad \forall i \in N, \tag{2.8}$$

which after some simplifications takes the form:

$$\Phi_i^{co}(v) = \frac{1}{2}\Phi^{Sh}(v) + \frac{1}{2}\Phi^E(v), \quad \forall i \in N.$$

(i) The per capita value Φ^{pc} due to [18] is given by

$$\Phi_i^{pc}(v) = n. \sum_{S \subseteq N \setminus i} \frac{s!(n-s-1)!}{n!} [\frac{v(S \cup i)}{s+1} - \frac{v(S)}{s}], \ i \in N. \tag{2.9}$$

(j) The per capita Shapley value $\Phi^{Sh^{pc}}$ introduced in example 4.4 of [8] is given by

$$\Phi_i^{Sh^{pc}}(v) = \frac{v(N)}{n} - \frac{v(N \setminus i)}{n(n-1)} + \sum_{S \subsetneq N \setminus i} \frac{s!(n-s-1)!}{n!} [\frac{v(S \cup i)}{s+1} - \frac{v(S)}{s}], \ \forall i \in N. \tag{2.10}$$

All these above mentioned values satisfy the axiom's efficiency, symmetry and linearity, and therefore, are members of the ESL values. In the next section, we discuss in detail the various characterizations and representations of the ESL values.

3 The ESL Values

We begin with the description of the ESL values given by Ruiz et al. [19].

Theorem 1 (Lemma 9 in [19], pp. 117) *A value $\Phi : G(N) \to \mathbb{R}^n$ satisfies efficiency, linearity, and symmetry if and only if there exists $\rho_s (s = 1, \cdots n-1)$ such that*

$$\Phi_i(v) = \frac{v(N)}{n} + \sum_{S: i \in S \neq N} \rho_s \frac{v(S)}{s} - \sum_{S: i \notin S} \rho_s \frac{v(S)}{n-s}, \qquad (\forall i \in N, \ \forall v \in G(N)) \tag{3.1}$$

Proof It is easy to check that the value Φ given by Eq(3.1) satisfies efficiency, symmetry, and linearity. Conversely, let Φ satisfy efficiency, symmetry, and linearity. Now, the class of standard games $(e_S)_{S \subseteq N}$ is a basis for $G(N)$ so that any game $v \in G(N)$ can be uniquely represented as $v = \sum_{S \subseteq N} v(S) e_S$. By linearity,

$$\Phi_i(v) = \Phi_i(\sum_{S \subseteq N} v(S).e_S) = \sum_{S \subseteq N} v(S) \Phi_i(e_S)$$

It follows from the definition of e_S that a pair of players i and j are symmetric if either $i, j \in S$ or $i, j \in N \setminus S$. Hence, for any value Φ satisfying symmetry, we have

$$\Phi_i(e_S) = \begin{cases} a_S \text{ if } i \in S \\ b_S \text{ if } i \notin S \end{cases}$$

Moreover, from efficiency, for all $S \subseteq N$, $\sum_{i \in N} \Phi_i(e_S) = e_S(N) \Leftrightarrow s a_S + (n - s) b_S = 0$ and $\sum_{i \in N} \Phi_i(e_N) = e_N(N) \Leftrightarrow \Phi_i(e_N) = \frac{1}{n}$ for all $i \in N$.
Putting together, we get

$$\Phi_i(v) = \frac{v(N)}{n} + \sum_{S: i \in S \neq N} a_S v(S) - \sum_{S: i \notin S} \frac{s a_S}{n-s} v(S), \qquad (\forall i \in N, \ \forall v \in G(N)).$$

By symmetry again, the real number a_S is same for all coalitions of size s. Putting $\rho_s = sa_s$ the last expression of $\Phi_i(v)$ is reduced to the form

$$\Phi_i(v) = \frac{v(N)}{n} + \sum_{S:i\in S\neq N} \rho_s \frac{v(S)}{s} - \sum_{S:i\notin S} \rho_s \frac{v(S)}{n-s}, \quad (\forall i \in N,\ \forall v \in G(N)) \tag{3.2}$$

This completes the proof. □

It is shown by Ruiz et al. [19] that the least square pre-nucleolus Φ^L is of the form Eq. (3.2) with constants $\rho_s = \frac{s(n-s)}{n \cdot 2^{n-2}}$, $s = 1, \cdots, n-1$.

Extending the representation of ESL values given by Ruiz et al. [19], Driessen and Radzik [9] propose the following theorem:

Theorem 2 (Theorem 3 in [9] pp. 155) *The following three statements for a value Φ on $G(N)$ are equivalent.*

(i) Φ verifies efficiency, linearity, and symmetry.
(ii) There exists a (unique) collection of constants $\rho_s (s = 1, \cdots n-1)$ such that

$$\Phi_i(v) = \frac{v(N)}{n} + \sum_{S:i\in S\neq N} \rho_s \frac{v(S)}{s} - \sum_{S:i\notin S} \rho_s \frac{v(S)}{n-s}, \quad (\forall i \in N,\ \forall v \in G(N)) \tag{3.3}$$

(iii) There exists a (unique) collection of constants $\{b_{n,s} | n \in N \setminus \{l\},\ s \in \{1, 2, ..., n\}\}$, with $b_{n,n} := 1$, so that, for every n-person game v with at least two players, the value payoff vector $(\Phi_i(v))_{i=1}^n \in \mathbb{R}^n$ is of the following form:

$$\Phi_i(v) = \sum_{S\subseteq N\setminus\{i\}} \frac{1}{n.\binom{n-1}{s}} \left[b_{n,s+1} v(S \cup i) - b_{n,s} v(S)\right], \quad \forall i \in N. \tag{3.4}$$

Proof The proof follows immediately by taking $b_{n,s} = \binom{n}{s}\rho_s$. □

Remark 1 Observe that Φ given by Eq. (3.4 becomes Φ^{Sh} for $b_{n,s} := 1$ for all $s \in \{1, 2, ..., n\}$. Define a game $v^d : 2^N \to \mathbb{R}$ by $v^d(S) = b_{n,s} v(S)$ and call it the disrupted game [9], we can immediately verify that Φ given by Eq(3.4) of any game v is the Shapley value Φ^{Sh} of its disrupted game. Formally, we have $\Phi(v) = \Phi^{Sh}(v^d)$.

Hernandez-Lemoneda et al. [10] propose a class of continuous values that satisfy efficiency, symmetry, and additivity (instead of linearity) having a form similar to [9, 19]. However, for any continuous value which is additive, the values are also linear. The following theorem gives a characterization of this class of values.

Theorem 3 (Proposition 1 in [10], pp. 3) *The continuous value Φ satisfies efficiency, symmetry and additivity if and only if it is of the form*

$$\Phi_i^{\beta}(v) = \frac{v(N)}{n} + \sum_{i\in S, S\neq N} (n-s)[\beta_s v(S) - \beta_{n-s} v(N\backslash S)] \tag{3.5}$$

for a set of $n-1$ *real numbers* $\{\beta_s\}_{s=1}^{n-1}$.

Remark 2 In [10], an interpretation of the expression of Φ^{β} is given as follows. Each player initially assigns the equal share of the grand coalition $v(N)$ and then a linear transaction of wealth occurs from $N\backslash S$ to S for each $S \subset N$. Every player in $N\backslash S$ pays an amount $s\beta_s v(S)$ and every player in S receives an amount $(n-s)\beta_s v(S)$ where β_s depends on the coalition size and not on the worth generated by the coalition S (by symmetry). After the completion of this process, the payoff to player i is $\Phi_i^{\beta}(v)$. A particular set of real numbers $\{\beta_s\}_{s=1}^{n-1}$ uniquely determines all the members of this class of values.

Let us now take a quick look at the proof of Theorem 3

Proof First, we show that a value Φ satisfies additivity and symmetry if and only if there exists a set of real numbers $\{a_s\}_{s=1}^{n} \cup \{a'_s\}_{s=1}^{n-1}$ such that

$$\Phi_i(v) = \sum_{S:i\in S, S\subseteq N} a_s v(S) + \sum_{S:i\notin S, S\subset N} a'_s v(S), \quad \forall\, i \in N$$

In view of the fact that the value being continuous and additive is linear, we have, $\Phi_i(v) = \Phi_i\left(\sum_{S\subseteq N} v(S).e_S\right) = \sum_{S\subseteq N} v(S)\Phi_i(e_S) = \sum_{S\subseteq N} v(S)a^i_S$, where $\Phi_i(e_S) = a^i_S$.

Suppose S and S' are two sub-coalitions of N with $s = s'$. Consider a permutation π on N, such that $\pi(S) = S'$ and $\pi(m) = n$ for $m \in S,\ n \in S'$.

Therefore, by symmetry,

$$\Phi_{\pi(m)}(\pi(e_S)) = \Phi_m(e_S) \Leftrightarrow \Phi_n(e_{S'}) = \Phi_m(e_S) \Leftrightarrow a^m_S = a^n_{S'}.$$

Similarly, for $\hat{m} \notin S, \hat{n} \notin S'$ with $\pi(\hat{m}) = \hat{n}$, $\left[\because \pi(S) = S' \Rightarrow \pi(N\backslash S) = N\backslash S'\right]$

$$\Phi_{\pi(\hat{m})}(\pi(e_S)) = \Phi_{\hat{m}}(e_S) \Leftrightarrow \Phi_{\hat{n}}(e_{S'}) = \Phi_{\hat{m}}(e_S) \Leftrightarrow a^{\hat{m}}_S = a^{\hat{n}}_{S'}, \quad \forall\, s = s'$$

Therefore,

$$\Phi_i(v) = \sum_{S:i\in S, S\subseteq N} a_s v(S) + \sum_{S:i\notin S, S\subset N} a'_s v(S), \quad \forall\, i \in N.$$

By efficiency, $\sum_{i\in N} \Phi_i(e_S) = sa_s + (n-s)a'_s = 0, \quad \forall\, S \subset N$, i.e., $a'_s = -\frac{s}{n-s}a_s$ for $1 \le s \le n-1$. Moreover, $\sum_{i\in N} \Phi_i(e_N) = na_n = 1 \Rightarrow a_n = \frac{1}{n}$. Therefore

$$\begin{aligned}\Phi_i(v) &= \sum_{i\in S} a_s v(S) - \sum_{i\notin S} \frac{s}{n-s} a_s v(S)\\ &= \sum_{i\in S} a_s v(S) - \sum_{i\in S} \frac{n-s}{s} a_{n-s} v(N\backslash S)\\ &= \frac{v(N)}{n} + \sum_{i\in S\subset N} (n-s)[\beta_s v(S) - \beta_{n-s} v(N\backslash S)], \text{ where } \beta_s = \frac{a_s}{n-s}.\end{aligned}$$

Hence, an ESL value Φ can be uniquely determined over the basis $(e_S)_{S\subseteq N}$. The real numbers $\{\Phi_i(e_S)\}_{S\subseteq N}$ and equivalently $\{\beta_s\}_{s=1}^{n-1}$ refer to a unique representation of Φ given by

$$\Phi_i^{\beta}(v) = \frac{v(N)}{n} + \sum_{i\in S, S\subset N} (n-s)[\beta_s v(S) - \beta_{n-s} v(N\backslash S)]$$

□

Chameni-Nembua and Andjiga [4] independently make an analytical study of the ESL values and provide another unique representation of the class. They introduce a sequence of n real numbers $(A(k))_{k=1}^{n}$ with $A(n) = 1$ to characterize the class of ESL values on $G(N)$. The characterization theorem goes as follows:

Theorem 4 (Lemma 1 in [4], pp. 3) *A value Φ is linear, efficient, and symmetric if and only if there exists a set of n real numbers $A(k)_{k=1,\ldots,n}$ such that*

$$\Phi_i(v) = \frac{v(N)}{n} + \sum_{k=1}^{n-1}\left[\frac{(n-k)!(k-1)!}{n!} A(k) \sum_{i\in S, |S|=k} v(S) - \frac{(n-k-1)!(k)!}{n!} A(k) \sum_{i\in S, |S|\neq k} v(S)\right] \tag{3.6}$$

After some simplifications, Eq. (3.6) can be rewritten as

$$\Phi_i(v) = \sum_{k=1}^{n} \left\{ \sum_{i\in S, s\neq k} \frac{(n-k)!(k-1)!}{n!} [A(k)v(S) - A(k-1)v(S\backslash i)] \right\} \tag{3.7}$$

Proof Following the proof of Theorem 3, we have a value Φ satisfies additivity and symmetry if and only if there exists a set of real numbers $\{a_s\}_{s=1}^{n}$ such that

$$\Phi_i(v) = \frac{v(N)}{n} + \sum_{i\in S\subset N} a_s v(S) - \sum_{i\notin S\subset N} \frac{s}{n-s} a_s v(S), \tag{3.8}$$

Take $A(s) = \frac{n!}{(n-s)!(s-1)!} a_s$, Eq. (3.8) becomes

$$\Phi_i(v) = \frac{v(N)}{n} + \sum_{i \in S \subset N} \frac{(n-s)!(s-1)!}{n!} A(s)v(S) - \sum_{i \notin S} \frac{s}{n-s} \frac{(n-s)!(s-1)!}{n!} A(s)v(S),$$

$$= \frac{v(N)}{n} + \sum_{s=1}^{n-1} \left\{ \frac{(n-s)!(s-1)!}{n!} A(s) \sum_{i \in S} v(S) - \frac{(n-s-1)!(s)!}{n!} A(s) \sum_{i \notin S} v(S) \right\}$$

This completes the proof. □

In a follow-up paper, Chameni-Nembua [5] generalizes the notion of null player for the set of ESL values and characterizes each element of this class by efficiency, symmetry, and linearity, and a generalized null player axiom. The idea is very similar to Shapley value formulation, where the marginal contribution of the generalized null player is redefined to suit the given expression. The following proposition provides a unique representation of the class of ESL values in this regard.

Theorem 5 (Theorem 1 in [5], pp. 432) *A value Φ on $G(N)$ is an ESL value if and only if there exists a unique sequence $\alpha(s)_{s=1,2,\ldots,n}$ with $\alpha(1) = 1$ such that*

$$\Phi_i(v) = \sum_{s=1}^{n} \sum_{i \in S, |S|=s} \frac{(n-s)!(s-1)!}{n!} A_i^{\alpha(s)}(S) \tag{3.9}$$

where

$$A_i^{\alpha(s)}(S) = \alpha(s)[v(S) - v(S \backslash i)] + \frac{1-\alpha(s)}{s-1} \sum_{j \neq i} [v(S) - v(S \backslash j)]$$

Proof Note that both the expressions of Φ given in Theorems 4 and 5 are equivalent. The unique set of real number $\{\alpha(s)\}_{s=1}^{n}$ in Theorem 5 is obtained from Theorem 4 by the substitution $\alpha(s+1) = A(s)$. This completes the proof. □

Remark 3 Assume that in a coalition, each player has some effect on the contributions of the other players. It is then desirable to share the total profit among the players taking this effect into account (at least to some degree). In Eq. (3.9), the term $A_i^{\alpha(s)}(S)$ represents the total gain of player $i \in N$ after formation of the coalition $S \subseteq N$. The term $\alpha(s)[v(S) - v(S \backslash i)]$ is the fraction of her marginal contributions she retains for herself and the remaining portion is equally distributed among the other members of the coalition. Moreover, i gets a fraction of marginal contributions of each $j \in N \backslash i$. When $\alpha(s) = 1$ for all $s = 1, \ldots, n$, the marginal contributions $[v(S) - v(S \backslash i)]$ of player i are fully retained by herself. This profit-sharing scheme is given by the Shapley value. On the other hand, when each player decides to distribute all her marginal contributions to the other members of the coalition, *i.e.*, $\alpha(s) = 0$ for all $s = 1, \ldots, n$ then we have the ED. The idea of sharing the coalitional worth or profits, therefore, specifies an '$n-1$' dimensional path from individual productivity $((\alpha_s = 1)_{s=2}^{n})$ to equality $((\alpha_s = 0)_{s=2}^{n})$ within the class of ESL values. Chameni-Nembua [5] introduces the notion of generalized null player property to further characterize the class of ESL values.

Meanwhile, Driessen and Radzik [17] restate their earlier result from [9] in a simplified manner as follows:

Theorem 6 (Proposition 2 in [17], pp. 106) *A value Φ is an ESL value if and only if there exists a unique collection of real constants $\{b_s\}_{s=0,1\ldots,n}$ with $b_n = 1$ and $b_0 = 0$ such that for every game $v \in G(N)$ the value $(\Phi_i(v))_{i\in N}$ is of the following form:*

$$\Phi_i(v) = \sum_{S\subset N\setminus i} \frac{s!(n-s-1)!}{n!}[b_{s+1}v(S\cup i) - b_s v(S)],\ i \in N \tag{3.10}$$

The following table illustrates some of the standard values discussed above for different values of $\{b_s\}_{s=0,1\ldots,n}$ in Eq. (3.10).

4 Some Special ESL Values

In this section, we briefly study some ESL values of TU games, whose characterizations require one or more additional axioms to get specific forms of the coefficients in their representations.

4.1 The Procedural Values

Malawski [14] introduces the class of procedural values for TU games. In his model, a procedure is developed to share the marginal contributions of an incumbent player with the players already present in the coalition. The coalition formed before the incumbent player enters is called the predecessor set. The class of values is based on the procedures of redistributing the marginal contributions to a coalition formed by the players joining in random order. The model is confined to the set of procedures that allows a player to share her marginal contributions only among its predecessor set (not with the successor). The fraction of marginal contributions paid by an entrant player can be viewed as an "entrant fee" that she is obliged to pay to her predecessors. Note that Chameni-Nembua [5] also identifies a procedure to share the marginal contributions of a player among all her coalition members, and every time the coalition grows in size, they have to share the marginal contributions generated with respect to that coalition. On the contrary, Malawski [14] identifies the procedure of redistributing the marginal contributions of an entrant player with its predecessors presuming a particular order of entry by the players in forming the grand coalition. We formally describe the model as follows:

Given a particular ordering (permutation) π and player $j \in N$, let $P_{\pi,j}$ be the set of predecessors of player j following the ordering π and $N_{\pi,j}$ be the set of successors of player j. Thus, we have

$$P_{\pi,j} = \pi^{-1}(\{1, 2, \cdots, \pi(j)\}),$$

$$N_{\pi,j} = \pi^{-1}(\{\pi(j), j+1, \cdots, n\}).$$

Define $m_{j,\pi}(v) = v(P_{\pi,j}) - v(P_{\pi,j}\backslash j)$. Thus, $m_{j,\pi}(v)$ is the marginal contribution of player j to the set of his predecessors following the ordering π. Each entrant player $j \in N$ generates her marginal contribution $m_{j,\pi}(v)$ following the ordering π which is then divided among the players in $P_{\pi,k}$ according to a fixed procedure. Assuming that all possible orders on the player set N are equally probable, the expected gain of a player following this process is called the procedural value. Formally, we have the following definitions:

Definition 1 A procedure s on $G(N)$ is a family of non-negative coefficients $((s_{k,j})_{j=1}^{k}))_{k=1}^{n}$ such that $(\forall k)\ \sum_{j}^{k} s_{k,j} = 1$.

Definition 2 The procedural value Φ^s with respect to a procedure s on $G(N)$ is given by

$$\Phi_i^s(v) = \sum_{\pi\in\Pi} \sum_{j\in N_{\pi,i}} \frac{s_{\pi(j),\pi(i)} m_{j,\pi}(v)}{n!} \tag{4.1}$$

It follows that, given a permutation π, the procedural value Φ^s can be identified with a set of non-negative coefficients $((s_{k,j})_{j=1}^{k})_{k=1}^{n}$ such that $\sum_{j=1}^{k} s_{k,j} = 1,\ \forall k$. Clearly, $s_{1,1} = 1$. The procedure $s_{k,k} = 1$ for $k = j$ and $s_{k,j} = 0,\ \forall j < k$ describes the Shapley value. The solidarity value is given by the procedure $s_{k,j} = \frac{1}{k},\ \forall k \geq 1, j \leq k$. Further, it is interesting to note that the following procedures are all attributed to the ED value:

- $s_{1,1} = 1,\ \forall\, k > 1,\ s_{k,k-1} = 1$
- $s_{k,1} = 1,\ \forall\, k \geq 1$
- $s_{1,1} = 1,\ \forall\, k > 1$ and $\forall\, j < k,\ s_{k,j} = \frac{1}{k-1},$

In what follows next, we show how the procedure s is connected to the coefficients $(\rho_i)_{i=1}^{n}$ of the representation of the ESL values given in Eq.(3.1).

Theorem 7 (Lemma 2 in [14], pp. 314) *For the value Φ^s determined by a procedure $s = (s_1, s_2, \cdots s_n)$, the coefficients $\rho_1, \cdots, \rho_n$ are*

$$\rho_n = 1, \quad \rho_t = \frac{s_{t+1}}{\binom{n}{t}} \ \textit{for}\ t < n.$$

Theorem 8 (Corollary 2 in [14], pp. 314) *Every ESL value on $G(N)$ with coefficients $\rho_1, \cdots \rho_n$ given in Eq.(3.1) which satisfy*

$$\rho_n = 1,\ \ 0 \leq \frac{\rho_t}{\binom{n}{t}} \leq 1, \quad \textit{for}\ t = 1, 2, ...n-1$$

is procedural, and the coefficients of its procedure are given by:

$$s_1 = 1, \quad s_k = \rho_{k-1} \cdot \binom{n}{k-1} \; for \; k = 2, 3, \cdots n.$$

The following three additional properties are mentioned in [14] to characterize the procedural value apart from efficiency, symmetry, and linearity.

Definition 3 The value Φ on $G(N)$ is

- weakly monotonic if in every monotone game v (i.e., satisfying $S \subset T \Rightarrow v(S) \leq v(T)$), $\Phi_i(v) \geq 0$ for all $i \in N$;
- coalitionally monotonic if for every coalition T and every pair of games v, w such that ($v(T) > w(T)$ and $v(S) = w(S)$ for every $S \neq T$) we have $\Phi_i(v) \geq \Phi_i(w)$ for each $i \in T$;
- locally monotonic if for $v(S \cup i) \geq v(S \cup j)$ for all coalitions $S \subseteq N \setminus \{i, j\}$ we have $\Phi_i(v) \geq \Phi_j(v)$.

Theorem 9 *An ESL value on $G(N)$ is procedural if and only if it satisfies weak monotonicity and coalitional monotonicity.*

Theorem 10 *An ESL value on $G(N)$ is procedural if and only if it satisfies weak monotonicity and local monotonicity.*

Remark 4 The set of procedural values is a convex set where the extreme points of this convex set are the procedures $(s_i)_{i=1}^n$ with $s_i \in \{0, 1\}$, $\{i = 2, \ldots, n\}$. Since the Shapley value and the ED are the two extreme points, the class of egalitarian Shapley values also belongs to the class of procedural values. Further, any convex combination of the Shapley value and the solidarity value also belongs to the class of procedural values.

Driessen and Radzik in [17] call the property of local monotonicity: fair treatment and weak monotonicity: monotonicity. These two monotonicity properties are used to further specify the constants b_s's in their ESL value representation given in Eq. (3.10) as can be seen in the following theorem:

Theorem 11 *An ESL value Φ verifies fair treatment (local monotonicity), if and only if the constants b_s in its representation Eq. (3.10) satisfy:*

$$b_n = 1 \; and \; b_k \geq 0 \; for \; k = 1, 2, ..., n-1.$$

Theorem 12 *An ESL value Φ verifies fair treatment and monotonicity, if and only if the constants b_s in its representation Eq(3.10) satisfy:*

$$b_n = 1 \; and \; 0 \leq b_k \leq 1 \; for \; k = 1, 2, ..., n-1.$$

In view of Remark 4, Theorems 11 and 12 can be seen as particular cases of Theorem 10. Moreover, Driessen and Radzik [17] show that:

(a) The Shapley value and the solidarity value verify fair treatment and monotonicity.
(b) The least square pre-nucleolus and the consensus value verify fair treatment.
(c) The least square pre-nucleolus and the consensus value do not verify monotonicity for $n \geq 4$ and $n \geq 3$, respectively.

These results are in sync with Remark 4, Theorems 9 and 10.

4.2 The Egalitarian Shapley Value

Many values for TU games build on the principle of egalitarianism, marginalism or a combination of the two. Marginalism vows for distribution of the total profit according to the players' productivity, while egalitarianism prescribes equal share of the profit irrespective of who contributes how much. The Shapley value is an extreme case of marginalism, while the ED is the extreme case of egalitarianism. Recall from Sect. 2, that the egalitarian Shapley value Φ^{α} given by Eq. (2.4) is a convex combination of these two values with convexity parameter $\alpha \in [0, 1]$. The parameter α determines how much marginal and egalitarian a value should be. Table 1 shows that the class of egalitarian Shapley values is a subset of the class of ESL values. Casajus and Huttener [2] introduce the null player in productive environment property (NPE) to fully characterize this class of values. The NPE states that whenever the grand coalition has a non-negative worth, payoff to the null player should also be non-negative. This property can be termed as a solidarity property, and therefore, the

Table 1 Examples of ESL values

Value	Coefficient $\{b_{n,s}\}_{s=1}^{n}$ with $b_{n,0} = 0$ and $b_{n,n} = 1$
1. The Shapley value (Φ^{Sh})	$b_{n,s} = 1,\ \forall\, 1 \leq s \leq n-1$.
2. The ED (Φ^{ED})	$b_{n,s} = 0,\ \forall\, 1 \leq s \leq n-1$.
3. The Solidarity value (Φ^{sol})	$b_{n,s} = \frac{1}{s+1},\ \forall\, 1 \leq s \leq n-1$.
4. The Egaliterian Shapley value ($\Phi^{\alpha-ES}$)	$b_{n,s} = \alpha,\ \forall\, 1 \leq s \leq n-1$ with $\alpha \in [0, 1]$.
5. The δ-discounted shapley value (Φ^{δ})	$b_{n,s} = \delta^{n-s},\ \forall\, 1 \leq s \leq n-1$ with $\delta \in [0, 1]$.
6. The least square value (Φ^{L})	$b_{n,s} = \frac{(n-1)!}{(s-1)!(n-s-1)!} \frac{m(s)}{\sum_{s=1}^{n-1} m(s) \binom{n-2}{n-1}},\ \forall\, 1 \leq s \leq n-1$ where $m(s) \geq 0,\ \forall s \in \{1, ..., n-1\}$ represents weight on all the coalitions of size s.
7. The equal surplus division (Φ^{E})	$b_{n,1} = n-1$ and $b_{n,s} = 0,\ \forall\, 2 \leq s \leq n-1$
8. The consensus value (Φ^{co})	$b_{n,1} = \frac{n}{2}$ and $b_{n,s} = \frac{1}{2},\ \forall\, 2 \leq s \leq n-1$
9. The per capita value (Φ^{pc})	$b_{n,s} = \frac{n}{s},\ \forall\, 1 \leq s \leq n-1$
10. The per capita Shapley value ($\Phi^{Sh^{pc}}$)	$b_{n,s} = \frac{1}{s},\ \forall\, 1 \leq s \leq n-1$

egalitarian Shapley value can be considered as a solidarity value too. In the following, we give the NPE property and then we discuss the results found in [2].

- Null player in productive environment (NPE): For all $v \in G$ and $i \in N$ such that i is a null player in v and $v(N) \geq 0$, we have $\Phi_i(v) \geq 0$.

Theorem 13 *An ESL value Φ satisfies the null player in productive environment property if and only if there exists $\alpha \leq 1$ such that $\Phi = \Phi^\alpha$.*

The proof of Theorem 13 requires the ESL expression Eq. 3.9 given by Chameni-Nembua [5], however, it requires a number of additional concepts, and therefore, we think that the proof is beyond the scope of the present paper. Interested reader may refer to [2] for a long but ingenious proof. Note that Casajus and Huettner [2] call the local monotonicity property as desirability. The following theorem due to Driessen and Radzik [18] gives an interesting characterization of the ESL values.

Theorem 14 *An ESL value Φ of the form given by Eq. 3.10 satisfies desirability(local monotonicity) and monotonicity if and only if $b_n = 1$ and $0 \leq b_s \leq 1$ for all $1, 2, \ldots, n-1$.*

4.3 The Discounted Shapley Values

We have already mentioned that Joosten [11] introduces the class of δ-discounted Shapley values denoted by Φ^δ and given by Eq. (2.5). The parameter $\delta \in [0, 1]$ decides the degree of marginalism and egalitarianism of Φ^δ from a viewpoint different from the α-egalitarian Shapley value. Note that Φ^δ is the Shapley value of a δ reduced game w corresponding to the original game v, defined by $w(S) = \delta^{n-s}v(S),\ \forall\ S \subseteq N$ with $\delta \in [0, 1]$. The two extreme values of $\delta = 0$ and $\delta = 1$ correspond, respectively, to the ED and the Shapley value. Brink and Funaki [21] characterize the class of δ-discounted Shapley values as a unique class of ESL values that satisfy the $\delta-$reducing player property. Formally, we have the following:

- Given $\delta \in [0, 1]$ a value Φ satisfies the δ-reducing player property if $\Phi_i(v) = 0$ for each player $i \in N$, that satisfies $v(S \cup i) = \delta v(S),\ \forall\ S \subseteq N \backslash i$.

The corresponding characterization of the δ-discounted Shapley value proceeds as follows:

Theorem 15 ([21]) *Given $\delta \in [0, 1]$, the ESL value Φ satisfies the $\delta-$reducing player property if and only if $\Phi = \Phi^\delta$.*

Proof It is easy to prove that Φ^δ is an ESL value that satisfies the $\delta-$reducing player property. Conversely, let Φ satisfies these four axioms. Then by Driessen and Radzik [17], there exists a unique collection of constants $\{b_s | s = 1, 2, ..., n\}$ with $b_n = 1$ such that

$$\Phi_i(v) = \sum_{S\subseteq N\backslash i} \frac{s!(n-s-1)!}{n!}(b_{s+1}v(S\cup i) - b_s v(S)), \quad \text{for all } i \in N. \tag{4.2}$$

For $s \in \{1, ..., n\}$ and an arbitrarily fixed $S \subseteq N$ we define the game u by $u(S) = 1$, $u(S \cup i) = \delta$ and $u(T) = 0$ for all $T \in 2^N \setminus \{S, S \cup i\}$, we see that i is a δ-reducing player in u, and thus by the δ-reducing player property $\Phi_i(u) = 0$. Since $\Phi_i(u) = s!(n-s-1)!(b_{s+1}\delta - b_s)$, it follows that $b_{s+1}\delta = b_s$. This holds for all $s \in \{n-1, n-2, ..., 1\}$, and thus using $b_n = 1$, we get $b_s = \delta n - s$ for $s \in \{1, 2, ..., n\}$. This proves the existence and uniqueness of the δ-discounted value. □

4.4 The ξ-solidarity Values

Casajus and Huttener [3] suggest a new class of solidarity values which they denote by Sol^ξ. Akin to the class of egalitarian Shapley values and the class of $\delta-$discounted Shapley values, the class of solidarity values Sol^ξ depends on the parameter ξ. The class, however, differs from the class Φ^α, $\alpha \in [0, 1]$ as all the elements of Sol^ξ do not satisfy desirability (local monotonicity) and monotonicity (weak monotonicity). The key axiom used in the characterization of this class is the ξ-player out property.

Definition 4 For any sequence $\xi = (\xi_s)_{s\in\mathbb{N}} \in \mathbb{R}$, a player $i \in N$ is called a $\xi-$player in $v \in G(N)$ if $v(i) = 0$ and $v(S \cup i) - v(S) = \xi_s \frac{v(S)}{s}$, for all $\emptyset \neq S \subseteq N\backslash i$.

A sequence ξ specifies the amount by which the share of per capita worth of a coalition changes when a ξ-players enters the coalition.

Definition 5 A value Φ satisfies the ξ-player out property if $\Phi_j(v_{N\backslash i}) = \Phi_j(v)$ for a ξ-player $i \in N\backslash j$, where the game $v_{N\backslash i}$ is the restriction of game v on $2^{N\backslash i}$.

The following result specifies the set of admissible sequences for which the value Sol^ξ satisfies the ξ-player out property as well as efficiency.

Theorem 16 (Theorem 2 in [3], pp. 585) *There exists a value that satisfies efficiency and ξ-player out property if and only if $\xi_1 \in \mathbb{R} - \{\frac{-1}{n} | n \in N\}$ and $\xi_s = \frac{s\xi_1}{(s-1)\xi_1+1}$, for all $s \in N$.*

The next result associates Sol^ξ with the class of ESL values.

Theorem 17 (Theorem 3 of [3], pp. 585) *For every admissible ξ, Sol^ξ given by Eq(4.3) is the unique ESL value that satisfies the ξ-player out property which is given by*

$$Sol^\xi = \xi_n \frac{v(N)}{n} + \sum_{S\subseteq N\backslash i} \frac{s!(n-s-1)!}{n!}[(1-\xi_{s+1})v(S\cup i) - (1-\xi_s)v(S)] \tag{4.3}$$

for all $v \in G(N)$ and $i \in N$, where $\xi_s = \frac{s\xi_1}{(s-1)\xi_1+1}$ for all $s \in N$.

An alternative formula for Sol^{ξ} following Chameni-Nembua [5] is given as follows:

Theorem 18 *For every admissible ξ, the value Sol^{ξ} is given by*

$$Sol_i^{\xi}(v) = \frac{v(i)}{n} + \sum_{S \subseteq N: i \in S, s>1} \rho_{s-1}\Big((1-\xi_{s-1}).[v(S)-v(S\setminus i)] + \xi_{s-1}\frac{1}{s-1}\sum_{j \in S \setminus i}[v(S)-v(S\setminus j)]\Big) \tag{4.4}$$

for all $i \in N$, $v \in G(N)$ where $\xi_l = \frac{l.\xi}{(l-1)\xi+1}$, $l \in N$ and $\rho_{s-1} = \frac{1}{n}\binom{n-1}{s}^{-1}$.

Proof Since Sol^{ξ} is an ESL value, following the ESL representation of of Chameni-Nembua [5] given by Eq. (3.9), there is a sequence $\alpha = (\alpha_2, ..., \alpha_n) \in \mathbb{R}^{n-1}$ such that

$$\text{Sol}_i^{\xi}(v) = \frac{v(i)}{n} + \sum_{S \subseteq N: i \in S; s>1} \rho_{s-1} A_i^{\alpha}(v, S)$$

for all $i \in N$ where,

$$A_i^{\alpha}(v, S) = \alpha_s[v(S) - v(S \setminus i)] + \frac{1-\alpha_s}{s-1}\sum_{j \in S \setminus i}[v(S) - v(S \setminus j)].$$

In particular, when $v = e^T$ for some $T \subseteq N$, given by (ii), it can be easily shown that

$$\text{Sol}_i^{\xi}(e_T) = \rho_{t-1}.\alpha_{t+1} \quad \forall\, T \subset N \text{ and } i \in T.$$

By Eq. (4.3), $\text{Sol}_i^{\xi}(e^T) = \rho_{t-1}(1-\xi_t)$ for all $T \subset N$ and $i \in N$. Comparing the two expressions of Sol^{ξ}, we get the necessary and sufficient condition for Sol^{ξ} to have the Chameni-Nembua form is $\alpha_t = 1 - \xi_{t-1}$, $t \in \{2, ..., n\}$. □

Remark 5 Sol^{ξ} with ξ_1 restricted to [0, 1], contains the Shapley value and the ED as its extreme cases and for $\xi = \frac{1}{2}$ we get the solidarity value introduced by Nowak and Radzik [15]. The degree of solidarity expressed by Sol^{ξ} is the smallest when $\xi = 0$, i.e., $\text{Sol}^0 = \Phi^{Sh}$ and the largest when $\text{Sol}^1 = \Phi^{ED}$.

4.5 The Solidarity Allocation Rules

In a recent paper, Béal et al. [1] propose a subclass of the class of ESL values (which are called the solidarity allocation rule by the authors). This class also takes both egalitarian and marginalistic principles into account. The computation of the Shapley value is based on a model where the players enter one by one to form the grand coalition and each player is rewarded with its expected marginal contribution over all possible permutations on the player set. On the contrary, computing the collective contributions in forming the grand coalition is the new idea embedded in Béal et al. [1]. The solidarity allocation rule makes equal shares of some fraction of the worth $v(N)$ and the remaining fraction is shared according to their expected marginal productivities. To be precise, it takes into account individual productivity for smaller coalitions and coalitional productivity for larger coalitions. This requires a critical size p of the coalitions to be fixed ex ante.

The procedure of obtaining the solidarity allocation rule is given as follows. Given a permutation π on the player set N, a coalition is formed by the sequential gatherings of players. In this sequential gathering, before attaining size p, each player keeps on getting her marginal contributions. When the coalition exceeds its critical size p, all players entering afterwards get the per capita income of their cumulative payoffs up-till the grand coalition is formed. The solidarity allocation rule Sol^p is the average payoff obtained by the above process over all possible permutations on the player set. The expected solidarity of the values Sol^p's over a probability distribution β denoted by Sol^β. Two special cases Sol^0 and Sol^{n-1} are exactly the ED and the Shapley value, respectively.

The construction process of Sol^p and Sol^β is formally described as follows:

(i) Take a critical size $p \in \{0, 1, \dots, n-1\}$.
(ii) Consider a game v and a permutation $\pi \in \Pi$ so that the players would follow π to form the grand coalition $v(N)$.
(iii) Each player arriving at position $\pi(i) \le p$ guarantees her contribution $v(P_{\pi,p}) - v(P_{\pi,p} \backslash i)$ as her payoff.
(iv) Players arriving at position $\pi(i) > p$ get equal share of $v(N) - v(P_{\pi,p})$.
(v) The payoff vector of player i thus obtained is called the contribution vector $C_i^{P_{\pi,p}}$.

$$C_i^{P_{\pi,p}}(v) = \begin{cases} v(P_{\pi,p}) - v(P_{\pi,p} \backslash i), & \text{if } \pi(i) \le p \\ \dfrac{v(N) - v(P_{\pi,p} \backslash i)}{n-p}, & \text{if } \pi(i) > p \end{cases}$$

(vi) Sol^p is defined as the average of this contributions over all possible permutations.

$$\mathrm{Sol}^p(v) = \frac{1}{n!} \sum_{\pi \in \Pi} C_i^{P_{\pi,p}}$$
$$= \sum_{S \subset N, i \in S, s \le p} \frac{(n-s)!(s-1)!}{n!} [v(S) - v(S \backslash i)]$$
$$+ \sum_{S \subseteq N, i \notin S, s = p} \frac{(n-s-1)!s!}{n!} [v(N) - v(S)]$$

(vii) If p is drawn from $\{0, 1, \dots, n\}$ according to some discrete probability distribution β, $\beta = \{\beta_p : p \in \{0, 1, \dots, n-1\}\}$. Then the solidarity value generated by this particular distribution β is given by

$$\text{Sol}^{\alpha}(v) = \sum_{p=0}^{n-1} \beta_p \, \text{Sol}^{p}(v)$$

Observe that Sol^p is an ESL value. Next, we state the result of our interest

Theorem 19 *(Proposition 8 in [1], pp 73.) A value Φ is a solidarity allocation rule if and only if it can be represented by Eq(6), with constants $B^{\Phi} = (b_s^{\Phi} : s \in \{0, 1, 2, 3, ..., n\})$ such that, $0 = b_0^{\Phi} = 0 \leq b_{n-1}^{\Phi} \leq ... \leq b_2^{\Phi} \leq b_1^{\Phi} = 1$. Further, $\Phi = \Phi^{\alpha}$ where $\alpha = \{\alpha_s : s \in \{0, 1, \ldots, n-1\}\}$ is obtained from the transformation $B^{\Phi} \to \alpha$ such that, $\alpha_0 = 1 - b_1^{\Phi}$, $\alpha_{n-1} = b_{n-1}^{\Phi}$, $\alpha_s = b_s^{\Phi} - b_{s+1}^{\Phi}$, $\forall s \in \{1, 2, \ldots, n\}$.*

Since the proof of the existence and uniqueness of the ESL values are all similar, we skip the proof of this theorem. Interested reader may find the complete proof in [1].

4.6 The Generalized Egalitarian Shapley Value

In a recent work, Choudhury et al. [6] construct a class of values generalizing the egalitarian Shapley value Φ^{α} over a sequence of coalition size dependent TU games. Most of the values discussed so far, are equipped with some parameter that provides the planner flexibility to balance between marginalism and egalitarianism, but none of them has an explicit relationship with the coalition size. The class of values introduced in [6] bridges this gap. The guiding principle in this model is that when size of the coalitions increases the level of marginalism among players increases. For example, in small groups such as the startup or flat organizations, the profit is shared equally among the partners, however, when the organization grows into tiers and levels, it becomes a function of the productivities of the members of those tiers. Keeping this as the motivation, the generalized class of egalitarian Shapley values is defined. To start with, let $G(N)$ be decomposed into n subclasses based on its coalition size $k \in \{1, 2, \ldots, n\}$. The decomposition and the formation of the value can be described in the following steps.

Step 1. Decompose any game $v \in G(N)$ into n components $v_{<2}$, $v_{2\leq,<3}$, … $v_{n-1\leq,<n}$ and $v_{\geq n}$, where,

$$v_{<2}(S) = \begin{cases} v(S) & \text{for all } S \subseteq N \text{ suchthat } s < 2, \\ 0 & \text{otherwise,} \end{cases}$$

$$v_{\geq n}(S) = \begin{cases} v(S) & \text{for all } S \subseteq N \text{ suchthat } n-1 < s \leq n, \\ 0 & \text{otherwise,} \end{cases}$$

and for all $k \in \{3, ..., n\}$ we have,

$$v_{k-1\leq,<k}(S) = \begin{cases} v(S) & \text{for all } S \subseteq N \text{ suchthat } k-1 \leq s < k, \\ 0 & \text{otherwise.} \end{cases}$$

Then by the usual addition of functions, we can represent v as $v = v_{<2} + v_{2\leq,<3} + \cdots + v_{n-1\leq,<n} + v_{\geq n}$.

Step 2. On the basis of Step 1, decompose $G(N)$ as:

$$G(N) = G(N)_{<2} \oplus G(N)_{2\leq,<3} \oplus G(N)_{n-1\leq,<n} \oplus G(N)_{\geq n}$$

where for each $v \in G(N)$, we have $v_{<2} \in G(N)_{<2}$ $v_{2\leq,<3} \in G(N)_{2\leq,<3}$ etc.

Step 3. Given $\alpha \equiv (\alpha_1, \alpha_2, \ldots, \alpha_n) \in [0, 1]^n$ with $0 = \alpha_0 \leq \alpha_1 \leq \cdots \leq \alpha_n = 1$, the generalized egaliterian Shapley value is given by

$$\Phi_i^{\alpha-GES}(v) = \Phi_i^{\alpha_1-ES}(v < 2) + \sum_{k=2}^{k=n-1} \Phi_i^{\alpha_k-ES}(v_{\leq k,<k+1}) + \Phi_i^{\alpha_n-ES}(v_{\geq n}) \quad (4.5)$$

$$= \sum_{S:i\in S} \frac{(n-s)!(s-1)!}{n!}\{\alpha_s v(S) - \alpha_{s-1} v(S\backslash i)\} \quad (4.6)$$

where $0 = \alpha_0 \leq \alpha_1 \leq \ldots \leq \alpha_n = 1$.

The first characterization of this class is in the same line of Shapley [20]. A relatively weaker than the null player axiom, namely, null player in a non-negative environment property along with efficiency, symmetry and linearity characterize this generalized class. Formally we have the following: this generalized class.

- null player in a non-negative environment property: For $v \in G(N)$ and $i \in N$ such that i is a null player in v and $v(S) \geq 0$ for all $i \in S$, we have $\Phi_i(v) \geq 0$.

This assumption is weaker than the null player in a productive environment due to Casajus and Huettner [2]. The characterization theorem of the generalized egalitarian Shapley value goes as follows:

Theorem 20 *A value Φ satisfies efficiency, additivity, desirability and the null player in a non-negative environment property if and only if there exists an $\alpha = (\alpha_1, \alpha_2, \ldots, \alpha_n) \in [0, 1]^n$, where $0 = \alpha_0 \leq \alpha_1 \leq \alpha_2 \leq \ldots \leq \alpha_n = 1$ such that $\Phi = \Phi^{\alpha-GES}$.*

Remark 6 It is interesting to note that not all ESL values that satisfy the null player in a non-negative environment property are generalized egalitarian Shapley values. Casajus and Huettner [2] show that a value that satisfies efficiency, additivity, and desirability also satisfies linearity. Moreover, desirability implies symmetry. Thus, desirability cannot be avoided in this characterization. From now onwards, we call the generalized egalitarian Shapley value the GES value in short.

5 An Extension of the GES Value

The generalized egalitarian Shapley value proposed by Choudhury et al. [6] gives a monotonically increasing sequence of the parameters α_s's ($s = 0, ..., n$), i.e., $\frac{\alpha_s}{\alpha_{s+1}} \leq 1$ for $s = 0, ..., n$ such that with the increase of the coalition size, the value becomes more marginal. That is, the increase in marginality is determined by the parameters α_i's. In this section, we try to answer the question: can we obtain a class of ESL values where sharing of the profits in some of the smaller coalitions may also be less egalitarian than the larger coalitions. This may be the case when there is spillover or externalities in forming a coalition. If we can obtain such a value, the next question would be: how much flexibility can we impose on the choice of the parameters α_s's? To answer these questions, we obtain the extended GES value, which we denote by the EGES value. The procedure of obtaining the EGES value goes as follows:

Step (i) Decompose any game $v \in G(N)$ into n components $v_{<2}$, $v_{2\leq,<3}$, ... $v_{n-1\leq,<n}$ and $v_{\geq n}$, where

$$v_{<2}(S) = \begin{cases} v(S) & \text{for all } S \subseteq N \text{ suchthat } s < 2, \\ 0 & \text{otherwise,} \end{cases}$$

$$v_{\geq n}(S) = \begin{cases} v(S) & \text{for all } S \subseteq N \text{ suchthat } n-1 < s \leq n, \\ 0 & \text{otherwise,} \end{cases}$$

and for all $k \in \{3, ..., n\}$, we have

$$v_{k-1\leq,<k}(S) = \begin{cases} v(S) & \text{for all } S \subseteq N \text{ suchthat } k-1 \leq s < k, \\ 0 & \text{otherwise.} \end{cases}$$

Then by the usual addition of functions, we can represent v as

$$v = v_{<2} + v_{2\leq,<3} + \cdots + v_{n-1\leq,<n} + v_{\geq n}.$$

Step (ii) On the basis of Step 1, decompose $G(N)$ as:

$$G(N) = G(N)_{<2} \oplus G(N)_{2\leq,<3} \oplus G(N)_{n-1\leq,<n} \oplus G(N)_{\geq n}$$

where for each $v \in G(N)$ we have, $v_{<2} \in G(N)_{<2}$ $v_{2\leq,<3} \in G(N)_{2\leq,<3}$ etc.

Step (iii) Let $f : \{1, 2, \ldots, n\} \to \mathbb{R}$ be a strictly increasing function. Given $\alpha \equiv (\alpha_1, \alpha_2, \ldots, \alpha_n) \in [0, 1]^n$ such that $\alpha_0 = 0$, $\alpha_n = 1$ and $\frac{\alpha_s}{\alpha_{s+1}} \leq \frac{f(s+1)}{f(s)}$ for all $0 \leq s \leq n$. Note that, by taking a strictly monotonic increasing function, we include the possibility that for some s, $\alpha_s \geq \alpha_{s+1}$.

Step (iv) The extended GES value denoted by $\Phi_i^{\alpha-EGES}$ is given by,

$$\Phi_i^{\alpha-EGES}(v) = \Phi^{\alpha_1-ES}(v_{<2}) + \sum_{k=2}^{k=n-1} \Phi^{\alpha_k-ES}(v_{k\leq,<k+1}) + \Phi^{\alpha_n-ES}(v_{\geq n})$$

After simplifications

$$\Phi_i^{\alpha-EGES}(v) = \sum_{S\subseteq N:i\in S} \frac{(n-s)!(s-1)!}{n!}\{\alpha_s v(S) - \alpha_{s-1} v(S\backslash i)\} \quad (5.1)$$

$$= \sum_{S\subseteq N\backslash i} \frac{(n-s-1)!(s)!}{n!}\{\alpha_{s+1} v(S\cup i) - \alpha_s v(S)\} \quad (5.2)$$

where $0 \leq \frac{\alpha_s}{\alpha_{s+1}} \leq \frac{f(s+1)}{f(s)}$ for all $s \in \{0, 1, 2, \ldots, n-1\}$ and $\alpha_0 = 0, \alpha_n = 1$.

The characterization of the EGES value depends on the function f. In what follows, we consider $f(s) = s$, the identity function and obtain a characterization of the EGES values. It is interesting to observe that the EGES values under this assumption of the function f takes a form similar to the per capita values given by Eq. (2.10).

5.1 *Characterization*

For the initial characterization of this class, we follow [6]. Akin to the null player in a non-negative environment property, we introduce per capita null player in a non-negative environment property.

Definition 6 A player $i \in N$ is called the per capita null player if $\frac{v(S\cup i)}{s+1} = \frac{v(S)}{s}$ for all $S \subseteq N\backslash i,\ 0 \leq s \leq n-1$ (with the convention: $\frac{0}{0} = 0$).

- Per capita null player in a non-negative environment property: A value Φ satisfies the per capita null player in a non-negative environment property if for each per capita null player i in v, $\Phi_i(v) \geq 0$ whenever $v(S) \geq 0$ for all $S \subseteq N$ such that $i \subset S$.

Theorem 21 *A value Φ satisfies desirability, additivity, efficiency, monotonicity, and per capita null player in a non-negative environment property if and only if there exists an $\alpha \equiv (\alpha_1, \alpha_2, \ldots, \alpha_n)$ satisfying $\frac{\alpha_s}{\alpha_{s+1}} \leq \frac{s+1}{s}$ for $1 \leq s \leq n-1$ and $0 \leq \alpha_s \leq 1$ such that $\Phi = \Phi^{\alpha-EGES}$.*

Proof The value $\Phi^{\alpha-EGES}$ inherits desirability, additivity, efficiency, and monotonicity from the class of egalitarian Shapley values Φ^α.

Let $j \in N$ be a per capita null player player in $v \in G(N)$, i.e., $\frac{v(S\cup j)}{s+1} = \frac{v(S)}{s} \Leftrightarrow v(S\cup j) = \frac{s+1}{s} v(S)$, for all $\emptyset \neq S \subseteq N\backslash j$ and $v(S) \geq 0$, We have

$$\begin{aligned}
\Phi_j^{\alpha-EGES}(v) &= \sum_{\emptyset \neq S \subseteq N \setminus j} \frac{(n-s-1)!(s)!}{n!}\{\alpha_{s+1}v(S \cup j) - \alpha_s v(S)\} \\
&= \sum_{\emptyset \neq S \subseteq N \setminus j} \frac{(n-s-1)!(s)!}{n!}\{\alpha_{s+1}v(S)\frac{s+1}{s} - \alpha_s v(S)\} \\
&= \sum_{\emptyset \neq S \subseteq N \setminus j} \frac{(n-s-1)!(s)!}{n!}\{\alpha_{s+1}\frac{s+1}{s} - \alpha_s\}v(S) \\
&\geq 0, \quad \text{with the convention:} \frac{0}{0} = 0.
\end{aligned}$$

Therefore, $\Phi_j^{\alpha-EGES}$ satisfies per capita null player in a non-negative environment property.

Conversely, let Φ be an arbitrary value satisfying desirability, additivity, efficiency, monotonicity (weak monotonicity in our terminology) and per capita null player in a non-negative environment property. It follows that Φ satisfies linearity also. A value satisfying desirability also satisfies symmetry. Therefore, Φ is an ESL value, moreover it satisfies desirability and monotonicity. From Theorem 14, we conclude that the value is of the following form:

$$\Phi_i(v) = \sum_{S: S \subseteq N \setminus i} \frac{(n-s-1)!(s)!}{n!}\{\alpha_{s+1}v(S \cup i) - \alpha_s v(S)\} \text{ with } 0 \leq \alpha_s \leq 1.$$

Let $\frac{v(S \cup i)}{s+1} = \frac{v(S)}{s} \Leftrightarrow v(S \cup i) = \frac{s+1}{s}v(S)$ for $\emptyset \neq S \subseteq N \setminus i$ and $v(T) = 0$ for all $T \neq S,\ S \setminus i$ and $v(S \cup i) \geq 0$ and $v(S) \geq 0$. Since Φ satisfies per capita null player in a non-negative environment property, we have

$$\begin{aligned}
\Phi_i(v) &= \sum_{S \subseteq N \setminus i} \frac{(n-s-1)!(s)!}{n!}\{\alpha_{s+1}v(S \cup i) - \alpha_s v(S)\} \\
&= \frac{(n-s-1)!(s)!}{n!}\{\alpha_{s+1}v(S)\frac{s+1}{s} - \alpha_s v(S)\} \\
&= \{\alpha_{s+1}\frac{s+1}{s} - \alpha_s\}v(S) \geq 0 \\
&\Rightarrow \text{ either } \alpha_{s+1} = \alpha_s = 0 \text{ or } \alpha_s = 0, \alpha_{s+1} > 0 \text{ or } \frac{s+1}{s} \geq \frac{\alpha_s}{\alpha_{s+1}} \text{ for } 1 \leq s \leq n-1
\end{aligned}$$

Hence,

$$\Phi_i(v) = \sum_{S: S \subseteq N \setminus i} \frac{(n-s-1)!(s)!}{n!}\{\alpha_{s+1}v(S \cup i) - \alpha_s v(S)\}$$

where $\frac{s+1}{s} \geq \frac{\alpha_s}{\alpha_{s+1}}$ for all $s \in \{0, 1, 2, \ldots, n-1\}$ with $0 \leq \alpha_s \leq 1$ for all $s = \{0, \ldots, n\}$. It follows that $\Phi = \Phi^{\alpha-EGES}$. □

Remark 7 The axioms characterizing $\Phi^{\alpha-EGES}$ are independent as can be seen from the following:

(a) Φ^{pc} satisfies all but monotonicity (weak monotonicity).
(b) $\Phi_i^{EF}(v) = 0$ for all $i \in N$ satisfies all but efficiency.
(c) $\Phi_i^A(v) = \begin{cases} \Phi_i^{Sh}(v) \text{ if } v(N) < 1 \\ \Phi_i^{ED}(v) \text{ if } v(N) \geq 1 \end{cases}$ satisfies all but additivity.
(d) Φ^α with $\alpha_n = 1,\ \alpha_1 = -1,\ \alpha_s = 0$ for all $2 \leq s \leq n-1$ satisfies all but desirability.
(e) Φ^α with $\alpha_n = 1,\ \alpha_s = \frac{1}{(n+1)^s}$ for all $1 \leq s \leq n-1$ satisfies all but the per capita null player in a non-negative environment property.

We get another characterization of the EGES value by replacing desirability and monotonicity with a new axiom which we call bounded desirability. The bounded desirability is similar to the Shapley value proportionality axiom introduced by Nowak and Radzik in [16] that depends on Φ^{Sh}.

- A value Φ satisfies bounded desirability if $0 \leq [\Phi_i(v) - \Phi_j(v)] \leq [\Phi_i^{Sh}(v) - \Phi_j^{Sh}(v)]$ whenever $v(S \cup j) \geq v(S \cup i)$ for all $S \subseteq N \backslash \{i, j\}$.

Desirability ensures higher payoff to a highly productive player. Similarly, under the assumption of bounded desirability, the payoff difference of two players is always less than their payoff difference under Shapley value. Any value that establishes a trade-off between marginalism and egalitarianism satisfies bounded desirability as Shapley value is an extreme case of marginalism.

Proposition 1 *If a value Φ satisfies bounded desirability then it also satisfies desirability but the converse is not true. Moreover, an ESL value satisfies desirability and monotonicity if and only if it satisfies bounded desirability.*

Proof The first part is clear from the definition. For the converse part, consider $i, j \in N$ and the the per capita value

$$\Phi_i^{pc}(v) = \sum_{S:S \subseteq N \backslash i} \frac{(n-s-1)!s!}{n!} n[\frac{v(S \cup i)}{s+1} - \frac{v(S)}{s}]$$

$$\begin{aligned} \Phi_i^{pc}(v) - \Phi_j^{pc}(v) &= \sum_{S:S \subseteq N \backslash i,j} \frac{(n-s-2)!s!}{(n-1)!} \frac{n}{s+1} [v(S \cup i) - v(S \cup j)] \\ &\geq \sum_{S:S \subseteq N \backslash i,j} \frac{(n-s-2)!s!}{(n-1)!} [v(S \cup i) - v(S \cup j)] \\ &= \Phi_i^{Sh}(v) - \Phi_j^{Sh}(v) \geq 0 \end{aligned}$$

Hence, Φ^{pc} satisfies desirability but not bounded desirability. For the second part, let Φ be an arbitrary ESL value that satisfies desirability and monotonicity, hence (by Proposition 3 of [18]) there exists b_s with $b_0 = 0,\ b_n = 1,\ 0 \leq b_s \leq 1$ such that

$$\Phi_i^b(v) = \sum_{S:S\subseteq N\setminus i} \frac{(n-s-1)!(s)!}{n!}\{b_{s+1}v(S\cup i) - b_s v(S)\}$$

Now

$$\begin{aligned}
0 &\leq \Phi_i^b(v) - \Phi_j^b(v) \\
&= \sum_{S:S\subseteq N\setminus i,j} \frac{(n-s-2)!(s)!}{(n-1)!} b_{s+1}[v(S\cup i) - v(S\cup j)] \\
&\leq \sum_{S:S\subseteq N\setminus i,j} \frac{(n-s-2)!(s)!}{(n-1)!}[v(S\cup i) - v(S\cup j)] \\
&= \Phi_i^{Sh}(v) - \Phi_j^{Sh}(v)
\end{aligned}$$

Thus, Φ with $0 \leq b_s \leq 1$ satisfies bounded desirability. Conversely, let the ESL value Φ satisfy bounded desirability. Since, $1 = e_{T\cup i}(T\cup i) \geq e_{T\cup i}(T\cup j) = 0$ for all $T \subseteq N\setminus\{i, j\}$, we have

$$\begin{aligned}
&0 \leq \Phi_i(e_{T\cup i}) - \Phi_j(e_{T\cup i}) \leq \Phi_i^{Sh}(e_{T\cup i}) - \Phi_j^{Sh}(e_{T\cup i}) \\
&\Rightarrow 0 \leq \frac{(n-t-2)!(t)!}{(n-1)!} b_{t+1} \leq \frac{(n-t-2)!(t)!}{(n-1)!} \\
&\Rightarrow 0 \leq b_{t+1} \leq 1 \text{ for all } 0 \leq t \leq n-2.
\end{aligned}$$

Therefore, Φ satisfies desirability and monotonicity. □

Theorem 22 *A value Φ satisfies bounded desirability, efficiency, additivity, and per capita null player in non-negative environment if and only if there exists an $\alpha \equiv (\alpha_1, \alpha_2, \ldots, \alpha_n)$, where $0 \leq \alpha_s \leq 1$ and $\frac{\alpha_s}{\alpha_{s+1}} \leq \frac{s+1}{s}$ for all $s \in \{0, 1, 2, \ldots, n-1\}$ such that $\Phi = \Phi^{\alpha-EGES}$.*

Proof It is easy to check that the class $\Phi^{\alpha-EGES}$ satisfies bounded desirability, efficiency, additivity and per capita null player in non-negative environment property. Let an arbitrary value Φ satisfy these axioms. If Φ satisfies bounded desirability, then Φ also satisfies desirability. Since desirability, efficiency, and additivity all together implies linearity and we know that desirability implies symmetry, therefore, Φ is an ESL value. Again if an ESL value satisfies bounded desirability, then it satisfies desirability and monotonicity, and hence there exist a sequence b_s with $0 \leq b_s \leq 1$ such that

$$\Phi_i(v) = \Phi_i^b(v) = \sum_{S:S\subseteq N\setminus i} \frac{(n-s-1)!(s)!}{n!}\{b_{s+1}v(S\cup i) - b_s v(S)\}.$$

Again Φ satisfies per capita null player in non-negative environment property. Consider the game v such that $\frac{v(S\cup i)}{s+1} = \frac{v(S)}{s} \Rightarrow v(S\cup i) = \frac{s+1}{s}v(S)$ for a subset S such that $S \subseteq N\setminus i$ and $v(T) = 0$ for all $T \neq S, S\cup i$ and $v(S), v(S\cup i) \geq 0$ of Theorem 21 and the above expression $\Phi_i = \Phi_i^b$, we conclude that $\frac{b_s}{b_{s+1}} \leq \frac{s+1}{s}$ for all $1 \leq s \leq n-1$ (when $b_{s+1} > 0$, otherwise we assume $\frac{0}{0} = 0$).

Remark 8 (a) For $\alpha_s = 1 - \xi_s$, we have $\frac{\alpha_s}{\alpha_{s+1}} = \frac{s\xi+1}{(s-1)\xi+1} \leq \frac{s+1}{s}$, for all $\xi \in [0, 1]$, therefore, the class Sol^ξ [3] belongs to the $EGES$ class.

(b) The class Sol^β introduced by Béal et al. [1] is characterized by the sequence b such that $b_0 = 0,\ b_n = 1,\ 1 \geq b_1 \geq b_2 \geq, \ldots, b_{n-1} \geq 0$.
A value Φ^b of this class with $\frac{b_s}{b_{s+1}} \leq \frac{s+1}{s}$ for $1 \leq s \leq n-2$ belongs to the $EGES$ class.

6 Conclusion

We survey the recent developments on different ESL values and the corresponding characterizations following their representations proposed by Ruiz et al. [19], Driessen and Radzik [9] and Chameni-Nembua [5]. Then we obtain the characterization of a few special classes of games that require the representation of the ESL values given in [5, 9, 19]. Finally, an extended generalized egalitarian Shapley value is obtained and its characterization is given. Similar extensions and generalizations can also be made of those values where the planer has a choice to adjust egalitarianism and marginalism over a range of coalition size, viz., the δ-discounted Shapley value, the ξ-Sol value, etc., to name a few.

References

1. Béal, S., E. Rémila, and P. Solal. 2017. A class of solidarity allocation rules for TU-games. *Theory and Decision* 83: 61–94.
2. Casajus, A., and F. Huettner. 2013. Null players, solidarity, and the egalitarian Shapley values. *Journal of Mathematical Economics* 49: 58–61.
3. Casajus, A., and F. Huettner. 2014. On a class of solidarity values. *Eurupian Journal of Operation Research* 236: 583–591.
4. Chameni-Nembua, C., and N.G. Andjiga. 2008. Linear, efficient and symmetric values for TU-games. *Economics Bulletin* 3 (71): 1–10.
5. Chameni-Nembua, C. 2012. Linear efficient and symmetric values for TU-games: Sharing the joint gain of cooperation. *Games and Economic Behavior* 74: 431–433.
6. Choudhury D, Borkotokey S, Kumar R, Sarangi S (2020) Consolidating Marginalism with Egalitarianism: The Generalised Egalitarian Shapley value. *Annals of Operations Research* (Accepted).
7. Driessen TSH, Funaki Y (1991) Coincidence of and collinearity between game theoretic solutions. *Operations-Research-Spektrum* 13:15–30.
8. Driessen, T.S.H., and T. Radzik. 2002. A weighted pseudo-potential approach to values for TU-games. *International Transaction in Operation Research* 9: 303–320.
9. Driessen, T.S.H., and T. Radzik. 2003. Extensions of Hart and Mas-Colells consistency to efficient, linear, and symmetric values for TU-games. In: Petrosyan LA, Yeung DWK (eds), ICM Millennium Lectures on Games. Springer-Verlag, Heidelberg, Germany, 147–166. Volume

dedicated to the International Congress of Mathematicians, Game Theory and Applications Satellite Conference, August 14–17, 2002, Quindao, China.
10. Hernandez-Lamoneda, L., R. Juarez, and F. Sanchez-Sanchez. 2008. Solutions without dummy axiom for TU cooperative games. *Economics Bulletin* 3 (1): 1–9.
11. Jooston, R. 1996. Dynamics, Equilibira, and Values. Ph.D. Thesis, University of Maastrichit University (http://arno.unimaas.nl/show.cgi?did=10343).
12. Ju, Y., P. Borm, and P. Ruys. 2007. The consensus value: a new solution concept for cooperative games. *Social Choice and Welfare* 28: 685–703.
13. Malawski, M. 2007. A note on equal treatment and symmetry of values, report. Institute of Computer Science PAS.
14. Malawski, M. 2013. Procedural values for cooperative games. *International Journal of Game Theory* 42: 305–324.
15. Nowak, A.S., and T.S. Radzik. 1994. A solidarity value for n-person transferable utility games. *International Journal of Game Theory* 23: 43–48.
16. Nowak, A.S., and T.S. Radzik. 1996. On convex combination of two values. *Applicationes Mathematicae* 24 (1): 47–56.
17. Radzik, T., and T. Driessen. 2013. On a family of values for TU-games generalizing the Shapley value. *Mathematical Social Sciences* 65: 105–111.
18. Radzik, T., and T. Driessen. 2016. Modeling values for TU-games using generalized versions of consistency, standardness and the null player property. *Mathematical Operation Research* 83: 179–205.
19. Ruiz, L.M., F. Valenciano, and J.M. Zarzuelo. 1998. The family of least square values for transferable utility games. *Games Econ Behavior* 24: 109–130.
20. Shapley, L.S. 1953. A value for n-person games. In *Contributions to the Theory of Games*, vol. II, ed. H.W. Kuhn and A.W. Tucker. annals of mathematics studies, 307–317. Princeton: Princeton University Press.
21. van den Brink, R., and Y. Funaki. 2015. Implementation and axiomatization of discounted Shapley values. *Social Choice and Welfare* 45: 329–344.

New Characterizations of the Discounted Shapley Values

Parishmita Boruah

1 Introduction

Cooperative games model social and economic situations which are subject to some binding agreements among agents/players. Let $N = \{1, 2, ..., n\}$ denote the set of agents/players. A cooperative game with transferable utility or simply a TU game is a pair (N, v) where the function $v : 2^N \to \mathbb{R}$ is such that $v(\emptyset) = 0$. For each $S \subseteq N$, the value $v(S)$ represents the worth of S; a subset S in this terminology is called a coalition. Denote the cardinality of a coalition $S \subseteq N$ by the symbol $|S|$. If N is fixed, (N, v) is simply denoted by v. The class of all TU games with player set N, denoted by $\mathcal{G}(N)$, is a linear space of dimension $2^{|N|} - 1$ under standard addition and scalar multiplication of set functions. Since a basis for any linear space inherits most of the properties of its members, recognizing the basis elements is itself an important and interesting domain of study (for detailed description of various methods to obtain alternative basis vectors for the class $\mathcal{G}(N)$, we recommend [6, 9, 12]). A value is a function on the space $\mathcal{G}(N)$ that assigns each of its members an $|N|$-vector from $\mathbb{R}^{|N|}$. Among the various values proposed till date, the Shapley value [14] and the Equal division rule are the most popular values. Most of the popular values proposed till date satisfy Linearity, Symmetry, and Efficiency and therefore, these values are together called the ESL values [13]. The Shapley value, the δ-discounted Shapley value [8], the Solidarity value [10], the Equal division rule [16], the Egalitarian Shapley value [8], etc., are few examples of the ESL values. The Shapley value and the Equal division rule are respectively, based on the two extreme cases of marginalism and egalitarianism. On the other hand, the δ-discounted Shapley value ($\delta \in [0, 1]$) of the game (N, v) is defined as the Shapley value of the game (N, v') such that $v'(S) = \delta^{|N|-|S|} v(S)$, for each $S \subseteq N$ and hence can be considered as a dynamic way of distributing the resource with the total resource diminishes at each stage in the power of δ. The family of δ-discounted Shapley values makes a trade-off

P. Boruah (✉)
Department of Mathematics, Dibrugarh University, Dibrugarh 786004, Assam, India

S. Borkotokey et al. (eds.), *Game Theory and Networks*, Indian Statistical Institute Series,
https://doi.org/10.1007/978-981-16-4737-6_8

between marginalism (the Shapley value for $\delta = 1$) and egalitarianism (the Equal division rule for $\delta = 0$) and therefore, can be thought of as a solidarity value. Now onwards, we alternatively use the name "discounted Shapley value" instead of "δ-discounted Shapley value" where there is no ambiguity on δ.

Consolidation of marginalism and egalitarianism has been accepted as a more humane way of sharing resource among agents (see [8, 15, 17]). The discounted Shapley values have strong theoretical background in both cooperative and non-cooperative setup [17]. In this paper, we propose a new characterization of the discounted Shapley values. Our characterization goes along line of [2, 3, 18, 19] who have used similar procedure for the characterization of the Shapley value. For an arbitrarily fixed $\delta \in (0, 1]$, we obtain a basis for the kernel of the δ-discounted Shapley value which is a special game, we call it the δ-factious oligarchic game. The δ-factious oligarchic game is such that it cannot alter the solution of a game under the summation operation. Examples of such games can be seen in parliamentary system of democracy, where the treasury bench may have more members than it actually required to pass a bill. The remaining votes and respective contributions are therefore considered marginal from the utilitarian point of view. Thus, even if we add these extra contributions to the bill passing game, as it popularly known in the literature, their net contribution is zero. More such examples can be found in [2, 3, 18, 19].

The standard characterization of the δ-discounted Shapley value is done using the axioms Additivity (Linearity), Symmetry, Efficiency, and δ-reducing player property [17]. We show an alternative characterization using the δ-factious oligarchic axiom and the δ-inessential property, which we define in Sect. 3. Our approach builds on the solution procedure of the inverse problem discussed in [12], i.e., given an $|N|$-vector of payoffs, say $\mathbf{x}$, one needs to find all the games in $\mathcal{G}(N)$ of which the δ-discounted Shapley value equals $\mathbf{x}$. This is equivalent to describe the kernel of the δ-discounted Shapley value. Following this procedure, the commander basis (games) was proposed in [19] to characterize the Shapley value. The commander basis has two properties, namely (i) when a game is expressed as a linear combination of the commander basis, the coefficients related to the singleton coalitions coincide with the Shapley value of the game and (ii) the kernel of the Shapley value is induced by the commander basis. Similar to this, the δ-factious oligarchic games have the properties, namely (i) when a game is represented by the linear combination of these games, the coefficients related to the singleton coalitions coincide with the discounted Shapley value of the game. (ii) The class of δ-factious oligarchic games induces the null space of the discounted Shapley value.

The rest of the paper is organized as follows: Sect. 2 contains Preliminaries. In Sect. 3, we investigate the kernel of the discounted Shapley values. In Sect. 4, we discuss some new axioms and the bases for the kernel of the discounted Shapley values. These new axioms and bases lead to the natural axiomatizations of the discounted Shapley values, which we present in Sect. 5. Further, we extend the method adopted by us to characterize the discounted Shapley values for a subclass of the ESL values in Sect. 6.

2 Preliminaries

In this section, we compile and present the existing notions and subsequent results required for the development of our paper from [3, 8, 14, 17, 18], etc. Let the player set $N = \{1, 2, 3, ..., n\}$ with $|N| \geq 2$ be fixed. Recall that a value or a solution is a function $\Phi : \mathcal{G}(N) \rightarrow \mathbb{R}^{|N|}$ which uniquely determines for each game v a distribution $\Phi(v) = (\Phi_1(v), \Phi_2(v), \Phi_3(v), ..., \Phi_{|N|}(v))$, where $\Phi_i(v)$ is the payoff to player i, for all $i \in N$. Three well-known values on $\mathcal{G}(N)$ which are of importance to the development of our paper are given in the following.

(a) The **Shapley value** [14] is given by,

$$\Phi_i^{Sh}(N, v) = \sum_{S \subseteq N \setminus \{i\}} \frac{|S|!(|N| - |S| - 1)!}{|N|!}(v(S \cup \{i\}) - v(S)), \qquad \forall\, i \in N.$$

(b) The **Equal division rule** $ED : \mathcal{G}(N) \rightarrow \mathbb{R}^{|N|}$ is defined as

$$ED_i(v) = \frac{v(N)}{|N|}, \qquad \forall\, i \in N.$$

(c) For $\delta \in [0, 1]$, the δ-**discounted Shapley value** Φ^{δ}, introduced by Joosten [8] is given by

$$\Phi_i^{\delta}(N, v) = \sum_{S \subseteq N \setminus \{i\}} \frac{|S|!(|N| - |S| - 1)!}{|N|!} \delta^{|N|-|S|-1}(v(S \cup \{i\}) - \delta.v(S)), \qquad \forall\, i \in N.$$

Note that the class δ-discounted Shapley values contain the Equal division rule (for $\delta = 0$) and the Shapley value (for $\delta = 1$) as extreme cases.

Various axiomatizations of the Shapley value, the Equal division rule, and the discounted Shapley values are found in [2–4, 17], etc. In the following, we list some of the important axioms that characterize the discounted Shapley values and also are relevant to the present paper. Before that, we formally give some definitions as follows:

Definition 1 *(null and nullifying player)* Player $i \in N$ is a null player in $v \in \mathcal{G}(N)$ if for all $S \subseteq N \setminus \{i\}$, we have $v(S \cup \{i\}) = v(S)$. Player $i \in N$ is a nullifying player in $v \in \mathcal{G}(N)$ if for all $S \subseteq N \setminus \{i\}$, we have $v(S \cup \{i\}) = 0$.

It follows that the null player is non-productive in any coalition, whereas the nullifying player annihilates the productivities of all the coalitions to which it belongs.

Definition 2 *(δ-reducing player)* For $\delta \in [0, 1]$, player $i \in N$ is called a δ-reducing player in $v \in \mathcal{G}(N)$ if for all $S \subseteq N \setminus \{i\}$, we have $v(S \cup \{i\}) = \delta v(S)$.

The δ-reducing player reduces the productivity of a player by the fraction δ. For $\delta = 1$ and $\delta = 0$, the δ-reducing player becomes the null player and the nullifying player, respectively.

Definition 3 A game $v \in \mathcal{G}(N)$ is called inessential, if for all $\emptyset \neq S \subseteq N$, $v(S) = \sum_{i \in S} v(\{i\})$.

Definition 4 If a game $v \in \mathcal{G}(N)$ is such that $v(S) = 0, \;\; \forall\, S \subseteq N$, then it is called the null game.

The list of the standard axioms that characterize the Shapley value, the Equal division rule, and the discounted Shapley values is given below.

Axiom 1 Efficiency (EFF): The value Φ satisfies EFF if $\sum_{i \in N} \Phi_i(v) = v(N)\; \forall$ $v \in \mathcal{G}(N)$.

Axiom 2 Linearity (LIN): Φ satisfies LIN, if $\Phi(\alpha v + \beta w) = \alpha\Phi(v) + \beta\Phi(w)\; \forall$ $v, w \in \mathcal{G}(N)$, $\forall\; \alpha, \beta \in \mathbb{R}$. If $\alpha = \beta = 1$ then Φ is called additive and the corresponding axiom is abbreviated as ADD.

Axiom 3 Symmetry (SYM): Φ is symmetric if $\Phi_i(v) = \Phi_j(v)$, whenever i, j are such that $v(S \cup \{i\}) = v(S \cup \{j\})$, for $S \subseteq N \backslash \{i, j\}$.

Axiom 4 δ-reducing player property (δ-RP): A solution Φ will satisfy δ-reducing player property if $\Phi_i(v) = 0$, whenever $i \in N$ is a δ-reducing player.

Note that for $\delta = 0$, the δ-RP becomes the nullifying player property (NPP) due to [16] and for $\delta = 1$, it becomes the null player axiom due to [14]. In what follows next, Theorem 1 due to [17] gives a characterization of the δ-discounted Shapley value. The same set of axioms can be shown to characterize the Shapley value [14] and the Equal division rule [16] just by putting $\delta = 1$ and $\delta = 0$, respectively, in the axiom of δ-reducing player property.

Theorem 1 ([17], Theorem 4.2 on pp. 338) *For $\delta \in [0, 1]$, a solution satisfies EFF, SYM, LIN, and δ-RP iff it is the δ-discounted Shapley value.*

The null space or the kernel of a value Φ is the space of games of dimension $2^{|N|} - |N| - 1$ to which Φ assigns the zero vector. Therefore, the null space of the discounted Shapley value Φ^δ can be written as

$$ker(\Phi^\delta) = \{v \in \mathcal{G}(N) : \Phi_i^\delta(v) = 0\, \forall\, i \in N\}.$$

Remark 1 $\mathcal{G}(N)$ being a linear space, the games with identical values can be identified using the notion of the kernel (see [18]). Let v_1 and v_2 be two games with $|N|$ players. If $v_1 - v_2 \in ker(\Phi)$, then the two games v_1 and v_2 have identical values under Φ. For any $T \subseteq N, T \neq \emptyset$, the game $\overline{u}_T$ is defined as follows [18]:

$$\overline{u}_T(S) = \begin{cases} 1 & \text{if } |S \cap T| = k \\ 0 & \text{otherwise,} \end{cases}$$

where $1 \leq k \leq |T|,\ k \in \mathbb{N}$. The set $\{\overline{u}_T\}_{\emptyset \neq T \subseteq N}$ is a basis for $\mathcal{G}(N)$ which induces the null space of the Shapley value. In the next section, in a similar way, we define a basis for $\mathcal{G}(N)$ that induces the null space of the δ-discounted Shapley values.

3 Games in the Kernel of the Discounted Shapley Values

One of our principal targets in this paper is to introduce and identify those games which belong to the kernel of the discounted Shapley values. As a first step in this direction, in the following, we define the δ-inessential game.

Definition 5 For $\delta \in [0, 1]$, a game $v \in \mathcal{G}(N)$ is called a δ-inessential game if it satisfies

$$v(S) = \delta^{|S|-1} \sum_{i \in S} v(\{i\}), \quad \forall \emptyset \neq S \subseteq N.$$

Remark 2 For $\delta = 0$ and $\delta = 1$, the δ-inessential game coincides with the null game and the inessential game, respectively. This game is neither monotonic nor additive for $\delta \in (0, 1)$.

Given $\delta \in (0, 1]$, we now define the class of δ-factious oligarchic games, denoted by $\tilde{v}_k^T$ where the coalition $T \subseteq N$ is such that $|T| \geq 2$ with the parameter k $(1 \leq k \leq |T| - 1,\ k \in \mathbb{N})$ as follows:

$$\tilde{v}_k^T(S) = \begin{cases} \dfrac{1}{\delta^{|N|-|S|}} & \text{if } |S \cap T| = k \\ 0 & \text{otherwise.} \end{cases} \tag{1}$$

Remark 3 The term "Oligarchy" comes from the Greek word *oligarkhia*, which is a combination of two words *oligoi* (few) and *arkhein* (to rule), and it means a small group of players having the authority to control other players in the game and assign a value.[1]

Since the δ-factious oligarchic game $\tilde{v}_k^T$ given by Eq.(1) is defined for all $\delta > 0$, therefore we cannot obtain the Equal division rule for the class $\tilde{v}_k^T$ from the δ-discounted Shapley value simply by putting $\delta = 0$. Thus, while discussing the δ-discounted Shapley value for the parameter δ, we mean $0 < \delta \leq 1$. For $\delta = 1$, this game is nothing but the factious oligarchic game defined in [18], which belongs to the kernel of the Shapley value.

For $k = 1$ and $\delta \in (0, 1]$, the δ-factious oligarchic game takes the form

$$\overline{v}^T(S) = \begin{cases} \dfrac{1}{\delta^{|N|-|S|}} & \text{if } |S \cap T| = 1 \\ 0 & \text{otherwise.} \end{cases} \tag{2}$$

[1] https://en.wikipedia.org/wiki/Oligarchy.

We call this game the *δ-dog eat dog* game. For $\delta = 1$, the δ-dog eat dog game becomes the dog eat dog game (see [3, 19]).

For $k = |T| - 1$ and $|T| \geq 2$, we have a fourth class of games called *δ-scapegoat game* which is defined as

$$\underline{v}^T(S) = \begin{cases} \frac{1}{\delta^{|N|-|S|}} & \text{if } |S \cap T| = |T| - 1 \\ 0 & \text{otherwise.} \end{cases} \tag{3}$$

Note: For $\delta = 1$, the δ-scapegoat game is the scapegoat game (see [3, 18]).

Following theorem shows that all these δ-factious oligarchic games belong to the kernel of the δ-discounted Shapley value.

Theorem 2 *Given* $\delta \in (0, 1]$, *each δ-factious oligarchic game is in the kernel of the δ-discounted Shapley value.*

Proof Recall that for $T \subseteq N$ and $|T| \geq 2$ with parameter k $(1 \leq k \leq |T| - 1)$ the δ-factious oligarchic game $\tilde{v}_k^T$ is given by

$$\tilde{v}_k^T(S) = \begin{cases} \frac{1}{\delta^{|N|-|S|}} & \text{if } |S \cap T| = k \\ 0 & \text{otherwise.} \end{cases}$$

We first show that each member of $N \backslash T$ is a δ-reducing player in $\tilde{v}_k^T$. Let $i \in N \backslash T$.

Case I(a): Suppose $S \subseteq T$ and $|S| = k$, then $|S \cap T| = k$ and $\tilde{v}_k^T(S) = \frac{1}{\delta^{|N|-|S|}}$. Moreover, since $i \notin T$ we have $S \cup \{i\} \nsubseteq T$. It follows that $|(S \cup \{i\}) \cap T| = k$ and $\tilde{v}_k^T(S \cup \{i\}) = \frac{1}{\delta^{|N|-|S|-1}}$. Therefore, $\tilde{v}_k^T(S \cup \{i\}) - \delta\tilde{v}_k^T(S) = 0$.

Case I(b): suppose $S \subseteq T$ and $|S| < k$, then $|S \cap T| = |S| < k$ and $\tilde{v}_k^T(S) = 0$. Moreover, since $i \notin T$, we must have $S \cup \{i\} \nsubseteq T$. It follows that $|(S \cup \{i\}) \cap T| < k$ and $\tilde{v}_k^T(S \cup \{i\}) = 0$. Therefore, $\tilde{v}_k^T(S \cup \{i\}) - \delta.\tilde{v}_k^T(S) = 0$.

Case I(c): Let $S \subseteq T$ and $|S| > k$. It follows that $|S \cap T| = |S| > k$ and $\tilde{v}_k^T(S) = 0$. Moreover, since $i \notin T$, we must have $S \cup \{i\} \nsubseteq T$. It follows that $|(S \cup \{i\}) \cap T| > k$ and $\tilde{v}_k^T(S \cup \{i\}) = 0$. Therefore, $\tilde{v}_k^T(S \cup \{i\}) - \delta.\tilde{v}_k^T(S) = 0$.

Case II: Take $S \subseteq N \backslash T$. Then $|S \cap T| = 0$ and $\tilde{v}_k^T(S) = 0$. Moreover, $S \cup \{i\} \nsubseteq T$ as $i \notin T$. It follows that $|(S \cup \{i\}) \cap T| = 0$ and $\tilde{v}_k^T(S \cup \{i\}) = 0$. Therefore, $\tilde{v}_k^T(S \cup \{i\}) - \delta.\tilde{v}_k^T(S) = 0$. In all the above cases, i is a δ-reducing player. Since $i \in N \backslash T$ is arbitrary, the assertion follows.

The worths generated by the coalitions under $\tilde{v}_k^T$ depend only on the cardinality of $S \cap T$, not on the identity of the players. Since Φ^δ satisfies SYM, each player in T gets identical payoff. Moreover, $\Phi_i^\delta(\tilde{v}_k^T) = 0, \ \forall i \in N \backslash T$ as Φ^δ satisfies δ-RP. Now, $\tilde{v}_k^T(N) = 0$, as $|T \cap N| = |T| > k$. Hence by EFF,

$$\begin{aligned}
&\sum_{i\in N}\Phi^{\delta}(\tilde{v}_k^T) = \tilde{v}_k^T(N)\\
&\Rightarrow \sum_{i\in N\setminus T}\Phi_i^{\delta}(\tilde{v}_k^T) + \sum_{i\in T}\Phi_i^{\delta}(\tilde{v}_k^T) = 0\\
&\Rightarrow \sum_{i\in T}\Phi_i^{\delta}(\tilde{v}_k^T) = 0, \;\; \forall\; i\in T\\
&\Rightarrow \Phi_i^{\delta}(\tilde{v}_k^T) = 0, \;\; \forall\; i\in T.
\end{aligned}$$

Therefore, $\Phi_i^{\delta}(\tilde{v}_k^T) = 0, \quad \forall \;\; i \in N$ and this implies that every player has 0 δ-discounted Shapley value in all types of δ-factious oligarchic games. This completes the proof. □

In the next section, we propose some new bases for the space of TU games and accordingly obtain alternative axioms to characterize the δ-discounted Shapley value.

4 A New Basis for TU Games

We construct a new basis for the space of TU games expanding the family of δ-factious oligarchic games. Note that, the set of δ-factious oligarchic games given by Eq. (1) includes all games on coalitions T with $|T| \geq 2$ and therefore, it has exactly $2^{|N|} - |N| - 1$ games. To make it a basis for $\mathcal{G}(N)$ we add the δ-factious oligarchic games on the singleton coalitions with a slight modification on the definition as follows.

Definition 6 The family of δ-factious oligarchic games for any nonempty coalition T with parameter k $(1 \leq k \leq |T|)$ is defined as

$$\tilde{v}_k^T(S) = \begin{cases} \dfrac{1}{\delta^{|N|-|S|}} & \text{if } |S\cap T| = k\\ 0 & \text{otherwise.}\end{cases}$$

The following theorem shows that the set of δ-factious oligarchic games $\{\tilde{v}_k^T\}_{T\in 2^N\setminus\{\emptyset\}}$ forms a basis for the space of TU games.

Theorem 3 *The set of δ-factious oligarchic games $\{\tilde{v}_k^T\}_{T\in 2^N\setminus\{\emptyset\}}$ with parameter $1 \leq k \leq |T|$ forms a basis for the space $\mathcal{G}(N)$.*

Proof For any $T \subseteq N, T \neq \emptyset$, we define $\overline{u}_T^h$ as follows:

$$\overline{u}_T^h(S) = \begin{cases} \dfrac{1}{h(|S|)} & \text{if } |S\cap T| = k\\ 0 & \text{otherwise,}\end{cases}$$

where $h(|S|) = \delta^{|N|-|S|}$. It is easy to show that $(\overline{u}_T^h)^h(S) = \overline{u}_T^h(S)h(|S|) = \overline{u}_T(S)$ for all $T \subseteq N, T \neq \emptyset$ (see [19], pp. 24). Therefore, following Theorem 1 in [18], $\{\overline{u}_T\}$ is a basis for $\mathcal{G}(N)$. This further implies that the class $\{\tilde{v}_k^T\}_{T\in 2^N\setminus\{\emptyset\}}$ forms a basis for $\mathcal{G}(N)$. □

Following Theorem 3, we next show that the set of δ-factious oligarchic games $\{\tilde{v}_k^T\}_{T\in 2^N\setminus\{\emptyset\},|T|\geq 2}$ spans the kernel of the discounted Shapley values.

Theorem 4 *For any $1 \leq k \leq |T| - 1$, the set of δ-factious oligarchic games $\{\tilde{v}_k^T\}_{T\in 2^N\setminus\{\emptyset\},|T|\geq 2}$ constitutes a basis for the null space of the discounted Shapley value.*

Proof By Theorem 3, the family $\{\tilde{v}_k^T\}_{T\in 2^N\setminus\{\emptyset\},|T|\geq 2}$ is linearly independent. It also contains exactly $2^{|N|} - |N| - 1$ games, hence the family $\{\tilde{v}_k^T\}_{T\in 2^N\setminus\{\emptyset\},|T|\geq 2}$ spans a linear space of dimension $2^{|N|} - |N| - 1$. We have that $ker(\Phi^\delta) = \{v \in \mathcal{G}(N) | \Phi^\delta(v) = 0\}$ is a subspace of $\mathbb{R}^{2^{|N|}-1}$ of dimension $2^{|N|} - |N| - 1$ that contains all the δ-factious oligarchic games. Since the space spanned by $\{\tilde{v}_k^T\}_{T\in 2^N\setminus\{\emptyset\},|T|\geq 2}$ has dimension $2^{|N|} - |N| - 1$, it must coincide with $ker(\Phi^\delta)$. Therefore, $\{\tilde{v}_k^T\}_{T\in 2^N\setminus\{\emptyset\},|T|\geq 2}$ represents a basis for the null space of the discounted Shapley value. □

Note that the games v and w have identical discounted Shapley value if and only if $v - w$ is a game in the kernel of the discounted Shapley value. In view of Theorem 3, any game $v \in \mathcal{G}(N)$ can be expressed as follows:

$$v = \sum_{T\subseteq N,|T|\geq 1} a_T \tilde{v}_k^T = \sum_{i\in N} \beta_i \tilde{v}_k^{\{i\}} + \sum_{T\subseteq N,|T|\geq 2} a_T \tilde{v}_k^T, \tag{4}$$

where $\beta_i = a_T$ for $T = \{i\}$ and each $i \in N$.

The following proposition shows that when a game is expressed as a linear combination of the δ-factious oligarchic games as given in Eq. (4), the coefficients β_i are indeed the components of the discounted Shapley value of v.

Proposition 1 *If v is given by Eq. (4), then $\Phi_i^\delta(v) = \beta_i$ for each $i \in N$.*

Proof Consider, $v \in \mathcal{G}(N)$ and $\beta = (\beta_1, ..., \beta_n) \in \mathbb{R}^n$. Applying Φ^δ on v given by Eq.(4) and using LIN , we obtain the following.

$$\begin{aligned}\Phi^\delta(v) &= \sum_{j\in N} \beta_j \Phi^\delta(\tilde{v}_k^{\{j\}}) + \sum_{T\subseteq N,|T|\geq 2} a_T \Phi^\delta(\tilde{v}_k^T) \\ &= \sum_{j\in N} \beta_j \Phi^\delta(\tilde{v}_k^{\{j\}}) \text{ [since, } \Phi^\delta(\tilde{v}_k^T) = 0, \forall |T| \geq 2].\end{aligned}$$

Now, consider the game $(\tilde{v}_k^{\{i\}})_{i\in N}$: $\tilde{v}_k^{\{i\}}(S) = \begin{cases} \dfrac{1}{\delta^{|N|-|S|}} & \text{if } i \in S \\ 0 & \text{otherwise.} \end{cases}$

Therefore, for all $i \in N$, we have

$$\begin{aligned}\Phi_i^{\delta}(\tilde{v}_k^{\{i\}}) &= \sum_{S \subseteq N\setminus\{i\}} \frac{|S|!(|N|-|S|-1)!}{|N|!}\delta^{|N|-|S|-1}(\tilde{v}_k^{\{i\}}(S\cup\{i\}) - \delta.\tilde{v}_k^{\{i\}}(S)) \\ &= \sum_{S \subseteq N\setminus\{i\}} \frac{|S|!(|N|-|S|-1)!}{|N|!}\delta^{|N|-|S|-1}(\tilde{v}_k^{\{i\}}(S\cup\{i\}) \\ &= \sum_{S \subseteq N\setminus\{i\}} \frac{|S|!(|N|-|S|-1)!}{|N|!}\delta^{|N|-|S|-1}.\frac{1}{\delta^{|N|-|S|-1}} = 1.\end{aligned}$$

Again, $\tilde{v}_k^{\{i\}}(N) = 1$. Thus by EFF we have,

$$\sum_{j\in N}\Phi_j^{\delta}(\tilde{v}_k^{\{i\}}) = \sum_{j\in N\setminus\{i\}}\Phi_j^{\delta}(\tilde{v}_k^{\{i\}}) + \Phi_i^{\delta}(\tilde{v}_k^{\{i\}}) = \tilde{v}_k^{\{i\}}(N) = 1.$$

It follows that $\sum_{j\in N\setminus\{i\}}\Phi_j^{\delta}(\tilde{v}_k^{\{i\}}) + 1 = 1 \Rightarrow \sum_{j\in N\setminus\{i\}}\Phi_j^{\delta}(\tilde{v}_k^{\{i\}}) = 0$. Thus, $\Phi_j^{\delta}(\tilde{v}_k^{\{i\}}) = 0,\ \forall i \neq j$. Now,

$$\Phi_i^{\delta}(v) = \sum_{j\in N}\beta_j\Phi_i^{\delta}(\tilde{v}_k^{\{j\}}) = \sum_{j\in N\setminus\{i\}}\beta_j\Phi_i^{\delta}(\tilde{v}_k^{\{j\}}) + \beta_i\Phi_i^{\delta}(\tilde{v}_k^{\{i\}}) = \beta_i\Phi_i^{\delta}(\tilde{v}_k^{\{i\}}) = \beta_i.$$

Since, $i \in N$ is arbitrary, we have $\Phi_i^{\delta}(v) = \beta_i,\ \forall\, i \in N$. □

Proposition 1 clearly reflects the result that we have mentioned about our basis that if we represent a game using the δ-factious oligarchic basis, the coefficients related to the singleton coalitions coincide with the corresponding discounted Shapley value of the game. We provide an alternative characterization of the discounted Shapley values. The new characterization requires only 2 axioms instead of the axioms used in [17], namely, EFF, SYM, ADD, δ-RP.

5 An Alternative Characterization

We show that the three special games, namely, the δ-dog eat dog game, the δ-scapegoat, the δ-factious oligarchic games which we defined in Sect. 3 provide three intuitive axioms for the discounted Shapley values. The underlying idea is that changing the cooperation structure by adding games in the kernel of the discounted Shapley values should not affect the division of payoffs. Formally, we have the following.

Axiom 5 The δ-Dog Eat Dog Axiom (δ-DED): Φ satisfies the δ-Dog Eat Dog axiom if $\Phi(v) = \Phi(v + \alpha w)$ for every game $v \in \mathcal{G}(N)$, any δ-dog eat dog game w, and $\alpha \in \mathbb{R}$.

Axiom 6 δ-Scapegoat Axiom (δ-S): A value Φ satisfies the δ-Scapegoat axiom if $\Phi(v) = \Phi(v + \alpha w)$ for every game $v \in \mathcal{G}(N)$, any δ-scapegoat game w, and $\alpha \in \mathbb{R}$.

Axiom 7 δ-Factious Oligarchy (δ-FO): A value Φ satisfies the δ-Factious Oligarchy axiom if $\Phi(v) = \Phi(v + \alpha w)$ for every game $v \in \mathcal{G}(N)$, any δ-factious oligarchic game w, and $\alpha \in \mathbb{R}$.

Note that for $\delta = 1$, the δ-factious oligarchic game becomes the factious oligarchic game, so the axiom will be the Factious oligarchy axiom given in [3].

Proposition 2 *The δ-discounted Shapley value satisfies the δ-DED, δ-S and δ-FO.*

Proof The δ-dog eat dog games and the δ-scapegoat games are special types of the δ-factious oligarchic games. So, it is enough to show that the δ-discounted Shapley value satisfies the δ-Factious Oligarchy axiom. But it follows directly from LIN and Theorem 2. □

Corollary 1 *A solution Φ satisfies the δ-Dog Eat Dog, δ-Scapegoat, or δ-Factious Oligarchy axiom if and only if $\Phi(v) = \Phi(v + w)$ for every game v and all games w that are linear combinations of δ-dog eat dog, δ-scapegoat, or δ-factious oligarchic games, respectively.*

We now introduce the δ-inessential axiom as follows.

Axiom 8 δ-inessential axiom (δ-IA): A value Φ on $\mathcal{G}(N)$ satisfies the δ-inessential axiom (δ-IA) if given $\delta \in [0, 1]$, $\Phi_i(v) = \delta^{|N|-1}v(\{i\})$ for all $i \in N$ in every δ-inessential game $v \in \mathcal{G}(N)$.

For $\delta = 1$, **Axiom** 8 coincides with the inessential axiom.

Proposition 3 *The δ-discounted Shapley value satisfies δ-IA.*

Proof Let v be any δ-inessential game. We have

$$\begin{aligned}
\Phi_i^{\delta}(v) &= \sum_{S\subseteq N\setminus\{i\}} \frac{|S|!(|N|-|S|-1)!}{|N|!}\delta^{|N|-|S|-1}(v(S\cup\{i\}) - \delta.v(S)) \\
&= \sum_{S\subseteq N\setminus\{i\}} \frac{|S|!(|N|-|S|-1)!}{|N|!}\delta^{|N|-|S|-1}\Big(\delta^{|S|+1-1}\sum_{S\subseteq N,\ i\in S} v(\{j\}) \\
&\qquad\qquad - \delta.\delta^{|S|-1}\sum_{S\subseteq N} v(\{j\})\Big) \\
&= \sum_{S\subseteq N\setminus\{i\}} \frac{|S|!(|N|-|S|-1)!}{|N|!}\delta^{|S|}.\delta^{|N|-|S|-1}v(\{i\}) \\
&= \delta^{|N|-1}v(\{i\}).
\end{aligned}$$

Since, i is an arbitrary player, it follows that the discounted Shapley value satisfies the δ-IA. □

We next define the δ-discounted Shapley inessential game $\tilde{w}_v$ of v as follows:

$$\tilde{w}_v(S) = \frac{1}{\delta^{|N|-|S|}} \sum_{i \in S} \Phi_i^\delta(v), \ \forall \, S \subseteq N. \tag{5}$$

Then $\tilde{w}_v(\{i\}) = \frac{1}{\delta^{|N|-1}} \Phi_i^\delta(v) \Rightarrow \Phi_i^\delta(v) = \delta^{|N|-1} \tilde{w}_v(\{i\})$. Using this condition in Eq. (5), we get $\tilde{w}_v(S) = \delta^{|S|-1} \sum_{i \in S} \tilde{w}_v(\{i\}), \quad \forall \emptyset \neq S \subseteq N$. Thus, the δ-discounted Shapley inessential game $\tilde{w}_v$ of v given by Eq.(5) is also a δ-inessential game.

In line with Corollary 2 in [3], we present the following lemma which is latter used in the characterization of the δ-discounted Shapley value.

Lemma 1 *Every game is the sum of its δ-discounted Shapley inessential game and a game in the kernel of the δ-discounted Shapley value.*

Proof Let, $\tilde{w}_v$ be the δ-discounted Shapley inessential game of v. Then,

$$\begin{aligned}
\Phi_i^\delta(\tilde{w}_v) &= \sum_{S \subseteq N \setminus \{i\}} \frac{|S|!(|N|-|S|-1)!}{|N|!} \delta^{(|N|-|S|-1)} \big[\tilde{w}_v(S \cup \{i\}) - \delta . \tilde{w}_v(S)\big] \\
&= \sum_{S \subseteq N \setminus \{i\}} \frac{|S|!(|N|-|S|-1)!}{|N|!} \delta^{(|N|-|S|-1)} \Bigg[\frac{1}{\delta^{|N|-|S|-1}} \sum_{j \in S \cup \{i\}} \Phi_j^\delta(v) \\
&\qquad - \frac{1}{\delta^{|N|-|S|}} \sum_{j \in S} \Phi_j^\delta(v)\Bigg] \\
&= \sum_{S \subseteq N \setminus \{i\}} \frac{|S|!(|N|-|S|-1)!}{|N|!} \Phi_i^\delta(v) \\
&= \Phi_i^\delta(v).
\end{aligned}$$

Using LIN on Φ^δ, we obtain $\Phi^\delta(v - \tilde{w}_v) = 0$. This implies $v - \tilde{w}_v$ is a game in the kernel of the discounted Shapley value. Thus, v can be expressed as $v = (v - \tilde{w}_v) + \tilde{w}_v$. □

In view of Lemma 1, we observe that if a value Φ satisfies the δ-IA and the addition of games from the kernel of the discounted Shapley value does not affect Φ, then Φ must coincide with the discounted Shapley value. From this observation, along with Theorem 3 and Lemma 1, we obtain an alternative characterization of the discounted Shapley values using Axiom δ-FO and Axiom δ-IA as follows.

Theorem 5 *A value Φ satisfies the δ-inessential axiom and any one of the δ-Dog Eat Dog, δ-Scapegoat, δ-Factious Oligarchy axiom iff it is the δ-discounted Shapley value.*

Proof From Theorem 3 we have that each of the δ-dog eat dog games, the δ-scapegoat games and the δ-factious oligarchic games constitutes a basis for the space $\mathcal{G}(N)$ of all TU-games. Recall that the δ-discounted Shapley value is characterized by the axioms: δ-RP, EFF, SYM, and ADD. We deduce these axioms from δ-IA and δ-FO using Lemma 1.

Consider a solution Ψ which satisfies the δ-FO and the δ-IA. Let $v, w \in \mathcal{G}(N)$. By Lemma 1, we can write $v = v^{in} + v^f$ and $w = w^{in} + w^f$ where v^{in}, w^{in} are the δ-discounted Shapley inessential game of v and w and v^f, w^f are δ-factious oligarchic games of v and w, respectively.
ADD: Employing δ-FO and the δ-IA on Ψ, we obtain

$$\begin{aligned}\Psi_i(v+w) &= \Psi_i([v^{in}+v^f]+[w^{in}+w^f]) = \Psi_i(v^{in}+w^{in})\\ &= \delta^{|N|-1}(v^{in}+w^{in})(\{i\}) = \Psi_i(v^{in}) + \Psi_i(w^{in})\\ &= \Psi_i(v^{in}+v^f) + \Psi_i(w^{in}+w^f)\\ &= \Psi_i(v) + \Psi_i(w), \quad \forall i \in N.\end{aligned}$$

Thus, $\Psi(v+w) = \Psi(v) + \Psi(w)$. Therefore Ψ satisfies ADD.
δ-**RP:** Let i be a δ-reducing player in N. Therefore, $v(S \cup \{i\}) = \delta.v(S), \forall S \subseteq N\backslash\{i\}$. Following Lemma 1 we get, $(v^{in}+v^f)(S \cup \{i\}) = \delta.(v^{in}+v^f)(S)$. This implies that $\Psi((v^{in}+v^f)(S \cup \{i\})) = \delta.\Psi((v^{in}+v^f)(S))$. Consequently, $\Psi[\delta^{|S|}\sum_{k\in S\cup\{i\}} v^{in}(\{k\})] = \delta\Psi[\delta^{|S|-1}\sum_{k\in S} v^{in}(\{k\})]$. It follows from Lemma 1 that $\delta^{|S|}\sum_{k\in S\cup\{i\}}\Psi_k(v^{in}) = \delta^{|S|}\sum_{k\in S}\Psi_k(v^{in})$. Thus, $\Psi_i(v^{in}) = 0$. Using δ-FO, $\Psi_i(v) = 0$. Hence Ψ satisfies δ-RP.
SYM: Let player i and j be such that $v(S \cup i) = v(S \cup j)$ for all $S \subseteq N \setminus \{i, j\}$. Then, $(v^{in}+v^f)(S \cup \{i\}) = (v^{in}+v^f)(S \cup \{j\})$. This implies that $\Psi[\delta^{|S|}\sum_{k\in S\cup\{i\}} v^{in}(k)] = \Psi[\delta^{|S|}\sum_{k\in S\cup\{j\}} v^{in}(k)]$. Consequently,

$$\delta^{|S|}\sum_{k\in S\cup\{i\}}\Psi_k(v^{in}) = \delta^{|S|}\sum_{k\in S\cup\{j\}}\Psi_k(v^{in}).$$

It follows that $\Psi_i(v^{in}+v^f) = \Psi_j(v^{in}+v^f) \Rightarrow \Psi_i(v) = \Psi_j(v)$, where v^{in} is the δ-discounted Shapley inessential game of v and w is a δ-factious oligarchic game. Thus Ψ satisfies SYM.
EFF: We have again from δ-FO and δ-IA,

$$\begin{aligned}\sum_{i\in N}\Psi_i(v) = \sum_{i\in N}\Psi_i(v^{in}+v^f) = \sum_{i\in N}\Psi_i(v^{in}) = \sum_{i\in N}\delta^{|N|-1}v^{in}(\{i\})\\ = \delta^{|N|-1}\sum_{i\in N}v^{in}(\{i\}) = v^{in}(N) = (v^{in}+v^f)(N) = v(N).\end{aligned}$$

Thus Ψ satisfies EFF. It follows that Ψ is the discounted Shapley value Φ^δ.
Using the same procedure, we can show that the result is true if we replace the δ-Factious Oligarchy axiom by δ-Dog Eat Dog, or δ-Scapegoat axiom. Thus, a value satisfies the δ-IA and any one of the δ-Dog Eat Dog, δ-Scapegoat and δ-Factious Oligarchy axiom if and only if it is the discounted Shapley value. □

6 Generalization for the Set of ESL values

In this section, we generalize our method discussed in Sect. 5 to obtain an alternative characterization of a subclass of the ESL values. The following proposition due to [11] characterizes the class of ESL values.

Proposition 4 (Proposition 2 in [11], pp. 2) *A value Φ is an ESL value if and only if there exists a unique collection of constants $\{b_{|S|} : |S| = 1, 2, ..., |N|\}$ with $b_{|N|} = 1$ and $b_{|\emptyset|} = 0$ such that Φ is of the form:*

$$\Phi_i(N, v) = \sum_{S \subseteq N\setminus\{i\}} \frac{|S|!(|N| - |S| - 1)!}{|N|!}(b_{|S\cup\{i\}|}v(S \cup \{i\}) - b_{|S|}v(S)), \quad (6)$$

for $i \in N$.

Let $\mathcal{E}$ denote the class of ESL values given by Eq.(6) and $\mathcal{E}^+$ denote the class of ESL values given by Eq.(6) where for each $S \subseteq N$, $b_{|S|} \neq 0$. Let us call the members of $\mathcal{E}^+$ the ESL+ values. For $b_{|S|} : 1 \leq |S| \leq |N| - 1$, $b_{|S|} \neq 0$ and any coalition $T \subseteq N$ such that $|T| \geq 2$ with the parameter k $(1 \leq k \leq |T| - 1,\ k \in \mathbb{N})$, define the game μ_k^T by

$$\mu_k^T(S) = \begin{cases} \dfrac{1}{b_{|S|}} & \text{if } |S \cap T| = k \\ 0 & \text{otherwise.} \end{cases} \quad (7)$$

Proposition 5 *Given k such that $1 \leq k \leq |T| - 1$, the class $\{\mu_k^T : T \subseteq N, |T| \geq 2\}$ spans the null space of any ESL+ value determined by the coefficients $b_{|S|}$ and the class $\{\mu_k^{\{i\}}\}_{i\in N} \cup \{\mu_k^T\}_{|T|\geq 2}$ forms a basis for the space of all TU games.*

For the characterization of the ESL+ values, we define an ESL+-inessential game and the $b_{|S|}$-inessential game as follows:

Definition 7 *(ESL+-inessential game)* A game $v \in \mathcal{G}(N)$ is called ESL+-inessential, if for all $\emptyset \neq S \subseteq N$, $v(S) = \dfrac{b_1}{b_{|S|}} \sum_{i\in S} v(\{i\})$.

Definition 8 *($b_{|S|}$-inessential game)* Let Φ be an ESL+ value with the coefficients $b_{|S|},\ S \subseteq N$. The $b_{|S|}$-inessential game $\tilde{w}_v$ of any game $v \in \mathcal{G}(N)$ is defined as

$$\tilde{w}_v(S) = \frac{1}{b_{|S|}} \sum_{i\in S} \Phi_i(v),\ \forall\, S \subseteq N. \quad (8)$$

It is easy to check that every $b_{|S|}$-inessential game $\tilde{w}_v$ of $v \in \mathcal{G}(N)$ is an ESL+-inessential game. The two axioms to characterize the members of $\mathcal{E}^+$ are given below.

Axiom 9 ESL+-inessential axiom: A value Φ on $\mathcal{G}(N)$ satisfies the ESL+-inessential axiom if, $\Phi_i(v) = b_1 v(\{i\})$ for all $i \in N$ in every ESL+-inessential game $v \in \mathcal{G}(N)$.

Axiom 10 Zero-effect Axiom (Ze): A value Φ satisfies the Zero-effect axiom if $\Phi(v) = \Phi(v + \alpha w)$ for every game $v \in \mathcal{G}(N)$ and any game in the kernel of the ESL+ values.

All the results of Sect. 5 immediately follow for ESL+ values when we replace the δ-inessential game with ESL+-inessential game, δ-discounted Shapley inessential game with $b_{|S|}$-inessential game, δ-inessential axiom with ESL+-inessential axiom, and δ-Factious Oligarchic axiom with Zero-effect Axiom. We therefore list below these results without proofs.

Theorem 6 *(a) When $v \in \mathcal{G}(N)$ is expressed as a linear combination of the basis $(\mu_k^T)_{|T|\geq 1}$ with the parameter k $(1 \leq k \leq |T|,\ k \in \mathbb{N})$, the coefficients related to the singleton coalitions coincide with the ESL+ value of the game v.*
(b) The ESL+ values satisfy the ESL+-inessential axiom.
(c) Every game can be expressed as a sum of its $b_{|S|}$-inessential game and a game in the kernel of the ESL+ value.
(d) A value Φ satisfies the ESL+-inessential axiom and Zero-effect Axiom iff it is the ESL+ value.

7 Concluding Remarks

In this work, we have shown that the family of the δ-factious oligarchic games ($\delta \in (0, 1]$) spans the kernel of the δ-discounted Shapley value. This class also contains the family of factious oligarchic games for $\delta = 1$. Moreover, the extension of this family makes a basis for the space of all TU-games. When we express a game as a linear combination of the δ-factious oligarchic games, the coefficients related to the singleton coalitions are the payoffs under the δ-discounted Shapley value of the corresponding game. An alternative characterization of the δ-discounted Shapley value is presented. Our work is closely related to [3] where they have adopted a similar approach to characterize the Shapley value. Further, we have extended our results to the set of ESL+ values as well. A more general characterization to cover all the ESL values is kept for our future work.

References

1. Béal, S., and Rémilla, and Solal, P, . 2017. Axiomatization and implementation of a class of solidarity values for TU games. *Theory and Decisions* 83: 61–94.
2. Béal, S., E. Rémila, and P. Solal. 2016. Characterizations of three linear values for TU games by associated consistency: simple proofs using the Jordan normal form. *International Game Theory Review* 18 (1650003): 1–21.

3. Béal, S., Manea, M., Rémila, E. and Solal, P. (2018) Games with identical Shapley value, https://economics.mit.edu/files/11885
4. Chun, Y. 1989. A new axiomatization of the Shapley value, *Games and Economic Behavior*, 1 (2):119–130.
5. Driessen, T.S.H., and T. Radzik. 2013. On a family of values for TU-games generalizing the Shapley value. *Mathematical Social Sciences* 65: 105–111.
6. Faigle, U., and M. Grabisch. 2016. Bases and Linear Transforms of TU-games and Cooperation Systems. *International Journal of Game Theory* 45: 875–892.
7. Hamiache, G. 2001. Associated Consistency and the Shapley Value. *International Journal of Game Theory* 30: 279–289.
8. Joosten, R.A.M.G. 1996. Dynamics, equilibria, and values (Doctoral dissertation, Maastricht University).
9. Kleinberg, N.L., and J.H. Weiss. 1985. Equivalent n-person games and the null space of the Shapley value. *Mathematics of Operations Research* 10: 233–243.
10. Nowak, A.S., and T. Radzik. 1994. A solidarity value for n-person transferable utility games. *International Journal of Game Theory* 23 (1): 43–48.
11. Radzik, T., and T. Driessen. 2013. On a family of values for TU-games generalizing the Shapley value. *Mathematical Social science* 65: 105–111.
12. Rojas, J., and F. Sanchez. 2016. On the inverse problem for a subclass of linear, symmetric and efficient values of cooperative TU games. *Operations Research Letters* 44: 618–621.
13. Ruiz, L.M., F. Valenciano, and J.M. Zarzuelo. 1998. The family of least square values for transferable utility games. *Games and Economic Behavior* 24: 109–130.
14. Shapley, L.S. 1953. A Value for n-Person Games, *Contributions to the Theory of Games* (H.W. Kuhn and A.W. Tucker, eds.), 307–317, Annals of Mathematical Studies, 28, Princeton University Press.
15. van den Brink, R., Y. Funaki, and Y. Ju. 2011. Reconciling marginalism with egalitarianism: consistency, monotonicity, and implementation of egalitarian Shapley values. *Social Choice and Welfare*. https://doi.org/10.1007/s00355-011-0634-2.
16. van den Brink, R. 2007. Null or nullifying players: the difference between the Shapley value and equal division solutions. *Journal of Economic Theory* 136 (1): 767–775.
17. van den Brink, R., Funaki, Y. (2015) Implementation and axiomatization of discounted Shapley values, *Social choice welfare*, 45:329-344. https://doi.org/10.1007/s00355-015-0899-y.
18. Yokote, K. and Funaki, Y. 2015. Several bases of a game space and an application to the Shapley value, mimeo, https://ideas.repec.org/p/wap/wpaper/1419.html.
19. Yokote, K., Y. Funaki, and Y. Kamijo. 2016. A new basis and the Shapley value. *Mathematical Social Sciences* 80: 21–24.

No-envy in the Queueing Problem with Multiple Identical Machines

Manipushpak Mitra and Suresh Mutuswami

JEL Classifications: C72 and D63 and D82

1 Introduction

No-envy in the context of allocation problems was introduced by Foley [17] as a criterion of fair allocation. An allocation is said to satisfy no-envy if no agent wishes to swap his allocation with any other agent. Its implications have been studied in many different contexts (see Alkan, Demange and Gale [1], Bevia [2], Moulin [21], Svensson [27], Tadenuma and Thomson [28], Thomson [29, 31], Thomson Varian [32]).

In this paper, we identify and analyze no-envy allocations for queueing problems with multiple identical machines. In a queueing problem with multiple identical machines, there is a server, with many identical machines. The server has to process a finite number of jobs for a set of agents. All jobs take the same time to process and without loss of generality, we assume that it takes one unit of time to complete one job. Each agent has one job to process. Each machine takes one unit of time to process one job. We assume scarcity of resources: the number of jobs to be processed exceeds the number of machines available to the server. Hence, some agents have to wait in a queue. Waiting in a queue is costly for each agent, and we assume that agents have quasi-linear preferences.

M. Mitra (✉)
Economic Research Unit, Indian Statistical Institute, Kolkata, India
e-mail: mmitra@isical.ac.in

S. Mutuswami
Economics Division, School of Business, University of Leicester, LE1 7RH Leicester, UK
e-mail: sm403@le.ac.uk

S. Borkotokey et al. (eds.), *Game Theory and Networks*, Indian Statistical Institute Series,
https://doi.org/10.1007/978-981-16-4737-6_9

Mitra [20] analyzed the above problem under a general cost structure. In this paper, we analyze the same problem assuming a specific cost structure but addressing a different question. Specifically, we ask whether the server can allocate queue positions and provide compensatory monetary transfers to the agents so as to ensure no-envy. For the special case of queueing problems with single machine and with cost functions which is linear in time, this question was analyzed by Chun [3] and Chun, Mitra, and Mutuswami [7, 9, 10]. Our work adds to this literature by generalizing their results to the case of multiple identical machines. For two identical machines, queueing problem was addressed by Chun and Heo [6] from a cooperative game perspective.

In deriving our results, we also use some other properties of allocation rules: *outcome efficiency*, *equal treatment of equals*, *budget balance*, *Pareto efficiency*, and *Lorenz optimality*. Outcome efficiency requires that the queue minimizes the aggregate waiting cost. Equal treatment of equals requires that if two agents have identical waiting costs, then they also receive identical utilities from the allocation. In the absence of job or agent priority, these properties are uncontroversial basic requirements from any allocation rule. Budget balance requires that the sum of monetary transfers across agents must add up to zero. The definition of Pareto efficiency is standard and, in this context, boils down to identifying allocation rules that satisfy outcome efficiency and budget balance.

We first show that if an allocation rule satisfies no-envy, then it also satisfies outcome efficiency and equal treatment of equals. We then define a set of allocation rules that we call *fair allocation rules*. Our main theorem shows that an allocation rule satisfies no-envy if and only if it is a fair allocation rule. We proceed to identify the subset of fair allocation rules that also satisfy budget balance. We call these rules as PE-fair allocation rules. Our next proposition shows that an allocation rule satisfies no-envy and Pareto efficiency if and only if it is PE-fair.

We show that there is a large set of allocation rules satisfying no-envy and Pareto efficiency. This raises the issue of selecting an allocation rule from this set. Following Chun, Mitra, and Mutuswami [9], we use the egalitarian principle represented by the Lorenz criterion to derive a single allocation rule.[1] Our last proposition identifies the Lorenz optimal rule in the set of PE-fair allocation rules.

The paper is organized as follows. In Sect. 2, we provide the framework and introduced the relevant axioms on allocation rules. In Sect. 3, we provide our results. Finally, in Sect. 4, we discuss potential future research topics.

[1] Dutta [15] and Sen and Foster [25] used the Lorenz criterion to rank income distributions. The Lorenz criterion was also used to analyze the bankruptcy problem (see Chun, Schummer, and Thomson [11] and Thomson [30]).

2 The Framework

Let $N = \{1, \ldots, n\}$ be the finite set of agents and m be the number of identical machines. We assume that $n > m > 1$.[2] Let the non-negative real number θ_j represent the per-period waiting cost of agent j. The vector of waiting costs of all agents $\theta = (\theta_1, \ldots, \theta_n)$ is a *profile*. A *queueing problem* is $\Gamma = (n, m, \theta)$.

With multiple machines, two agents may occupy the same queue position (but in the queues for different machines). To represent this situation, we use the notion of a *multi-set* where all elements need not be distinct. For example, $X = \{1, 1, 1, 3, 6, 6, 9\}$ is a multi-set. For any positive real number x, let $\lceil x \rceil$ denote the least integer greater than or equal to x. With m machines and n agents, there are a total of $M = \lceil \frac{n}{m} \rceil > 1$ queue positions. The set of all possible queue positions is the multi-set

$$\mathcal{O}_{n;m} = \{\underbrace{1, \ldots, 1}_{m}, \underbrace{2, \ldots, 2}_{m}, \ldots, \underbrace{M-1, \ldots, M-1}_{m}, \underbrace{M, \ldots, M}_{n-m(M-1)}\}.$$

A *queue* σ is a one-to-one correspondence between N and $\mathcal{O}_{n;m}$. Thus, each queue σ is a permutation of the set $\mathcal{O}_{n;m}$. Let $P(\mathcal{O}_{n;m})$ be the set of all possible permutations of $\mathcal{O}_{n;m}$.[3]

For notational convenience, we represent $\sigma(j) = k$ as $\sigma_j = k$; this means that individual j has the k-th position in the queue.

Agents incur costs from waiting in the queue. If agent j gets the kth queue position, then she incurs a total waiting cost of $k\theta_j$. The preferences of all agents are quasi-linear; hence agent j's utility takes the form $U_j(k, t_j; \theta_j) = -k\theta_j + t_j$ where t_j is the monetary transfer that the agent receives. The associated total waiting cost of all agents is $C(\sigma; \theta) = \sum_{j \in N} \sigma_j \theta_j$.

An *allocation rule* $A = \langle \sigma, t \rangle$ associates a queue and a vector of monetary transfers to every queueing problem. In our setup, we keep the set of agents and the number of machines unchanged, so an allocation rule effectively associates a queue and a vector of transfers to every profile of waiting costs. Abusing notation, we write $A_j(\theta) = (\sigma_j(\theta), t_j(\theta))$ as the allocation of agent j when the profile of waiting costs is θ. The associated utility of the agent is $U_j(A_j(\theta); \theta_j) = -\sigma_j(\theta)\theta_j + t_j(\theta)$.

2.1 *Axioms on Allocation Rules*

We are primarily interested in allocation rules satisfying *no-envy* (Foley [17]).

---

[2] If $n \leq m$, then all agents can be served without delay and the queueing problem is non-existent. The case of $m = 1$ has been analyzed by Chun [3] and Kayi and Ramaekers [18] among others.

[3] Readers can verify that the cardinality of $P(\mathcal{O}_{n;m})$ is $|P(\mathcal{O}_{n;m})| = n!/\{(n - m(M - 1))!(m!)^{M-1}\}$.

Definition 2.1 An allocation rule $A = \langle \sigma, t \rangle$ satisfies *no-envy* if for all profiles $\theta \in \Theta^N$ and all $i, j \in N$, $U_i(A_i(\theta); \theta_i) \geq U_i(A_j(\theta); \theta_i)$.

An allocation satisfies *no-envy* if all agents prefer their own allocation over any other agent's allocation.

Definition 2.2 A queue σ^* is *order efficient for the profile* θ if $C(\sigma^*; \theta) \leq C(\sigma; \theta)$ for all $\sigma \in P(\mathcal{O}_{n;m})$.

Simply put, an order efficient queue (for the profile θ) minimizes aggregate waiting cost. It is easy to verify that σ^* is order efficient for the profile θ if and only if $\sigma_j^* \leq \sigma_i^*$ whenever $\theta_j > \theta_i$. Observe that there can be profiles with more than one order efficient queue since no conclusion can be drawn about the relative queue positions of i and j if $\theta_i = \theta_j$. Let $E(\theta)$ denote the set of all order efficient queues for the profile θ.

Example 2.3 Consider the queueing problem $\Gamma = (7, 2, \theta)$ where $\theta_5 > \theta_4 = \theta_1 > \theta_2 > \theta_3 = \theta_6 > \theta_7$. In this case, $M = 4$ and $\mathcal{O}_{7;2} = \{1, 1, 2, 2, 3, 3, 4\}$.[4] The profile θ has two order efficient queues: $\sigma^1 = (2, 2, 3, 1, 1, 3, 4)$ and $\sigma^2 = (1, 2, 3, 2, 1, 3, 4)$. In σ^1, agent 4 is served first and agent 1 served second while it is the other way around in σ^2. However, one can easily verify that $C(\sigma^1; \theta) - C(\sigma^2; \theta) = \theta_1 - \theta_4 = 0$. There are also profiles for which the order efficient queue is unique. For example, if $\theta_5' > \theta_1' > \theta_2' = \theta_4' > \theta_7' = \theta_6' > \theta_3'$ then the order efficient queue is uniquely given by $(1, 2, 4, 2, 1, 3, 3)$.

The next definition extends outcome efficiency from profiles to allocation rules. It requires that the allocation rule picks an order efficient queue at every profile.

Definition 2.4 An allocation rule $A = \langle \sigma, t \rangle$ is *outcome efficient* if for all profiles θ, $\sigma(\theta) \in E(\theta)$.

Since order efficient queues are not unique at all profiles, our definition implicitly assumes a *tie-breaking rule* for selecting an order efficient queue whenever such queues are not unique. For instance, in Example 2.3, there are two order efficient queues for the profile θ. The tie-breaking rule needs to select either σ^1 or σ^2. There are any number of ways of breaking ties. For example, one can use the following rule: Fix the linear order $1 \succ 2 \succ \ldots \succ n$ on the set of agents N. For any profile θ with $\theta_i = \theta_j$, we pick the queue σ such that $\sigma_i < \sigma_j$ if and only if $i \succ j$.

The next definition requires that there be no net transfers into or out of the system.

Definition 2.5 An allocation rule $A = \langle \sigma, t \rangle$ is *budget balanced* if for all profiles θ, $\sum_{i \in N} t_i(\theta) = 0$.

The last basic property that we discuss is *equal treatment of equals*. It is a basic equity requirement: if two agents have the same waiting cost, then there is no reason to favor either one and therefore, they both should receive the same net utility from the allocation.

---

[4] The cardinality of the set $P(\mathcal{O}_{7;2})$ is $|P(\mathcal{O}_{7;2})| = 7!/\{(7 - 2.3)!(2!)^3\} = 630$.

Definition 2.6 An allocation rule $A = \langle \sigma, t \rangle$ satisfies *equal treatment of equals* if for all profiles θ, $\theta_i = \theta_j$ implies $U_i(A_i(\theta); \theta_i) = U_j(A_j(\theta); \theta_j)$.

3 Results

As mentioned before, outcome efficiency and equal treatment of equals are two basic desirable properties that one would expect from any allocation rule. Our first result shows that if an allocation rule satisfies no-envy, then it also satisfies these two properties.

Proposition 3.1 *If an allocation rule $A = \langle \sigma, t \rangle$ satisfies no-envy, then it also satisfies outcome efficiency and equal treatment of equals.*

Proof Suppose A satisfies no-envy. Let θ be a profile and let $i, j \in N$. No-envy implies that neither i envies j nor j envies i. Therefore,

(1) $U_i(A_i(\theta); \theta_i) \geq U_i(A_j(\theta); \theta_i)$, and,
(2) $U_j(A_j(\theta); \theta_j) \geq U_j(A_i(\theta); \theta_j)$.

Simplifying (1) and (2), we get $t_i(\theta) - t_j(\theta) \geq \big(\sigma_i(\theta) - \sigma_j(\theta)\big)\,\theta_i$ and $t_i(\theta) - t_j(\theta) \leq \big(\sigma_i(\theta) - \sigma_j(\theta)\big)\,\theta_j$ respectively. Combining the two inequalities, we obtain

$$\big(\sigma_i(\theta) - \sigma_j(\theta)\big)\,\theta_i \leq t_i(\theta) - t_j(\theta) \leq \big(\sigma_i(\theta) - \sigma_j(\theta)\big)\,\theta_j. \tag{3.1}$$

Note that (3.1) can be satisfied only if $\big(\sigma_i(\theta) - \sigma_j(\theta)\big)\big(\theta_j - \theta_i\big) \geq 0$. Or, $\theta_i > \theta_j$ implies $\sigma_i(\theta) \leq \sigma_j(\theta)$. This is the condition for outcome efficiency of a queue and establishes the first part of the proposition.

To prove the second part, suppose $\theta_i = \theta_j$. Then from (3.1) it follows that $t_i(\theta) = t_j(\theta) + \{\sigma_i(\theta) - \sigma_j(\theta)\}\theta_j$. Therefore,

$$\begin{aligned} U_i(A_i(\theta); \theta_i) &= -\,\sigma_i(\theta)\theta_i + t_i(\theta) \\ &= -\sigma_i(\theta)\theta_j + t_j(\theta) + \big(\sigma_i(\theta) - \sigma_j(\theta)\big)\,\theta_j \\ &= U_j(A_j(\theta); \theta_j). \end{aligned} \tag{3.2}$$

Remark 3.2 Proposition 3.1 generalizes the results of Chun, Mitra, and Mutuswami [7, 9] for the case of queueing problems with a single machine. The former shows that no-envy implies outcome efficiency and the latter shows that no-envy implies equal treatment of equals. Proposition 3.1 establishes that these two implications of no envy is true even in multiple identical machine queue problems.

Before going to our main result, we provide some more definitions. For any profile θ with $\sigma(\theta) \in E(\theta)$ and any $k \in \{1, \ldots, M\}$, define $I(k; \theta) = \{j \in N \mid \sigma_j(\theta) = k\}$. Moreover, let $l_k, h_k \in I(k; \theta)$ be such that $\theta_{l_k} \leq \theta_j \leq \theta_{h_k}$ for all $j \in I(k; \theta)$. Therefore, for any $\theta \in \Theta^n$, any order efficient queue $\sigma(\theta) \in E(\theta)$, and, any $k \in$

$\{1, \ldots, M\}$, l_k (h_k) is that agent whose waiting cost is lowest (highest) among the set of all agents getting the k-th queue position under the order efficient queue $\sigma(\theta) \in E(\theta)$. Observe that given $\sigma(\theta) \in E(\theta)$, from the definition of order efficiency of queue we also have $\theta_{h_{k+1}} \leq \theta_{l_k}$ for all $k \in \{1, \ldots, M-1\}$.

Definition 3.3 An allocation rule $A = \langle \sigma, t \rangle$ is *fair* if for all profiles θ,

(EF1) $\sigma(\theta) \in E(\theta)$, and
(EF2) there exists $\lambda_k(\theta) \in [0, 1]$ for all $k \in \{1, \ldots, M-1\}$ and a real number $R_1(\theta)$ such that for each $j \in N$,

$$t_j(\theta) = R_1(\theta) + \sum_{k=1}^{\sigma_j(\theta)-1} \lambda_k(\theta)\theta_{l_k} + (1 - \lambda_k(\theta))\theta_{h_{k+1}}. \tag{3.3}$$

Fair allocation rules have the following features. For any profile θ, it always picks a queue which is order efficient. The transfers depend on the queue position. All agents in the first queue position get $R_1(\theta)$ as their transfer. All agents in queue positions $k, k > 1$ get a common transfer which is the transfer of any agent in the $(k-1)$th queue position plus a convex combination of the *lowest* waiting cost of the agents in the $k-1$th queue position and the *highest* waiting cost of the agents in the kth position. Hence, the transfers to agent i are

$$t_i(\theta) = \begin{cases} R_1(\theta) & \text{if } \sigma_i(\theta) = 1, \\ R_1(\theta) + \lambda_1(\theta)\theta_{l_1} + (1 - \lambda_1(\theta))\theta_{h_2} & \text{if } \sigma_i(\theta) = 2. \\ R_1(\theta) + \lambda_1(\theta)\theta_{l_1} + (1 - \lambda_1(\theta))\theta_{h_2} & \\ + \ \lambda_2(\theta)\theta_{l_2} + (1 - \lambda_2(\theta))\theta_{h_3} & \text{if } \sigma_i(\theta) = 3, \end{cases}$$

and so on.

To obtain a fair allocation rule A, we need to fix an order efficient queue $\sigma(\theta)$, a vector $\lambda(\theta) \in [0, 1]^{M-1}$ of queue position specific weights and a real number $R_1(\theta)$ for every profile θ. Given these selections, (3.3) fixes the transfers of all the agents. Thus, one can construct many allocation rules that are fair. Let $\mathcal{A}(n, m)$ be the set of all fair allocation rules. The next result shows that these are the only rules that satisfy no-envy.

Theorem 3.4 *An allocation rule $A = \langle \sigma, t \rangle$ satisfies no-envy if and only if it is a fair allocation rule.*

Proof Let A be an allocation rule satisfying no-envy and let θ be a profile. We first show necessity. The necessity of (EF1) follows from Proposition 3.1 and hence we only prove the necessity of (EF2). From Proposition 3.1 we know that if A satisfies no-envy, then for any $i, j \in N$, (3.1) must hold. It follows from this that (a) if $\sigma_i(\theta) = \sigma_j(\theta)$, then $t_i(\theta) = t_j(\theta)$ and (b) if $\sigma_i(\theta) = \sigma_j(\theta) + 1$, then $\theta_i \leq t_i(\theta) - t_j(\theta) \leq \theta_j$. The first conclusion is that all agents getting the same queue position get the same transfer. The second conclusion is that the difference in transfer between two agents in consecutive queue positions is bounded between their respective waiting costs. Since this must hold for any two agents in consecutive positions, it follows that the difference $t_{k+1}(\theta) - t_k(\theta)$ is bounded below by the *highest* waiting cost of agents in the $(k+1)$th position and above by the *lowest* waiting cost of agents in the kth position. Therefore, there exists $\lambda_k(\theta) \in [0, 1]$ such that

$$t_k(\theta) - t_{k+1}(\theta) = \lambda_k(\theta)\theta_{l_k} + (1 - \lambda_k(\theta))\theta_{h_{k+1}}. \tag{3.4}$$

It follows from (3.4) that $t_{k+1}(\theta)$ is determined once $t_k(\theta)$ is determined. The recursion process obviously terminates at $k = 1$. By selecting any number $R_1(\theta)$ and setting $t_j(\theta) = R_1(\theta)$ for all j such that $\sigma_j(\theta) = 1$ and then recursively solving for the transfers of agents in other queue positions using (3.4), we get the transfers given by (3.3).[5]

To prove the converse, let A be a fair allocation rule, let θ be a profile, and let $i, j \in N$. Let $A(\theta) = (\sigma(\theta), t(\theta))$ be the allocation at the profile θ. We have the following mutually exclusive and exhaustive possibilities: (a) $\sigma_i(\theta) = \sigma_j(\theta)$, (b) $\sigma_i(\theta) > \sigma_j(\theta)$, and (c) $\sigma_i(\theta) < \sigma_j(\theta)$. We show that $U_i(A_i(\theta); \theta_i) - U_i(A_j(\theta); \theta_i) \geq 0$ for each of these three cases.

If (a) holds, then $t_i(\theta) = t_j(\theta)$ from (3.3); hence, $A_i(\theta) = A_j(\theta)$. It follows now that $U_i(A_i(\theta); \theta_i) - U_i(A_j(\theta); \theta_i) = 0$.

If (b) holds, then by (3.3), we get

$$\begin{aligned} U_i(A_i(\theta); \theta_i) - U_i(A_j(\theta); \theta_i) &= \sum_{k=\sigma_j(\theta)}^{\sigma_i(\theta)-1} \left[\lambda_k(\theta)\theta_{l_k} + (1 - \lambda_k(\theta))\theta_{h_{k+1}}\right] \\ &\quad - \left[\sigma_i(\theta) - \sigma_j(\theta)\right]\theta_i \\ &= \sum_{k=\sigma_j(\theta)}^{\sigma_i(\theta)-1} \left[\lambda_k(\theta)\theta_{l_k} + (1 - \lambda_k(\theta))\theta_{h_{k+1}}\right] \\ &\quad - \sum_{k=\sigma_j(\theta)}^{\sigma_i(\theta)-1} \theta_i \\ &= \sum_{k=\sigma_j(\theta)}^{\sigma_i(\theta)-1} \left[\lambda_k(\theta)\theta_{l_k} + (1 - \lambda_k(\theta))\theta_{h_{k+1}} - \theta_i\right] \\ &\geq 0. \end{aligned}$$

[5] Actually, (3.4) says that $n - 1$ transfers can be determined in terms of one transfer. We choose $t_1(\theta)$ as the free variable here but we could have chosen any one of the other $n - 1$ transfers. The choice of the free variable is immaterial.

The last inequality follows due to order efficiency of $\sigma(\theta)$ and from the inequality $\theta_{h_{\sigma_i(\theta)}} \geq \theta_i$. Specifically, $\theta_{h_{\sigma_i(\theta)}} \geq \theta_i$ along with order efficiency ensures that $\lambda_k(\theta)\theta_{l_k} + (1-\lambda_k(\theta))\theta_{h_{k+1}} \geq \theta_i$ for $k = \sigma_j(\theta), \ldots, \sigma_i(\theta) - 1$.

Finally, if (c) holds, then by (3.3), we get

$$\begin{aligned}
U_i(A_i(\theta);\theta_i) - U_i(A_j(\theta);\theta_i) &= \left[\sigma_j(\theta) - \sigma_i(\theta)\right]\theta_i \\
&\quad - \sum_{k=\sigma_i(\theta)}^{\sigma_j(\theta)-1} \left[\lambda_k(\theta)\theta_{l_k} + (1-\lambda_k(\theta))\theta_{h_{k+1}}\right] \\
&= \sum_{k=\sigma_i(\theta)}^{\sigma_j(\theta)-1} \theta_i - \sum_{k=\sigma_i(\theta)}^{\sigma_j(\theta)-1} \left[\lambda_k(\theta)\theta_{l_k} + (1-\lambda_k(\theta))\theta_{h_{k+1}}\right] \\
&= \sum_{k=\sigma_i(\theta)}^{\sigma_j(\theta)-1} \left[\theta_i - \left(\lambda_k(\theta)\theta_{l_k} + (1-\lambda_k(\theta))\theta_{h_{k+1}}\right)\right] \\
&\geq 0.
\end{aligned}$$

The last inequality follows due to order efficiency and the inequality $\theta_{l_{\sigma_i(\theta)}} \leq \theta_i$. The argument is similar to the last case and is omitted.

Remark 3.5 Theorem 3.4 generalizes the results of Chun, Mitra, and Mutuswami [9] for the single machine queueing problem (see Proposition 3.11 in their paper) to the case of multiple identical machines. The difference between the two results is driven solely by the fact that there may be more than one agent in any given queue position with multiple machines. To deduce fair allocation rules, we required the set of agent that gets the kth queue position for any profile θ, and specifically, the agents with the highest and lowest waiting costs in this set. In single machine queueing problems, this distinction is not needed since there is only one agent associated with a given queueing position.

3.1 Pareto Efficiency

Theorem 3.4 shows that in queueing problem with multiple identical machines, the set of fair allocation rules coincides with the set of rules satisfying no-envy. Given Proposition 3.1, these rules also satisfy outcome efficiency and equal treatment of equals. However, these rules may not be budget balanced. Budget balanced fair allocation rule are desirable because they guarantee no-envy allocations without incurring any budget deficit or budget surplus.

Definition 3.6 An allocation rule $A = \langle \sigma, t \rangle$ is *Pareto efficient* if it satisfies outcome efficiency and budget balance.

We now define a set of rules that we call *PE-fair* allocation rules.

Definition 3.7 An allocation rule $A = \langle \sigma, t \rangle$ is said to be *PE-fair* if it satisfies the following two properties.

(PEF1) For each profile θ, $\sigma(\theta) \in E(\theta)$.
(PEF2) For each profile θ, there exists $\lambda_k(\theta) \in [0, 1]$, $k = 1, \dots, M-1$ such that for each $j \in N$,

$$t_j(\theta) = \sum_{k=1}^{\sigma_j(\theta)-1} \frac{km}{n} \left[\lambda_k(\theta)\theta_{l_k} + (1-\lambda_k(\theta))\theta_{h_{k+1}}\right] - \sum_{k=\sigma_j(\theta)}^{M-1} \left[1 - \frac{km}{n}\right] \left[\lambda_k(\theta)\theta_{l_k} + (1-\lambda_k(\theta))\theta_{h_{k+1}}\right]. \tag{3.5}$$

The transfer given by (3.5) says that the transfer of agent j is a weighted sum of the lowest and the highest waiting costs in each queue position. The weights given to each queueing position is km/n if $\sigma_j(\theta) > k$ and $-(1 - km/n)$ if $\sigma_j(\theta) < k$.

Remark 3.8 Equation (EF1) of Definition 3.3 is same as (PEF1) of Definition 3.7 as both sets of allocation rules require outcome efficiency. We obtain (PEF2) from (EF2) by substituting $R_1(\theta) = -\sum_{k=1}^{M-1}\{1 - (km/n)\}\{\lambda_k(\theta)\theta_{l_k} + (1-\lambda_k(\theta))\theta_{h_{k+1}}\}$ for each profile θ in (3.3) and then simplifying the resulting expression. The result of this substitution and simplification is reflected in the expression of each agent's transfer given by condition (3.5) of Definition 3.7. The next result shows that this substitution leads to budget balance.

Proposition 3.9 *An allocation rule $A \in \mathcal{A}(n, m)$ is budget balanced if and only if it is PE-fair.*

Proof We proceed by imposing budget balance on the transfers for a fair allocation rule. Using (3.3), the sum of transfers across all agents is given by

$$\sum_{j \in N} t_j(\theta) = nR_1(\theta) + \sum_{k=1}^{M-1} (n - km) \left[\lambda_k(\theta)\theta_{l_k} + (1-\lambda_k(\theta))\theta_{h_{k+1}}\right]. \tag{3.6}$$

We have budget balance if and only if $\sum_{j \in N} t_j(\theta) = 0$. Putting the left-hand side of (3.6) equal to zero, we solve for $R_1(\theta)$. This gives

$$R_1(\theta) = -\sum_{k=1}^{M-1} \left(1 - \frac{km}{n}\right) \left[\lambda_k(\theta)\theta_{l_k} + (1-\lambda_k(\theta))\theta_{h_{k+1}}\right]. \tag{3.7}$$

Substituting (3.7) in (3.3) we get that for any $j \in N$,

$$t_j(\theta) = -\sum_{k=1}^{M-1}\left(1-\frac{km}{n}\right)\left[\lambda_k(\theta)\theta_{l_k} + (1-\lambda_k(\theta))\theta_{h_{k+1}}\right] + \sum_{k=1}^{\sigma_j(\theta)-1}\left[\lambda_k(\theta)\theta_{l_k} + (1-\lambda_k(\theta))\theta_{h_{k+1}}\right]. \quad (3.8)$$

Note that for j such that $\sigma_j(\theta) > 1$ and for any $k = 1, \ldots, \sigma_j(\theta) - 1$, the coefficient of the term $\lambda_k(\theta)\theta_{l_k} + (1 - \lambda_k(\theta))\theta_{h_{k+1}}$ in (3.8) is $1 - (1 - (km/n)) = km/n$. For any $k = \sigma_j(\theta), \ldots, M - 1$, the coefficient is $-(1 - (km/n))$ and hence we get the transfer given by (3.5).

Let us denote $\mathcal{A}^*(n, m)$ as the set of all PE-fair allocation rules. From Proposition 3.9, it follows that $\mathcal{A}^*(n, m) \subset \mathcal{A}(n, m)$. In choosing a fair allocation rule, we have M degrees of freedom to determine the transfers: the position specific weights $(\lambda_1(\theta), \ldots, \lambda_{M-1}(\theta))$ and the transfers to the agents in the first queue position $R_1(\theta)$. With budget balance, we only have $M - 1$ degrees of freedom: once the queue position specific weights are chosen, $R_1(\theta)$ is automatically determined by budget balance.

The proof of the following corollary is obvious and is left as an exercise for the reader.

Corollary 3.10 *An allocation rule $A \in \mathcal{A}(n, m)$ satisfies no-envy and Pareto efficiency if and only if it is PE-fair.*

3.2 Lorenz Optimality

The set $\mathcal{A}^*(n, m)$ is large since we have the freedom to select the variables $\lambda_k(\theta)$.[6] Hence, the server faces the problem of deciding which allocation rule to select from this set. Can we use some criterion to select an allocation rule from $\mathcal{A}^*(n, m)$? In this subsection, we use the egalitarian principle embodied in the Lorenz criterion to address this issue.

Let $x = (x_1, \ldots, x_n)$ be a vector in $\mathbb{R}^n$ and define $\bar{x}$ to be the permutation of x such that $\bar{x}_1 \leq \bar{x}_2 \leq \ldots \leq \bar{x}_n$.

Definition 3.11 Let $\mathbf{x} = (x_1, \ldots, x_n)$ and $\mathbf{y} = (y_1, \ldots, y_n)$ be vectors in $\mathbb{R}^n$. We say that $\mathbf{x}$ *Lorenz dominates* $\mathbf{y}$ if for all $k = 1, \ldots, n$, $\sum_{j=1}^{k} \bar{x}_j \geq \sum_{j=1}^{k} \bar{y}_j$.

It is well known that Lorenz domination is a partial order since two vectors may not be Lorenz comparable. For instance, $(20, 35, 45)$ and $(15, 42, 43)$ are not Lorenz comparable since $20 > 15$ but $15 + 42 = 57 > 20 + 35 = 55$.

Let A be an allocation rule and θ be a profile. For each $j \in N$, define the agent's utility corresponding to his allocation $A_j(\theta)$ as $U_j^A(\theta) = U_j(A_j(\theta); \theta_j)$. The utility of all agents is given by the vector $U^A(\theta) = (U_1^A(\theta), \ldots, U_n^A(\theta))$.

---

[6] Note that we can select these variables differently at each profile.

Definition 3.12 Let $A, A' \in \mathcal{A}^*(n,m)$. We say that A *Lorenz dominates* A' if for all profiles θ, $U^A(\theta)$ Lorenz dominates $U^{A'}(\theta)$.

This is a weak criterion since not only might $U^A(\theta)$ and $U^{A'}(\theta)$ be Lorenz non-comparable but the domination relationships could also go in opposite directions for two profiles. That is, we might have $U^A(\theta)$ Lorenz dominates $U^{A'}(\theta)$ but $U^{A'}(\theta')$ Lorenz dominates $U^A(\theta')$.)

We write $A \succ_{LD} A'$ to denote that the allocation rule A Lorenz dominates the allocation rule A'.

Definition 3.13 An allocation rule $A \in \mathcal{A}^*(n,m)$ is *Lorenz optimal* if $A \succ_{LD} A'$ for all $A' \in \mathcal{A}^*(n,m)$.

It is clear from the above discussion that it is not obvious that a Lorenz optimal allocation rule exists on $\mathcal{A}^*(n,m)$.[7] Our final result shows that we do have a Lorenz optimal allocation rule on $\mathcal{A}^*(n,m)$.

Definition 3.14 The allocation rule $A^* = \langle \sigma, t^* \rangle \in \mathcal{A}^*(n,m)$ is *LPE-fair* if it satisfies the following two properties.

(LPN1) For all profiles θ, $\sigma(\theta) \in E(\theta)$.
(LPN2) For each profile θ and for each $j \in N$,

$$t_j^*(\theta) = \sum_{k=2}^{\sigma_j(\theta)} \left\lceil \frac{(k-1)m}{n} \right\rceil \theta_{h_k} - \sum_{k=\sigma_j(\theta)+1}^{M} \left[1 - \frac{(k-1)m}{n} \right] \theta_{h_k}. \tag{3.9}$$

The transfers in (3.9) are derived from (3.5) by substituting $\lambda_k(\theta) = 0, k = 1, \ldots, M-1$. The LPE-fair allocation rule is a generalization of the Lorenz no-envy rule defined in Chun, Mitra, and Mutuswami [9] for the single machine queueing problem. The following proposition generalizes Proposition 3.14 in their paper.

Proposition 3.15 *The LPE-fair allocation rule is Lorenz optimal in* $\mathcal{A}^*(n,m)$.

Proof Let A be a fair allocation rule. Let θ be a profile and let the rank based reordering of waiting costs be $\theta_{(1)} \geq \ldots \geq \theta_{(n)}$. For any $r = 1, \ldots, n-1$, we have

(1) $U^A_{(r+1)}(0) \quad U^A_{(r)}(\theta) = q(\theta_{(r)} - \theta_{(r+1)})$ if $\lceil \frac{r}{m} \rceil = \lceil \frac{r+1}{m} \rceil = q$,
(2) $U^A_{(r+1)}(\theta) - U^A_{(r)}(\theta) = q(\theta_{(r)} - \theta_{(r+1)}) + \lambda_q(\theta)(\theta_{l(q)} - \theta_{h(q+1)}) + (\theta_{h(q+1)} - \theta_{(r+1)})$ if $\lceil \frac{r}{m} \rceil = q < \lceil \frac{r+1}{m} \rceil = 1 + q$.

From (3.2) and (3.2), we have two observations. First, if $\theta_{(r)} = \theta_{(r+1)}$, then $U^A_{(r+1)}(\theta) = U^A_{(r)}(\theta)$. This is clear in (3.2) so let us consider (3.2). Since $\theta(r) = \theta_{(r+1)}$ and $\lceil \frac{r+1}{m} \rceil = 1 + \lceil \frac{r}{m} \rceil$, we must have $\theta_{l(q)} = \theta_{h(q+1)} = \theta_{(r)} = \theta_{(r+1)}$. It follows from this that $U^A_{(r+1)}(\theta) = U^A_{(r)}(\theta)$. Second, if $\theta_{(r)} > \theta_{(r+1)}$, then $U^A_{(r+1)}(\theta) > U^A_{(r)}(\theta)$.

[7] See Chun, Mitra, and Mutuswami [9] for more discussion on Lorenz optimal allocation rules.

Again, this is clear in (3.2). For (3.2), note that $\theta_{(r)} = \theta_{l(q)} > \theta_{h(q+1)} = \theta_{(r+1)}$ and the result follows from this observation.

We conclude that $U^A_{(1)}(\theta) \leq \ldots \leq U^A_{(n)}(\theta)$ implying that $\overline{U}^A_r(\theta) = U^A_{(r)}(\theta)$ for all $r = 1, \ldots, n$.[8] Hence, the agent having the highest waiting cost will have the lowest utility, the agent having the second highest waiting cost will have the next lowest level of utility and so on.

If A is Lorenz optimal, then $\overline{U}^A_1(\theta) \geq \overline{U}^{A'}_1(\theta)$ for all $A' \in \mathcal{A}^*(n, m)$. Given the above discussion and (3.9), we have

$$\overline{U}^{A'}_1(\theta) = -\theta_{(1)} - \sum_{k=1}^{M-1}\left(1 - \frac{km}{n}\right)\left[\lambda_k(\theta)\theta_{l_k} + (1 - \lambda_k(\theta))\theta_{h_{k+1}}\right]. \qquad (3.10)$$

If $\theta_{l_k} > \theta_{h_{k+1}}$, it follows from (3.10) that $\overline{U}^A_1(\theta)$ is minimized only if $\lambda_k(\theta) = 0$. If $\theta_{l_k} = \theta_{h_{k+1}}$, then $\lambda_k(\theta) = 0$ is a minimizer but no longer uniquely so. This proves the necessity of the LPE-fair allocation rule A^* for Lorenz optimality.

To prove sufficiency, let A be an allocation rule and θ a profile. Using (3.2) and (3.2), it follows that

$$\begin{aligned}\sum_{q=1}^{r}\overline{U}^A_q(\theta)) = &- m\left(1 - \frac{r}{n}\right)\sum_{k=1}^{\lceil \frac{r}{m}\rceil - 1} k\left[\lambda_k(\theta)\theta_{l_k} + (1 - \lambda_k(\theta))\theta_{h_{k+1}}\right] \\ &- r\sum_{k=\lceil \frac{r}{m}\rceil}^{M-1}\left(1 - \frac{km}{n}\right)\left[\lambda_k(\theta)\theta_{l_k} + (1 - \lambda_k(\theta))\theta_{h_{k+1}}\right] \qquad (3.11) \\ &- \sum_{q=1}^{r}\left\lceil \frac{q}{m}\right\rceil \theta_{(q)}.\end{aligned}$$

It is clear from (3.11) that the sum $\sum_{q=1}^{r}\overline{U}^A_q(\theta)$ is maximized by setting $\lambda_k(\theta) = 0$. Hence, $\sum_{q=1}^{r}\overline{U}^{A^*}_q(\theta) \geq \sum_{q=1}^{r}\overline{U}^A_q(\theta)$ for every $r = 1, \ldots, n-1$ and this establishes sufficiency.

The following corollary follows straightforwardly from Proposition 3.15.

Corollary 3.16 *An allocation rule* $A \in \mathcal{A}^*(n, m)$ *satisfies no-envy, Pareto efficiency, and Lorenz optimality if and only if it the LPE-fair allocation rule* A^*.

Proof Follows from Corollary 3.10 and Proposition 3.15.

Example 3.17 Consider the queueing problem with seven agents and three machines $\Gamma = (7, 3, \theta)$. Then, the list of all queue positions is given by the multi-set $\mathcal{O}_{7;3} =$

---

[8] Recall that for any vector x, $\overline{x}$ is the rank-order based reordering such that $\bar{x}_1 \leq \ldots \leq \bar{x}_n$.

$\{1, 1, 1, 2, 2, 2, 3\}$. A queue σ is a one-to-one mapping from $\{1, \ldots, 7\}$ to $\mathcal{O}_{7;3}$.[9] Let A be a fair allocation rule let $\theta_5 > \theta_4 = \theta_1 > \theta_2 > \theta_3 = \theta_6 > \theta_7$. The unique order efficient queue for this profile is $\sigma(\theta) = (1, 2, 2, 1, 1, 2, 3)$. The monetary transfers to the agents is $t(\theta)$ such that the position specific transfers are $T_1(\theta) = t_1(\theta) = t_4(\theta) = t_5(\theta)$, $T_2(\theta) = t_2(\theta) = t_3(\theta) = t_6(\theta)$, and $T_3(\theta) = t_7(\theta)$. Using (3.3), we get

$$\begin{aligned} T_1(\theta) &= R_1(\theta) \\ T_2(\theta) &= T_1(\theta) + \lambda_1(\theta)\min\{\theta_1, \theta_4, \theta_5\} + (1 - \lambda_1(\theta))\max\{\theta_2, \theta_3, \theta_6\} \\ &= R_1(\theta) + \lambda_1(\theta)\theta_1 + (1 - \lambda_1(\theta))\theta_2 \\ T_3(\theta) &= T_2(\theta) + \lambda_2(\theta)\min\{\theta_2, \theta_3, \theta_6\} + (1 - \lambda_2(\theta))\theta_7 \\ &= R_1(\theta) + \lambda_1(\theta)\theta_1 + (1 - \lambda_1(\theta))\theta_2 + \lambda_2(\theta)\theta_6 + (1 - \lambda_2(\theta))\theta_7. \end{aligned}$$

To obtain Pareto efficiency, we need budget balance. Hence, we have $7R_1(\theta) + 4\lambda_1(\theta)\theta_1 + 4(1 - \lambda_1(\theta))\theta_2 + \lambda_2(\theta)\theta_6 + (1 - \lambda_2(\theta))\theta_7 = 0$. Solving for $R_1(\theta)$ gives

$$R_1(\theta) = -\frac{4}{7}\lambda_1(\theta)\theta_1 - \frac{4}{7}(1 - \lambda_1(\theta))\theta_2 - \frac{1}{7}\lambda_2(\theta)\theta_6 - \frac{1}{7}(1 - \lambda_2(\theta))\theta_7.$$

One can now get the position specific transfers by appropriate substitution.

Finally, by substituting $\lambda_1(\theta) = \lambda_2(\theta) = 0$, we obtain the LPE-fair allocation rule A^*. This gives the position specific transfers:

$$T_1(\theta) = -\frac{4}{7}\theta_2 - \frac{1}{7}\theta_7,\ T_2(\theta) = \frac{3}{7}\theta_2 - \frac{1}{7}\theta_7,\ T_3(\theta) = \frac{3}{7}\theta_2 + \frac{6}{7}\theta_7.$$

4 Future Research Prospects

In this concluding section, we discuss other fairness criteria that can be analyzed in the context of queueing problems with multiple machines.

4.1 Egalitarian Equivalence

Egalitarian equivalence was introduced by Pazner and Schmeidler [24] as a solution to the deficiencies of the no-envy criterion. An allocation rule is egalitarian equivalent if there is a reference bundle for each profile such that each agent is indifferent between his allocation and the reference bundle.

[9] The number of possible queues is $|P(\mathcal{O}_{7;3})| = 7!/\{(7 - 3 \times 2)!(3!)^2\} = 140$.

Definition 4.1 Let $\Gamma = (n, m, \theta)$ be a queueing problem. An allocation $(\sigma(\theta), t(\theta))$ is *egalitarian equivalent for the profile* θ if there is a *reference bundle* $(\sigma^R(\theta), t^R(\theta))$ such that for all $j \in N$, $(\sigma_j(\theta), t_j(\theta)) \sim (\sigma^R(\theta), t^R(\theta))$.[10] An allocation rule $A = \langle\sigma, t\rangle$ is egalitarian equivalent if for all profiles θ, $(\sigma(\theta), t(\theta))$ is an egalitarian equivalent allocation.

Like no-envy, an attractive feature of egalitarian equivalence is that it is an ordinal concept, makes no inter-personal utility comparisons and satisfies equal treatment of equals. It has also been studied in many contexts (see Demange [13]; Dutta and Vohra [16]; Ohseto [23]; Thomson [29]; Yengin [33]); and in the single machine queueing context by Chun, Mitra, and Mutuswami [7, 9]. However, to the best of our knowledge, egalitarian equivalence for the multiple machine queueing context has not been studied so far.

One can address the same questions as those addressed in this paper by replacing no-envy with egalitarian equivalence and then compare those results with the ones obtained in this paper. For the single machine queueing problem (Chun, Mitra, and Mutuswami [7]) and, in general allocation problems with indivisibilities (Thomson [29]), one can show that no allocation rule satisfies both no-envy and egalitarian equivalence. Therefore, one would hope to get the same impossibility result for multiple machine queueing problems while identifying allocation rules that satisfy both no-envy and egalitarian equivalence.

4.2 *Identical Preferences Lower Bound*

Equal Division Lower Bound is one of the oldest fairness concepts (see Dubins and Spanier [14] and Steinhaus [26]). It requires that an agents utility is at least that of consuming his equal share of the resources. It has been addressed in many different contexts (see Bevia [2], Moulin [21, 22], Thomson [31], Yengin [34] among others).

In the queueing problem, though, this concept cannot be directly applied as there is no sense in which resources can be divided equally. Maniquet [19] in his important paper on the single machine queueing problem therefore applied a related concept, *identical preferences lower bound*. This requires that each agent gets at least as much utility assuming all agents *have the same preferences as herself.* Note that this is a hypothetical scenario because, of course, waiting costs are likely to differ across agents. Following Maniquet, further contributions were made by Chun [3–5]; Chun, Mitra, and Mutuswami [8–10]; and Chun and Yengin [12].

For the queueing problem with multiple identical machines, identical preferences lower bound implies that the utility of any agent must be no less than his expected cost assuming that all queues and all positions have an equal chance of being chosen. It is left as an exercise for the reader to show that under identical preferences lower bound that for each profile θ, agent j's utility must be at least $-\frac{M}{n}\left[n - \frac{m(M-1)}{2}\right]\theta_j$.

[10] Here, the symbol "$\sim$" stands for indifference between the two allocations.

Again, one can address the same questions as those addressed in this paper by replacing no-envy with identical preferences lower bound and compare those results with the ones obtained in this paper.

Remark 4.2 There is no allocation rule satisfying no-envy and egalitarian equivalence for the single machine queueing problem. Hence, one would expect to get the same impossibility result for the queueing problem with multiple identical machines. On the other hand, it would be an interesting question to ask whether budget balance and no-envy together imply the identical preferences lower bound as in the case for the single machine queueing problem.

Acknowledgements We would like to thank Surajit Borkotokey, Rajnish Kumar, K. S. Mallikarjun Rao, Diganta Mukherjee and Sudipta Sarangi for encouraging us to write this chapter. We are also thankful to Reviewer 1 for helpful comments and suggestions.

References

1. Alkan, A., G. Demange, and D. Gale. 1991. Fair allocation of indivisible goods and criteria of justice. *Econometrica* 59: 1023–1039.
2. Bevia, C. 1996. Identical preferences lower bound solution and consistency in economies with indivisible goods. *Social Choice and Welfare* 13: 113–126.
3. Chun, Y. 2006. No-envy in queueing problems. *Economic Theory* 29: 151–162.
4. Chun, Y. 2006. A pessimistic approach to the queueing problem. *Mathematical Social Sciences* 51: 171–181.
5. Chun, Y. 2016. *Fair queueing*. Berlin: Springer.
6. Chun, Y., and E.J. Heo. 2008. Queueing problems with two parallel servers. *International Journal of Economic Theory* 4: 299–315.
7. Chun, Y., M. Mitra, and S. Mutuswami. 2014. Egalitarian equivalence and strategyproofness in the queueing problem. *Economic Theory* 56 (2): 425–442.
8. Chun, Y., M. Mitra, and S. Mutuswami. 2019. A characterization of the symmetrically balanced VCG rule in the queueing problem. *Games and Economic Behavior* 118: 486–490.
9. Chun, Y., M. Mitra, and S. Mutuswami. 2019. Egalitarianism in the queueing problem. *Journal of Mathematical Economics* 81: 48–56.
10. Chun, Y., M. Mitra, and S. Mutuswami. 2019. Recent developments in the queueing problem. *TOP* 27: 1–23.
11. Chun, Y., J. Schummer, and W. Thomson. 2001. Constrained egalitarianism: A new solution to bankruptcy problems. *Seoul Journal of Economics* 14: 269–297.
12. Chun, Y., and D. Yengin. 2017. Welfare lower bounds and strategy-proofness in the queueing problem. *Games and Economic Behavior* 102: 462–476.
13. Demange, G. 1984. Implementing efficient egalitarian equivalent allocations. *Econometrica* 52: 1167–1177.
14. Dubins, L., and E. Sapnier. 1961. How to cut a cake fairly. *American Mathematical Monthly* 68: 17.
15. Dutta, B. 2002. Inequality, poverty and welfare. In *Handbook of social choice and welfare*, vol. 1, ed. K.J. Arrow, A.K. Sen, and K. Suzumura, 597–633. The Netherlands: Elsevier.
16. Dutta, B., and R. Vohra. 1993. A characterization of egalitarian equivalence. *Economic Theory* 3: 465–479.
17. Foley, D. 1967. Resource allocation and the public sector. *Yale Economic Essays* 7: 45–98.
18. Kayi, C., and E. Ramaekers. 2010. Characterizations of Pareto-efficient, fair, and strategy-proof allocation rules in queueing problems. *Games and Economic Behavior* 68: 220–232.

19. Maniquet, F. 2003. A characterization of the Shapley value in queueing problems. *Journal of Economic Theory* 109: 90–103.
20. Mitra, M. 2005. Incomplete information and multiple machine queueing problems. *European Journal of Operational Research* 165: 251–266.
21. Moulin, H. 1990. Fair division under joint ownership: recent results and open problems. *Social Choice and Welfare* 7: 149–170.
22. Moulin, H. 1991. Welfare bounds in the fair division problem. *Journal of Economic Theory* 54: 321–337.
23. Ohseto, S. 2004. Implementing egalitarian-equivalent allocation of indivisible goods on restricted domains. *Economic Theory*, 659–670.
24. Pazner, E., and D. Schmeidler. 1978. Egalitarian equivalent allocations: A new concept of equity. *Quarterly Journal of Economics* 92: 671–687.
25. Sen, A., and J. Foster. 1997. *On economic inequality*. Oxford: Claredon Press.
26. Steinhaus, H. 1948. The problem of fair division. *Econometrica* 16: 101–104.
27. Svensson, L.G. 1983. Large indivisibiities: an analysis with respect to price equilibrium and fairness. *Econometrica* 51: 939–954.
28. Tadenuma, K., and W. Thomson. 1991. No-envy and consistency in economies with indivisible goods. *Econometrica* 59: 1755–1767.
29. Thomson, W. 1990. On the non-existence of envy-free and egalitarian-equivalent allocations in economies with indivisibilities. *Economics Letters* 34: 227–229.
30. Thomson, W. 2012. Lorenz rankings of rules for the adjudication of conflicting claims. *Economic Theory* 50: 547–569.
31. Thomson, W. 2013. *The theory of fair allocation*. Book Manuscript: University of Rochester.
32. Thomson, W., and H. Varian, 1985. Theories of justice based on symmetry, Chapter 4. In: Hurwicz, L., Schmeidler, D., Sonnenschein, H. (eds.) Social Goals and Social Organization: Essays in memory of E. Pazner, 107–129. Cambridge: Cambridge University Press.
33. Yengin, D. 2012. Egalitarian-equivalent Groves mechanisms in the allocation of heterogenous objects. *Social Choice Welfare* 38: 137–160.
34. Yengin, W. 2013. Identical preferences lower bound for allocation of heterogenous tasks and NIMBY problems. *Journal of Public Economic Theory* 15: 580–601.

Rationing Rules Under Uncertain Claims: A Survey

Sinan Ertemel and Rajnish Kumar

JEL code: C71 · D63 · D81

1 Introduction

Working from home under the lockdown situation in our respective countries when we are writing this chapter, the whole world is struggling to deal with the pandemic of COVID-19—the most serious challenge the world has faced in a long time. Jared Kushner, senior advisor and son-in-law of the US president Donald Trump has been widely criticized for his statement about the federal stockpile of life-saving pharmaceuticals and medical supplies on April 3, 2020 (see, e.g., Dale, April [4]), regarding the federal stockpile of life-saving pharmaceuticals and medical supplies. Jared Kushner was trying to make the point that the states should first assess their needs and their stockpiles before making a request from the federal stockpile. Although the criticism is well-founded based on the words used by Kushner,[1] "the problem

[1]Kushner's statement included "the notion of the federal stockpile was it's supposed to be our stockpile; it's not supposed to be states' stockpiles that they then use".

We are grateful to Jingyi Xue for her many valuable comments and suggestions. We also would like to thank the two referees whose suggestions have helped improve this chapter. Rajnish Kumar acknowledges the financial support provided by British Council, UK, Grant UGC-UKIERI 2016-17-059.

S. Ertemel (✉)
Management School, Queen's University Belfast, 185 Stranmillis Road, BT9 5EE Belfast, UK
e-mail: rajnish.kumar@qub.ac.uk

R. Kumar
Department of Economics, Istanbul Technical University, 34367, Macka, Istanbul, Turkey
e-mail: ertemels@itu.edu.tr

S. Borkotokey et al. (eds.), *Game Theory and Networks*, Indian Statistical Institute Series,
https://doi.org/10.1007/978-981-16-4737-6_10

of dividing when there isn't enough" is not trivial (see, e.g., Thomson [15], Moulin [13]). It becomes even more challenging when the problem is of dividing today when the shortage tomorrow is unknown. This is the situation that corresponds to the division of the federal stockpile of medical supplies to the various states when these states are preparing for their unknown peaks of the pandemic. This is the situation that we address in this chapter.

A rationing problem involves a resource to be divided among a set of agents each of whom has a claim on the resource. The resource is not enough to meet all the claims. Claims of creditors on a bankrupt firm and inheritance claims on the estate of a deceased person are examples of such situations (as first formal studies, see O'Neill [14], and Aumann and Maschler [1]). A rationing problem under uncertain claims consists of a rationing problem where the claims are state-contingent or follow a probability distribution. Agents may submit a vector of claims one for each state of the world to be realized in the second stage, or they may submit a claim distribution. The resource must be divided into stage one before the realization of the state of the world.

Apart from the pandemic situation mentioned above, such a situation may arise in the allocation of the fiscal budget of a country. Different ministries of a government may require different resources in different states of the world to be realized in the next fiscal year. For example, the Ministry of Defense may have different requirements depending on its relations with other countries as well as the situations of disasters that may happen in the following year. The Ministry of Agriculture may have requirements based on factors like rainfall next year. The Ministry of Health may have requirements that depend on factors like the incidence of epidemics and the weather. However, the federal budget must be allocated at the beginning of the fiscal year.

Another example of our setting is the distribution of research funds (or travel grants) among graduate students of a school in a university who expect travel or research expenses contingent on the state of the world (e.g., expenses based on the results of their research and travels plans based on the conferences accepting their papers). A situation like our setting also arises in the allocation of university funds among different schools based on their performance, or need, or government research funds to researchers from various universities.

The underlying ethical nature of the rationing problem rules out the market mechanism or any other traditional mechanism. Thus, one must resort to an axiomatic approach to find a solution to this situation. A solution may be universally acceptable to the agents if it is deemed fair by the agents. The notion of fairness is captured by the axioms used to select a rule. Axioms considered in this chapter belong to the following four categories: Axioms on equity, axioms on incentives, structural invariance axioms, and axioms on uncertainty.

We consider two setups and provide two classes of rules. Under the state-contingent claims setup, we introduce classes of ex-ante and ex-post rules. We provide axiomatic characterizations of these rules.

Under the setup where claims are probability distributions, we introduce the equal-quantile rules and expected-waste-constrained uniform gains rules and we also provide an axiomatic characterization of these rules.

In Sect. 2, we introduce the formal model for the two setups mentioned above. In Sect. 3, we introduce the rules along with their axiomatic characterization. In Sect. 4, we conclude with some open questions.

2 The Setups and Rationing Rules

The rationing problem under uncertain claims can be defined as an extension of the standard rationing problem to a stochastic setting where the claims are uncertain and shares are to be distributed before the uncertainty is resolved. There are essentially two approaches to model claim uncertainty: (1) state-contingent claims and (2) random claims represented by probability distribution functions. First, we will present the state-contingent claims setup defined for the proportional and parametric rules. Next, we follow with the setup where claims are probability distributions and introduce equal-quantile rules along with the expected-waste-constrained uniform gains rules.

2.1 *State-Contingent Claims Setup*

The rationing problem under the state-contingent framework is defined as the tuple $(N, \mathbf{c}, E, \mathbf{p})$. The set of *individuals* is a nonempty finite set N and the set of *states* is a nonempty finite set S. A *profile of a state-contingent claims matrix*, $\mathbf{c} \equiv \langle c_{is} : i \in N; s \in S\rangle$, is a map, $(i, s) \in N \times S \mapsto c_{is} \in \mathbb{R}_+$. The set of profile of state-contingent claims shall be denoted by $\mathscr{C}$. For any individual i, $\mathbf{c}_i \equiv \langle c_{is} : s \in S\rangle$ is *individual* i*'s state-contingent claim vector*, *i.e.*, the map $s \in S \mapsto c_{is} \in \mathbb{R}_+$ as obtained by restriction of the map $\boldsymbol{c}$ to the set $\{i\} \times S$. For any state $s \in S$, $\mathbf{c}_s \equiv \langle c_{is} : i \in N\rangle$ is the *profile of claims in state* s, *i.e.*, the map $i \in N \mapsto c_{is} \in \mathbb{R}_+$ as obtained by restriction of the map $\mathbf{c}$ to the set $N \times \{s\}$. An *estate* is any element of $\mathscr{E}{:}{-}\mathbb{R}_+$ typically denoted by $E, E', \dots$ or $E_1, E_2, \dots$ and so on. An *assessment of state probabilities*, denoted by $\mathbf{p} = \langle p_s : s \in S\rangle$, is a map, $s \in S \mapsto p_s \in [0, 1]$ such that $\sum_{s \in S} p_s = 1$. Thus, the set of assessments of state probabilities is the $|S| - 1$-dimensional simplex $\Delta(S)$, *i.e.*, $\mathbf{p} \in \Delta(S)$. For any $s \in S$, $\boldsymbol{\delta}_s \in \Delta(S)$ shall denote the lottery which is degenerate at the state s. For the map $\mathbf{c}_i$ and $\mathbf{p} \in \Delta(S)$, define $\bar{c}_i(\mathbf{p}){:=}\sum_{s \in S}(p_s \cdot c_{is})$, *i.e.*, the *expected claim of individual* i. We shall follow the convention that for any set K, and $\mathbf{x}, \mathbf{y} \in \mathbb{R}^K$, $\mathbf{x} \geq \mathbf{y} \iff x_k \geq y_k$ for all $k \in K$. Vector of zeroes, $\mathbf{0}_K$, shall denote the map, $k \in K \mapsto 0 \in \mathbb{R}_+$, and vector of ones, $\mathbf{1}_K$, shall denote the map, $k \in K \mapsto 1 \in \mathbb{R}_+$.

For a fixed population N, a *rationing problem under state- contingent claims* is an ordered triple $(\mathbf{c}, E, \mathbf{p}) \in \mathscr{C} \times \mathscr{E} \times \Delta(S)$ such that for every $s \in S$, $\sum_{i \in N} c_{is} \geq E$.

The set of rationing problems, $\mathscr{D}:=\{(\mathbf{c}, E, \mathbf{p}) \in \mathscr{C} \times \mathscr{E} \times \Delta(S) : (\forall s \in S)[\sum_{i\in N} c_{is} \geq E]\}$ shall be called the *domain*. A *rule* is a map, $\phi : \mathscr{D} \rightarrow \mathbb{R}^N_+$, such that, for any $(\mathbf{c}, E, \mathbf{p}) \in \mathscr{D}$, $\sum_{i\in N} \phi_i(\mathbf{c}, E, \mathbf{p}) = E$ where $\phi_i(\mathbf{c}, E, \mathbf{p})$ shall be called the *share* of individual i in the rationing problem $(\mathbf{c}, E, \mathbf{p})$ according to the rule ϕ. Denote by $\mathscr{D}^*$ the set $\{(\mathbf{x}, t) \in \mathbb{R}^N_+ \times \mathbb{R}_+ : \sum_{i\in N} x_i \geq t\}$. We define any rule ϕ to be *ex-ante*, if and only if there exists a corresponding function $\psi : \mathscr{D}^* \rightarrow \mathbb{R}^N_+$ such that $\sum_{i\in N} \psi_i(\mathbf{x}, t) = t$ for any $(\mathbf{x}, t) \in \mathscr{D}^*$, and $\phi(\mathbf{c}, E, \mathbf{p}) = \psi(\bar{c}(\mathbf{p}), E)$ for any $(\mathbf{c}, E, \mathbf{p}) \in \mathscr{D}$ where $\bar{c}(\mathbf{p}):=\langle \bar{c}_i(\mathbf{p}) : i \in N\rangle$. A rule ϕ is defined to be *ex-post*, if and only if, for any $(\mathbf{c}, E, \mathbf{p}) \in \mathscr{D}$, $\phi(\mathbf{c}, E, \mathbf{p}) = \sum_{s\in S}[p_s \cdot \phi(\mathbf{c}, E, \boldsymbol{\delta}_s)]$.

2.2 Proportional Rules

In the state-contingent claims setting, ex-ante and ex-post proportional rules are defined as follows.

The ex-ante proportional rule ($\overline{pr}$) is defined as applying the proportional rule to the expectation of the state-contingent claims:

$$\overline{pr_i}(\mathbf{c}, E, \mathbf{p}) = \frac{\bar{c}_i(\mathbf{p})}{\sum\limits_{i\in N} \bar{c}_i(\mathbf{p})} E, \text{ for all } i \in N.$$

The ex-post proportional rule ($\widetilde{pr}$) is defined as the expectation of the shares found by applying the proportional rule on the state-contingent claims:

$$\widetilde{pr}_i(\mathbf{c}, E, \mathbf{p}) = \sum_{s\in S} \left(p_s \frac{c_{is}}{\sum\limits_{i\in N} c_{is}} \right) E, \text{ for all } i \in N.$$

2.3 Parametric Rationing Rules

In this section, we define a class of rules over the domain $\mathscr{D}$, which shall be denoted by Φ, such that any rule $\phi \in \Phi$ satisfies each of the standard axioms which will be given in the next section. These rules have some similarities with a class of rules studied in Juarez and Kumar [9]. Members of the class Φ are constructed by the composition of rules, from a family contained in the class of Young's "parametric rules", (Young [20]), with a profile of $\mathbb{R}$-valued functions, which map pairs $(\mathbf{c}, \mathbf{p}) \in \mathscr{C} \times \Delta(S)$ satisfying some conditions. Typical profiles shall be denoted by $T \equiv \langle T_i : i \in N\rangle$, and the class of all such profiles shall be denoted by $\mathscr{T}$.

Let $T \equiv \langle T_i : i \in N\rangle \in \mathscr{T}$, if and only if, for every $i \in N$, $T_i : \mathscr{C} \times \Delta(S) \rightarrow \mathbb{R}$ is a map, and, for any $i \in N$ and any $(\mathbf{c}, \mathbf{p}) \in \mathscr{C} \times \Delta(S)$, each of the following conditions hold:

$R.1$ $\sum_{i\in N} T_i(\mathbf{c}, \mathbf{p}) \geq \min_{s\in S} \sum_{i\in N} c_{is}$.
$R.2$ $T_i(\mathbf{c}, \mathbf{p}) \geq \min_{s\in S} c_{is}$.
$R.3$ If $\mathbf{c}'_i > \mathbf{c}_i$, then $T_i((\mathbf{c}'_i, \mathbf{c}_{-i}), \mathbf{p}) > T_i(\mathbf{c}, \mathbf{p})$.

Now, we shall specify a certain subclass of rules inspired by Young's "parametric rules". Let $h \equiv \langle h_i : i \in N\rangle \in \mathscr{H}$, if and only if there exists $\theta_*, \theta^* \in \mathbb{R}$ with $\theta_* < \theta^*$ such that, for every $i \in N$, $h_i : [\theta_*, \theta^*] \times \mathbb{R}_+ \to \mathbb{R}_+$ satisfies each of the following:

$H.1$ For any $x \in \mathbb{R}_+$, $h_i(\theta_*, x) = 0$ and $h_i(\theta^*, x) = x$.
$H.2$ For any $x \in \mathbb{R}_+$, the map $\theta \in [\theta_*, \theta^*] \mapsto h_i(\theta, x)$ is continuous.
$H.3$ For any $x \in \mathbb{R}_+$, the map $\theta \in [\theta_*, \theta^*] \mapsto h_i(\theta, x)$ is strictly increasing.
$H.4$ For any $x, x' \in \mathbb{R}_+$ with $x < x'$ and any $\theta \in [\theta_*, \theta^*]$, $h_i(\theta, x) < h_i(\theta, x')$.

For any $h \in \mathscr{H}$, we define a corresponding map $\psi^h : \mathscr{D}^* \to \mathbb{R}^N_+$ as follows. For every $i \in N$, and for any $(\mathbf{x}, t) \in \mathscr{D}^*$, let $\psi^h_i(\mathbf{x}, t) := h_i(\theta, \min\{x_i, t\})$ where $\theta \in [\theta_*, \theta^*]$ solves $\sum_{i\in N} h_i(\theta, \min\{x_i, t\}) = t$. Set $\psi^h := \langle \psi^h_i : i \in N\rangle$. Observe, for any problem $(\mathbf{x}, t) \in \mathscr{D}^*$, the resulting profile of "truncated claims" $\mathbf{x}^t := \langle \min\{x_i, t\} : i \in N\rangle$ defines some $\theta \in [\theta_*, \theta^*]$ that solves $\sum_{i\in N} h_i(\theta, x^t_i) = t$ by the properties $H.1$ and $H.2$. That such a solution is unique follows from property $F.3$. That is, ψ^h is indeed a rule over the domain $\mathscr{D}^*$ if $h \in \mathscr{H}$. For any $T \in \mathscr{T}$ and any $h \in \mathscr{H}$, the corresponding map $\phi^{h,T} : \mathscr{D} \to \mathbb{R}^N_+$ is defined by

$$\phi^{h,T}(\mathbf{c}, E, \mathbf{p}) := \psi^h(T(\mathbf{c}, \mathbf{p}), E), \text{ for every } (\mathbf{c}, E, \mathbf{p}) \in \mathscr{D}.$$

Let $M \in \mathbb{N}$. For every $\mathbf{p} \in \Delta(S)$, let $\beta_1(\mathbf{p}), \beta_2(\mathbf{p}), \ldots, \beta_M(\mathbf{p}) \in [0, 1]$ such that $\sum_{m=1}^M \beta_m(\mathbf{p}) = 1$. Also, for each $m \in \{1, 2, \ldots, M\}$, let $h_m \in \mathscr{H}$ and $T_m \in \mathscr{T}$. Define $\phi : \mathscr{D} \to \mathbb{R}^N_+$ as follows:

$$\phi(\mathbf{c}, E, \mathbf{p}) := \sum_{m=1}^{M} \beta_m(\mathbf{p}) \cdot \phi^{h_m, T_m}(\mathbf{c}, E, \mathbf{p}), \text{ for every } (\mathbf{c}, E, \mathbf{p}) \in \mathscr{D}.$$

We introduce two particular subclasses of Φ. The first one is a subclass of rules that have the ex-ante form. To see that this is indeed the case, we consider $T \in \mathscr{T}$ as follows. Let $T \equiv \langle T_i : i \in N\rangle$ where, for every individual $i \in N$, the map $T_i : \mathscr{C} \times \Delta(S) \to \mathbb{R}_+$ is defined as $T_i(\mathbf{c}, \mathbf{p}) := \|(\mathbf{c}_i, \mathbf{p})\|_1$. Recall, $\|(\mathbf{c}_i, \mathbf{p})\|_1 = \sum_{s\in S}(p_s \cdot c_{is})$ by definition. Thus, for any choice of $h \in \mathscr{H}$, the resulting rule $\phi^{h,T}$ has the ex-ante form. We shall call this rule the *ex-ante rule defined by* h. Given the definition of the class $\mathscr{H}$, the ex-ante versions of many prominent rules such as Talmud rule, minimal overlap rule, and the family of priority-augmented weighted uniform gains rules studied in Flores-Szwagrzak [6] which includes uniform gains rule and weighted uniform rules are contained in the class Φ. Some rules are, however, not in Φ. For instance, proportional rule, uniform losses rule, reverse Talmud rules (van den Brink et al. [16], van den Brink and Moreno Ternero [17]), and random order of arrival rule do not belong to this class.

Next, we observe that a subclass of rules, having the ex-post form, are also contained in Φ. Define $M := |S|$. For every $s \in S$, we define $T_s \in \mathscr{T}$ as follows.

Fix $s \in S$. Let $T_s \equiv \langle T_{s,i} : i \in N \rangle$, where, for any $i \in N$, $T_{s,i}(\mathbf{c}, \mathbf{p}) := c_{is}$ for every $(\mathbf{c}, \mathbf{p}) \in \mathscr{C} \times \Delta(S)$. Also, for any $s \in S$, define $\beta_s(\mathbf{p}) := p_s$ for every $\mathbf{p} \in \Delta(S)$. Clearly, $\sum_{s \in S} \beta_s(\mathbf{p}) = 1$. Fix any $h \in \mathscr{H}$, and define the map $\phi : \mathscr{D} \to \mathbb{R}^N_+$ as follows:

$$\phi(\mathbf{c}, E, \mathbf{p}) := \sum_{s \in S} \beta_s(\mathbf{p}) \cdot \psi^{h, T_s}(T(\mathbf{c}, \mathbf{p}), E), \text{ for every } (\mathbf{c}, E, \mathbf{p}) \in \mathscr{D}.$$

We shall call this rule the *ex-post rule defined by* h. In particular, the ex-post versions of the Talmud rule, minimal overlap rule, and any member of the priority-augmented weighted uniform gains rules are elements of this class Φ. By the same token, the uniform losses rule and the random order of arrival rule do not belong to this class.

2.4 Claims as Probability Distributions Setup

Let $\mathbb{R}$ be the set of real numbers, $\mathbb{R}_+$ non-negative real numbers, $\mathbb{R}_{++}$ positive real numbers, and $\mathbb{N}$ positive integers. We now move to a changing-population environment. Let $\mathbb{N}$ be the set of potential individuals. A population is a finite subset of $\mathbb{N}$. Let $\mathcal{N}$ be the set of all finite subsets of $\mathbb{N}$.

The claim of an individual is modeled as a probability measure over non-negative real numbers which denote quantities of a resource. It is represented by a cumulative distribution function (CDF), and assumed to have finite interval support. Let $\mathcal{F}$ be the set of such claims. A typical claim is denoted by F_i and its support by $[c_i, C_i]$. For each $N \in \mathcal{N}$, a problem for population N is a pair $(F, E) \in \mathcal{F}^N \times \mathbb{R}_+$ where $F = (F_i)_{i \in N}$ is a profile of individual claims and E is an endowment of the resource. A rule is a function that specifies for each $N \in \mathcal{N}$ and each $(F, E) \in \mathcal{F}^N \times \mathbb{R}_+$ an allocation $\phi(F, E) \in \mathbb{R}^N_+$ such that for each $i \in N$, $\phi_i(F, E) \leq C_i$, and $\sum \phi_i(F, E) \leq E$.

For each $F_i \in \mathcal{F}$, a value in its support is interpreted as a realized need of individual i, with c_i being his sure need and C_i his maximal need. The need of an individual is understood as his satiation point. That is, when his assignment is less than his need, he is better off from getting more of the resource, and when his assignment exceeds his need, he is indifferent to any increase of the resource. Thus, when individual i is assigned t_i units of the resource and his realized need is x_i, $\max\{t_i - x_i, 0\}$ is the induced waste and $\max\{x_i - t_i, 0\}$ his deficit.

The current setup differs from the previous one in two aspects. First, in the previous setup, uncertainty is modeled in terms of an exogenous state space and a probability measure on the states, and a claim is modeled as a state-contingent non-negative real vector. The advantage of modeling uncertainty in this way is to capture the underlying joint distribution of agents' claims. In contrast, the current setup only contains the data of the marginal distributions. Therefore, it applies only to the cases in which agents' claims are subject to idiosyncratic risk or the joint distribution is irrelevant to resource allocation.[2]

Second, in the previous setup, the endowment is assumed to be no larger than the sum of individual claims and must be fully allocated, whereas in the current setup, it can be arbitrarily small or large and is only an upper bound on the amount of the resource that can be allocated to the individuals. In comparison, the current setup accommodates the case where full use of the resource may not be efficient due to an opportunity cost brought about by the claim uncertainty (Long, Sethuraman, and Xue [11]).

2.5 Equal-Quantile Rules

Long, Sethuraman, and Xue introduce the class of equal-quantile rules. Each rule in this class is parameterized by a number $\lambda \in (0, 1]$. When the endowment is no larger than the sum of individuals' sure needs, the equal-quantile rule with λ fully allocates the endowment by applying the uniform gains rule to the profile of sure needs. When the endowment exceeds the sum of individuals' sure needs, roughly speaking, the rule assigns to each individual an amount such that all individuals have the same probability of having a need no more than the assignment. Moreover, the common probability of satiation is maximized under the constraint that it does not exceed λ.

Formally, for each $F_i \in \mathcal{F}$, define the quantile function $Q_{F_i} : (0, 1] \to \mathbb{R}$ by setting for each $\alpha \in (0, 1]$,

$$Q_{F_i}(\alpha) := \min\{x_i \in \mathbb{R} : F_i(x_i) \geq \alpha\}.$$

Then for each $\lambda \in (0, 1]$, the **equal-quantile rule with parameter λ**, denoted by ϕ^λ, is defined as follows. For each $N \in \mathcal{N}$ and each $(F, E) \in \mathcal{F}^N \times \mathbb{R}_+$, when $E \leq \sum c_i$, for each $j \in N$,

$$\phi_j^\lambda(F, E) := \min\{c^*, c_j\}, \text{ where } c^* \in \mathbb{R}_+ \text{ satisfies } \sum \min\{c^*, c_i\} = E;$$

when $E > \sum c_i$, for each $j \in N$,

$$\phi_j^\lambda(F, E) := Q_{F_j}(\alpha^*), \text{ where } \alpha^* \in (0, \lambda] \text{ satisfies } \sum Q_{F_i}(\alpha^*) = \min\left\{E, \sum Q_{F_i}(\lambda)\right\}.$$

The parameter λ determines when the endowment should be fully allocated to the individuals. In particular, $Q_{F_i}(\lambda)$ is the maximal amount of the resource that can be assigned to an individual who claims F_i. The endowment is fully allocated whenever it does not exceed the sum of individuals' maximal assignments. Otherwise, each

---

[2] See Long, Sethuraman, and Xue [11] for more discussion.

individual just receives his maximal assignment, and the endowment is only partially allocated.

2.6 Expected-Waste-Constrained Uniform Gains Rules

Xue [19] introduces the class of expected-waste uniform gains rules. To understand this class, imagine that an assignment to an individual induces two components of cost to a society: the resource assigned to him and the expected waste generated by him. Each expected-waste-constrained uniform gains rule is associated with a cost function that aggregates the two components and allocates the resource in a way that makes the costs of individuals as equal as possible. More precisely, for each problem, the rule would set a common cost. If assigning the maximal need to an individual induces a cost smaller than the common cost, then the individual receives his maximal need. Otherwise, he receives the amount that induces exactly the common cost. The common cost is set by the binding feasibility constraint.

Relaxing the interval support assumption, assume now that a claim is a probability measure over $\mathbb{R}_+$ that has a compact support. As in the previous setup, claims are represented by CDFs. Let $\mathcal{F}'$ be the set of all such claims. For each $F_i \in \mathcal{F}'$, let c_i and C_i denote, respectively, the minimal value and the maximal value in the support of F_i. For each $N \in \mathcal{N}$, a problem for population N is a pair (F, E) where $F \in \mathcal{F}'^N$ is a claim profile and $E \in [0, \sum C_i]$ is an endowment. Let $\mathcal{P}^N$ be the set of all problems for population N. For each $N \in \mathcal{N}$ and each $(F, E) \in \mathcal{P}^N$, an allocation is a vector $t \in \mathbb{R}^N_+$ such that for each $i \in N$, $t_i \leq C_i$, and $\sum t_i = E$. Note that in this setup, the endowment is always assumed to fall short of the sum of the maximal needs and is required to be fully allocated. A rule is a function ϕ that specifies for each population $N \in \mathcal{N}$ and each problem $(F, E) \in \mathcal{P}^N$ an allocation.

Formally, let $D := \{(t_i, \int_0^{t_i}(t_i - x_i)dF_i(x_i)) : F_i \in \mathcal{F}', t_i \in [0, C_i]\}$. It can be readily seen that equivalently, $D = \{(x_1, x_2) \in \mathbb{R}^2_+ : x_1 > x_2\} \cup \{(0, 0)\}$. A cost function is a continuous function $U : D \to \mathbb{R}$ satisfying that (i) for each pair $(x_1, x_2), (x_1', x_2') \in D$ such that $x_1 < x_1'$ and either $x_2 < x_2'$ or $x_2 = x_2' = 0$, $U(x_1, x_2) < U(x_1', x_2')$, and (ii) $U(0, 0) = 0$. Let $\mathcal{U}$ be the set of all cost functions. For each $F_i \in \mathcal{F}'$ and each $t_i \in [0, C_i]$, let $w(F_i, t_i) := \int_0^{t_i}(t_i - x_i)dF_i(x_i)$ denote the expected waste induced by an individual who has a claim F_i and is assigned t_i. For each $U \in \mathcal{U}$ and each $F_i \in \mathcal{F}'$, define $U_{F_i}^{-1} : U(D) \to [0, C_i]$ by setting for each $c \in U(D)$,

$$U_{F_i}^{-1}(c) := \begin{cases} C_i & c > U(C_i, w(F_i, C_i)) \\ t_i & c = U(t_i, w(F_i, t_i)) \text{ where } t_i \in [0, C_i] \end{cases}.$$

Notice that $U_{F_i}^{-1}$ is continuous and increasing on $[0, C_i]$. Then for each $U \in \mathcal{U}$, the **expected-waste-constrained uniform gains rule with U**, denoted by ϕ^U, is defined as follows. For each $N \in \mathcal{N}$, each $(F, E) \in \mathcal{P}^N$, and each $i \in N$,

$$\phi_i^U(F, E) = U_{F_i}^{-1}(c^*), \text{ where } c^* \text{ satisfies } \sum U_{F_i}^{-1}(c^*) = E.$$

3 Axiomatic Characterizations

3.1 Proportional Rules

Before giving the characterization of the parametric rules defined in the previous section, we focus our attention to the most prominent member of the parametric rules family, that is, proportional rules. As the proportional rule doesn't satisfy independence of claim truncation axiom, it is not included in our parametric rules domain. Accordingly, we provide a separate characterization for these rules in the rich domain.[3] For a more detailed account of proportional rules under uncertainty, the reader is referred to Ertemel and Kumar [5].

First, we will provide the list of axioms that will be used in the characterization of the ex-ante proportional rules.

Continuity: For all $(\mathbf{c}, E, \mathbf{p})) \in \mathcal{D}$ and for all sequences $(\mathbf{c}^k, E^k, \mathbf{p}^k) \in \mathcal{D}$, if $(\mathbf{c}^k, E^k, \mathbf{p}^k) \to (\mathbf{c}, E, \mathbf{p})$, then $\phi(\mathbf{c}^k, E^k, \mathbf{p}^k) \to \phi(\mathbf{c}, E, \mathbf{p})$.

Continuity tells us that small changes in the parameters of the problem do not bring big jumps in the allocations. Continuity is desirable because we do not want small errors (e.g., measurement errors) to lead to big changes in the allocations.

No award for null: For all $(\mathbf{c}, E, \mathbf{p}) \in \mathcal{D}$ and for all $i \in N$, if $c_{is} = 0$ for all $s \in S$, then $\phi_i(\mathbf{c}, E, \mathbf{p}) = 0$.

No award for null says that an individual with zero claim for each state should get zero share. This axiom is also called the dummy axiom in the literature.

Moulin [12] defined a non-advantageous reallocation axiom to characterize the egalitarian and utilitarian solutions in quasi-linear social choice problems. We will define two axioms on invariance to reallocation in a similar manner where transfers are made either across individuals or across states.

Non-advantageous reallocation across individuals: For all $(\mathbf{c}, E, \mathbf{p}), (\mathbf{c}', E, \mathbf{p}) \in \mathcal{D}$ and for all $i \in N$, if $\sum\limits_{j \in N \setminus \{i\}} c_{js} = \sum\limits_{j \in N \setminus \{i\}} c'_{js}$ and $c_{is} = c'_{is}$ for all $s \in S$, then $\phi_i(\mathbf{c}, E, \mathbf{p}) = \phi_i(\mathbf{c}', E, \mathbf{p})$.

[3] A domain $\mathcal{D}$ is rich if for all $(\mathbf{c}, E, \mathbf{p}), (\mathbf{c}', E, \mathbf{p}) \in \mathscr{C} \times \mathscr{E} \times \Delta(S)$ if $\sum\limits_{i \in N} c_{is} = \sum\limits_{i \in N} c'_{is}$ for all $s \in S$ then $(\mathbf{c}, E, \mathbf{p}) \in \mathcal{D}$ implies $(\mathbf{c}', E, \mathbf{p}) \in \mathcal{D}$.

Non-advantageous reallocation across individuals states that the share of individual i depends on the sum of the claims of the individuals other than himself. In other words, individuals other than i cannot affect the share of i by reallocating their claims among themselves.

Non-advantageous reallocation across states: For all $(\mathbf{c}, E, \mathbf{p}), (\mathbf{c}', E, \mathbf{p}) \in \mathcal{D}$ and for all $i \in N$, if $\bar{c}_i(\mathbf{p}) = \bar{c}'_i(\mathbf{p})$ and $c_{js} = c'_{js}$ for all $j \in N \backslash \{i\}$ and for all $s \in S$, then $\phi_j(\mathbf{c}, E, \mathbf{p}) = \phi_j(\mathbf{c}', E, \mathbf{p})$ for all $j \in N \backslash \{i\}$.

Non-advantageous reallocation across states implies that if individual i reallocates his claim across all the states given his expected claim is constant, then other individuals' share (hence his own share) would not change.

The theorem below provides a characterization for ex-ante proportional rules.

Theorem 1 (Ertemel and Kumar [5]) *Let* $|N| \geq 3$ *and* $|S| \geq 3$. *A rationing rule* ϕ *satisfies Continuity, No award for null, Non-advantageous reallocation across individuals, and Non-advantageous reallocation across states if and only if* ϕ *is ex-ante proportional rule.*

Now we will characterize the ex-post proportional rule. The functional form of ex-post proportional rule is additively separable with respect to the states. This is similar to the *expected utility form* due to von Neumann and Morgenstern [18]. Therefore, in the spirit of Expected Utility Theory, we will utilize the *Independence* axiom which is defined below.

Independence: For all $(\mathbf{c}, E, \mathbf{p}), (\mathbf{c}, E, \mathbf{q}), (\mathbf{c}, E, \mathbf{r}) \in \mathcal{D}$, for all $i \in N$, and for all $\lambda \in (0, 1)$, we have $\phi_i(\mathbf{c}, E, \mathbf{p}) \geq \phi_i(\mathbf{c}, E, \mathbf{q})$ if and only if $\phi_i(\mathbf{c}, E, \lambda\mathbf{p} + (1 - \lambda)\mathbf{r}) \geq \phi_i(\mathbf{c}, E, \lambda\mathbf{q} + (1 - \lambda)\mathbf{r})$.

Independence implies that the ordering of an individual's share with respect to two different state probabilities is preserved if these two state probabilities are mixed with any other state probability.

We further introduce *Symmetry* axiom which implies that the names of the states do not matter. This is a very natural axiom and is central to the literature on fairness.

Symmetry: For all $(\mathbf{c}, E, \mathbf{p}) \in \mathcal{D}$, for all permutations $\rho : S \to S$, and for all $i \in N$, $\phi_i(\mathbf{c}, E, \mathbf{p}) = \phi_i(\mathbf{c}_\rho, E, \mathbf{p}_\rho)$, where $\mathbf{p}_\rho = (p_{\rho(1)}, p_{\rho(2)}, ..., p_{\rho(|S|)})$ and $\mathbf{c}_\rho = (c_{\rho(1)}, c_{\rho(2)}, ..., c_{\rho(|S|)})$.

By keeping all the axioms of Theorem 1 and replacing Non-advantageous reallocation across states axiom with the Independence axiom and adding Symmetry axiom, we get characterization of the ex-post proportional rule in Theorem 2.

Theorem 2 (Ertemel and Kumar [5]) *Let $|N| \geq 3$. A rationing rule ϕ satisfies Continuity, No award for null, Non-advantageous reallocation across individuals, Independence, and Symmetry if and only if ϕ is an ex-post proportional rule.*

3.2 Parametric Rules

Now we move to our family of parametric rules with the corresponding axioms given below. For more details on these rules, the reader is referred to Chatterjee, Ertemel, and Kumar [3].

Claim Monotonicity: If $(\mathbf{c}, E, \mathbf{p})$, $((\mathbf{c}_i', \mathbf{c}_{-i}), E, \mathbf{p}) \in \mathscr{D}$ and $i \in N$ such that $\mathbf{c}_i \leq \mathbf{c}_i'$, then $\phi_i(\mathbf{c}, E, \mathbf{p}) \leq \phi_i((\mathbf{c}_i', \mathbf{c}_{-i}), E, \mathbf{p})$.

The above axiom is the most obvious adaptation of the "claim monotonicity" axiom from the rationing literature with deterministic claims. Consider two profiles of state-contingent claims such that the profiles differ only in one specific individual's state-contingent claim and that too in the sense that the claims, of the individual, in the second profile are greater than that in the first state-wise. The rule satisfies the axiom, if and only if it provides that individual at least as much in the second profile as it does in the first. The "claim monotonicity" property, in the deterministic setting, holds for many major rules such as priority-augmented weighted uniform gains rules, the Talmud rule, the proportional rule, and so on. However, formally there exist several rules which do not satisfy this property as can be seen from the definitions of two particularly wide classes of rules which are the class of *fixed path rules* and Young's class of *parametric rules*. The first class is important in the characterization of the rules that satisfy the property of "independence of irrelevant claims" while the second class characterizes the property of "consistency". In particular, both the classes admit as nonempty proper subclasses of rules that either do satisfy "claim monotonicity" or do not.

Weak Consistency: If $(\mathbf{c}, E, \mathbf{p})$, $((\mathbf{c}_i', \mathbf{c}_{-i}), E, \mathbf{p}) \in \mathscr{D}$ and $i \in N$, then $\sum_{j \in N \setminus \{i\}} \phi_j (\mathbf{c}, E, \mathbf{p}) = \sum_{j \in N \setminus \{i\}} \phi_j((\mathbf{c}_i', \mathbf{c}_{-i}), E, \mathbf{p})$ implies $\phi_j(\mathbf{c}, E, \mathbf{p}) = \phi_j((\mathbf{c}_i', \mathbf{c}_{-i}), E, \mathbf{p})$ for every $j \in N \setminus \{i\}$.

The property called "consistency" is of wide appeal in the rationing literature with deterministic claims. Many of the major rules satisfy the property of "consistency". For instance, the class of priority-augmented weighted uniform gains rules, the proportional rule, and the Talmud rule. In fact, every rule in Young's class of *parametric rules* satisfies "consistency". However, the random order of arrival rule does not satisfy this property. To briefly recall the idea of "consistency", consider the shares computed by the rule for a problem involving some set of individuals. Next, a group of some of the individuals leaves having obtained their shares. The property demands that the rule allocate the same shares from the sum of the shares of the remaining individuals as it had computed initially. The above axiom is not a direct adaptation of the "consistency" property to the setting involving uncertainty.

No Reward for More Irrelevant Claims: If $(\mathbf{c}, E, \mathbf{p}) \in \mathscr{D}$ and $i \in N$ such that $\min_{s \in S} c_{is} \geq E$, then $\delta\mathbf{c}_i \geq \mathbf{0}_S$ implies $\phi_i(\mathbf{c}, E, \mathbf{p}) \geq \phi_i((\mathbf{c}_i + \delta\mathbf{c}_i, \mathbf{c}_{-i}), E, \mathbf{p})$.

The above axiom is an adaptation to the setting with an uncertainty of a slight weakening of the "independence of irrelevant claims" property which is also known as "truncation of irrelevant claims". The idea is, given individual claims and a resource, any claim matters only as long as it does not exceed the resource. If an individual's claim does exceed the resource, then her claim is "truncated" in the sense that the rule considers her claim to be the level of the resource itself. Many important rules satisfy "truncation of irrelevant claims".

First, we present the axioms that are relevant in the characterization theorems of rules that have the ex-ante form. The following two axioms below describe how uncertainty inherent in claims may be treated.

No Penalty for Risk: If $(\mathbf{c}, E, \mathbf{p}), ((\mathbf{c}'_i, \mathbf{c}_{-i}), E, \mathbf{p}) \in \mathscr{D}$ and $i \in N$ such that $\mathbf{c}'_i = \bar{c}_i(\mathbf{p}) \cdot \mathbf{1}_S$, then $\phi_i(\mathbf{c}, E, \mathbf{p}) \geq \phi_i((\mathbf{c}'_i, \mathbf{c}_{-i}), E, \mathbf{p})$.

To see the interpretation of the above axiom, fix the claims of every other individual and consider two state-contingent claims of the individual that differ only in that the first claim is "risky" while the second is not. In particular, the first claim is a mean-preserving spread of the second. Note that the notion of "riskier claim" is equivalent to second-order stochastic dominance which in turn is much weaker than the notion of mean-preserving spread. This is so as second-order stochastic dominance is obtained by any *sequence* of mean-preserving spreads.

No Sudden Response to Uncertainty: If $\mathbf{c}, \mathbf{c}' \in \mathscr{C}$ and $E^\dagger \in \mathbb{R}_{++}$ such that $\mathbf{c}'_i = \bar{c}_i(\mathbf{p}) \cdot \mathbf{1}_S$ for every $i \in N$, and $(\mathbf{c}, E^\dagger, \mathbf{p}), (\mathbf{c}', E^\dagger, \mathbf{p}) \in \mathscr{D}$, then $\phi(\mathbf{c}, E, \mathbf{p}) = \phi(\mathbf{c}', E, \mathbf{p})$ for every $E \leq E^\dagger$ implies that $\phi(\mathbf{c}, E, \mathbf{p}) = \phi(\mathbf{c}', E, \mathbf{p})$ for any E in some neighborhood of $E^\dagger$.

Consider two state-contingent profiles of individual claims as follows. The second of the two is the "equivalent deterministic profile" in that every individual's claim is the same across every state of nature equalling the risk-free mean claim. Now, suppose it is a case that the two profiles are treated identically by the rule in that every individual gets the same in both the profiles under the rule as long as the resource is up to some strictly positive level of the resource. Then the rule satisfies the above axiom, if and only if the rule continues to process the two profiles identically in some neighborhood of that level.

Now we state characterization theorems for the parametric rules. We begin with Theorem 3 which states that the class of parametric rules Φ satisfy the three axioms:

claim monotonicity, *weak consistency*, and *no reward for more irrelevant claims*.

Theorem 3 (Chatterjee, Ertemel, and Kumar [3]) *Any rule $\phi \in \Phi$ satisfies claim monotonicity, weak consistency, and no reward for more irrelevant claims.*

Within the class of rules that satisfy the axioms of *claim monotonicity*, *weak consistency*, and *no reward for more irrelevant claims*, Theorem 4 is the main result that characterizes those rules that have the ex-ante form. Since the class Φ is a subclass of such rules according to Theorem 3, we conclude that the ex-ante rules characterization below pins down the critical feature about how any rule processes the inherent riskiness that manifests in the state-contingent claims of the individuals.

Theorem 4 (Chatterjee, Ertemel, and Kumar [3]) *Consider any rule that satisfies claim monotonicity, weak consistency, and no reward for more irrelevant claims. The rule satisfies no penalty for risk and no sudden response to uncertainty, if and only if it is an ex-ante rule.*

In the light of Theorems 3 and 4, the following corollary is immediate.

Corollary 1 (Chatterjee, Ertemel, and Kumar [3]) *Any rule $\phi \in \Phi$ has the ex-ante form, if and only if ϕ satisfies no penalty for risk and no sudden response to uncertainty.*

Now, we present the only axiom that characterizes the ex-post form of rules. Before defining our main axiom, *Indifference to independent combinations*, we give the following preliminaries for the combination of games of rationing problems.

For any $K \in \mathbb{N}$, let $\pi_1, \pi_2, \ldots, \pi_K \in [0, 1]$ such that $\sum_{k=1}^{K} \pi_k = 1$, and consider any K rationing problems $(\mathbf{c}_1, E_1, \mathbf{p}_1), \ldots, (\mathbf{c}_K, E_K, \mathbf{p}_K) \in \mathscr{D}$. Then $[\bigoplus_{k=1}^{K} \pi_k \centerdot (\mathbf{c}_K, E_K, \mathbf{p}_k)]$ shall denote the *gamble* with outcomes $(\mathbf{c}_1, E_1, \mathbf{p}_1), \ldots, (\mathbf{c}_K, E_K, \mathbf{p}_K)$ in $\mathscr{D}$ having probabilities $\pi_1, \ldots, \pi_K$, respectively. The gamble $[\bigoplus_{k=1}^{K} \pi_k \centerdot (\mathbf{c}_K, E_K, \mathbf{p}_k)]$, under the rule ϕ, induces the *money lottery* $[\bigoplus_{k=1}^{K} \pi_k \centerdot \phi_i(\mathbf{c}_k, E_k, \mathbf{p}_k)]$ for each individual i. For any $M \in \mathbb{N}$, let $\mu_1, \ldots, \mu_M \in \Delta(\mathscr{D})$, let $\bigotimes_{m=1}^{M} \mu_m$ denote *independent combination of gambles*. Let μ_m be the gamble $[\bigoplus_{k=1}^{K_m} \pi_k^m \centerdot (\mathbf{c}_k^m, E_k^m, \mathbf{p}_k^m)]$ for each $m \in \{1, \ldots, M\}$. Then the independent combination of gambles $\bigotimes_{m=1}^{M} \mu_m$ gives each individual i the share $\sum_{m=1}^{M} \phi_i(\mathbf{c}_{k_m}^m, E_{k_m}^m, \mathbf{p}_{k_m}^m)$ with probability $\prod_{m=1}^{M} \pi_{k_m}^m$. The class of all independent combinations of gambles shall be denoted by $\mathscr{I}_{\mathscr{D}}$. For each individual i, driven by von Neumann–Morgenstern preferences over money lotteries, let $\succsim_i$ be the complete and transitive binary relation over $\mathscr{I}_{\mathscr{D}}$.

Indifference to Independent Combinations: If $(\mathbf{c}, E, \mathbf{p}) \in \mathscr{D}$ and $\mathbf{q} \in \Delta(S)$, then $\bigotimes_{s \in S}[p_s \centerdot (\mathbf{c}, E, \boldsymbol{\delta}_s) \bigoplus (1 - p_s) \centerdot (\mathbf{c}, 0, \mathbf{q})] \sim_i (\mathbf{c}, E, \mathbf{p})$.

Indifference to Independent Combination axiom forces a risk neutral individual to regard independent combination of the "ex-post" problem $(\mathbf{c}, E, \boldsymbol{\delta}_s)$ with probability p_s and zero estate problem $(\mathbf{c}, 0, \mathbf{p})$ with probability $1 - p_s$ equivalent to the original rationing problem $(\mathbf{c}, E, \mathbf{p})$.

Theorem 5 (Chatterjee, Ertemel, and Kumar [3]) *The rule ϕ has the ex-post form, if and only if the induced $\succsim_i$ satisfies indifference to independent combinations for every risk neutral $i \in N$.*

Here, we point out that the theorem above is not a mere statement of definitional equivalence of two notions: "ex-post rule" and "Indifference to Independent Combinations". A rule is required to make an individual indifferent between the original problem and its corresponding "independent combination of gambles" version *only if* that individual is risk-neutral. The rule is silent about how an individual compares these two versions of a problem if he is *not* risk-neutral. Further, it is *not* required that at least one or more individuals are risk-neutral.

Theorems 4 and 5 provide a very top-level characterization of the ex-ante and the ex-post forms of rules to resolve rationing problems. One way to think of the relevance of these theorems is that whenever a rule—or a class of rules—from the standard rationing literature in the "deterministic" setting is considered for extension to the setting involving "uncertainty", then standard characterizations of the corresponding "deterministic" versions are adaptable to characterizations of the corresponding "ex-ante" and "ex- post" forms. The key idea is whether the rule being proposed should be chosen so as to satisfy either the property of *no penalty for risk* or that of *indifference to independence combinations*. Within the class Φ, these two choices are not compatible.

3.3 *Equal-Quantile Rules*

The class of equal-quantile rules admits an axiomatic justification. All the axioms are extensions, studied by Long, Sethuraman, and Xue [11], of their counterparts in the deterministic claims problems.

Symmetry says that individuals who have equal claims should receive equal amounts.

Symmetry: For each $N \in \mathcal{N}$, each $(F, E) \in \mathcal{F}^N \times \mathbb{R}_+$, and each pair $i, j \in N$, if $F_i = F_j$, then $\phi_i(F, E) = \phi_j(F, E)$.

Ranking says that if an individual has no smaller claim than another individual in the sense of first-order stochastic dominance (FSD), then the former should receive a no smaller amount than the latter. For each pair $F_i, F_j \in \mathcal{F}$, F_i is no smaller than F_j in the FSD sense, denoted by $F_i \succsim_{FSD} F_j$, if for each $c \in \mathbb{R}$, $F_i(c) \leq F_j(c)$.

Ranking: For each $N \in \mathcal{N}$, each $(F, E) \in \mathcal{F}^N \times \mathbb{R}_+$, and each pair $i, j \in N$, if $F_i \succsim_{FSD} F_j$, then $\phi_i(F, E) \geq \phi_j(F, E)$.

Strict ranking says that if an individual has a larger claim than another individual in a strict FSD sense, then the former should receive a larger amount than the latter. For each pair $F_i, F_j \in \mathcal{F}$, F_i is larger than F_j in the strict FSD sense, denoted by $F_i \succ_{FSD} F_j$, if for each $c \in [0, C_j]$, $F_i(c) < F_j(c)$, or equivalently, if $C_i > C_j$ and for each $c \in [0, C_i)$, $F_i(c) < F_j(c)$.

Strict ranking: For each $N \in \mathcal{N}$, each $(F, E) \in \mathcal{F}^N \times \mathbb{R}_+$, and each pair $i, j \in N$, if $F_i \succ_{FSD} F_j$ and $E > 0$, then $\phi_i(F, E) > \phi_j(F, E)$.

Continuity says that if the data of a problem does not change too much, the allocation should not change too much. Here, the topology adopted to evaluate changes of a claim is based on the notion of weak convergence of CDFs and the convergence of sure needs and maximal needs.

Continuity: For each $N \in \mathcal{N}$, each $(F, E) \in \mathcal{F}^N \times \mathbb{R}_+$, and each sequence $\{(F^n, E^n)\}_{n=1}^{\infty}$ of elements of $\mathcal{F}^N$, if for each $i \in N$, F_i^n converges weakly to F_i, $\lim c_i^n = c_i$, $\lim C_i^n = C_i$, and $\lim E^n = E$, then $\lim \phi(F^n, E^n) = \phi(F, E)$.

Endowment continuity requires only that when the endowment does not change too much, the allocation does not change too much.

Endowment continuity: For each $N \in \mathcal{N}$, each $(F, E) \in \mathcal{F}^N \times \mathbb{R}_+$, and each sequence $\{(F, E^n)\}_{n=1}^{\infty}$ of elements of $\mathcal{F}^N \times \mathbb{R}_+$, if $\lim E^n = E$, then $\lim \phi(F, E^n) = \phi(F, E)$.

Consistency says that after an allocation has been chosen for a problem and some individuals leave with their assignments, if the rest of the endowment is redivided among the remaining individuals, then each of them should receive the same amount as initially. Since in the current setup, the resource is not required to be fully allocated, depending on whether the unassigned resource has been disposed of or not, what is left for the remaining individuals is either the sum of their assignments or the difference between the endowment and the sum of the assignments taken away by those who leave. Here, *consistency* requires the invariance of the remaining individuals' assignments in both cases.

Consistency: For each $N \in \mathcal{N}$, each $(F, E) \in \mathcal{F}^N \times \mathbb{R}_+$, and each $N' \subseteq N$, $\phi_{N'}(F, E) = \phi(F_{N'}, \sum_{j \in N'} \phi_j(F, E)) = \phi(F_{N'}, E - \sum_{j \in N \setminus N'} \phi_j(F, E))$, where $\phi_{N'}(F, E)$ and $F_{N'}$ are, respectively, the restrictions of $\phi(F, E)$ and F onto N'.

Lastly, *ordinality* says that if a problem is "transformed", the allocation should be "transformed" accordingly. Precisely, a transformation is a function from $\mathbb{R}_+$ to $\mathbb{R}_+$.

Let $\mathcal{T}$ be the set of all increasing and continuous transformations. For each $F_i \in \mathcal{F}$ and each $T \in \mathcal{T}$, define $F_i^T : \mathbb{R} \to [0, 1]$ by setting for each $x_i \in \mathbb{R}$,

$$F_i^T(x_i) := \begin{cases} 0 & x_i \in (-\infty, T(0)) \\ F_i(T^{-1}(x_i)) & x_i \in T([0, \infty)) \\ 1 & x_i \in [\lim_{x\to\infty} T(x), \infty) \end{cases}.$$

It can be readily seen that $F_i^T \in \mathcal{F}$. For each $N \in \mathcal{N}$, each $F \in \mathcal{F}^N$, and each $T \in \mathcal{T}$, let F^T denote the transformed claim profile in $\mathcal{F}^N$, namely for each $i \in N$, $(F^T)_i = F_i^T$.

Ordinality: For each $I \in \mathcal{N}$, each $(F, E) \in \mathcal{F}^N \times \mathbb{R}_+$, each $T \in \mathcal{T}$, and each $i \in N$, $\phi_i(F^\phi, \sum T(\phi_j(F, E))) = T(\phi_i(F, E))$.

It turns out that as long as a rule satisfies some basic axioms, it must associate each claim with a maximal assignment. Moreover, the resource is fully allocated whenever it does not exceed the sum of the maximal assignments, and otherwise, each individual just receives his maximal assignment.

Theorem 6 (Long, Sethuraman, and Xue [11]) *Let r be symmetric, endowment-continuous, and consistent. There is $M : \mathcal{F} \to \mathbb{R}_+$ such that for each $N \in \mathcal{N}$, each $(F, E) \in \mathcal{F}^N \times \mathbb{R}_+$, and each $i \in N$, (1) $E < \sum M(F_j) \Rightarrow \sum r_j(F, E) = E$ and $r_i(F, E) \leq M(F_i)$, and (2) $E \geq \sum M(F_j) \Rightarrow r_i(F, E) = M(F_i)$.*

With more axioms being imposed, *symmetry* becomes redundant, and the class of equal-quantile rules is characterized.

Theorem 7 (Long, Sethuraman, and Xue [11]) *A rule satisfies strict ranking, continuity, consistency, and ordinality if and only if it is an equal-quantile rule.*

The class of equal-quantile rules not only admits an axiomatic justification but also is optimal with respect to a utilitarian social welfare function. Imagine that each individual obtains a common and constant marginal utility $u > 0$ from each unit of the assigned resource that does not exceed his realized need. Thus, if an individual is assigned t_i and his realized need is x_i, then his utility is $u \cdot \min\{x_i, t_i\}$. Imagine also that there is an alternative way of using the resource outside the model, which could be thought of as an outside individual. The outside individual obtains a constant marginal utility $v \in [0, u)$ for each unit of the unassigned resource. For each $N \in \mathcal{N}$ and each $(F, E) \in \mathcal{F}^N \times \mathbb{R}_+$, a utilitarian planner who cares about the welfare of all individuals including the outside individual sums up their utilities and chooses an allocation t to maximize

$$\int \left[\sum u \min\{x_i, t_i\} + v(T - \sum t_i) \right] dF. \tag{1}$$

Note that with abuse of notation, F in (1) denotes a joint distribution of individuals' needs whose marginal distributions are their claims. The choice of such a joint distribution can be arbitrary since the value of (1) depends only on the marginal distributions.

Proposition 1 (Long, Sethuraman, and Xue [11]) *Let $u, v \in \mathbb{R}_+$ and $\lambda \in (0, 1]$ be such that $u > v$ and $\lambda = \frac{u-v}{u}$. For each $N \in \mathcal{N}$ and each $(F, E) \in \mathcal{F}^N \times \mathbb{R}_+$, $\phi^\lambda(F, E)$ maximizes the utilitarian social welfare function (1).*

Alternatively, each equal-quantile rule minimizes a utilitarian social cost function. Imagine that each unit of waste generated by an individual incurs a constant marginal cost $c^w \geq 0$ and each unit of deficit borne by an individual incurs a constant marginal cost $c^d > 0$. Then the sum of the individual costs is

$$c^w \sum \int_{x_i < t_i} (t_i - x_i) dF_i + c^d \sum \int_{x_i > t_i} (x_i - t_i) dF_i. \tag{2}$$

Proposition 2 (Long, Sethuraman, and Xue [11]) *Let $c^w \geq 0$, $c^d > 0$, and $\lambda \in (0, 1]$ be such that $\lambda = \frac{c^d}{c^w + c^d}$. For each $N \in \mathbb{N}$ and each $(F, E) \in \mathcal{F}^N \times \mathbb{R}_+$, $\phi^\lambda(F, E)$ minimizes the utilitarian social cost function (2).*

The two propositions provide respective welfare interpretations for the parameter of an equal-quantile rule. It is the optimal upper bound on the probability of satiation, which is determined by the trade-off between the expected utilities obtained by the individuals within the model and by the outside individual, or the trade-off between waste and deficit. Long, Sethuraman, and Xue [11] further establish a link between the two utilitarian objective functions by showing that the waste and deficit cost can be viewed as an opportunity cost generated by resource allocation uncertainty.

3.4 Expected-Waste-Constrained Uniform Gains Rule

Xue [19] proposes three new axioms that address, in particular, how to deal with the issue of waste in resource allocation under uncertainty.

The first is *no domination*. It says that no agent should dominate another agent by receiving more of the resource and generating more expected waste, unless the second agent is fully satiated.

No domination: For each $N \in \mathcal{N}$, each $(F, E) \in \mathcal{P}^N$, and each pair $i, j \in N$, if $\phi_i(F, E) > \phi_j(F, E)$ and $w(F_i, \phi_i(F, E)) > w(F_j, \phi_j(F, E))$, then $\phi_j(F, E) = C_j$.

The second is *risk aversion*. It says that in each problem with two individuals, if the claim of one of them is riskier than that of the other, then the former should receive no larger amount than the latter. For each pair $F_i, F_j \in \mathcal{F}$, F_i is riskier than F_j if they have the same mean, for each $c \in \mathbb{R}$,

$$\int_{-\infty}^{c} F_i(x_i)dx_i \geq \int_{-\infty}^{c} F_j(x_j)dx_j, \tag{3}$$

and the inequality (3) is strict at some $c \in \mathbb{R}$. It can be readily seen that if F_i is riskier than F_j and $C_i = C_j$, then the expected waste induced by assigning an amount to individual i is always no smaller than that induced by assigning the same amount to individual j. Thus, *risk aversion* imposes punishment on the more wasteful individual.

Risk aversion: For each pair $i, j \in \mathbb{N}$ and each $(F, E) \in \mathcal{P}^{\{i,j\}}$, if $C_i = C_j$ and F_i is riskier than F_j, then $\phi_i(F, E) \leq \phi_j(F, E)$.

The third axiom is *No reversal*. It complements *risk aversion* by requiring a rule not to be too sensitive to risk—it should not impose an overly harsh punishment on an individual who has a riskier claim. In particular, the riskier individual should not be assigned so little as to generate a smaller expected waste than the more deterministic individual.

No reversal: For each pair $i, j \in \mathbb{N}$ and each $(F, E) \in \mathcal{P}^{\{i,j\}}$, if $C_i = C_j$ and F_i is riskier than F_j, then $w(F_i, \phi_i(F, E)) \geq w(F_j, \phi_j(F, E))$.

Besides the new axioms, Xue [19] also studies extensions of some existing axioms in the deterministic claims problems. *Symmetry* can be straightforwardly extended to the current setup.

Symmetry: For each $N \in \mathcal{N}$, each $(F, E) \in \mathcal{P}^N$, and each pair $i, j \in N$, if $F_i = F_j$, then $\phi_i(F, E) = \phi_j(F, E)$.

Endowment monotonicity says that no individual should receive less when the endowment increases.

Endowment monotonicity: For each $N \in \mathcal{N}$, each $(F, E) \in \mathcal{P}^N$, and each $E' \in [0, E)$, $\phi(F, E') \leq \phi_j(F, E)$.

Conditionally strict endowment monotonicity says that when the endowment increases, an individual should receive more only if he has not been fully compensated.

Conditionally strict endowment monotonicity: For each $N \in \mathcal{N}$, each $(F, E) \in \mathcal{P}^N$, each $E' \in [0, E)$, and each $i \in N$, if $\phi_i(F, E') < C_i$, then $\phi_i(F, E') < \phi_i(F, E)$.

A property that is weaker than *conditionally strict endowment monotonicity* is *positivity*. It says that if an individual has a non-zero claim and if the endowment is positive, then he should receive a positive amount.

Positivity: For each $N \in \mathcal{N}$, each $(F, E) \in \mathcal{P}^N$, and each $i \in N$, if $C_i > 0$ and $E > 0$, then $\phi_i(F, E) > 0$.

The usual *consistency* axiom can be directly extended to the current setup.

Consistency: For each $N \in \mathcal{N}$, each $(F, E) \in \mathcal{P}^N$, and each $N' \subseteq N$, $\phi_{N'}(F, E) = \phi(F_{N'}, \sum_{j \in N'} \phi_j(F, E))$, where $\phi_{N'}(F, E)$ and $F_{N'}$ are, respectively, the restrictions of $\phi(F, E)$ and F onto N'.

Strong upper composition pertains to the possibility that after an allocation has been chosen for a problem, the endowment is found to be overestimated. In this case, the initial assignment to an individual could be used as an upper bound on his claim, and the axiom requires that truncating his claim at this upper bound does not affect the allocation. In other words, dividing the smaller endowment based on the initial claim profile should be the same as based on any revised claim profile in which the claims of a subset of agents are truncated at their initial assignments.

Strong upper composition: For each $N \in \mathcal{N}$, each $(F, E) \in \mathcal{P}^N$, each $E' \in [0, E)$, and each $N' \subseteq N$, if $F' \in \mathcal{F}'^{N'}$ is such that for each $i \in N'$ and each $x_i \in \mathbb{R}$,

$$F_i'(x_i) = \begin{cases} F_i(x_i) & x_i < \phi_i(F, E) \\ 1 & x_i \geq \phi_i(F, E) \end{cases},$$

then $\phi(F, E') = \phi((F', F_{N \setminus N'}), E')$.

Lower composition pertains to the possibility that after an allocation has been chosen for a problem, the endowment is found to be underestimated. There are two ways of dealing with the situation. The first is to cancel the initial allocation and divide the larger endowment based on the initial claim profile. The second is to keep the initial allocation and then divide the increment in the endowment based on the profile of the claims reduced by the initial assignments. The axiom requires that both ways of dealing with the situation lead to the same final allocation.

Lower composition: For each $N \in \mathcal{N}$, each $(F, E) \in \mathcal{P}^N$, and each $E' \in [0, E)$, if $F' \in \mathcal{F}'^N$ is such that for each $i \in N$ and each $x_i \in \mathbb{R}$,

$$F_i'(x_i) = \begin{cases} 0 & x_i < 0 \\ F_i(x_i + \phi_i(F, E')) & x_i \geq 0 \end{cases},$$

then $\phi(F, E) = \phi(F, E') + \phi(F', E - E')$.

Claims truncation invariance says that the part of a claim that exceeds what is available should be regarded as irrelevant, so truncating claims at the endowment should not affect the allocation.

Claims truncation invariance: For each $N \in \mathcal{N}$ and each $(F, E) \in \mathcal{P}^N$, if $F' \in \mathcal{F}'^N$ is such that for each $i \in N$ and each $x_i \in \mathbb{R}$,

$$F_i'(x_i) = \begin{cases} F_i(x_i) & x_i < E \\ 1 & x_i \geq E \end{cases},$$

then $\phi(F, E) = \phi(F', E)$.

Scale invariance says that if claims and the endowment are rescaled by a common factor, then assignments should be rescaled by the same factor.

Scale invariance: For each $N \in \mathcal{N}$, each $(F, E) \in \mathcal{P}^N$, and each $c > 0$, if $F' \in \mathcal{F}'^N$ is such that for each $i \in N$ and each $x_i \in \mathbb{R}$, $F_i'(cx_i) = F_i(x_i)$, then $c\phi(F, E) = \phi(F', cE)$.

The class of expected-waste-constrained uniform gains rules admits an axiomatic justification.

Theorem 8 (Xue [19]) *A rule satisfies no domination, conditionally strict endowment monotonicity, consistency, and strong upper composition if and only if it is an expected-waste-constrained uniform gains rule.*

A subclass of expected-waste-constrained uniform gains rules is further pinned down by *scale invariance*.

Proposition 3 (Xue [19]) *An expected-waste-constrained uniform gains rule satisfies scale invariance if and only if it is associated with a homogeneous cost function.*

The specific class of expected-waste-constrained uniform gains rules associated with linear cost functions can be characterized by two different sets of axioms.

Theorem 9 (Xue [19]) *A rule satisfies (i) no domination, positivity, consistency, strong upper composition, and lower composition, or (ii) no domination, positivity, consistency, lower composition, and claims truncation invariance if and only if it is an expected-waste-constrained uniform gains rule associated with a linear cost function.*

It turns out that under some axioms, *no domination* is equivalent to the combination of *risk aversion* and *no reversal*.

Proposition 4 (Xue [19]) *Let a rule satisfy either (i) conditionally strict endowment monotonicity, consistency, and strong upper composition, or (ii) positivity, consistency, lower composition, and claims truncation invariance. Then, it satisfies no domination if and only if it satisfies risk aversion and no reversal.*

Corollary 2 (Xue [19]) *Theorem 8 and Theorem 9 hold with no domination replaced by risk aversion and no reversal.*

4 Conclusion and Open Questions

We survey a recent line of literature on the fair division under the rationing framework. This is an important and growing literature which considers uncertainty in the claims. We study four classes of rules in this framework; the proportional rules, the parametric rules, equal-quantile rules, and expected-waste-constrained uniform gains rules. We provide axiomatic characterizations of these rules.

Theories of distributive justice in general should be developed to incorporate the uncertainty of individual characteristics. Familiar axioms may well have new implications under uncertainty. For example, consider the efficiency requirement defined in terms of maximizing a utilitarian social welfare function. In the deterministic rationing/bankruptcy model where the resource always falls short of the sum of the agents' (deterministic) satiation points, efficiency implies that the resource should be fully allocated to the agents. When their satiation points are uncertain, Long, Sethuraman, and Xue [11] showed, by bringing an outside individual into the picture, that it may no longer be efficient to do so, whether the resource is limited or not. Moreover, by allowing the resource not to be fully allocated, Long, Sethuraman, and Xue [11] discover new implications of familiar axioms on maximal assignments and partial resource allocation. This suggests that more work is needed to extend our understanding of existing efficiency and fairness axioms in a more general framework.

Possible directions for future research may consider a situation when the resource is also uncertain. Koster and Boonen [10] consider a situation when the resource is state-contingent. However, the claims are not state-contingent in their framework. It will be challenging and yet interesting to consider this framework.

Another possible direction for future research is to relax the structural invariance axioms which will open the doors for rules that may satisfy more axioms pertaining to equity or incentives. One such possibility is to study the parametric rules when we relax the axiom of "No Reward for more irrelevant claims" which will allow for more rules to be included such as proportional.

Further research may also explore the allocation rules for state-contingent claims when the individuals have subjective probabilities over the states. Also, when considering the probability distribution of claims, one may want to consider joint density rather than independent distribution.

Another open question can be about the relationship between an objective function that rationalizes a division rule and the axiomatic foundation of the rule. In the deterministic rationing/bankruptcy model, Young's parametric rules, characterized by *continuity*, *symmetry*, and *pairwise consistency*, are rationalizable by the minimization of additively separable and convex cost functions (Young [20]). Long, Sethuraman, and Xue [11] show, in the uncertain context, that maximizing/minimizing a particular utilitarian social welfare/cost function is equivalent to imposing some extensions of standard axioms. It will be interesting to find optimal rules with respect to some other reasonable objective functions and understand the connection between a specific objective function and the requirements that it imposes on a rule.

There are some issues specific to the uncertain context. For example, Xue [19] proposes three new axioms dealing with the issue of waste that is specific to uncertainty. More axioms should be formulated to address, for example, the trade-off between waste and deficit in particular and the balance between efficiency and fairness under uncertainty in general.

"No-envy" axiom first introduced by Foley [7] has been of interest in recent literature (see, e.g., Bhardwaj, Kumar, and Ortega [2] and the references therein). It will be interesting to have a version of "No-envy" in the context of uncertainty.

One may also study situations where the uncertainty is revealed sequentially, rather than knowing the probability distribution of the claims. Juarez, Ko, and Xue [8] study such a setup where the shares are allocated sequentially.

Last but not the least, one may consider incentive axioms for the uncertain setting. This can include risk dominance, pessimistic consideration, or optimistic consideration. It is clear that the incentives of agents will be quite different in the uncertain setting. Thus, it will be an opportunity to exploit this property.

Given the multitude of directions in which this research can be extended and the scope of applications to real life problems, we feel that much more is needed to be explored in this literature.

References

1. Aumann, R., and M. Maschler. 1985. Game theoretic analysis of a bankruptcy problem from the Talmud. *Journal of Economic Theory* 36:195–213.
2. Bhardwaj, B., Kumar, R., and J. Ortega. 2020. Fairness and efficiency in cake-cutting with single peaked preferences. *Economics Letters*. https://doi.org/10.1016/j.econlet.2020.109064.
3. Chatterjee, S., Ertemel, S., and R. Kumar. 2020. Parametric rules for state contingent claims. *Working paper*. http://www.sertemel.com/wp-content/uploads/2019/11/Parametric.pdf.
4. Dale, D. 2020. Trump administration edits national stockpile website a day after it contradicted Jared Kushner. https://edition.cnn.com/2020/04/03/politics/stockpile-website-edited-kushner-claim/index.html.
5. Ertemel, S., and R. Kumar. 2018. Proportional rules for state contingent claims. *International Journal of Game Theory* 47:229–246.
6. Flores-Szwagrzak, K. 2015. Priority classes and weighted constrained equal awards rules for the claims problem. *Journal of Economic Theory* 160:36–55.
7. Foley, D. 1967. Resource allocation and the public sector. *Yale Economic Essays* 7:45–98.

8. Juarez, R., Ko, C. Y. and J. Xue. 2018. Sharing sequential values in a network. *Journal of Economic Theory* 177:734–779.
9. Juarez, R. and R. Kumar. 2013. Implementing efficient graphs in connection networks. *Economic Theory* 54(2):359–403.
10. Koster, M., and T.J Boonen. 2019. Constrained stochastic cost allocation. *Mathematical Social Sciences* 101:20–30.
11. Long, Y., Sethuraman, J., and J. Xue. 2019. Equal– quantiles rule in rationing with claim uncertainty. Research Collection School Of Economics. https://ink.library.smu.edu.sg/soe_research/2274/.
12. Moulin, H. 1985. Egalitarianism and utilitarianism in quasi– linear bargaining. *Econometrica* 53:49–67.
13. Moulin, H. 2002. Axiomatic Cost and Surplus– Sharing. K. Arrow, A. Sen, and K. Suzumura (Eds.) *The Handbook of Social Choice and Welfare (Vol. 1)* Netherlands: North– Holland.
14. O'Neill, B. 1982. A problem of rights arbitration from the Talmud. *Mathematical Social Sciences* 2:345–371.
15. Thomson, W. 2019. How to divide when there isn't enough from aristotle, the talmud, and maimonides to the axiomatics of resource allocation, Vol. 62. Cambridge University Press.
16. van den Brink, R., Funaki, Y., and G. van der Laan. 2013. Characterization of the reverse Talmud bankruptcy rule by exemption and exclusion properties. *European Journal of Operational Research* 228(2):413–417.
17. van den Brink, R., and J.D. Moreno-Ternero. 2017. The reverse TAL-family of rules for bankruptcy problems. *Annals of Operations Research* 254(1–2):449–465.
18. von Neumann, J., and O. Morgenstern. 1944. Theory of games and economic behavior. Princeton University Press: Princeton.
19. Xue, J. 2018. Fair division with uncertain needs. *Social Choice and Welfare* 51:105–136.
20. Young, P. 1987. On dividing an amount according to individual claims or liabilities. *Mathematics of Operations Research* 12:398–414.

Network Theory and Applications

Building Social Networks Under Consent: A Survey

Robert P. Gilles

1 Mutual Consent in Network Formation

During the past two decades there has emerged an extensive literature on game-theoretic models of network formation. Seminally, the fundamentals of such a game-theoretic perspective were set out by [3] in which players are guided by the Myerson value of corresponding communication situations. This contribution explored network formation under mutual consent through a non-cooperative signalling game: A link between two players is formed if and only if both players signal to each other their willingness to form this relationship. The main insight of the *Myerson model* [36] is that the network without any links is always supported through a Nash equilibrium of this signalling game. This theoretical result leads to the conclusion that network formation under mutual consent has to be considered as difficult, even impossible. This would contradict the well-established understanding of human nature as that of a social networker [21, 41].[1] Nevertheless, paradoxically, the Myerson model is and remains the most natural, straightforward and convincing non-cooperative model of network formation under mutual consent.

[1]The main conclusion is strengthened in the case of costly link formation, in which the empty network is a *strong* Nash equilibrium, indicating that starting from an empty network it seems unlikely that rational agents would be able to establish non-trivial networks. I also refer to [30] for a dynamic model of such non-trivial network formation.

I am very grateful for the extensive and helpful comments and suggestions of an anonymous referee. I also thank my co-authors Sudipta Sarangi, Subhadip Chakrabarti and Owen Sims for their contributions and support to explore the paradoxes of social network formation under the hypothesis of mutual consent. The work summarised here could not have been developed without their contributions, insights and help.

R. P. Gilles (✉)
Economics Group, Management School, The Queen's University of Belfast, Riddel Hall, 185 Stranmillis Road, Belfast BT9 5EE, UK
e-mail: r.gilles@qub.ac.uk

S. Borkotokey et al. (eds.), *Game Theory and Networks*, Indian Statistical Institute Series,
https://doi.org/10.1007/978-981-16-4737-6_11

The relative failure of this natural non-cooperative approach induced [29] to introduce an alternative approach, which is founded on a bilateral cooperative consideration.[2] In their approach, Jackson and Wolinsky allow pairs of players to cooperatively deviate from an existing network to modify it. The equilibrium networks under such pairwise modification are denoted as *pairwise stable* networks. Pairwise stability provided a fertile foundation for further exploration of network formation under cooperative consent. This resulted in the development and study of variations of pairwise stability.

Although the Jackson–Wolinsky approach founded on pairwise stability has been very successful in explaining the emergence of non-trivial networks, there remained a gap in our understanding concerning a purely non-cooperative approach to the modelling of mutual consent in network formation. This has been more recently explored through the design of bespoke equilibrium concept applied in the Myerson model. In particular, [18, 43] introduced models of trusting behaviour in network formation through trust-based belief systems. The equilibrium concepts that are developed from these models have very strong properties, showing that trust in network formation leads to non-trivial equilibrium networks. For example, [18]s notion of monadic stability results in equilibrium networks that form a specific subclass of pairwise stable networks—denoted as the *strictly* pairwise stable networks.

Overview of this survey. This survey explores the various methodologies to properly model mutual consent in network formation. I compare the different classes of equilibrium networks that emerge from these different methodologies. After discussing the principles of link formation under mutual consent and Myerson's seminal model, I turn to the exploration of Jackson–Wolinsky type stability concepts based on pairwise cooperative behaviour. I distinguish different subclasses of stable network based on hypotheses about how coalitions of certain sizes can modify the current network. This mainly pertains to pairs of players, but also extends to coalitions of players of arbitrary size—resulting in the notion of a *strongly stable* network [27].

Subsequently, I turn to the main non-cooperative theory of network formation under mutual consent, namely, extensions of the Myerson model [36]. I survey the results from the literature that categorise the various classes of equilibrium networks in the Myerson model with two- as well as one-sided link formation costs. There emerges a close link to certain classes of stable networks in the Jackson–Wolinsky framework.

Subsequently, I discuss the idea of equilibrium refinement in the Myerson model to reflect considerations of mutual trust in link formation. Indeed, links are representations of socio-economic relationships that are founded on mutual trust between the interacting parties. This results in the unilateral [43] and monadic stability [18] concepts in the Myerson model. I explore monadic stability further, which is founded

[2] An alternative mathematical model emerged with [4] based on one-sided link formation: One assumes *ex-ante*, or implicit, consent among players in the network formation game. The resulting equilibrium networks are denoted as *Nash networks* in the subsequently developed literature. This approach is unsatisfactory due to its unnatural social foundations with rather limited applicability to explain social and economic phenomena.

on a conception of mutual trust through a belief system in the Myerson model. The properties of these monadically stable networks as well as their existence, using Monderer and Shapley [33]s theory of game-theoretic potentials, are also reviewed.

I conclude this survey by looking at an alternative method to modelling mutual consent in link- and network formation. This refers to the introduction of correlated strategies in the Myerson model as a tool to represent coordinated interaction. The resulting class of "correlated equilibrium networks" still needs to be explored in future research.

2 Introducing Mutual Consent: Modelling Principles

Throughout this survey, I use a broad class of game-theoretic techniques to model how relationships—or "links"—between pairs of socio-economic agents come about. We refer to these socio-economic agents as *players* in the context of these models. Each player is assumed to be a fully rational individual decision-maker that acts according to a set of behavioural rules described in the developed equilibrium concept.

Besides the specific behavioural hypotheses on which these equilibrium concepts are based, it is important to realise that there are some fundamental broad axioms made. These fundamental axioms introduce a few fundamental limitations of the approach that is surveyed here:

(i) This game-theoretic approach is purely *static* in nature. This implies that we start from a zero state in which no links exist and in which these socio-economic agents decide whether and which links to build. The end result is a fully formed network in which certain value-generating activities are achieved. It would be more realistic to model the formation of a network as a dynamic building process. However, in the static conception followed throughout this survey, one network does *not* evolve into another.

This has major consequences for how we view network formation and which networks actually are identified in these constructions. Indeed, the identified equilibrium networks do not exhibit the features of large social networks identified in the literature quoted on social networks [5, 38]. So, these equilibrium networks are usually neither scale-free nor small world networks nor satisfying the basic property of assortative mixing. This is a severe limitation of such a static approach.[3]

On the other hand, the static approach highlights certain properties of rational decision-making in the context of pairwise cooperation, required for building value-generating relationships under mutual consent. Rather contradictorily, the main theorem in Myerson's non-cooperative model shows that rational decision-

[3] In my discussion in this survey I omit the recent development of incentive-based stochastic models of network formation. This approach focuses not only on game-theoretic incentives in network formation—as the subject matter of this survey—but combines this concept with stochastic processes that describe random meetings. This approach was seminally developed in Jackson and Rogers [25, 26] and further addressed in, e.g. Golub and Livne [19].

making does actually not result in any sensible network formation—the empty network is always supported through a Nash equilibrium in the Myerson model. So, starting from an empty network, fully rational players have no mechanism to create a meaningful interaction structure. Only if we impose that the decision-makers are *boundedly rational*—and, thus, use animal spirits rather than optimisation in decision-making—we arrive at the conclusion that non-trivial and sensible networks emerge under mutual consent.[4] This important insight is the main conclusion presented in this survey.

(ii) The game-theoretic approach explored in this survey is founded on a *negative* stability methodology. Hence, a network is called "stable" if there are no incentives to change the network. This is the standard methodology in game theory. Rather than constructing an actual building process, this methodology only looks at which networks *cannot* emerge due to the existing incentives to change the network that the players are endowed with. We thus arrive at a class of equilibrium networks that describe configurations in which such incentives for deviation are absent.

The consequence of the application of this standard game-theoretic methodology is that reality is only approximated. This approach, for example, does not allow the mixing of modes of incentives, which is common in real-life interaction. This, therefore, is another reason why the theoretically derived networks do not have the desired features discussed in the literature on large social networks as surveyed by Barabási [5].

The next section sets out the basic framework of modelling mutual consent in the formation of a relationship between two players.

2.1 Players, Links and Networks

We use the basic concepts from the theory of social networks set out in the literature. Following the accepted symbolism, the set $N = \{1, \ldots, n\}$ represents a set of *players*. The fundamental issue addressed here is how these players will build pairwise or binary relationships with other players and ultimately construct a socio-economic network consisting of such binary relationships.

Each player $i \in N$ is explicitly endowed with the social ability to build such pairwise relationships or *links* with other players, provided that consent is given by the other party. Again following the accepted terminology in the literature [24], the pairwise subset $\{i, j\} \subset N$ with $i \neq j$ denotes a pairwise relationship between players $i \in N$ and $j \in N$. We follow convention to use shorthand notation and define a *link* between players i and j as $ij = \{i, j\} \in g^N$, where

[4] This is captured in the notion of a monadically stable network that is founded on trusting behaviour by the players. Such trusting behaviour is fundamentally boundedly rational. Indeed, to trust another player is not founded on calculation, but on a leap of faith.

$$g^N = \{\, \{i, j\} \mid i, j \in N \text{ and } i \neq j\} = \{ij \mid i, j \in N\} \tag{1}$$

denotes the set of all potential links on the player set N. As such the set g^N acts as the universal set of all potential links on player set N.

A *network* on N is now an arbitrary subset of links, i.e. any subset $g \subset g^N$ is a network on N. In particular, $g = g^0 = \varnothing$ is the *empty network* on N which describes a situation where no links are formed. Furthermore, $g = g^N$ is the *complete network* on N, which is the largest network consisting of all potential links among players in N. We introduce $\mathbb{G}^N = \{g \mid g \subset g^N\}$ as the collection of all networks on N.

The *neighbourhood* of player $i \in N$ in network $g \in \mathbb{G}^N$ is given by $N_i(g) = \{j \in N \mid ij \in g\}$. The collection of corresponding neighbouring relationships or links is denoted by $L_i(g) = \{ij \in g \mid j \in N_i(g)\}$. The complete collection of all potential links that involve player $i \in N$—or that can be formed by player i—is denoted by $L_i = L_i(g^N) = \{ij \mid j \neq i\}$.

Adding and deleting links to a network. In formal models of network formation we consider the deletion and addition of links to given networks. For this I introduce some well-accepted notation [24]. Consider a network $g \in \mathbb{G}^N$. For every pair of players $i, j \in N$ with $ij \notin g$ we now denote by $g + ij$ the network that results from g by adding the link $ij \notin g$, i.e. $g + ij = g \cup \{ij\} \in \mathbb{G}^N$. Similarly, for some collection of links $h \subset g^N$ with $g \cap h = \varnothing$, we denote $g + h = g \cup h$ the network that results from adding link collection h to the network g.

Next, consider two players $i, j \in N$ with $ij \in g$. We denote by $g - ij = g \setminus \{ij\} \in \mathbb{G}^N$ the network that results from removing the link ij from the network g. Again, for any collection of links $h \subset g$ we denote $g - h = g \setminus h$ the network that results from removing the links in h from the network g.

Example 2.1 With these notational conventions we are now equipped to address link formation processes. To illustrate this notation, consider the network $g = \{12, 13, 24, 34, 35\}$ on $N = \{1, 2, 3, 4, 5\}$ as depicted in Fig. 1 below consisting of the red and black links. Considering the green link $45 \notin g$, then $g' = g + 45 = \{12, 13, 24, 34, 35, 45\}$ is depicted in Fig. 1 as the network consisting of all coloured links. Finally, removing the red link set $h = \{13, 35\} \subset g$ from g results into $g'' = \{12, 24, 34\}$, depicted by collection of the black links only in Fig. 1. ♦

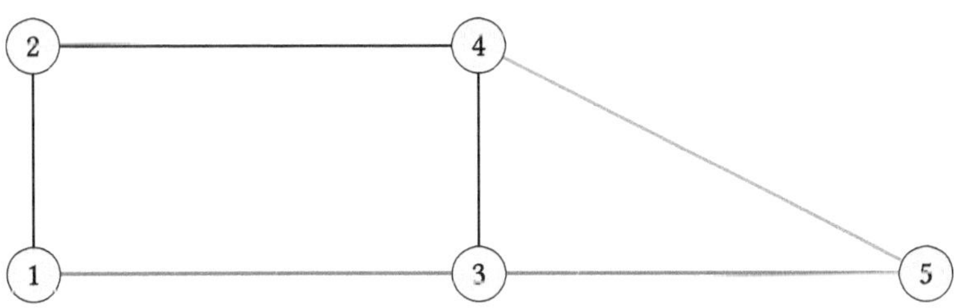

Fig. 1 Illustration for link addition and deletion

Payoffs. Throughout the literature on game-theoretic approaches to network formation, players are assumed to be fully incentivised in their drive to build and maintain links as well as delete links in existing networks. These incentives are introduced as an individualised payoff function. Indeed, for every player $i \in N$ we introduce *player i's network payoff function* as $\varphi_i : \mathbb{G}^N \to \mathbb{R}$, which assigns to every network $g \in \mathbb{G}^N$ a value $\varphi_i(g)$ that evaluates i's situation as a member of the networked community described by g.

We can now capture all payoff information on the population N of players in the *network payoff function* given by $\varphi = (\varphi_1, \ldots, \varphi_n) : \mathbb{G}^N \to \mathbb{R}^N$. In particular, I emphasise that the function φ indeed captures all incentives for the decision-makers in N in the network formation processes to be considered next.[5]

A network payoff for a player captures *all* values emanating in the structured community that is perceived or received by that player. This includes all perceived *externalities* of third parties. In this regard, the network payoff function can capture widespread externalities from relationship and network formation in that community. The addition of network externalities in the payoff structure differentiates this inclusive network payoff approach from the more classical cooperative game-theoretic payoff structure employed by Myerson [34, 35], Dutta and Mutuswami [12] and van den Nouweland [44, 45]. The payoff function including widespread externalities has been seminally introduced in network theory by Jackson and Wolinsky [29].

Example 2.2 I illustrate this concept by revisiting the networks depicted in Fig. 1. For example, player 1 can be assigned $\varphi_1(g) = 1$ as well as $\varphi_1(g') = 5$ even though her neighbours in both networks are exactly the same, i.e. $N_1(g) = N_1(g') = \{2, 3\}$. This, therefore, captures widespread externalities from the creation of the link 45 in the network g from the perspective of player 1. ♦

2.2 Myerson's Approach to Network Formation

The most fundamental and basic model of how networks form under mutual consent was seminally introduced as an example in Myerson [36, p. 448]. He pointed out that in a very simple network formation game—known as the *Myerson model*—, the resulting networks that are supported by Nash equilibria in this game *always* include the empty network g^0. Hence, building no links at all is an equilibrium in any incentive structure generated by player benefits to network formation.

Myerson presented this as a negative insight, since it indicates that purely non-cooperative game theory cannot provide a fertile basis for a debate of how non-trivial networks between players emerge. However, what this really expresses is that networks are not forming if players act purely selfishly. My contention is throughout that it actually has to be expected that pure selfishness would undermine cooperative acts such as forming links between pairs of players.

[5] We might refer to the multi-dimensional function φ also as representing the *network payoff structure*.

Here I initially explore the seminal Myerson model itself. In subsequent sections I turn to extensions of this basic model with added consideration of link formation costs. For the proper development of the Myerson model we need to review some basic non-cooperative game theory.

Preliminaries: Some game theory. This section relies heavily on standard non-cooperative game theory. Again we let $N = \{1, \dots, n\}$ be the set of players. A *game* on N is a pair $(\mathcal{A}, \pi)$ with $\mathcal{A} = (A_1, \dots, A_n)$ an ordered collection of *strategy sets* such that each player $i \in N$ is assigned her individual strategy set A_i and a game-theoretic *payoff function* $\pi = (\pi_1, \dots, \pi_n) \colon A \to \mathbb{R}^N$ where $A = \prod_{i \in N} A_i$ is the set of all *strategy tuples* generated in $\mathcal{A}$.

Hence, in a non-cooperative game, each player $i \in N$ is endowed with her individual strategy set A_i and a payoff function $\pi_i \colon A \to \mathbb{R}$. The fundamental idea is that every player selects a strategy that optimises her payoffs, provided that other players also select strategies that affect this payoff. As such, a game is a mathematical representation of a social interaction situation. Game theory is now a collection of rules and tools that model how players make decisions in the context of such social interaction situations.

A strategy tuple is a list $a = (a_1, \dots, a_n) \in A$. We use the convention that the list of strategies of players other than $i \in N$ are indicated by $a_{-i} = (a_1, \dots, a_{i-1}, a_{i+1}, \dots, a_n) \in \prod_{j \in N \colon j \neq i} A_j$. Hence, $a = (a_i, a_{-i})$.

Definition 2.3 A strategy tuple $a^* \in A$ is a **Nash equilibrium** in the game $(\mathcal{A}, \pi)$ if for every player $i \in N$ and any strategy $b_i \in A_i$ it holds that $\pi_I(a^*) \geqslant \pi_i(b_i, a^*_{-i})$. In a Nash equilibrium, every player optimises her strategy, *given the strategic choices of all other players*.

A Nash equilibrium can also be expressed in terms of "best responses". Formally, a strategy $a_i \in A_i$ is a *best response* to strategy tuple $a_{-i} \in \prod_{j \in N \colon j \neq i} A_j$ if for every strategy $a_i' \in A_i$ it holds that $\pi_i(a_i, a_{-i}) \geqslant \pi_i(a_i', a_{-i})$. Hence, a best response is the strategy for player i that optimises her payoffs given that all other players $j \neq i$ select the strategy $a_j \in A_j$.

Now a strategy tuple $a^* \in A$ is a Nash equilibrium if and only if for every player $i \in N$ it holds that a_i^* is a best response to a^*_{-i}. As such a Nash equilibrium is a fixed point of the best response correspondence that is generated by the game. Furthermore, it can be shown that in this respect a Nash equilibrium usually can be interpreted as a saddle point in a well-constructed geometric representation of the game.

The Myerson model. Myerson [36] introduced his approach to modelling the formation of networks as an illustration of the underlying processes that determine the Nash equilibria in a non-cooperative strategic form game. Myerson's framework is *the* quintessential model of mutual consent in link formation. The Myerson model encompasses a basic signalling game in which players send each other messages about whether they want to form a link or not. Due to its very fundamental and basic nature, it is a model that acts as the benchmark in any discussion on consent in link formation.

In Myerson's framework, players costlessly signal to each other whether they want to form links. Now, a link is established if and only if the two players signal both that they would like to form the link. Formally, the *Myerson model* Γ^m_φ on player set N under network payoff function $\varphi \colon \mathbb{G}^N \to \mathbb{R}^N$ is a non-cooperative game $\Gamma^m_\varphi = (\mathcal{A}^m, \pi^m)$ given as follows:

- For every player $i \in N$, her strategy set is given by all vectors of signals to other players in N:

$$A^m_i = \left\{ \ell_i = (\ell_{i1}, \ell_{i2}, \dots, \ell_{in}) \,\middle|\, \ell_{ij} \in \{0, 1\} \text{ and } \ell_{ii} = 1 \right\} ; \tag{2}$$

  Here, ℓ_{ij} is a signal that player i communicates to player j about her intentions to form a link with j. If $\ell_{ij} = 1$, player i indicates that she is interested in forming the link with player j; if $\ell_{ij} = 0$, player i signals that she wants to remain unattached to player j.
- A link ij is now formed if both players i and j signal to each other they want to form the link, i.e. if $\ell_{ij} = \ell_{ji} = 1$. If we denote by $\ell = (\ell_1, \dots, \ell_n) \in A^m = A^m_1 \times \cdots \times A^m_n$ a strategy profile, then the resulting network can be identified as

$$g(\ell) = \{ij \in g^N \mid \ell_{ij} = \ell_{ji} = 1\}. \tag{3}$$

  We say that $g(\ell)$ is the network *supported* by the strategy profile ℓ in the Myerson model.
- The Myerson model is completed by the game-theoretic payoff function $\pi^m \colon A^m \to \mathbb{R}^N$ defined by

$$\pi^m_i(\ell) = \varphi_i\left(g(\ell)\right). \tag{4}$$

  Clearly, the payoff function π^m reflects the property that signalling is costless and that there are no costs incurred in the formation of a link between any pair of players.

In the next discussion, I investigate the networks that are supported through Nash equilibria in the Myerson model.

M-networks. The Nash equilibria in the basic Myerson model form a class of signalling profiles that support networks on N that are stable against unilateral modification. We denote these Nash equilibrium networks as "M-networks" to distinguish this class of networks from other classes of networks.

Definition 2.4 Let φ be a network payoff function on player set N and let $\Gamma^m_\varphi = (\mathcal{A}^m, \pi^m)$ be the corresponding basic Myerson model. A network $g \in \mathbb{G}^N$ is an **M-network** if there exists a Nash equilibrium strategy tuple $\ell^g \in A^m$ in Γ^m_φ such that $g(\ell^g) = g$.

Clearly, using the Nash equilibrium conditions and the definition of π^m, we get the following M-network requirement: For every player $i \in N$ and every signal vector $\ell_i \in A^m_i$ it holds that

$$\varphi_i\left(g(\ell_i, \ell^g_{-i})\right) \leqslant \varphi_i\left(g(\ell)\right).$$

The concept of M-network is at the core of the assessment of network formation itself, since it describes the stable outcomes of the basic signalling framework represented in the Myerson model. Crucially, Myerson [36] already pointed out that the empty network is always supported as an M-network. Formally, this can be expressed as follows.

Proposition 2.5 (Myerson's Lemma) *In the Myerson model* $\Gamma^m_\varphi = (\mathcal{A}^m, \pi^m)$ *the "no-link" signal profile* $\ell^0 = (0, \dots, 0) \in A^m$ *is a Nash equilibrium. Consequently, the empty network* $g^0 = g(\ell^0)$ *is an M-network.*

Proof Let $\ell^0_{ij} = 0$ for all $i, j \in N$, making up the strategy profile ℓ^0. Then, for any player $i \in N$, any signal vector $\ell_i \in A^m_i$ is a best response to ℓ^0_{-i}, since $g(\ell^0_{-i}, \ell_i) = g^0$ irrespective of the selected signal vector ℓ_i. Therefore, ℓ^0_i itself is a best response to ℓ^0_{-i}, showing that ℓ^0 is a Nash equilibrium in $\Gamma^m_\varphi = (\mathcal{A}^m, \pi^m)$. ■

This property points out that non-trivial M-networks are very hard to form; rational self-interest easily results in complete failure and no cooperation might emerge. In this case, Myerson's Lemma indicates that, without some supporting mechanism, there simply are no incentives to justify that any links are formed at all. So, Myerson's Lemma points to the very fundamental issue of human cooperation: Why would rational human beings be cooperative? In this regard, Myerson's Lemma is a very succinct expression of this major question in social science and economics.

3 Jackson–Wolinsky Stability Concepts

The challenge of modelling non-trivial network formation stated in the discussion of the Myerson model as Myerson's Lemma was taken on by Jackson and Wolinsky [29]. They formulated *cooperative* equilibrium concepts that are tailored to the specific demands of modelling bilateral link formation. This resulted in the notion of a "pairwise stable" network.

I first discuss a class of cooperative or pairwise concepts of network stability from a link-based perspective as explored in Gilles et al. [15, 17]. This concerns four fundamental link-stability principles, each founding a particular form of cooperative stability, and three further derived stability notions—including the seminal pairwise stability concept introduced by Jackson and Wolinsky [29].

Central to this approach is that while mutual consent is required for establishing a link, a player is able to delete her links unilaterally. Here, we focus on link-centred considerations. Hence, how would the deletion of one or more links affects the players' payoffs? Similarly, how would the addition of one or more links affect payoffs? These mutual considerations are brought together into a link- or network-based notion of stability.

Deleting links from networks. Throughout it is assumed that players have full autonomy or sovereignty over the decision to delete one or more of her links. Indeed, the principle of mutual consent requires that players control which links they participate in. This implies that every player can veto her participation in any link or relationship. Based on this consideration, I introduce two fundamental stability concepts concerning the deletion of links.

As before, let $\varphi\colon \mathbb{G}^N \to \mathbb{R}^N$ be a network payoff function on the player set N.

(i) A network $g \in \mathbb{G}^N$ is **link deletion proof** (LDP) for φ if for every player $i \in N$ and every neighbour $j \in N_i(g)$, it holds that $\varphi_i(g - ij) \leqslant \varphi_i(g)$.
Link deletion proofness requires that no player has an incentive to sever an existing link with one of her neighbours.
We denote by $\mathcal{D}(\varphi) \subset \mathbb{G}^N$ the class of all link deletion proof networks for the given payoff function φ [29].

(ii) A network $g \in \mathbb{G}^N$ is **strong link deletion proof** (SLDP) for φ if for every player $i \in N$ and every set of her direct links $h \subset L_i(g)$, it holds that $\varphi_i(g - h) \leqslant \varphi_i(g)$.
Strong link deletion proofness requires that no player has incentives to sever links with one or more of her neighbours simultaneously.
We denote by $\mathcal{D}_s(\varphi) \subset \mathbb{G}^N$ the class of all strong link deletion proof networks for the given payoff function φ [15].

From the definition it is clear that any SLDP network is always LDP and, therefore, strong link deletion proofness is indeed a stronger notion than (regular) link deletion proofness. As indicated, LDP was seminally introduced in Jackson and Wolinsky [29], while SLDP was only introduced as a stand-alone concept in early drafts of Gilles et al. [15].

Second, the empty network $g^0 = \varnothing$ on any set of players N is trivially strong link deletion proof. Indeed, this network does not contain any links and, therefore, the deletion of links is vacuously satisfied. We can therefore summarise that:

Proposition 3.1 *For any network payoff function $\varphi\colon \mathbb{G}^N \to \mathbb{R}^N$ it holds that*

$$g^0 \in \mathcal{D}_s(\varphi) \subset \mathcal{D}(\varphi) \subset \mathbb{G}^N. \tag{5}$$

The first question that I consider is under which conditions link deletion proofness is exactly the same as strong link deletion proofness. This seems a rather innocuous question, since SLDP is so much stronger a concept than LDP. Nevertheless, it is enlightening to identify the exact property on the network payoff structure φ that allows this equivalence.

Theorem 3.2 *Strong link deletion proofness and link deletion proofness are equivalent for network payoff structure φ in the sense that $\mathcal{D}(\varphi) = \mathcal{D}_s(\varphi)$ if and only if the network payoff structure φ is* ***convex*** *on the class of link deletion proof networks $\mathcal{D}(\varphi) \subset G^N$ in the sense that for every LDP network $g \in \mathcal{D}(\varphi)$, every player $i \in N$, every neighbour $j \in N_i(g)$ and every link set $h \subset L_i$ with $h \cap L_i(g) = \varnothing$ it holds that*

$$\sum_{ij \in h} [\, \varphi_i(g + ij) - \varphi_i(g)\,] \geqslant 0 \text{ implies that } \varphi_i(g + h) \geqslant \varphi_i(g). \tag{6}$$

For a proof of Theorem 3.2 I refer to Appendix A.1 of this survey.

The convexity property on the payoff structure φ requires that the sign of the sum of values from adding one link to a network from a set of links fully determines whether adding all links is beneficial or not. Hence, looking at links one-by-one gives complete information about whether it is beneficial to add all links to the network or not.

Adding links to networks. Next I consider how players assess the addition of a link to an existing network. Again we take the idea of consent in link formation as central into our reasoning here. This implies that both parties in the formation of a new link have to agree that adding this link is beneficial.

(iii) A network $g \in \mathbb{G}^N$ is **link addition proof** (LAP) for φ if for all $i, j \in N$ with $ij \notin g$, it holds that $\varphi_i(g + ij) > \varphi_i(g)$ implies $\varphi_j(g + ij) < \varphi_j(g)$.
Link addition proofness states that there are no incentives for any pair of players to form an additional link. This is based on the requirement of mutual consent in link formation. Indeed, if one player would like to add a link, the other player would have strong objections. In this case this is formulated as that if one player has benefits from forming the link, the other (consenting) party has losses and, thus, would withhold her consent.
We denote by $\mathcal{A}(\varphi) \subset \mathbb{G}^N$ the class of all link addition proof networks for the given payoff function φ [29].

(iv) A network $g \in \mathbb{G}^N$ is **strict link addition proof** (SLAP) for φ if for all $i, j \in N$, it holds that $ij \notin g$ if and only if $\varphi_i(g + ij) < \varphi_i(g)$ as well as $\varphi_j(g + ij) < \varphi_j(g)$.
Strict link addition proofness is a far stronger notion that LAP. Indeed, it requires that both players agree that forming an additional link between them is not beneficial for either of them. This agreement is imposed and only a certain very specific type of network payoff structures would support such networks to exist. Consequently, it has to be expected that, for an arbitrary regular network payoff function, only a rather small class of networks actually satisfies this property.
We denote by $\mathcal{A}_s(\varphi) \subset \mathbb{G}^N$ the class of all strict link addition proof networks for the given payoff structure φ [18].

The introduced notions of link addition proofness require some clarification. These two notions indeed only seem to partially cover the idea that a network is stable if it satisfies the property that "if i has an incentive to form an additional link with j, then j has no incentive to form a link with i". This is subject to the next discussion.

To understand link addition proofness in more detail, we can reformulate it. Indeed, a network g is link addition proof if and only if for all players $i, j \in N$ with $ij \notin g$:

$$\varphi_i(g + ij) \geqslant \varphi_i(g) \text{ implies } \varphi_j(g + ij) \leqslant \varphi_j(g). \tag{7}$$

This has some interesting consequences regarding the interpretation of the LAP property. First, a link $ij \notin g$ for some $i, j \in N$ is *non-discerning* if it holds that

$$\varphi_i(g + ij) = \varphi_i(g) \text{ as well as } \varphi_j(g + ij) = \varphi_j(g). \tag{8}$$

From the formulation above, the definition of link addition proofness is indeed ambiguous whether any non-discerning link ij should be in the network for it to be LAP or not. Hence, such non-discerning links can arbitrarily be added to or deleted from networks without the LAP property being affected. Thus, the class of non-discerning links makes the determination of LAP networks "fuzzy".

To address this issue of the addition or deletion of non-discerning links, I introduce a third type of link addition proofness:

(v) A network $g \in \mathbb{G}^N$ is **$\star$-link addition proof** ($\star$-LAP) for φ if for all players $i, j \in N$, it holds that if $ij \notin g$, then $\varphi_i(g + ij) \geqslant \varphi_i(g)$ implies $\varphi_j(g + ij) < \varphi_j(g)$.
We denote by $\mathcal{A}_\star(\varphi) \subset \mathbb{G}^N$ the class of all $\star$-link addition proof networks for the given payoff structure φ.

This minor modification of the definition of link addition proofness simply requires that all non-discerning links should be part of a $\star$-link addition proof network. This makes the definition unambiguous.

Example 3.3 To delineate the three-link addition proofness concepts introduced here, we can explore an example of a network payoff function in which these concepts result in different classes of networks. We consider three players and all possible networks, i.e. $N = \{1, 2, 3\}$ and $\mathbb{G}^N = \{g \mid g \subset g^N\}$ where $g^N = \{12, 23, 13\}$. Note that there are exactly eight possible networks on N, i.e. $\#\mathbb{G}^N = 8$.

We now consider a particular network payoff function φ on the generated class of networks $\mathbb{G}^N$ on N. All potential network payoffs represented by φ can be represented in an appropriately constructed table:

Network g	$\varphi_1(g)$	$\varphi_2(g)$	$\varphi_3(g)$		**Stability**
$g^0 = \varnothing$	0	0	0		LAP
$g^1 = \{12\}$	0	0	1		$\star$-LAP
$g^2 = \{13\}$	0	0	0		
$g^3 = \{23\}$	0	0	0		
$g^4 = \{12, 13\}$	2	1	0		
$g^5 = \{12, 23\}$	1	2	0		
$g^6 = \{13, 23\}$	0	1	0		
$g^7 = g^N$	3	3	3		SLAP

First, note that g^0 is link addition proof, but not $\star$-link addition proof. Indeed, if any link is added to the empty network, no payoffs are changed for any of the players involved. On the other hand, there are no losses, thus precluding that g^0 is $\star$-link addition proof.

Next, g^1 is $\star$-link addition proof, but not strong link addition proof. Indeed, any addition of a link to g^1 results in a loss for player 3. However, adding link 13 results in a strict gain for player 1, implying that g^1 is not strong link addition proof.

Third, the complete network g^N is strong link addition proof by tautology. Indeed, there are no links to be added to this network, and therefore vacuously the property of strong link addition proofness is satisfied.

I remark that none of the other networks have any link addition proofness properties. ♦

Next I explore the equivalence of these link addition proofness concepts. In order to explore these equivalences effectively, I introduce two auxiliary properties of the network payoff structure.

Definition 3.4 Consider a network payoff structure φ on $\mathbb{G}^N$. Then:

- The structure φ is said to be **discerning** on the class of networks $\mathbb{G} \subset \mathbb{G}^N$ if for every network $g \in \mathbb{G}$ it holds that for any pair $i, j \in N$ with $ij \notin g$ either $\varphi_i(g+ij) \neq \varphi_i(g)$ or $\varphi_j(g+ij) \neq \varphi_j(g)$ or both.
- The structure φ is said to be **uniform** on the class of networks $\mathbb{G} \subset \mathbb{G}^N$ if for every network $g \in \mathbb{G}$ and for any pair $i, j \in N$ with $ij \notin g$ it holds that

$$\varphi_i(g+ij) \geqslant \varphi_i(g) \text{ implies } \varphi_j(g+ij) \geqslant \varphi_j(g). \tag{9}$$

Using these auxiliary concepts we can now show the following equivalences:

Theorem 3.5 *Let φ be some network payoff structure on the class of all networks $\mathbb{G}^N$ on the set of players N. Then the following properties hold:*

(a) $g^N \in \mathcal{A}_s(\varphi) \subset \mathcal{A}_\star(\varphi) \subset \mathcal{A}(\varphi)$*;*
(b) It holds that $\mathcal{A}_\star(\varphi) = \mathcal{A}(\varphi)$ if and only if φ is discerning on $\mathcal{A}(\varphi)$, and;
(c) It holds that $\mathcal{A}_s(\varphi) = \mathcal{A}_\star(\varphi)$ if and only if φ is uniform on $\mathcal{A}_\star(\varphi)$.

For a proof of Theorem 3.5 I refer to Appendix A.2 in this survey. Furthermore, from Theorem 3.5 it is easily concluded that the following equivalence also holds:

Corollary 3.6 *SLAP and LAP are equivalent concepts for payoff structure φ in the sense that $\mathcal{A}_s(\varphi) = \mathcal{A}(\varphi)$ if and only if the payoff structure φ is discerning and uniform on $\mathcal{A}(\varphi)$.*

3.1 *Notions of Pairwise Stability*

In the previous discussion, I introduced four fundamental stability concepts on adding links to and deleting links from a network. These four basic notions can be combined to define derived concepts. The first concept—known as *pairwise stability* [29]—combines the weakest link-stability notions and has been the subject of extensive discussion in the literature. This notion implicitly assumes that players only consider the deletion and addition of one specific link at a time.

(vi) Network g is **pairwise stable** (PS) for φ if g is link deletion proof as well as link addition proof. We denote by $\mathcal{P}(\varphi) = \mathcal{D}(\varphi) \cap \mathcal{A}(\varphi)$ the family of pairwise stable networks for the payoff function φ.

This notion implicitly assumes that the original pairwise stability concept—introduced by Jackson and Wolinsky [29]—only concerns itself with the contemplation of adding a single link to or deleting a single link from a given network. If there are no incentives for players to either add a link to the existing network or delete a link from the network, then the network is "pairwise stable": There are no incentives present under the hypothesis of mutual consent in link formation that anybody wants to change a single link in this network.

Two further derived stability concepts, which strengthen the notion of pairwise stability, have particular relevance in the theory of consent in link formation. Strong pairwise stability [15, 17] assumes that players can delete an arbitrary collection of links under their control. Hence, they can veto any link in which they participate. On the other hand, the contemplation of adding links remains confined to adding a single link.

Strict pairwise stability [18] is the strongest notion in this framework. It not only considers that players can delete any number of their existing links, but also that they are assumed to be in agreement regarding the addition of a link to an existing network. It is clear that for an arbitrary network payoff structure, the collection of such strictly pairwise stable networks might well be empty. Only for certain network payoff structures such networks might emerge.

(vii) Network g is **strongly pairwise stable** (SPS) for φ if it is strong link deletion proof as well as link addition proof.
We denote by $\mathcal{P}_\star(\varphi) = \mathcal{D}_s(\varphi) \cap \mathcal{A}(\varphi)$ the family of strongly pairwise stable networks for the payoff function φ.

(viii) Network g is **strictly pairwise stable** (SPS*) for φ if it is strong link deletion proof as well as strict link addition proof.
We denote by $\mathcal{P}_s(\varphi) = \mathcal{D}_s(\varphi) \cap \mathcal{A}_s(\varphi)$ the family of strictly pairwise stable networks for the payoff function φ.

These three pairwise stability concepts generate different classes of networks in most cases. I consider an example to illustrate this.

Example 3.7 Again consider three players and all potentially generated networks, i.e. $N = \{1, 2, 3\}$ with $g^N = \{12, 23, 13\}$. Now, consider a network payoff function φ on the generated class of networks $\mathbb{G}^N$ on N represented in the following table:

Network g	$\varphi_1(g)$	$\varphi_2(g)$	$\varphi_3(g)$		**Stability**
$g^0 = \varnothing$	0	0	0		Strongly PS
$g^1 = \{12\}$	0	0	5		Strictly PS
$g^2 = \{13\}$	0	0	0		
$g^3 = \{23\}$	0	0	0		
$g^4 = \{12, 13\}$	-1	0	0		
$g^5 = \{12, 23\}$	0	-1	0		
$g^6 = \{13, 23\}$	0	1	1		
$g^7 = g^N$	3	3	3		PS

Again we discuss the properties of these networks.

First, note that the empty network g^0 is trivially SLDP and in this case as well LAP. Therefore, it is indeed strongly pairwise stable.[6]

Second, g^1 is LDP and, therefore, SLDP. Moreover, g^1 is SLAP. Indeed, adding link 13 to g^1 results in strict losses for both players 1 and 3. Similarly, for link 23. Thus, we conclude that g^1 is strictly pairwise stable.

Finally, the complete network g^N is SLAP due to being the maximal network. Furthermore, g^N is LDP. However, g^N is not SLDP. player 3 has the strict incentive to delete both her links and revert to network g^1.

We conclude from this discussion that this simple network payoff example induces three distinct classes of pairwise stable networks. ♦

Using the equivalence results stated in Theorems 3.2 and 3.5, we can now conclude the following equivalences between the formulated pairwise stability concepts. The proofs are rather transparent and therefore omitted.

Corollary 3.8 *Consider a network payoff structure φ on the class of all networks $\mathbb{G}^N$ on set of players N. Then the following relationships hold:*

(a) $\mathcal{P}_s(\varphi) \subset \mathcal{P}_\star(\varphi) \subset \mathcal{P}(\varphi)$;
(b) Pairwise stability and strong pairwise stability are equivalent concepts for φ in the sense that $\mathcal{P}(\varphi) = \mathcal{P}_\star(\varphi)$ if and only if φ is convex on $\mathcal{P}(\varphi)$;
(c) Strong pairwise stability and strict pairwise stability are equivalent concepts for φ in the sense that $\mathcal{P}_\star(\varphi) = \mathcal{P}_s(\varphi)$ if and only if φ is discerning and uniform on $\mathcal{P}_\star(\varphi)$, and;
(d) Pairwise stability and strict pairwise stability are equivalent concepts for φ in the sense that $\mathcal{P}(\varphi) = \mathcal{P}_s(\varphi)$ if and only if φ is convex, discerning as well as uniform on $\mathcal{P}(\varphi)$.

3.2 Strong Stability

Next I discuss some of the ideas put forward by Jackson and van den Nouweland [27]. They investigated networks that emerge if coalitions of arbitrary size can make changes to the network in a coordinated fashion to the coalition's overall benefit.[7] As such strong stability is an extension of the pairwise stability concept to allow arbitrary coalitions to adjust the network structure under their control.

As a preliminary we denote a *coalition* as any subset S of players in N; hence, a coalition is any $S \subset N$. This includes the empty coalition $\varnothing$ as well as the "grand" coalition N itself. In a non-cooperative game $(\mathcal{A}, \pi)$, for any coalition $S \subset N$ and

[6] It should be remarked that networks with at most one link are SLDP if they are LDP. Therefore, they are strongly pairwise stable if they are link addition proof and link deletion proof.

[7] This approach is akin to the strong equilibrium concept proposed by Aumann [1] in non-cooperative game theory. Jackson–Nouweland's concept of strong stability can be viewed as a network theoretical implementation of the ideas behind Aumann's strong equilibrium concept.

strategy profile $a \in A$ we denote by a_S the S-restriction of a defined by $(a_j)_{j \in S}$ and by $a_{N \setminus S}$ its complement $(a_k)_{k \notin S}$.

Now, in a non-cooperative game $(\mathcal{A}, \pi)$ a strategy tuple $a \in A$ is a *strong equilibrium* if for every (non-empty) coalition of players $S \subset N$ and every coordinated strategic deviation $b_S = (b_i)_{i \in S} \in A_S = \prod_{i \in S} A_i$ it holds that

$$\pi_i \left(a_{N \setminus S}, b_S\right) \leqslant \pi_i(a) \qquad \text{for all } i \in S. \tag{10}$$

Next we introduce the strong stability concept put forward by Jackson and van den Nouweland [27]. The next definition essentially transposes strong equilibrium conditions to network formation situations.

Definition 3.9 Let φ be a network payoff function on N and consider the corresponding Myerson model $\Gamma^m_\varphi = (\mathcal{A}^m, \pi^m)$.

(i) A network $g' \in \mathbb{G}^N$ **can be obtained from** network $g \in \mathbb{G}^N$ through the coordinated actions of coalition $S \subset N$ if $g' = g + h^+ - h^-$, where $h^+ \subset g^S = \{ij \mid i, j \in S\}$ and $h^- \subset \cup_{i \in S} L_i(g)$.
(ii) A network $g \in G^N$ is **strongly stable** if for every coalition $S \subset N$ and every network g' that is obtainable from network g through coordinated actions from coalition S it holds that $\varphi_i(g') > \varphi_i(g)$ for some player $i \in S$ implies that there exists some other player $j \in S$ with $\varphi_j(g') < \varphi_j(g)$.

It should be remarked that Dutta and Mutuswami [12] introduced a slightly different definition of "strong stability". They consider that all members of S need to be made strictly better off for a deviation to be successful.[8]

Strong equilibrium is a very demanding concept and these equilibria do not exist in many game-theoretic decision situations. Similarly, the notion of strong stability is equally demanding, implying that such networks rather unlikely exist. The next example illustrates these issues and introduces the notion of *costly* link formation that will be explored further in the next two subsections.

Example 3.10 (*Costly trade networks*) This example of a Walrasian trade network has been introduced seminally in Jackson and Watts [28] and further developed in Jackson and van den Nouweland [27] and Gilles et al. [16]. It considers an economy of n players who trade goods through connection paths. There are two commodities X and Y and all players are endowed with a Cobb–Douglas utility function $u(x, y) = \sqrt{xy}$. All players are assumed to have a commodity endowment of either $(1, 0)$ or $(0, 1)$ with an equal probability of $\frac{1}{2}$.

Players can trade with any other player that they are connected with, directly or indirectly. Hence, there emerge complete markets in each of the components. So, for $n = 5$ a network $g = \{12, 23, 45\}$ generates two components and two markets, namely, 123 separated from 45. Additional links, therefore, not always contribute to

[8] In the definition used by Jackson and van den Nouweland [27] a deviation needs to make all members of S to be at least as well off and making one member strictly better off.

the extent of these markets: $g' = \{12, 23, 13, 45\}$ results in exactly the same markets 123 and 45.

The cost c of forming any link ij is uniform and set at $c > \frac{1}{2}$. The costs of the formation of the trade network are divided equally among the members of a market, being a component of the network.

The network payoff function φ is now defined as the expected net benefits from participating in the generated market structure. This can be developed as follows.

First, consider the case of a market of the size two. There is a probability of $\frac{1}{2}$ that these two players have opposite endowments and a probability of $\frac{1}{2}$ that they have the same endowment. Hence, the probability of trade is $\frac{1}{2}$ resulting in a Walrasian allocation of $(\frac{1}{2}, \frac{1}{2})$ resulting in $\varphi = \frac{1}{2} \cdot \sqrt{\frac{1}{4}} - \frac{1}{2}c = \frac{1}{4} - \frac{1}{2}c < 0$.

More generally consider a market (component) of k players. The probability of r players having endowment $(0, 1)$ and $(k - r)$ players having endowment $(1, 0)$ is now

$$^{k}C_r \left(\tfrac{1}{2}\right)^{k} - r \cdot \left(\tfrac{1}{2}\right)^{r} = {}^{k}C_r \left(\tfrac{1}{2}\right)^{k}.$$

The expected gross payoff from trade is now given by

$$\tfrac{r}{2k} \cdot \left(\frac{k-r}{r}\right)^{\frac{1}{2}} + \tfrac{k-r}{2k} \cdot \left(\frac{r}{k-r}\right)^{\frac{1}{2}} = \frac{\sqrt{r(k-r)}}{k}.$$

Hence, taking into account that there are exactly $k - 1$ links required to build a market for k players, the resulting net payoff from this trade network is given by

$$\varphi = \frac{1}{k \cdot 2^k} \left[\sum_{r=1}^{k-1} {}^{k}C_r \sqrt{r(k-r)} \right] - \frac{(k-1)c}{k}.$$

Turning to $n = k = 3$ it can easily be computed that the net benefits to each player are given by

$$\varphi = \frac{\sqrt{2}}{4} - \frac{2c}{3} > 0 \text{ for } \tfrac{1}{2} < c < \tfrac{3\sqrt{2}}{8}.$$

For $n = k = 3$ and the given link formation cost range there are two pairwise stable networks, namely, the connected network and the (inefficient) empty network. The empty network is bilaterally stable, since creating a single link between two players is not beneficial for the given link formation cost range. On the other hand, the empty network is not strongly stable. Indeed, if all three players coordinate they would create two links to make a beneficial market among them.

This also shows that the connected component based on two links among the three players is strongly stable. ♦

In this section I discussed the pairwise stability concept and its variants in the link-based cooperative framework as seminally set out by Jackson and Wolinsky [29]. It is clear that these concepts are rather limited in their scope, since they are link-based only. Individual and collective incentives are not truly taken into account. Indeed, considerations are founded on adding and deleting links; the players' incentives are assumed to coincide with the (marginal) benefits generated from these links rather than the individualised payoffs. Next, I return to Myerson's original non-cooperative framework founded on the direct benefits to players to the formation of links.

4 Refinements of M-Networks

In this section I review stability and equilibrium concepts that refine the class of M-networks that emerges from the Myerson approach to non-cooperative network formation under mutual consent. This literature is founded on the insight that the class of M-networks is very large. This is the subject of the next theorem, which states the equivalence of the class of M-networks with the set of strong link deletion proof networks.

Theorem 4.1 *Let φ be a network payoff function on N and consider the corresponding Myerson model $\Gamma^m_\varphi = (\mathcal{A}^m, \pi^m)$.*

(a) A network $g \in \mathbb{G}^N$ is an M-network for φ if and only if g is strong link deletion proof for φ.
(b) Suppose that the network payoff structure φ is ***link monotone*** *in the sense that for every player $i \in N$, every network $g \in \mathbb{G}^N$ and every link $ij \notin L_i(g)$ it holds that $\varphi_i(g + ij) \geqslant \varphi_i(g)$. Then every network $g \in \mathbb{G}^N$ is supported as an M-network.*

For a proof of this theorem I refer to Appendix A.3.

The fundamental insights presented as Myerson's Lemma and Theorem 4.1 have motivated economists and social scientists to look into "refinements" of the Nash equilibrium concept in the Myerson model. These refinement equilibrium concepts have been developed particularly for addressing link formation issues from the perspective of consent. These attempts can be divided into two classes.

First, the standard approach in game-theoretic models of network formation is to strictly apply methodological individualistic perspectives. Thus, all motivations emanate from the player decision makers and are not considered to be external to the rational decision-making process. This has resulted in a number of equilibrium concepts that simply assume that decision-makers have a natural ability to cooperate if the incentives are in favour of such cooperation. Below I present the refinements considered by Bloch and Jackson [6] and Gilles et al. [16, 17].

The second approach is to explicitly assume that decision-makers are *not* fully individualistic, but adhere to some institutional or trusting norms of behaviour. Van de Rijt and Buskens [43] and Gilles and Sarangi [18] explicitly introduce a model of trusting behaviour through the introduction of an individualised belief or conjecture that other decision-makers will form links if they benefit from that. Thus, the trust in network formation is internalised into the player decision-makers; all such decision makers adhere to a well-defined norm of decision-making that expresses trusting behaviour. This is fully developed in Sect. 5.

Similarly, certain equilibrium concepts in non-cooperative game theory are founded on institutional signalling systems. The main such concept is Aumann's *correlated equilibrium*, which can be used to introduce institutional arrangements in the decision-making processes of players [2]. Here these institutions are explicitly modelled as external to these players. They adhere to these institutions since they benefit from applying these institutional behavioural rules instead of acting purely selfish. This is explored fully in Sect. 6.

4.1 Pairwise Nash Equilibrium and Bilateral Stability

Goyal and Joshi [20] introduced a refinement of the M-network concept that implements the idea of cooperation between players to modify the network through coordinated actions. Thus, it is assumed that decision-makers can implement bilateral or pairwise coordinated network modification. So, we consider any pair of players $i, j \in N$ who consider how to modify their strategic signals ℓ_i and ℓ_j to modify the resulting network in their favour.

This bilaterally coordinated action can be modelled in two different fashion. First, within the Myerson model as the so-called "pairwise" Nash equilibrium [20] and, second, as a network stability notion, denoted as "bilateral" stability [16].[9] This is introduced in the next definition.

Definition 4.2 Let φ be a network payoff function on N and consider the corresponding Myerson model $\Gamma^m_\varphi = (\mathcal{A}^m, \pi^m)$.

(i) A signal profile $\ell \in A^m$ is a **pairwise Nash equilibrium** in Γ^m_φ if ℓ is a Nash equilibrium in Γ^m_φ and for every pair of players $i, j \in N$ it holds that

$$\pi^m_i\left(\ell'_i, \ell'_j, \ell_{-i,j}\right) > \pi^m_i(\ell) \text{ implies that } \pi^m_j\left(\ell'_i, \ell'_j, \ell_{-i,j}\right) < \pi^m_j(\ell) \qquad (11)$$

for all deviations $\ell'_i \in A^m_i$ and $\ell'_j \in A^m_j$. (Here, $\ell_{-i,j}$ refers to the restricted signal profile $(\ell_h)_{h \neq i,j}$.)

(ii) A network $g \in \mathbb{G}^N$ is **bilaterally stable** for φ if g is strong deletion proof for φ and for every pair of players $i, j \in N$ and network $g' = g + \mathring{h} - h_i - h_j$ with $\mathring{h} \in \{\{ij\}, \varnothing\}$, $h_i \subset L_i(g)$ and $h_j \subset L_j(g)$ it holds that

$$\varphi_i(g') > \varphi_i(g) \text{ implies that } \varphi_j(g') < \varphi_j(g). \qquad (12)$$

[9] I remark here that I use a terminology that deviates from the literature. Indeed, the pairwise Nash equilibrium concept in the Myerson model was seminally introduced in Goyal and Joshi [20] and explored further by Bloch and Jackson [6] and Joshi et al. [30]. It refers to M-networks that are additionally link addition proof. Therefore, I use the notion of pairwise Nash equilibrium here in a slightly different way as introduced in Gilles et al. [16].

It is not hard to see that in the Myerson model there is a complete equivalence between these two concepts. The pairwise Nash equilibrium is simply a strategic formulation of bilateral stability. I give the following proposition therefore without proof.

Proposition 4.3 *Let φ be a network payoff function on N and consider the corresponding Myerson model $\Gamma^m_\varphi = (\mathcal{A}^m, \pi^m)$. A network $g \in \mathbb{G}^N$ is supported through a pairwise Nash equilibrium $\ell \in A^m$ with $g(\ell) = g$ if and only if g is bilaterally stable for φ.*

Although these concepts are quite natural within the context of network formation, the additional benefits are rather limited. Coordinated pairwise activity is well captured by the three pairwise stability concepts that have been introduced in this survey. The notion of unilateral stability (See Sect. 5) also captures coordinated action in the sense that it is assumed that players respond positively to a player's proposal to change the network if that is to their benefit. Bilateral stability does not extend this to pairs of players but reverts back to the normal best response rationality principle that others keep their actions unchanged.

Stability of higher orders. The notion of bilateral stability can easily be extended to the stability of higher orders. Indeed, under bilateral stability it is assumed that coalitions of two players can modify the network as proposed above. This can be extended to coalitions of at most r members, where $r \in \mathbb{N}$ is the assumed maximum size of the coalition under consideration. This is referred to as "stability of order r" in Gilles et al. [16]. In particular, if $r = n$, we arrive at the strong stability notion of Jackson and van den Nouweland [27]. This shows that these concepts represent the intermediate stability notion between M-networks and strongly stable networks.[10]

4.2 Two-Sided Link Formation Costs

Example 3.10 introduced the idea that there are normally link formation costs. In this particular case the costs of network formation are borne equally among all players that participate in the network. This signifies a collective approach to the allocation of network formation costs. It is more natural to assume that players only bear the costs of the links that they participate in. Next, I develop the idea of link formation costs further and refine the notion of M-networks to capture this.

In particular, I consider a modification of the Myerson model where the "intent to form links" is costly in the sense that approaching another player to form a link involves *explicit* investment of time, effort and energy. Hence, the act of sending a signal is costly. However, if the other player does not reciprocate and the link does not materialise, the player choosing to "reach out" still incurs this cost.[11] This means that if player $i \in N$ contemplates building a link ij with player $j \in N$ and sends a

[10] For results concerning these intermediate stability concepts, I refer to the quoted papers.

[11] This model of two-sided link formation costs was introduced in Gilles et al. [15] and developed further by Gilles and Sarangi [18], Gilles et al. [17].

message $\ell_{ij} = 1$, she incurs a cost of $c_{ij} > 0$. On the other hand, $\ell_{ij} = 0$ signifies no link is attempted to be made, which imposes *no* costs on player i.

Formally, a *link formation cost structure* can therefore be represented by a function $c\colon N \times N \to \mathbb{R}_+$ where $c(i, j) = c_{ij} \geqslant 0$ is the cost that player $i \in N$ incurs for sending a message to player $j \in N$, using the convention that $c(i, i) = 0$ for all $i \in N$. Hence, player i incurs a cost $c_{ij} \geqslant 0$ when communicating to player j that she wants to form a link. In particular, this cost refers to the effort to respond to messages sent by others. Obviously, if $c_{ij} = 0$, then there is no cost to communicating and sending messages from i to j.

This construction introduces the *consent model with two-sided link formation costs* as a modification of the (basic) Myerson model Γ^m_φ given as a non-cooperative game $\Gamma^a_\varphi(c) = (\mathcal{A}^a, \pi^a)$, where player i's strategy set is given by $A^a_i = A^m_i$ and player i's payoff for any strategy tuple $\ell \in A^a$ is given by

$$\pi^a_i(\ell) = \varphi_i(g(\ell)) - \sum_{j \neq i} \ell_{ij} \cdot c_{ij} = \pi^m_i(\ell) - \sum_{j \neq i} \ell_{ij} \cdot c_{ij}, \tag{13}$$

where $\varphi\colon \mathbb{G}^N \to \mathbb{R}^N$ is the network payoff function representing the gross benefits from network formation without taking into account the costs of link formation.

Our first result develops a complete characterisation of the Nash equilibria in the consent model with two-sided link formation costs. Part of this equivalence theorem was already stated without proof in Gilles and Sarangi [18] and as stated here is taken from Gilles et al. [17]. There are some preliminaries that need to be developed before stating the main assertion.

Definition 4.4 Let φ be a network payoff function on player set N and let $c\colon N \times N \to \mathbb{R}_+$ a link formation cost structure on N. Furthermore, let $\Gamma^a_\varphi(c) = (\mathcal{A}^a, \pi^a)$ be the associated consent model with two-sided link formation costs.

A strategy tuple $\ell \in A^a = A^m$ is **non-superfluous** in the consent model with two-sided link formation costs $\Gamma^a_\varphi = (\mathcal{A}^a, \pi^a)$ if for all pairs of players $i, j \in N$, $\ell_{ij} = 1$ if and only if $\ell_{ji} = 1$.

We call a non-superfluous strategy tuple $\ell \in A^a$ that is a Nash equilibrium a **non-superfluous Nash equilibrium**.

The main theorem states that in $\Gamma^a_\varphi(c)$ the networks that are supported by Nash equilibria are exactly the strong link deletion proof networks for a network payoff function that takes account of the link formation costs. For a proof of the next theorem I refer to Appendix A.4.

Theorem 4.5 *Let φ be a network payoff function on player set N and let $c\colon N \times N \to \mathbb{R}_+$ be a link formation cost structure on N. Furthermore, let $\Gamma^a_\varphi(c) = (\mathcal{A}^a, \pi^a)$ be the associated consent model with two-sided link formation costs.*

Then for every network $g \in \mathbb{G}^N$ the following three statements are equivalent:

(a) *Network g is supported by a Nash equilibrium of the consent model with two-sided link formation costs $\Gamma^a_\varphi(c)$.*

(b) *Network g is supported by a non-superfluous Nash equilibrium of the consent model with two-sided link formation costs* $\Gamma^a_\varphi(c)$.
(c) *Network g is strong link deletion proof with regard to the network payoff function* $\varphi^a : \mathbb{G}^N \to \mathbb{R}^N$ *given by*

$$\varphi^a_i(g) = \varphi_i(g) - \sum_{j \in N_i(g)} c_{ij}. \tag{14}$$

Theorem 4.5 provides a complete and detailed characterisation of the set of all Nash equilibria of the consent model with two-sided link formation costs. Furthermore, Theorem 4.5 clearly generalises the insight that the class of M-networks in the basic Myerson model is exactly the class of strong deletion proof networks under network payoff function φ.

In particular, each Nash equilibrium network is actually supported by a *unique* non-superfluous strategy profile if the cost structure is non-trivial in the sense that all link formation costs are positive. Gilles et al. [17] also discuss that there actually exist superfluous Nash equilibria if costs of link formation are zero for one of the players.

Example 4.6 ([17]) Consider the binary network formation situation with $N = \{1, 2\}$ and the network payoff function given by $\varphi_1(g^0) = \varphi_2(g^0) = \varphi_1(g^N) = 0$ and $\varphi_2(g^N) = 1$. Link formation costs are given by $c_{12} = 0$ and $c_{21} = 1$. Hence, we can derive that under two-sided link formation costs that $\varphi^a_i(g^0) = 0$ as well as $\varphi^a_i(g^N) = 0$, for $i = 1, 2$.

Clearly, the empty network g^0 is both (strong) link deletion proof for the net payoff function φ^a and supported by the superfluous Nash equilibrium characterised by $\ell_{12} = 1$ and $\ell_{21} = 0$. Of course, g^0 is also supported as a Nash equilibrium through its non-superfluous strategy profile $\ell^0_{12} = \ell^0_{21} = 0$ in (A^a, π^a). ♦

4.3 One-Sided Link Formation Costs

It is a natural extension to consider a network formation process under a one-sided cost structure. In this approach, one of the two linking players acts as the *initiator* and sends an initiation message to the other. If the other player, called the *responder*, chooses to reciprocate positively, the link materialises; otherwise, not. This link formation process has a similar nature as the process considered in Bala and Goyal [4], except that here the responder has to consent to the formation of the link, while in Bala–Goyal's model this is not required. There the initiator can create a link with the respondent in the absence of consent.

The decision-making process is more complex than that under two-sided link formation costs. Consequently, the action set has to be constructed differently. Following Gilles et al. [17], for each player i, we introduce a strategy set given by

$$A_i^b = \left\{ (l_{ij}, r_{ij})_{j \neq i} \;\middle|\; l_{ij}, r_{ij} \in \{0, 1\} \right\}. \tag{15}$$

This means that player i chooses to act as an initiator in forming a link with j if she initiates a message to j indicated as $l_{ij} = 1$. In this case, player j acts as the respondent and responds positively to this initiative if $r_{ji} = 1$. On the other hand, player j rejects the initiated link with i if $r_{ji} = 0$. Therefore, a link is only established if the initiated link is accepted, i.e. if $l_{ij} = r_{ji} = 1$. This is formalised as follows.

Let $A^b = \prod_{i \in N} A_i^b$ be the set of such communication profiles. Given the link formation process set out above, for any profile $(l, r) \in A^b$, the resulting network is now given by

$$g^b(l, r) = \{ij \in g^N \mid l_{ij} = r_{ji} = 1\}. \tag{16}$$

To delineate the one-sided model from the two-sided model, it is preferred to use a different notation for the incurred link formation costs. Instead, I introduce the function $\gamma\colon N \times N \to \mathbb{R}_+$ as the one-sided link formation cost structure. Here, when i initiates a link with j—represented by $l_{ij} = 1$—i incurs a cost of $\gamma_{ij} \geqslant 0$, regardless of whether the initialised link is accepted by j or not. On the other hand, responding to a link initialisation message is costless, i.e. j incurs no cost in responding to any message ℓ_{ij} sent by i in the link formation process.

For a given network payoff function φ on N this now results in the following net payoff function for player i:

$$\pi_i^b(l, r) = \varphi_i\left(g^b(l, r)\right) - \sum_{j \neq i} l_{ij} \cdot \gamma_{ij}. \tag{17}$$

Formally, let φ be a network payoff function on N and let $\gamma\colon N \times N \to \mathbb{R}_+$ be a given one-sided link formation cost structure. Then we refer to the non-cooperative game in strategic form $\Gamma_\varphi^b(\gamma) = (\mathcal{A}^b, \pi^b)$ as the *consent model of network formation with one-sided link formation costs*.

Nash equilibria of the consent model with one-sided link formation costs. As before, we can now introduce non-superfluous strategy tuples in the consent model with one-sided link formation costs:

Definition 4.7 Let φ be a network payoff function on N and let $\gamma\colon N \times N \to \mathbb{R}_+$ be a given one-sided link formation cost structure. Consider the corresponding consent model with one-sided link formation costs $\Gamma_\varphi^b(\gamma) = (\mathcal{A}^b, \pi^b)$.

Then a strategy profile $(l, r) \in A^b$ is **non-superfluous** if for all pairs $i, j \in N$ it holds that

$$l_{ij} = 1 \text{ implies that } r_{ji} = 1 \text{ as well as } l_{ji} = r_{ij} = 0, \text{ and} \tag{18}$$

$$r_{ij} = 1 \text{ implies that } l_{ji} = 1 \text{ as well as } l_{ij} = r_{ji} = 0. \tag{19}$$

Unlike for the consent model with two-sided link formation costs, each network is no longer supported by a unique non-superfluous strategy profile. Indeed, it depends on who of the two players involved initiates and who responds in the link formation process.

On the other hand, under a non-superfluous strategy profile, only one player bears the establishment cost of each existing link, and every initialisation is responded to positively. As a first step in the analysis of this one-sided approach, I explore the relationship between the Nash equilibria of the two-sided and the one-sided model. Secondly, I present a full characterisation of the Nash equilibria of the one-sided model in terms of network stability properties. These results are taken from Gilles et al. [16].

The main question to be considered here is whether there is a network payoff function which would provide equivalence between Nash equilibria of the one-sided model and strong link deletion proofness with regard to a payoff function in a similar fashion as Theorem 4.5 for two-sided link formation costs. In particular, I follow efficiency logic and consider a payoff function which only assigns link formation costs to the player with the lower cost of link formation. If link formation costs are equal, a tie-breaking rule is applied.

Let $M_i(g) = \{j \in N_i(g) \mid \gamma_{ij} < \gamma_{ji} \text{ or } \gamma_{ij} = \gamma_{ji},\ i < j\} \subset N_i(g)$ be the potential links that player i should finance based on incurring the lowest link formation costs. The corresponding payoff function φ^b is defined for $i \in N$ by

$$\varphi_i^b(g) = \varphi_i(g) - \sum_{j \in M_i(g)} \gamma_{ij}$$

given the network payoff function φ representing benefits without taking into account costs of link formation. We can show the following implication, which proof is relegated to Appendix A.5.

Theorem 4.8 *Let φ be a network payoff function on N and let $\gamma\colon N \times N \to \mathbb{R}_+$ be a given one-sided link formation cost structure. If network $g \in \mathbb{G}^N$ is strong link deletion proof for the net payoff function φ^b, then g can be supported by a non-superfluous Nash equilibrium in the consent model with one-sided link formation costs $\Gamma_\varphi^b(\gamma) = (A^b, \pi^b)$.*

The converse of Theorem 4.8 does not hold as shown by the following counter-example.

Example 4.9 Consider the minimal binary network formation situation with $N = \{1, 2\}$ and network payoffs given by $\varphi_1(g^0) = \varphi_2(g^0) = 0, \varphi_1(g^N) = 2$ and $\varphi_2(g^N) = 10$. Link formation costs are given by $\gamma_{12} = 5$ and $\gamma_{21} = 7$.

Hence for $i = 1, 2, \varphi_i^b(g^0) = 0, \varphi_1^b(g^N) = -3$ and $\varphi_2^b(g^N) = 3$. Clearly, the complete network g^N is not link deletion proof for the network payoff function φ^b, since player 1 would benefit from severing the unique link 12.

However, there is a Nash equilibrium of the one-sided consent model $\Gamma^b_\varphi(\gamma) = (A^b, \pi^b)$ that supports the complete network g^N: $l_{12} = 0$; $r_{12} = 1$; $l_{21} = 1$; $r_{21} = 0$.[12] ♦

One might expect that a network payoff function that assigns a link initiator role to the player with the higher marginal net benefits as a result of formation of the link in question might resolve the issue of characterising the supported equilibrium networks in $\Gamma^b_\varphi(\gamma) = (A^b, \pi^b)$. Below it is shown that this is actually not the case.

Example 4.10 Consider a situation with three players, $N = \{1, 2, 3\}$. The following table gives the benefits for each of the three players in the case of the formation of one of only three relevant networks:

Network g	$\varphi_1(g)$	$\varphi_2(g)$	$\varphi_3(g)$
$\{12\}$	10	10	0
$\{13\}$	10	0	10
$\{12, 13\}$	15	20	20

All other networks generate no benefits to any of the three players, i.e. $\varphi_i(g) = 0$ for all other networks g not listed in the table.

Consider the following one-sided link formation cost structure: $\gamma_{12} = \gamma_{13} = 9$, $\gamma_{21} = 10$, $\gamma_{31} = 10$, and $\gamma_{23} = \gamma_{32} = 10$. Within this context, player 1 has the highest marginal net benefit from forming links 12 as well as 13, namely $\varphi_1(\{12\}) - \gamma_{12} = \varphi_1(\{13\}) - \gamma_{13} = 1$, while the other players have no positive marginal benefits from forming links 12 and 13.

Now, the network $\{12, 13\}$ is not link deletion proof for the network payoff function that is based on the property that the player with the highest net marginal benefit is assumed to finance the formation of a link. Indeed, player 1—who has the highest net marginal benefits from both links—has a negative net return from forming network $\{12, 13\}$ and would prefer to sever one of the two links to increase her net benefit to 1.

On the other hand, $\{12, 13\}$ is supported by a non-superfluous Nash equilibrium strategy profile under one-sided link formation costs with $l_{21} = r_{12} = 1$ and $l_{31} = r_{13} = 1$. ♦

These examples show that the problem of finding a reasonable payoff function that completely characterises all Nash equilibria of the one-sided consent model in terms of network stability remains open. The issues are such that it can be argued that there is actually no reasonable network payoff function that characterises all supported equilibrium networks in the consent model under one-sided link formation costs.

[12] Note that in the case of two-sided link formation costs, the cost of link formation is a total of $\gamma_{12} + \gamma_{21} = 7 + 5 = 12$, which clearly makes the complete network g^N not being supported by a Nash equilibrium in $\Gamma^a_\varphi(\gamma)$. This indicates the underlying reason why two-sided link formation costs shrink the set of supported networks in comparison with the case of one-sided link formation costs.

Multi-stage network formation under one-sided link formation costs. One can ask whether certain other approaches can resolve the coordination and free-riding issues that are indicated in the discussion of the converse of Theorem 4.8 above.[13]

Here, I consider a two-stage network formation process to restore equivalence between equilibria of that model under one-sided costs and strong link deletion proofness with respect to some well-constructed network payoff function. This is motivated by the fact that often sequential decision-making solves coordination problems. With this in mind, consider the following natural two-stage process:

(i) In the first stage, every players $i \in N$ initiates links by selecting initiation messages $(l_{ij})_{j \neq i}$.
(ii) In the second stage, all players respond to links initiated in the first stage and select $(r_{ij} : l_{ji} = 1)_{j \neq i}$.

The question is whether the subgame perfect Nash equilibria of this game are strong link deletion proof with regard to φ^b. We show that this is not necessarily the case.

Example 4.11 Reconsider the simple binary linking situation in Example 4.9. We showed earlier that the complete network $g^N = \{12\}$ is not (strong) link deletion proof for the net payoff function φ^b but that there is a Nash equilibrium communication profile of the one-sided model that supports it, namely, $l_{12} = 0$; $r_{12} = 1$; $l_{21} = 1$; $r_{21} = 0$.

We now show that in the two-stage network formation process described above, this communication profile is subgame perfect as well. Consider the reduced game in the second stage, given that $l_{12} = 0$ and $l_{21} = 1$ has been chosen in the first stage. In normal form it can now be represented as the matrix game

	0	**1**
0	0, −7	0, −7
1	2, 3	2, 3

There are two Nash equilibria in this game, one of which is $r_{12} = 1$ and $r_{21} = 0$. This is exactly the second part of the indicated communication profile. Thus, the given communication profile is indeed a subgame perfect equilibrium in the two-stage link formation process. ♦

The reason why sequential decision-making cannot resolve the coordination problem is that here the problem stems from costs not being transferable. Complete transferability of costs and benefits would take us into the framework of Jackson and Wolinsky [29] and, in particular, Bloch and Jackson [6, 7].

A formal comparison of one-sided and two-sided link formation costs. Since the two models that we considered in this section have different philosophical bases, we must make some simplifying assumptions to enable a more formal comparison. In

[13] This discussion requires knowledge of multi-stage, sequential games and the notion of subgame perfection. This discussion can be skipped without any difficulty. For more elaborate discussion of multi-stage and sequential games I refer to Osborne [39], Harrington [22] and Maschler et al. [32].

particular, we have to address how the two different link formation cost formulations are related. This simply requires us to formulate the one-sided cost structure γ in terms of the two-sided cost structure c. Hence, we consider γ to be a particular functional form of c.

I look at two simplified cases that facilitate this comparison.

Case A: The initiator bears all. Suppose that the initiator in the model with one-sided costs bears both his cost and the cost of the responder in the context of the two-sided consent model. So, initiation is tantamount to bearing the total cost of link formation, i.e. $\gamma_{ij} = c_{ij} + c_{ji}$ for all $i \neq j$. Benefits described by φ remain individualised and are not transferable.

In this case, it is quite obvious that the Nash equilibria of the two models are not comparable, which is shown in the next simple example.

Example 4.12 Consider again a binary link formation situation with $N = \{1, 2\}$ and $\varphi_i(g^N) = 51$, $\varphi_i(g^0) = 0$, $i = 1, 2$. Moreover, let $c_{12} = c_{21} = 50$. Hence, $\gamma_{12} = \gamma_{21} = 100$. Then, $g^N = \{12\}$ is supported by a Nash equilibria of the two-sided model, namely through $\ell_{12} = \ell_{21} = 1$. But there is no Nash equilibrium in the one-sided model that would support it because no one would be willing to pay a cost of 100 in order to sustain this link.

Next, modify the situation to let $\varphi_1(g^N) = 12$, $\varphi_2(g^N) = 2$, $\varphi_i(g^0) = 0$, $i = 1, 2$ and $c_{12} = c_{21} = 5$. Hence, $\gamma_{12} = \gamma_{21} = 10$. Then, $g^N = \{12\}$ is now supported by a Nash equilibrium of the one-sided model, namely, through $l_{12} = r_{21} = 1$, $l_{21} = r_{12} = 0$. The strategy supporting this network is not a Nash equilibrium in the two-sided model. ♦

Case B: A sunk cost formulation. Next, we consider the case in which the link formation costs are not transferable and that the initiator has to bear only his own cost. This corresponds to a scenario where the costs of the responding party are sunk and, thus, not relevant to the decision-making process.

Hence, we assume that $\gamma_{ij} = c_{ij}$ for all $i \neq j$. In this case, it can be shown that networks supported by Nash equilibria of the two two-sided model are also supported by some Nash equilibrium of the one-sided model, while the converse does not hold. For a proof of the next theorem I refer to Appendix A.6.

Theorem 4.13 *Let φ be a network payoff function on player set N and let $c\colon N \times N \to \mathbb{R}_+$ a two-sided link formation cost structure on N.*

If a network $g \in \mathbb{G}^N$ is supported by a Nash equilibrium of the consent model with two-sided link formation costs $\Gamma^a_\varphi(c)$, then there exists a non-superfluous Nash equilibrium supporting network g in the consent model with one-sided link formation costs $\Gamma^b_\varphi(c)$, i.e. for one-sided link formation cost structure γ given by $\gamma_{ij} = c_{ij}$ for all $i, j \in N$.

We show that the converse of Theorem 4.13 does not hold.

Example 4.14 Consider again the binary link formation situation with $N = \{1, 2\}$. Furthermore, assume now that $\varphi_1(g^0) = \varphi_2(g^0) = 0$, $\varphi_1(g^N) = 6$ and $\varphi_2(g^N) = 4$. Let two-sided costs of link formation be uniform, given by $c_{ij} = 5$ for all $i, j \in N$.

The complete network $g^N = \{12\}$ initiated by player 1 is supported by a Nash equilibrium in the one-sided model for $\gamma_{ij} = c_{ij}$. But the strategy tuple $\ell_{12} = \ell_{21} = 1$ in the two-sided model that supports this network is not a Nash equilibrium in that model. ♦

This discussion shows that one-sided link formation processes require a very careful analysis and do not necessarily result in very delineated conclusions.

5 Trust and Network Formation

In this section I review some concepts that try to capture the fundamental idea that "trust builds networks". These concepts go beyond the approaches that I have reviewed thus far, being Myerson's model and its variations as well as the Jackson–Wolinsky approach to incorporate cooperative conceptions into a network formation setting.

I discuss two different implementations of trusting behaviour into network formation. First, van de Rijt and Buskens [43] consider the notion of *unilateral stability* that is founded on the principle that players attempt the formation of links even if their correspondents did not signal that they would necessarily agree to the formation of these links. Thus, players follow the rule that one should certainly try to form links if one expects the correspondent to benefit from its formation. This leads to a refinement of the class of M-networks.

A similar conception has been developed by Gilles and Sarangi [18]. Within the consent model under two-sided link formation costs Gilles and Sarangi [18] developed a belief-based stability concept denoted as *monadic stability* for understanding a purely non-cooperative process of network formation based on trusting behaviour. Again players are assumed to pursue the formation of links if they perceive the correspondents to benefit from their creation. However, monadic stability is defined as a self-confirming equilibrium [14] based on these belief systems, deviating considerably from van de Rijt and Buskens [43]s conception of trusting behaviour.

5.1 Unilateral Stability

The mathematical sociologists van de Rijt and Buskens [43] proposed a refinement of the Nash equilibrium concept that considers expanding a player's ability to affect the network that is formed in a broader way than allowed through best response rationality underlying the Nash equilibrium concept. They recognised that the multitude of Nash equilibria in the Myerson model is due to a simple (mis-)coordination problem: Players are indifferent between proposing or not proposing a link if the other player actually does not propose the link herself already. This resulted in a refinement of the Nash equilibrium concept that takes account of the idea that players trust that mutually beneficial link formation will indeed be pursued by other players.

Definition 5.1 Let φ be a network payoff function on N and consider the corresponding Myerson model $\Gamma^m_\varphi = (\mathcal{A}^m, \pi^m)$. A network $g \in \mathbb{G}^N$ is **unilaterally stable** if there exists a strategy profile $\ell \in A^m$ in the Myerson model with $g(\ell) = g$ such that

(i) for all $i \in N$ and $\ell'_i \in A^m_i : \pi^m_i(\ell) \geqslant \pi^m_i(\ell'_i, \ell_{-i})$ *(Nash equilibrium condition)*, and
(ii) for every $i \in N$ and every alternative strategy $\ell'_i \in A^m_i$, it holds that

$$\pi^m_i(\ell^\star) > \pi^m_i(\ell)$$

implies that there is some $j \in N$ with $\ell'_{ij} = 1$ and $\ell_{ij} = 0$ for whom

$$\pi^m_j(\ell^\star) < \pi^m_j(\ell),$$

where $\ell^\star \in A^m$ is given by $\ell^\star_i = \ell'_i$, $\ell^\star_{jk} = \ell_{jk}$ for $j \neq i \neq k$ and $\ell^\star_{ji} = \ell'_{ij} = 1$ for $j \neq i$.

A network is unilaterally stable if it is supported through a Nash equilibrium in the Myerson model under the additional provision that every player can modify her direct neighbourhood provided that this modification can be constructed with the consent of her chosen neighbours. So, if i's proposal would make herself better off, then all newly selected neighbours would have no objections and would not receive lower payoffs as a consequence of this modification of the network.

Unilateral stability introduces a form of trusting behaviour into the Myerson approach to network formation under mutual consent. The consent of any player's neighbours is reasoned by that player is conducted in such a way that it reflects trusting behaviour by that particular player. In some sense it introduces a *bounded* form of rationality of any player in her consideration of how other players respond to changes in her behaviour. As such the notion of unilateral stability can be categorised as a model of trusting behaviour in network formation under mutual consent.

An alternative definition of unilateral stability is also possible as captured in the proposition below. It reflects the idea to add trusting behaviour to the M-network concept.

Proposition 5.2 (An alternative definition of unilaterally stable networks) *A network $g \in \mathbb{G}^N$ is unilaterally stability if and only if g is an M-network such that for every player $i \in N$ and all link sets $h^-_i \subset L_i(g)$ and $h^+_i \subset L_i(g^N \setminus g)$ it holds that* either $\varphi_i(g - h^-_i + h^+_i) \leqslant \varphi_i(g)$ or $\varphi_i(g - h^-_i + h^+_i) > \varphi_i(g)$ *implies there is some $j \in N$ such that $ij \in h^+_i$ and $\varphi_j(g - h^-_i + h^+_i) < \varphi_j(g)$.*

Unilateral stability is the strongest individualistic or "monadic" network formation concept that has been proposed in the literature. Indeed, going beyond the unilateral formation of links under consent as formulated here would actually involve active participation of multiple players.

Next, we turn to discussing some simple properties of unilateral stability.

Proposition 5.3 *Let φ be a network payoff function on N and consider the corresponding Myerson model $\Gamma^m_\varphi = (\mathcal{A}^m, \pi^m)$. Then the following properties hold:*

(a) Every unilaterally stable network is strongly pairwise stable.
(b) There exist strictly pairwise stable networks that are not unilaterally stable.
(c) If the network payoff structure φ is link monotone, then $g^N \in \mathbb{G}^N$ is the unique unilaterally stable network for φ.

I prove all three assertions in Proposition 5.3 in an informal fashion, rather than rigorously.

First, from Proposition 4.1 it follows that every M-network g is strong link deletion proof. Furthermore, applying the unilateral stability condition to a single link $ij \in g$ reduces to the LAP property. This immediately shows Proposition 5.3(a).

Next, if the network payoff structure is link monotone, then there are no objections of any player to add more links to an existing network. Hence, the complete network g^N is the only M-network that satisfies the unilateral stability condition, implying the assertion stated as Proposition 5.3(c).

Finally, to show Proposition 5.3(b), I device an example for the case of three players. This example also has an important role to assess the relationship between unilateral stability and other stability concepts, introduced further down in this chapter.

Example 5.4 Consider three players $N = \{1, 2, 3\}$ and a network payoff structure φ given in the next table.

Network g	$\varphi_1(g)$	$\varphi_2(g)$	$\varphi_3(g)$	**Stability**
$g^0 = \varnothing$	0	0	0	Strongly PS
$g^1 = \{12\}$	0	0	2	Strictly PS
$g^2 = \{13\}$	0	0	0	
$g^3 = \{23\}$	0	0	0	
$g^4 = \{12, 13\}$	−1	0	0	
$g^5 = \{12, 23\}$	0	−1	0	
$g^6 = \{13, 23\}$	0	1	1	
$g^7 = g^N$	3	3	3	U-stable

Here, g^0 is strongly pairwise stable but is not unilaterally stable. Indeed, player 3 can add both links 13 and 23 to make g^6 without objection of the other players.

Furthermore, g^1 is strictly pairwise stable and again not unilaterally stable. As before, player 3 can add links 13 and 23 to move to g^N without any objections of the other two players. This shows assertion Proposition 5.3(b).

Also, it is clear from the table that the complete network g^N is unilaterally stable, since it is strong link deletion proof. Note that in this case g^N is strictly pairwise stable as well.

Finally, I refer to Example 5.10 for a detailed discussion of an example in which assertion of Proposition 5.3(b) is strengthened in the sense that the class of strictly pairwise stable networks is completely disjoint from the class of unilaterally stable networks. ♦

To assess unilateral stability, it is clear that [43] introduce it as an expression of firmly methodological individualistic behavioural principles: Players act selfishly only, but conjecture that other players will consent to the creation of links that directly benefit them. It builds on the hypothesis that players offer no objections to the formation of links that directly benefit them.

However, an alternative interpretation can easily be applied here as well. Indeed, the unilateral stability concept can be interpreted to be an application of a principle of trusting behaviour: players trust others to consent to forming links if it does not hurt them. This is closely akin to the model of trusting behaviour. An alternative model of trusting behaviour founded on belief systems in Myerson's framework is discussed next.

5.2 *Monadic Stability*

Gilles and Sarangi [18] introduced a belief-based conception of trusting behaviour in the setting of the consent model with two-sided link formation costs. Their approach imposes minimal informational requirements. Unlike other models of strategic network formation, players need not be aware of the payoffs associated with every network. For any given network $g \in \mathbb{G}^N$ to emerge in such a setting, a player is required to know the payoffs associated with any change (creation or deletion) only involving their own direct links $ij \in L_i(g)$.

This results in an amendment of Myerson's consent game such that, based on their information, players form simple, myopic beliefs about the direct benefits other players will receive from establishing links with them. According to these myopic beliefs, each player $i \in N$ assumes that another player $j \in N$ is willing to form a new link with i if j stands to benefit from it in the prevailing network. Similarly i also assumes that j will break an existing link ij in the prevailing network if j does not benefit from having this link. Thus, in this process player i assumes that all other links in the prevailing network remain unchanged.

Therefore, these monadic beliefs are indeed "myopic" in the sense that they only pertain to direct effects of the addition or removal of a link in the network. Hence, these beliefs disregard higher order effects on the payoffs of all players in the network due to the addition or removal of such a link. As such these behavioural standards reflect a *bounded* form of rationality in decision-making, implying that the boundedly rational foundation of monadic stability is fundamentally different from the rational standard imposed by unilateral stability.

Such myopic beliefs essentially capture the idea that network formation primarily occurs between acquaintances with sufficiently large an amount of information about each other to assess first-order effects of network changes.[14] This concept is a normal form implementation of the self-confirming equilibrium concept introduced by Fudenberg and Levine [14] within the setting of the Myerson model and its variations.

One can assess these myopic belief systems as reflecting a certain form of "confidence" on the part of each player to engage in communication to form links with other players that have an obvious (first-order) benefit from the addition of such a link. This confidence suffices to form non-trivial social networks. As stated, a certain commonality is assumed among the players in order to formulate such common priors and beliefs on which this confidence is founded. In this regard we assume that players are acquaintances and build relationships through beliefs about actions undertaken by other players.[15]

We now formalise these myopic belief systems for the consent model under two-sided link formation costs.

Defining monadic stability. Throughout we assume there is a given network payoff function $\varphi\colon \mathbb{G}^N \to \mathbb{R}^N$ and we impose a two-sided link formation cost structure $c = (c_{ij})_{i,j \in N}$. Based on this data, consider the corresponding consent model under two-sided link formation costs $\Gamma^a_\varphi(c) = (\mathcal{A}^a, \pi^a)$. We can introduce specific belief systems in this setting that represent the trusting behavioural principle as discussed above.

Definition 5.5 Let $\ell \in A^a$ be an arbitrary communication profile resulting in network $g = g(\ell)$. For every player $i \in N$ we define i's **monadic belief system** concerning ℓ as a communication profile $\ell^{i\star} \in A^a$ given by

(i) for every $j \neq i$ with $ij \in g$ let

- $\ell^{i\star}_{ji} = 0$ if $\varphi_j(g - ij) + c_{ji} > \varphi_j(g)$ and
- $\ell^{i\star}_{ji} = 1$ if $\varphi_j(g - ij) + c_{ji} \leqslant \varphi_j(g)$;

(ii) for every $j \neq i$ with $ij \notin g$ let

- $\ell^{i\star}_{ji} = 0$ if $\varphi_j(g + ij) - c_{ji} < \varphi_j(g)$ and
- $\ell^{i\star}_{ji} = 1$ if $\varphi_j(g + ij) - c_{ji} \geqslant \varphi_j(g)$;

(iii) and for all $j, k \in N$ with $j \neq i \neq k$ let $\ell^{i\star}_{jk} = \ell_{jk}$.

A monadic belief system reflects that a player believes that other players are myopically selfish and will act in their myopic self-interest. Hence, links are consented to if that directly benefits the other player and are refused if deleting that link benefits the other player.

---

[14] That social relations are mainly formed between acquaintances is confirmed empirically by Wellman et al. [46] using data from the East York area. This principle also forms the foundation of the model in Brueckner [8], who models friendship as building links between players chosen from a given set of acquaintances.

[15] It is clear that this approach is akin to the notion of unilateral stability introduced before. A comparison of monadic stability with unilateral stability is, therefore, called for. This is further developed here as well.

Now monadic stability simply requires that each player acts rationally in view of these beliefs.

Definition 5.6 Let φ and c be given with the corresponding consent model under two-sided link formation costs $\Gamma^a_\varphi(c) = (\mathcal{A}^a, \pi^a)$.

(a) A network $g \in \mathbb{G}^N$ is **weakly monadically stable** for (φ, c) if there exists some communication profile $\ell \in A^a$ with $g = g(\ell)$ such that for every $i \in N$: $\ell_i \in A^a_i$ is a best response to her monadic beliefs $\ell^{i\star}_{-i} \in A^a_{-i}$ for payoff function π^a; thus,

$$\pi^a_i \left(g(\ell'_i, \ell^{i\star}_{-i}) \right) \leqslant \pi^a_i \left(g(\ell_i, \ell^{i\star}_{-i}) \right) \tag{20}$$

for all $\ell'_i \in A^a_i$.

(b) A network $g \in \mathbb{G}^N$ is **monadically stable** for (φ, c) if there exists some communication profile $\ell \in A^a$ with $g = g(\ell)$ such that for every $i \in N$: $\ell_i \in A^a_i$ is a best response to her monadic beliefs $\ell^{i\star}_{-i} \in A^a_{-i}$ for payoff function π^a and player i's monadic belief system $\ell^{i\star}$ is confirmed in the sense that for every $j \neq i$ it holds that $\ell^{i\star}_{ji} = \ell_{ji}$.

Weak monadic stability of a network is founded on the principle that every player $i \in N$ anticipates—as captured by her (monadic) expectations about direct links—that other players will respond myopically selfishly to her attempts to form a link with them. Note that ℓ_{-i} is fully replaced by the player's belief system $\ell^{i\star}_{-i}$ in the standard best response formulation of Nash equilibrium for player i and is therefore irrelevant for the decision-making process of i.

Monadic stability strengthens the above concept by requiring that the beliefs of each player are confirmed in the resulting equilibrium. Hence, monadic stability imposes a self-confirming condition on the weakly monadic equilibrium. This describes the situation that all players are fully satisfied with their beliefs; the observations that they make about the resulting network confirm their beliefs about the other players' payoffs. This amounts to updating one's initial beliefs. As such, monadic stability is an implementation of a *self-confirming equilibrium* based on the monadic belief system in the context of consent model with two-sided link formation costs [14].

To delineate the two monadic stability concepts for networks, we discuss a three-player example. This example shows that the class of monadically stable networks is usually strictly larger than the class of the weakly monadically stable networks.

Example 5.7 Consider $N = \{1, 2, 3\}$ and assume uniform link formation costs with $c_{ij} = 1$ for all $i, j \in N$. Let the network payoff function φ be given in the table below:

Network g	$\varphi_1(g)$	$\varphi_2(g)$	$\varphi_3(g)$	Stability
$g^0 = \varnothing$	0	0	0	M_w
$g^1 = \{12\}$	0	1	0	
$g^2 = \{13\}$	0	0	3	
$g^3 = \{23\}$	0	0	0	
$g^4 = \{12, 13\}$	3	0	0	
$g^5 = \{12, 23\}$	1	3	3	
$g^6 = \{13, 23\}$	2	2	5	M_w
$g^7 = g^N$	3	5	6	M_w and M

This table identifies whether the network in question is weak monadically stable—indicated by M_w—or whether it is monadically stable—indicated by M.

Within this example we now consider some of the networks given and analyse their stability properties.

Network g^0 : We show that this network is weakly monadically stable for a supporting communication profile that is superfluous. Indeed, select $\ell_0 = (\,(1, 1), (0, 0), (0, 0)\,) \in A^a$ with $g(\ell_0) = g^0 = \varnothing$. Observe here that player 1 incurs link formation costs with $\pi_1^a(\ell_0) = -2$, while $\pi_2^a(\ell_0) = \pi_3^a(\ell_0) = 0$. Then we can determine the monadic belief systems for all players as

$$\ell_0^{1\star} = (-, (1, 0), (1, 0)\,)$$
$$\ell_0^{2\star} = ((0, 1), -, (0, 0)\,)$$
$$\ell_0^{3\star} = ((1, 0), (0, 0), -\,).$$

It should be emphasised that in this case player 1 believes that both other players are willing to make links with her, because there are direct benefits from forming such links. However, the other players believe that player 1 will not attempt to make a link with them, because she has no direct (net) benefits from doing so. This refers to a classical coordination problem.

Now we determine that the best responses for all players are given by

- $\beta_1\left(\ell_0^{1\star}\right) = (1, 1)$ is the unique best response to $\ell_0^{1\star}$ for player 1.
- $\beta_2\left(\ell_0^{2\star}\right) = (0, 0)$ is the unique best response to $\ell_0^{2\star}$ for player 2.
- $\beta_3\left(\ell_0^{3\star}\right) = (0, 0)$ is the unique best response to $\ell_0^{3\star}$ for player 3.

This confirms that g^0 is indeed weakly monadically stable for ℓ_0. However, g^0 is not monadically stable, since in the communication profile ℓ_0, player 1's beliefs are not confirmed. She expects the other two players to be willing to form links with her, although they do not do so.

Network g^5 : This network is neither weakly monadically stable, nor monadically stable. The non-superfluous communication profile $\ell_5 = (\,(1, 0), (1, 1), (0, 1)\,)$ is an obvious candidate to support this network. For this profile we compute that

$$\ell_5^{1\star} = (-, (1, 1), (1, 1))$$
$$\ell_5^{2\star} = ((1, 0), -, (0, 1))$$
$$\ell_5^{3\star} = ((1, 1), (1, 1), -).$$

This results into the following best response configuration:

- $\beta_1\left(\ell_5^{1\star}\right) = (1, 1)$ is the unique best response to $\ell_5^{1\star}$ for player 1.
- $\beta_2\left(\ell_5^{2\star}\right) = (1, 1)$ is the unique best response to $\ell_5^{2\star}$ for player 2.
- $\beta_3\left(\ell_5^{3\star}\right) = (1, 1)$ is the unique best response to $\ell_5^{3\star}$ for player 3.

From this it is clear that g^5 cannot be supported by ℓ_5. This illustrates that weak monadic stability requires selecting a best response to a *specific* set of beliefs for each player $i \in N$. Without such a restriction on the beliefs it would be possible to support any strategy as weakly monadic stable. Moreover, observe that players only form beliefs about the behaviour of their acquaintances with regard to direct links, making it myopic but realistic. In fact, because of this, it is possible that monadically stable equilibria do not exist.

Finally, we can complete the argument by checking that other communication profiles can be ruled out in similar fashion.

Network g^6 : We argue that this network is weakly monadically stable as well. We can show that g^6 is supported by the action tuple $\ell_6 = ((0, 1), (1, 1,), (1, 1))$. Again we compute

$$\ell_6^{1\star} = (-, (1, 1), (1, 1))$$
$$\ell_6^{2\star} = ((1, 1), -, (1, 1))$$
$$\ell_6^{3\star} = ((0, 1), (1, 1), -).$$

Note here that player 1 is indifferent between g^6 and g^7 in terms of her net payoff π^a. Thus, in the computation of $\ell_6^{2\star}$ we use the bias of player 1 towards having more links rather than fewer in player 2's belief system.

This results in the following best response configuration:

- $\beta_1\left(\ell_6^{1\star}\right) = \{(0, 1), (1, 1)\}$ is the set of best responses to $\ell_6^{1\star}$ for player 1, i.e. (0, 1) and (1, 1) are both best responses for this player.
- $\beta_2\left(\ell_6^{2\star}\right) = (1, 1)$ is the unique best response to $\ell_6^{2\star}$ for player 2.
- $\beta_3\left(\ell_6^{3\star}\right) = (1, 1)$ is the unique best response to $\ell_6^{3\star}$ for player 3.

This shows that ℓ_6 is indeed supported as a weak monadically stable communication profile. On the other hand, g^6 is not monadically stable, since the beliefs of player 2 are not confirmed.

Network g^7: First, we claim that this network is strictly pairwise stable. Strong link deletion proofness follows trivially from the payoffs listed. Indeed, the net payoffs in other networks $(g^0, \dots, g^6)$ are at most the net payoff in g^7 for all players. Second, strict link addition proofness is trivially satisfied since there are no links that are not part of $g^7 = g^N$.

Furthermore, the complete network $g^7 = g^N$ is weakly monadically stable. We claim that g^7 is supported by the only communication profile supporting this network, $\ell_7 = ((1,1),(1,1,),(1,1))$. We can determine that the monadic belief systems are given by

$$\ell_7^{1\star} = (-, (1,1), (1,1))$$
$$\ell_7^{2\star} = ((1,1), -, (1,1))$$
$$\ell_7^{3\star} = ((1,1), (1,1), -).$$

From this we conclude that

- $\beta_1\left(\ell_7^{1\star}\right) = \{(0,1),(1,1)\}$ is the set of best responses to $\ell_7^{1\star}$ for player 1.
- $\beta_2\left(\ell_7^{2\star}\right) = (1,1)$ is the unique best response to $\ell_7^{2\star}$ for player 2.
- $\beta_3\left(\ell_7^{3\star}\right) = (1,1)$ is the unique best response to $\ell_7^{3\star}$ for player 3.

So, ℓ_7 is indeed a best response profile with regard to the generated monadic belief systems. Hence, g^7 is indeed weakly monadically stable.
Finally, all players' monadic belief systems are confirmed here. So, in fact, g^7 is monadically stable.

In this example, it is made clear that the introduced monadic belief systems require only that players use minimal information about each other's payoffs to formulate appropriate expectations about each other's linking behaviour. Indeed, monadic stability only considers players to use first-order effects of forming new links and deleting existing links to formulate their monadic beliefs. ♦

This example clarifies the relationship between the notion of weak monadic stability and the monadic stability concept. Next, I provide a more general characterisation.

Proposition 5.8 *Let the network payoff function φ and the link formation cost structure c be given. Every monadically stable network $g \in \mathbb{G}^N$ for (φ, c) satisfies the following two properties:*

(i) g is weakly monadically stable, and
(ii) g is supported by a monadic belief system ℓ^g that is non-superfluous in the sense that $\ell_{ij}^g = \ell_{ji}^g$ for all pairs $i, j \in N$.

Proof Let $g \in \mathbb{G}^N$ be monadically stable and let action tuple $\ell^g \in A^a$ support g as such. Suppose that $ij \notin g$ with $\ell_{ij}^g = 1$ and $\ell_{ji}^g = 0$. Then from the property that $\ell_i^g \in A_i^a$ is a best response to the belief system $\ell_{-i}^{g\,i\star}$ it can be concluded that $\ell_{ij}^g = 1$ implies that $\ell_{ji}^{g\,i\star} = 1$. But this would then imply that $\ell_{ji}^g \neq \ell_{ji}^{g\,i\star}$, violating the monadic stability self-confirmation condition. ■

The reverse of the assertion of Proposition 5.8 is not true. Simple examples can be constructed in which weakly monadically stable networks exist that satisfy the stated property, but which are not monadically stable.

A few comments regarding the relationship between weak monadic stability and network-based stability concepts are in order here. First, weakly monadically stable networks are not necessarily strong link deletion proof or link addition proof. Second, a network that is strong link deletion proof as well as link addition proof is not necessarily weakly monadically stable. We refer to network g^6 in Example 5.7, which is weakly monadically stable, but not link addition proof. The other comparisons can also be shown by properly constructed counterexamples.

An equivalence result. The main insight from this approach is that trust indeed builds very strong networks. This is exemplified by the equivalence of the class of monadically stable and strictly pairwise stable networks. For a proof I refer to Appendix A.7.

Theorem 5.9 *Let the network payoff function* φ *and the link formation cost structure* $c = (c_{ij})_{i,j \in N}$ *be given such that* $c_{ij} > 0$ *for all* $i, j \in N$ *with* $i \neq j$. *Then a network* $g \in \mathbb{G}^N$ *is monadically stable for* (φ, c) *if and only if* g *is strictly pairwise stable for the network payoff function* φ^a *given by*

$$\varphi_i^a(g) = \varphi_i(g) - \sum_{ij \in L_i(g)} c_{ij}. \tag{21}$$

Through the monadic stability concept we have considered the notion of confidence—as a form of mutual trust—into an advanced equilibrium concept, specifically designed for network formation. Confidence is introduced as an *internalised* feature into the behaviour of the players in network formation. Thus, trusting behaviour is as such a individualised feature rather than a social normative phenomenon.

The strength as well as the weakness of the monadic stability approach is the myopic nature of the belief systems. Players do not apply very sophisticated reasoning; they only look at the first order effects of link formation. Natural future extensions of this line of theoretical research should explore the possibility of introducing forward-looking behaviour to understand how farsightedly stable networks arise.[16]

5.3 A Comparison of Unilateral and Monadic Stability

As mentioned in the introduction to this section, unilateral and monadic stability seem to be founded on the same principles of trusting behaviour: Players attempt to form links with other players if they perceive these players to benefit from these links.

---

[16] This can be compared with existing models of farsighted network formation developed in [10, 11, 13, 23, 31, 37, 40, 42].

Recall that a network is unilaterally stable if there is no player who can induce changes to the network based on the belief that other players will consent to these changes if they are not harmful to them. Note here that unilateral stability assumes a fully rational form of farsightedness in the decision-making process: All proposed changes to the network—as made by a single player—are fully taken into account by all involved players before consent is granted. Thus, unilateral stability assumes a sophisticated level of rational forecasting by all players, who need to consent to the proposed changes to the network.

This implies that unilateral stability is indeed founded on the principle of trusting behaviour. Implicitly, players are indeed acting on beliefs that other players will act in their self-interest when confronted with proposed changes to their link sets. As such, unilateral stability is a trust equilibrium concept.

On the other hand, monadic stability assumes a much less sophisticated form of rational decision-making. Indeed, players are actually assumed to be boundedly rational: Players form monadic beliefs that only take first-order changes to the payoffs of other individualised into account. So, if a player proposes to add multiple links, her beliefs are founded on payoff changes per addition of a single link rather than the complete set of links. Beliefs are, thus, founded on a bounded form of reasoning by these players.

Moreover, only after beliefs are formed, all players base their actions on maximising their payoffs given these boundedly rational monadic beliefs. There can arise a build-in mismatch of beliefs and actual outcomes in the form of realised changes to the network. However, actual actions need to confirm the monadic beliefs of players. This pushes the decision-making process from unrealistic to justified, since these beliefs are observed by the player decision makers.

Therefore, monadic stability is a trust equilibrium concept as well and is designed explicitly to be based on an embedded form of trusting behaviour in the disguise of belief formation on trusting principles. These trusting principles are not violated due to the confirmation condition in the monadic stability concept—in contrast to the weak monadic stability notion.

In summary, monadic stability is founded on a boundedly rational form of trusting behaviour. This contrasts with unilateral stability in which all decisions are based on a more farsighted, rational implementation of similar ideas.

A formal comparison. Next I consider a more technical comparison of the two concepts. From the discussion above it cannot be expected that the application of monadic stability and unilateral stability results in exactly the same class of stable network. The next example shows that these two conceptions can lead to completely different sets of stable networks.

Example 5.10 Again consider the by-now familiar case of three players $N = \{1, 2, 3\}$. Let the network payoff function φ be given in the table below and assume that link formation is costless, i.e. $c_{ij} = 0$ for all $i, j \in N$.

Network g	$\varphi_1(g)$	$\varphi_2(g)$	$\varphi_3(g)$	**Stability**
$g^0 = \varnothing$	0	0	0	
$g^1 = \{12\}$	1	1	2	M-stable
$g^2 = \{13\}$	0	0	0	
$g^3 = \{23\}$	0	0	0	
$g^4 = \{12, 13\}$	0	0	1	
$g^5 = \{12, 23\}$	0	0	1	
$g^6 = \{13, 23\}$	3	3	3	U-stable
$g^7 = g^N$	4	2	4	

The table reports the stability properties of the various networks. There emerge three interesting networks to be investigated, namely, g^1, g^6 and $g^7 = g^N$. I discuss these in detail below:

Network g^1: We investigate the stability properties of this network. First, note that g^1 is not unilaterally stable. Indeed, player 3 prefers to propose the formation of links 13 and 23 to create network g^N, which represents a strict Pareto improvement for all players in N.
Second, network g^1 is supported by a non-superfluous communication profile that is represented as $\ell^1 = (\,(1,0), (1,0), (0,0)\,)$. This results into a monadic belief system given by

$$\ell_1^{1\star} = (-, (1,0), (0,0)\,)$$
$$\ell_1^{2\star} = ((1,0), -, (0,0)\,)$$
$$\ell_1^{3\star} = ((1,0), (1,0), -\,).$$

Clearly ℓ^1 constitutes a best response profile to the given monadic belief system and the monadic belief system is confirmed through ℓ^1, showing that g^1 is supported as a monadically stable network.[17]

Network g^6: First, note that g^6 is strongly pairwise stable as well as unilaterally stable. Indeed, only player 1 has an incentive to add link 12 to form the complete network $g^7 = g^N$, which is rejected by player 2 due to a loss in payoff. There are no players who have incentives to sever any of the two existing links.
Next, g^6 is not monadically stable. Indeed, take the non-superfluous communication profile that supports it, given by $\ell^6 = (\,(0,1), (0,1), (1,1)\,)$. Then the corresponding monadic belief system is

[17] Similarly, note that g^1 is actually a strictly pairwise stable network. The equivalence theorem shows that, therefore, g^1 has to be monadically stable.

$$\ell_6^{1\star} = (-, (0,1), (1,1))$$
$$\ell_6^{2\star} = ((1,1), -, (1,1))$$
$$\ell_6^{3\star} = ((0,1), (0,1), -).$$

Obviously, the communication profile ℓ^6 is a best response to the monadic belief system above. This implies that g^6 is weakly monadically stable. However, it is *not* monadically stable. Indeed, player 2 believes that player 1 would pursue the creation of a link with her—as represented by $\ell_{12}^{2\star} = 1$. This is not as described by ℓ^6; player 1 does not propose a link to player 2 and, as such, the belief system of player 2 is not confirmed in the equilibrium communication profile.

Network $g^7 = g^N$: To conclude the discussion of the situation described in this example, we consider the complete network $g^7 = g^N$, which is uniquely supported by the communication profile $\ell^7 = ((1,1), (1,1), (1,1))$. The resulting monadic belief systems can now be represented by

$$\ell_7^{1\star} = (-, (0,1), (1,1))$$
$$\ell_7^{2\star} = ((1,1), -, (1,1))$$
$$\ell_7^{3\star} = ((1,1), (1,1), -).$$

Obviously, the communication strategy $\ell_1^7 = (1,1)$ is not a best response to $\ell_7^{1\star}$, since player 1 expects player 2 not to form a link with her. Therefore, ℓ^7 is not supported as a monadically stable communication profile. Thus, g^7 is not weakly monadically stable.
Furthermore, this network is neither unilaterally stable; in particular, it is not link deletion proof. Indeed, player 2 has an incentive to break the link with player 1 to move to network g^6.

This example clearly shows that the class of unilaterally stable networks can be completely disjoint from the class of monadically stable networks. In this example, however, the unilaterally stable network is weakly monadically stable. This implies that in a unilaterally stable network monadic beliefs can destabilise the network, leading to unending improvement attempts by the players in the network. Thus, boundedly rational belief formation can undermine a farsightedly rational foundation for the network; as such, it represents an example of a direct conflict between farsighted or full and boundedly rational behaviour. ♦

5.4 Existence of Monadically Stable Networks

The question of existence of monadically stable networks is an important one. The previous discussion already identified the class of monadically stable networks to be exactly equal to the class of strictly pairwise stable networks. Obviously, this class is empty for a large collection of network payoff structures. Here I investigate certain conditions under which the class of such networks is non-empty.

These conditions are related to the notion of a network potential as seminally developed by Chakrabarti and Gilles [9]. There it is explored what the consequences are of founding network payoffs on an underlying link-based payoff function—denoted as a *network potential*. Network payoff functions that admit a potential impose a payoff structure in which players assess the value of links in a similar fashion. It can be shown that for network payoff structures that are founded on such potentials, there exist strictly pairwise stable networks.

In the subsequent discussion, I summarise the main insights from Chakrabarti and Gilles [9]. For details of the proofs of the main theorems I also refer to that paper and its appendices. Before stating the main definitions and the resulting properties, I recall the definition of two potential concepts in the context of a non-cooperative game $(\mathcal{A}, \pi)$ on the player set N as seminally introduced by Monderer and Shapley [33].

Definition 5.11 Let $(\mathcal{A}, \pi)$ be a non-cooperative game on player set N. Then:

(a) The game $(\mathcal{A}, \pi)$ **admits an exact potential** in the sense of Monderer and Shapley [33] if there exists a function $P\colon A \to \mathbb{R}$ such that

$$\pi_i(a) - \pi_i(b_i, a_{-i}) = P(a) - P(b_i, a_{-i}) \tag{22}$$

for every player $i \in N$, every strategy tuple $a \in A$ and every strategy $b_i \in A_i$.

(b) The game $(\mathcal{A}, \pi)$ **admits an ordinal potential** in the sense of Monderer and Shapley [33] if there exists a function $P\colon A \to \mathbb{R}$ such that

$$\pi_i(a) > \pi_i(b_i, a_{-i}) \quad \text{if and only if} \quad P(a) > P(b_i, a_{-i}) \tag{23}$$

for every player $i \in N$, every strategy tuple $a \in A$ and every strategy $b_i \in A_i$.

Based on these two notions of game-theoretic potentials, we can now consider how network payoff structures might be founded on similar constructs.

Network potentials. There are two main conceptions of the notion of a potential as a founding device in the determination of network payoffs. Again we refer to these notions as an "exact potential" and an "ordinal potential", following the accepted terminology in the literature. The next definition introduces these two notions.

Definition 5.12 Let $\varphi\colon \mathbb{G}^N \to \mathbb{R}^N$ be a network payoff function.

(a) The network payoff function φ **admits an exact potential** if there exists a function $\Lambda\colon \mathbb{G}^N \to \mathbb{R}$ such that

$$\varphi_i(g) - \varphi_i(g - ij) = \Lambda(g) - \Lambda(g - ij) \tag{24}$$

for every network $g \in \mathbb{G}^N$, every player $i \in N$ and every link $ij \in L_i(g)$.

(b) The network payoff function φ **admits an ordinal potential** if there exists a function $\Lambda \colon \mathbb{G}^N \to \mathbb{R}$ such that the following conditions hold:

$$\varphi_i(g) > \varphi_i(g-ij) \text{ if and only if } \Lambda(g) > \Lambda(g-ij) \tag{25}$$
$$\varphi_i(g) < \varphi_i(g-ij) \text{ if and only if } \Lambda(g) < \Lambda(g-ij) \tag{26}$$
$$\varphi_i(g) = \varphi_i(g-ij) \text{ if and only if } \Lambda(g) = \Lambda(g-ij) \tag{27}$$

for every network $g \in \mathbb{G}^N$, every player $i \in N$ and every link $ij \in L_i(g)$.

An exact potential imposes that the network payoff structure exhibits a *cardinally uniform* way of how players assess the addition or deletion of a link to a network. It is clear that the admittance of an exact potential is a very strong condition on the network payoff structure. This is confirmed by the following insight from Chakrabarti and Gilles [9, Theorem 3.3]:

Lemma 5.13 *A network payoff function φ admits an exact potential if and only if the corresponding Myerson model Γ^m_φ admits an exact potential in the sense of Monderer and Shapley [33].*

The admittance of an ordinal potential in a network payoff structure imposes a uniform assessment of deleting and adding links to networks by all players in purely *ordinal* terms. Although this property is significantly weaker than the admittance of an exact potential, it remains a rather demanding condition on the network payoff structure. The next lemma makes clear that there is again a relationship with the notion of an ordinal potential in the sense of Monderer and Shapley [33]. The next lemma is stated as Theorem 4.3 in Chakrabarti and Gilles [9]. For a proof I refer to that source.

Lemma 5.14 *Let φ be some network payoff structure. If the corresponding Myerson model Γ^m_φ admits an ordinal potential in the sense of Monderer and Shapley [33], then φ admits an ordinal potential.*

The reverse of the assertion stated in Lemma 5.14 is not true, as shown in Chakrabarti and Gilles [9, Example 4.4].

Properties of network payoff structures that admit potentials. Using the introduced notions of game-theoretic and network potentials, we can now distinguish three essential classes of network payoff structures. First, those network payoff structures that admit an exact potential; second, those network payoff structures for which the corresponding Myerson game admits an ordinal potential; and, finally, those network payoff structures that admit an ordinal potential. Each of these classes is larger than the previous.

The next propositions collect some properties of the third class, namely those network payoff structures that admit an ordinal potential. For proofs of these assertions I again refer to Chakrabarti and Gilles [9].

Proposition 5.15 *Let φ be some network payoff structure that admits an ordinal potential Λ. Then the following properties hold:*

(i) There exists at least one pairwise stable network.
(ii) The sets of strongly pairwise stable and strictly pairwise stable networks coincide.

The class of network payoff structures for which the corresponding Myerson game admits an ordinal potential is particularly interesting. Indeed, Chakrabarti and Gilles [9, Theorem 5.7] show that for this class of network payoff structures there exist strictly pairwise stable networks. I state for completeness the complete assertion:

Proposition 5.16 *Let φ be a network payoff function for which the corresponding Myerson model Γ^m_φ admits an ordinal potential in the sense of Monderer and Shapley [33]. Then there exists at least one strictly pairwise stable network for φ.*

This property gives rise to the main conclusion regarding the existence of a monadically stable network. Indeed, the admittance of an ordinal potential in the Myerson model gives rise to the existence of a strictly pairwise stable network, which in turn is monadically stable due to the fundamental equivalence theorem. As a consequence, we can formulate the following main existence theorem:

Theorem 5.17 *Let $\varphi\colon \mathbb{G}^N \to \mathbb{R}^N$ be a network payoff structure and let $c\colon N \times N \to \mathbb{R}_+$ be a link formation cost structure. If the corresponding consent model with two-sided link formation costs $\Gamma^a_\varphi(c)$ admits an ordinal potential in the sense of Monderer and Shapley [33], then there exists at least one monadically stable network for (φ, c).*

6 Correlated Network Formation

The previous section focussed mainly on the internalisation of trust in the behaviour of players to result in so-called "trusting behaviour" in link formation. We chose to internalise trusting behaviour in the form of belief systems (monadic stability) or through stability concepts themselves (unilateral stability). However, there is rather different an approach possible in which trusting behaviour is explicitly modelled through an externally determined institutional arrangement. These institutional arrangements are implemented collectively and are endowed with a form of collectively accepted self-enforcement.

In my discussion I mainly considered behavioural rules that can be viewed as being part of a trusted governance system. All players are assumed to be embedded in such a governance system, expressing this in the formulated monadic stability concept as embedded monadic belief systems. Hence, we use game-theoretic concepts to give this embeddedness an explicit, institutional form as a generally accepted behavioural rule, to behave according to the stated monadic belief system.

Correlation devices. Here, I turn to a much more explicit conception of behavioural sociality. One can model guiding behavioural norms also as being *external* to the players, rather than fully internalised—as is the case for the notion of monadic belief systems. This refers to the possibility to let external "devices" guide and coordinate decision-making in a game-theoretic setting. In particular, one can consider the question: "Can external guidance let decision-makers achieve a higher payoff than that is achieved through the set of supported Nash equilibria?"

A seminal study by Aumann [2] introduced an innovative way to exactly introduce a formal way to establish mutually beneficial coordination among players. These external arrangements are denoted as *correlation devices*. The basic idea is that the decisions made by players are influenced by things that are external to the decision problem itself but are situated in their immediate surrounding. The classical example is that of a traffic light.[18]

The game-theoretic representation is a form of the *Game of Chicken* as explored extensively in the literature. Two drivers approach a road crossing. At the crossing, each driver can either "stop" (action S) or "continue" (action C). If both continue there will result a crash; if both stop, both look foolish and need to coordinate their passing through prolonged negotiation (with hand gestures); and if one stops and the other continues, there is regret of the stopper and maximal payoff to the one who continues. The resulting payoffs can be captured by the following game-theoretic payoff matrix:

	S	C
S	5, 5	2, 7
C	7, 2	0, 0

There result in three Nash equilibria in mixed strategies here, namely, one driver stops and the other continues—resulting in payoffs $(7, 2)$ and $(2, 7)$ depending on who actually stops—and the case in which both players stop or continue with equal probability—resulting in the expected payoff vector $\left(3\frac{1}{2}, 3\frac{1}{2}\right)$. The latter includes a probability of $\frac{1}{4}$ of a crash, due to both players continuing.

Now consider that there is an outside regulator—represented as a correlation device—added to this situation in the form of a traffic light. The most important assumption of this arrangement is that both drivers are fully informed about what fraction of time the traffic light is in what colour. Hence, both drivers know the probability distribution that is implemented through the traffic light. We investigate two traffic light arrangements:

- First, consider that with equal probability the traffic light gives a red light to one player and a green light to the other. Adopting the normal rule to stop for red and to continue for green, we actually coordinate between the two Nash equilibria (S, C) and (C, S), resulting in an expected payoff computed as

$$\mathbb{E}\,\pi_1 = \tfrac{1}{2}(7, 2) + \tfrac{1}{2}(2, 7) = \left(4\tfrac{1}{2}, 4\tfrac{1}{2}\right).$$

[18] The following discussion is mainly based on the excellent account of correlated equilibrium in Chap. 9 of Maschler et al. [32]. I recommend the interested reader to look at their presentation.

Here there no positive probability of a crash and both drivers are reasonably content with their expected payoff.

Would this traffic light be *self-enforceable* within the given social decision situation? We need to check whether this traffic light arrangement is indeed beneficial to both player drivers if it is implemented as suggested by these two drivers. Obviously, if any of these drivers deviate from the recommendation, while the other follows it, there is a crash—resulting in zero payoffs. So, the suggested arrangement is indeed self-enforcing.

- In comparison with our regular traffic light, we can even increase the expected payoff by introducing a more complicated coordination device. Indeed, consider a traffic light that can stop both drivers simultaneously with a given probability. In that case, the drivers negotiate themselves and proceed with caution. So, the traffic light can give both drivers simultaneously the signal "red", at which both drivers are suggested to stop and proceed with caution.

  This allows the mixing of three outcomes in this decision situation. Suppose now that the traffic light gives both drivers simultaneously "red" with probability $\frac{1}{2}$ and one driver "red" and the other driver "green" with equal probabilities $\frac{1}{4}$.[19] We can depict the resulting probability distribution over all outcomes in a probability matrix:

	S	C
S	$\frac{1}{2}$	$\frac{1}{4}$
C	$\frac{1}{4}$	0

  The resulting expected payoffs can now be computed as

$$\mathbb{E}\,\pi_2 = \tfrac{1}{2}(5,5) + \tfrac{1}{4}(7,2) + \tfrac{1}{4}(2,7) = \left(4\tfrac{3}{4}, 4\tfrac{3}{4}\right) \gg \left(4\tfrac{1}{2}, 4\tfrac{1}{2}\right) = \mathbb{E}\,\pi_1.$$

  Again we can ask whether this traffic light is self-enforcing. If one driver receives "red", he knows that the other driver receives "red" with probability $\frac{2}{3}$ and "green" with probability $\frac{1}{3}$. So, if he continues there is a crash with probability $\frac{1}{3}$ and he receives an expected payoff of $\frac{1}{3} \cdot 0 + \frac{2}{3} \cdot 7 = 4\frac{2}{3} < 4\frac{3}{4}$, the latter being the expected payoff if he follows the recommendation of the traffic light. Again, we conclude that the traffic light arrangement is indeed self-enforcing; no player has an incentive to deviate from the provided arrangement and recommendations.

One can ask whether this reasoning can be extended to even higher payoffs. Indeed, Aumann showed that this is the case up to payoff level 5. The arrangement that both drivers always face a red light—that is, "red" with probability 1—is, of course, not self-enforcing.

Using correlation devices in network formation. Correlation devices can also be introduced in the processes of network formation. I return to the network formation

[19] This means that both drivers get *private* recommendations from the traffic light; they do not know what the colour to the other driver is. This is the usual arrangement in modern traffic law.

process under consent that we discussed thus far and consider how external correlation devices in the form of external recommender systems can guide players to form "good" networks. We first take a look at a by-now familiar network formation situation with three players.

Example 6.1 As before, let $N = \{1, 2, 3\}$ be the set of three players. Also, we choose φ to be a minor modification of the network payoff function studied in Example 5.10, given in the table below, and again we assume that link formation is costless, i.e. $c_{ij} = 0$ for all $i, j \in N$.

As reported in the table below, there are actually five M-networks, namely all strong link deletion proof networks given by $\mathcal{M} = \{g^0, g^1, g^2, g^3, g^6\}$. These five M-networks correspond only to three payoff vectors, namely, $(0, 0, 0)$, $(1, 1, 2)$ and $(3, 3, 3)$.

Network g	$\varphi_1(g)$	$\varphi_2(g)$	$\varphi_3(g)$	**M-network**
$g^0 = \varnothing$	0	0	0	M
$g^1 = \{12\}$	1	1	2	M
$g^2 = \{13\}$	0	0	0	M
$g^3 = \{23\}$	0	0	0	M
$g^4 = \{12, 13\}$	8	8	1	
$g^5 = \{12, 23\}$	0	0	1	
$g^6 = \{13, 23\}$	3	3	3	M
$g^7 = g^N$	4	2	4	

The main question I consider here is: Can we introduce a correlation device in this network formation situation that results in higher expected payoffs than those from the high-paying M-network g^6? Indeed, g^6 is the most obvious M-network that the players can aim for. Therefore, the payoff vector $(3, 3, 3)$ acts as a benchmark in relationship to any correlation device.

Consider an external recommender system based on the three networks g^4, g^6, and $g^7 = g^N$. In particular, assume that this correlation device recommends (i) all three players to execute signalling strategy $\ell^a = (\,(1, 1), (1, 0), (1, 0)\,)$ resulting in network g^4 with probability $\alpha = \frac{1}{12}$; (ii) the signalling strategy $\ell^b = (\,(0, 1), (0, 1), (1, 1)\,)$ resulting in network g^6 with probability $\beta = \frac{2}{3}$; and (iii) the signalling strategy $\ell^c = (\,(1, 1), (1, 1), (1, 1)\,)$ resulting in network $g^7 = g^N$ with probability $\gamma = \frac{1}{4}$. The expected payoffs under this system are now given by

$$\begin{aligned}\mathbb{E}\,\pi\,(\ell) &= \alpha \cdot \varphi(g^4) + \beta \cdot \varphi(g^6) + \gamma \cdot \varphi(g^7)\\ &= \tfrac{1}{12} \cdot \varphi(g^4) + \tfrac{2}{3} \cdot \varphi(g^6) + \tfrac{1}{4} \cdot \varphi(g^7) = \left(3\tfrac{2}{3}, 3\tfrac{1}{6}, 3\tfrac{1}{12}\right) \gg (3, 3, 3) = \varphi(g^6).\end{aligned}$$

Hence, coordinating the link building actions through this recommender system results into a *strict* Pareto improvement over the best M-network. It remains to show that all three players have no incentives to deviate from the recommended correlated strategy:

- Player 1: The only plausible alternative signalling strategy is to play $\ell_1 = (1, 1)$ to achieve the high-paying network g^4. This results actually in no changes to the

recommended networks, due to the recommended strategies executed by the two other players under the selected correlation device. Hence, player 1 has no gain from deviating from the recommended strategy.
- Player 2: The only plausible alternative signalling strategy for this player is to execute $\ell_2' = (1, 0)$ to establish network g^4. But this results in a lower expected payoff for player 2 if the other players follow the recommended strategies in ℓ:

$$\mathbb{E}\,\varphi_2\left(\ell_2'\right) = \tfrac{1}{12}\cdot\varphi_2(g^4) + \tfrac{2}{3}\cdot\varphi_2(g^2) + \tfrac{1}{4}\cdot\varphi_2(g^4) = \tfrac{1}{3}\cdot 8 + \tfrac{2}{3}\cdot 0 = 2\tfrac{2}{3} < 3\tfrac{1}{6} = \mathbb{E}\,\pi_2\left(\ell\right).$$

This recommender system uses two non-M-networks, g^4 and $g^7 = g^N$. Therefore, this correlation device is founded on considerations outside the realm of the stability concepts that we have considered thus far. It shows that inefficient networks and non-stable networks play a role in network formation processes. ♦

The example above shows just a single application of the correlated equilibrium concept to network formation analysis. The application of this concept opens the way to further exploration, even though the multitude of correlated equilibria is discouraging. Indeed, Aumann showed that the collection of expected payoff vectors supported by correlated equilibria includes the convex hull of all Nash equilibrium payoff vectors. This is rather daunting and discouraging from the perspective that correlation will not lead to a smaller class of supported networks.

However, the main research question that is still open is whether there exists a specific class of correlation devices that could guide players to highly productive networks. Throughout our history, humans have in fact found ways to implement very effective correlation devices to build effective and high-paying networks. This includes recommender systems such as job recommendation referrals and socio-economic recommendations through friendship networks. Further exploration of these systems from a Aumannian perspective is required to develop a theory that interprets these practical systems as correlation devices.

A Proofs of the Main Theorems

A.1 Proof of Theorem 3.2

If: Let φ be convex on $\mathcal{D}(\varphi)$. Obviously from the definitions and the discussions it follows that $\mathcal{D}_s(\varphi) \subset \mathcal{D}(\varphi)$. Thus, we only have to show that $\mathcal{D}(\varphi) \subset \mathcal{D}_s(\varphi)$.

Now let $g \in \mathcal{D}(\varphi)$. Then for every player $i \in N$ and link $ij \in L_i(g)$ it has to hold that $\varphi_i(g) \geqslant \varphi_i(g - ij)$ due to link deletion proofness of g. In particular, for any link set $h \subset L_i(g)$: $\sum_{ij \in h}[\varphi_i(g) - \varphi_i(g - ij)] \geqslant 0$. Since φ is convex on $\mathcal{D}(\varphi)$ and $g \in \mathcal{D}(\varphi)$, it follows that $\varphi_i(g) \geqslant \varphi_i(g - h)$ for every link set $h \subset L_i(g)$. In other words, g is strong link deletion proof, i.e. $g \in \mathcal{D}_s(\varphi)$.

Only if: Assume that $\mathcal{D}(\varphi) = \mathcal{D}_s(\varphi)$. Suppose further to the contrary that the payoff structure φ is not convex on $\mathcal{D}(\varphi)$. Then there exists some network $g \in \mathcal{D}(\varphi)$ and some player $i \in N$ such that for some link set $h \subset L_i(g)$ we have that $\sum_{ij \in h}[\varphi_i(g) - \varphi_i(g - ij)] \geqslant 0$ as well as $\varphi_i(g) < \varphi_i(g - h)$. But then this implies straightforwardly that player i would prefer to sever all links in h, i.e. $g \notin \mathcal{D}_s(\varphi)$. Thus, g cannot be strong link deletion proof giving us the necessary contradiction.

This completes the proof of the assertion of Theorem 3.2.

A.2 Proof of Theorem 3.5

Assertion (a) is trivial and a proof is therefore omitted.

Proof of (b).
If: Let φ be discerning on $\mathcal{A}(\varphi)$. Suppose that g is LAP. Furthermore, assume that $i, j \in N$ with $ij \notin g$ are such that $\varphi_i(g + ij) \geqslant \varphi_i(g)$. Now, if $\varphi_j(g + ij) = \varphi_j(g)$, then by definition of φ being discerning, $\varphi_i(g + ij) > \varphi_i(g)$. This contradicts the hypothesis that g is LAP. Thus, $\varphi_j(g + ij) < \varphi_j(g)$, confirming that g is indeed $\star$-LAP.

Only if: Suppose that φ is not discerning on $\mathcal{A}(\varphi)$. Then there exists some network g that is LAP and for some $i, j \in N$ with $ij \notin g$ it holds that $\varphi_i(g + ij) = \varphi_i(g)$ as well as $\varphi_j(g + ij) = \varphi_j(g)$. But this immediately implies that g can in fact not be $\star$-LAP, since the link ij should be in g for it to be $\star$-LAP. This is a contradiction.

Proof of (c).
If: Suppose that φ is uniform on $\mathcal{A}_\star(\varphi)$ and take some $g \in \mathcal{A}_\star(\varphi)$. Assume that $i, j \in N$ with $ij \notin g$. Then first suppose that

$$\varphi_i(g) \leqslant \varphi_i(g + ij). \tag{28}$$

Then by g being $\star$-LAP it has to hold that

$$\varphi_j(g) > \varphi_j(g + ij). \tag{29}$$

But also by uniformity of φ it has to hold that

$$\varphi_j(g) \leqslant \varphi_j(g + ij). \tag{30}$$

But (29) is in direct contradiction to (30). Thus, we conclude that (28) cannot hold. Therefore, for any $ij \notin g$ it has to hold that $\varphi_i(g) > \varphi_i(g + ij)$ as well as $\varphi_j(g) > \varphi_j(g + ij)$. Hence, we conclude that g is actually SLAP, i.e. $g \in \mathcal{A}_s(\varphi)$.

Only if: Assume that $\mathcal{A}_s(\varphi) = \mathcal{A}_\star(\varphi)$. Now take $g \in \mathcal{A}_\star(\varphi)$ to be $\star$-LAP. Then from g being SLAP, it follows that $\varphi_i(g) > \varphi_i(g + ij)$ as well as $\varphi_j(g) > \varphi_j(g + ij)$. This implies that φ indeed has to be uniform for g.

This proves the assertion of Theorem 3.5.

A.3 Proof of Theorem 4.1

First, we show assertion (a).

Suppose that there is an M-network $g \in \mathbb{G}^N$ supported by a Nash equilibrium strategy profile $\ell \in A^m$ that is not strong link deletion proof. Then there is some $i \in N$ and $h_i \subset L_i(g)$ with $\varphi_i(g - h_i) > \varphi_i(g)$. But then player i can modify his linking strategy as $\ell'_{ij} = 0$ if $ij \in h_i$ and $\ell'_{ij} = \ell_{ij}$. Then $g(\ell'_i, \ell_{-i}) = g - h_i$ implying that $\pi^m_i(\ell'_i, \ell_{-i}) > \pi^m(\ell)$. Therefore, ℓ cannot be a Nash equilibrium in $(\mathcal{A}^m, \pi^m)$. This is a contradiction, showing that M-networks are strong link deletion proof.

Next, let $g \in \mathcal{D}_s(\varphi)$ be a strong link deletion proof network for the network payoff function φ on N. Suppose that g is not an M-network. Then the corresponding signalling tuple ℓ^g—where $\ell^g_{ij} = 1$ if $ij \in g$ and $\ell^g_{ij} = 0$ otherwise—is not a Nash equilibrium strategy tuple in the Myerson model Γ^m_φ. Hence, there is a player $i \in N$ and an alternative strategy $\ell_i \in A_i$ with $\ell_i \neq \ell^g_i$ such that $\pi^m_i(\ell^g) < \pi^m_i(\ell_i, \ell^g_{-i})$. If we denote by $h + i = \{ij \mid \ell^g_{ij} = 1 \text{ and } \ell_{ij} = 0\}$, then it is clear that $g(\ell_i, \ell^g_{-i}) = g - h_i \subset L_i(g)$. Using the definition of the Myerson payoff function π^m, we have established that $\varphi_i(g) < \varphi_i(g - h_i)$, which contradicts the hypothesis that g is strong link deletion proof.

To show assertion (b), suppose that φ is link monotone. Take any network $g \in \mathbb{G}^N$ and construct a strategy profile $\ell^g \in A^m$ by $\ell^g_{ij} = 1$ if and only if $ij \in g$, for all $i, j \in N$. It is easy to see that ℓ^g is indeed a Nash equilibrium in $(\mathcal{A}^m, \pi^m)$ due to φ being link monotone: For any $i \in N$, any deviation ℓ_i from ℓ^g_i induces the link set $L_i\left(g(\ell_i, \ell^g_{-i})\right) \subseteq L_i(g)$ for i. This implies by link monotonicity that $\pi^m_i(\ell_i, \ell^g_{-i}) = \varphi_i\left(g(\ell_i, \ell^g_{-i})\right) \leqslant \varphi_i(g) = \pi^m(\ell^g)$.

A.4 Proof of Theorem 4.5

(a) implies (c): Let $\ell^\star$ be an arbitrary Nash equilibrium in $(\mathcal{A}^a, \pi^a)$. Then denote $g^\star = g^m(\ell^\star) = \{ij \in g^N \mid \ell^\star_{ij} = \ell^\star_{ji} = 1\}$. We show that $g^\star$ is strong link deletion proof for the derived network payoff function φ^a.

Suppose player i deletes a certain link set $h_i \subset L_i(g^\star)$. Define $\ell_i \in A^a_i$ as $\ell_{ij} = 1$ if $ij \in g^\star - h_i$ and $\ell_{ij} = 0$ for $ij \notin g^\star - h_i$. Then by $\ell^\star$ being a Nash equilibrium in (A^a, π^a) it follows that $g^m(\ell_i, \ell^\star_{-i}) = g^\star - h_i$ and $\pi^a_i(\ell^\star) \geqslant \pi^a_i(\ell_i, \ell^\star_{-i})$. Hence,

$$\begin{aligned}\varphi_i^a(g^\star) &= \varphi_i(g^\star) - \sum_{j \in N_i(g^\star)} c_{ij} = \pi_i^a(\ell^\star) + \sum_{k:\, \ell_{ik}^\star=1, \ell_{ki}^\star=0} c_{ik} \\ &\geqslant \pi_i^a(\ell^\star) \geqslant \pi_i^a(\ell_i, \ell_{-i}^\star) = \varphi_i(g^m(\ell_i, \ell_{-i}^\star)) - \sum_{k \neq i} \ell_{ik} \cdot c_{ik} \\ &= \varphi_i(g^\star - h_i) - \sum_{k \in N_i(g^\star - h_i)} c_{ik} = \varphi_i^a(g^\star - h_i).\end{aligned}$$

This proves that $g^\star$ is strong link deletion proof for φ^a.

(c) implies (b): Suppose that $g^\star \subset g^N$ is a strong link deletion proof network for φ^a. We show that it is supported by a non-superfluous Nash equilibrium strategy in $(\mathcal{A}^a, \pi^a)$. Consider the unique non-superfluous strategy profile $\ell^\star \in A^a$ such that $g^m(\ell^\star) = g^\star$. We proceed to show that $\ell^\star$ is a Nash equilibrium in $(\mathcal{A}^a, \pi^a)$ and $\ell_{ij}^\star = 1$ if and only if $ij \in g^\star$. Indeed,

$$\pi_i^a(\ell^\star) = \varphi_i(g^m(\ell^\star)) - \sum_{k \neq i} \ell_{ik}^\star \cdot c_{ik} = \varphi_i(g^\star) - \sum_{k \in N_i(g^\star)} c_{ik} = \varphi_i^a(g^\star).$$

Next, for some player i consider some deviation $\ell_i \neq \ell_i^\star$. Define $h_i = \{ik \in g^\star \mid \ell_{ik} = 0\}$. Then, $g^m(\ell_i, \ell_{-i}^\star) = g^\star - h_i$. Since $g^\star$ is strong link deletion proof with respect to φ^a, it follows that $\varphi_i^a(g^\star - h_i) \leqslant \varphi_i^a(g^\star)$. Thus,

$$\begin{aligned}\pi_i^a(\ell_i, \ell_{-i}^\star) &= \varphi_i(g^m(\ell_i, \ell_{-i}^\star)) - \sum_{k \neq i} \ell_{ik} \cdot c_{ik} \\ &= \varphi_i(g^\star - h_i) - \sum_{k \in N_i(g^\star - h_i)} c_{ik} - \sum_{k:\, \ell_{ik}=1, \ell_{ki}^\star=0} c_{ik} \\ &\leqslant \varphi_i(g^\star - h_i) - \sum_{k \in N_i(g^\star - h_i)} c_{ik} \\ &= \varphi_i^a(g^\star - h_i) \leqslant \varphi_i^a(g^\star) = \pi_i^a(l^\star).\end{aligned}$$

This proves that the non-superfluous signal profile $\ell^\star$ is indeed a Nash equilibrium. Trivially (b) implies (a), which proves the assertion and completes the proof of Theorem 4.5.

A.5 Proof of Theorem 4.8

Let $g^\star$ be strong link deletion proof under the net payoff function φ^b. For $g^\star$, define a non-superfluous communication profile $\lambda^\star = (l^\star, r^\star) \in A^b$ as follows:

(i) $l_{ij}^\star = r_{ji}^\star = 1$ if $ij \in g^\star$ and $\gamma_{ij} < \gamma_{ji}$, or
(ii) $l_{ij}^\star = r_{ji}^\star = 1$ if $ij \in g^\star$, $\gamma_{ij} = \gamma_{ji}$ and $i < j$, or
(iii) $l_{ij}^\star = r_{ji}^\star = 0$ if $ij \notin g^\star$.

Obviously, $g^b(l^\star, r^\star) = g^\star$ and

$$\pi_i^b(\lambda^\star) = \varphi_i\left(g^b(\lambda^\star)\right) - \sum_{j \neq i} l_{ij}^\star \cdot \gamma_{ij} = \varphi_i(g^\star) - \sum_{j \in M_i(g^\star)} \gamma_{ij} = \varphi_i^b(g^\star).$$

Now, for player $i \in N$ consider an arbitrary deviation $\widehat{\lambda}_i = \left(\widehat{l}_i, \widehat{r}_i\right) \neq \left(l_i^\star, r_i^\star\right) = \lambda_i^\star$. In any such deviation, no new links will be formed because if $ij \notin g^\star$, it follows that $l_{ji}^\star = r_{ji}^\star = 0$. However, links in i's neighbourhood link set $L_i(g^\star)$ can be deleted. Hence, let $g^b\left(\widehat{\lambda}_i, \lambda_{-i}^\star\right) = g^\star - h_i$ where $h_i \subset L_i(g^\star)$.

We prove that $j \in N_i(g^\star - h_i)$ and $\left[\gamma_{ij} < \gamma_{ji} \text{ or } \gamma_{ij} = \gamma_{ji},\ i < j\right]$ implies that $\widehat{l}_{ij} = 1$. In other words, $j \in M_i(g^\star - h_i) \subset N_i(g^\star - h_i)$ implies that $\widehat{l}_{ij} = 1$.

Now, assume by contradiction that for some $j \in M_i(g^\star - h_i)$: $\widehat{l}_{ij} = 0$. Now,

$$j \in N_i(g^\star - h_i) \Leftrightarrow \widehat{l}_{ij} = 1 \text{ and } r_{ji}^\star = 1 \text{ or } \widehat{r}_{ij} = 1 \text{ and } l_{ji}^\star = 1. \tag{31}$$

But $l_{ji}^\star = 1$ implies by construction that $\gamma_{ij} > \gamma_{ji}$ or $\gamma_{ij} = \gamma_{ji},\ i > j$. Furthermore, $r_{ji}^\star = 1$ implies by construction that $\gamma_{ij} < \gamma_{ji}$ or $\gamma_{ij} = \gamma_{ji}$, $i < j$. Since $\widehat{l}_{ij} = 0$, by (31), it follows that $\widehat{r}_{ij} = l_{ji}^\star = 1$ which implies that $\gamma_{ij} > \gamma_{ji}$ or $\gamma_{ij} = \gamma_{ji}$ with $i > j$. This contradicts $j \in M_i(g^\star - h_i)$ completing the proof of the claim stated above.

Now, the proven claim implies that

$$\sum_{j \in M_i(g^\star - h_i)} \gamma_{ij} \leqslant \sum_{j \in N_i(g^\star - h_i)} \widehat{l}_{ij} \cdot \gamma_{ij} \leqslant \sum_{j \neq i} \widehat{l}_{ij} \cdot \gamma_{ij}. \tag{32}$$

Hence,

$$\begin{aligned}
\pi_i^b\left(\widehat{\lambda}_i, \lambda_{-i}^\star\right) &= \varphi_i\left(g^b(\widehat{\lambda}_i, \lambda_{-i}^\star)\right) - \sum_{j \neq i} \widehat{l}_{ij} \cdot \gamma_{ij} = \varphi_i\left(g^\star - h_i\right) - \sum_{j \neq i} \widehat{l}_{ij} \cdot \gamma_{ij} \\
&\leqslant \varphi_i\left(g^\star - h_i\right) - \sum_{j \in M_i(g^\star - h_i)} \gamma_{ij} = \varphi_i^b(g^\star - h_i) \\
&\leqslant \varphi_i^b(g^\star) = \pi_i^b\left(l^\star, r^\star\right).
\end{aligned}$$

The first inequality follows from (32) and the second follows from the fact that $g^\star$ is strong link deletion proof with respect to φ^b. This completes the proof of Theorem 4.8.

A.6 Proof of Theorem 4.13

Let $g^\star$ be supported by a Nash equilibrium signalling profile $\ell^\star \in A^a$ in the consent model with two-sided link formation costs $(\mathcal{A}^a, \pi^a)$. We now construct a non-superfluous strategy tuple $\left(\widehat{l}, \widehat{r}\right) \in A^b$ in the consent model with one-sided link formation costs such that $g^b\left(\widehat{l}, \widehat{r}\right) = g^\star$ and $\left(\widehat{l}, \widehat{r}\right)$ is a Nash equilibrium in $(\mathcal{A}^b, \pi^b)$.

From Theorem 4.5, we can assume without loss of generality that $\ell^\star \in A^a$ is non-superfluous. Given $\ell^\star$, we define $\widehat{\lambda} = \left(\widehat{l}, \widehat{r}\right) \in A^b$ by

(i) $\widehat{l}_{ij} = \widehat{r}_{ji} = 1$ and $\widehat{l}_{ji} = \widehat{r}_{ij} = 0$ if and only if $\ell^\star_{ij} = \ell^\star_{ji} = 1$, and either $c_{ij} < c_{ji}$, or $c_{ij} = c_{ji}$ with $i < j$.
(ii) $\widehat{l}_{ij} = \widehat{l}_{ji} = \widehat{r}_{ij} = \widehat{r}_{ji} = 0$ if and only if $\ell^\star_{ij} = \ell^\star_{ji} = 0$.

It follows immediately that $\widehat{\lambda} = \left(\widehat{l}, \widehat{r}\right)$ is a non-superfluous communication profile in A^b supporting $g^b\left(\widehat{l}, \widehat{r}\right) = g^\star$.

It remains to be shown that $\widehat{\lambda}$ is a Nash equilibrium of the consent model with one-sided link formation costs. We sketch the proof of this assertion.

Now, if $\widehat{\lambda}$ is not a Nash equilibrium, then it has to be because some player prefers to delete one or more of her links. Also, any link delivers the same benefit to the player as under two-sided link formation costs, while it would cost no more to establish the link. Thus, preferring to keep a link under two-sided link formation costs implies that the player would prefer to keep the link under one-sided link formation costs. Mathematical details of this argument are left to the reader.

This completes the proof of Theorem 4.13.

A.7 Proof of Theorem 5.9

We first develop some simple auxiliary insights for weakly monadically stable networks. Suppose that $g \in \mathbb{G}^N$ is weakly monadically stable relative to the data φ and $c = (c_{ij})_{i,j \in N}$. Then there exists some action tuple $\hat{\ell} \in A^a$ such that $g = g(\hat{\ell})$ and for every player $i \in N$: $\hat{\ell}_i \in A^a_i$ is a best response to the monadic belief system $\hat{\ell}^{i\star}_{-i} \in A^a_{-i}$ for the payoff function π^a.

For this setting we state two auxiliary results.

Lemma 1 *If $\hat{\ell}^{i\star}_{ji} = 0$ and $c_{ij} > 0$, then $\ell_{ij} = 0$ is the unique best response to $\hat{\ell}^{i\star}_{-i}$.*

Proof Clearly, if player i selects $\ell_{ij} = 1$, i only incurs strictly positive costs $c_{ij} > 0$ and no benefits. This implies that player i makes a loss from trying to establish link ij. Hence, $\ell_{ij} = 0$ is the unique best response to $\hat{\ell}^{i\star}_{-i}$. ■

Lemma 2 *If $ij \in g(\hat{\ell})$ with $c_{ij} > 0$ as well as $c_{ji} > 0$, then $\hat{\ell}^{i\star}_{ji} = \hat{\ell}^{j\star}_{ij} = 1$.*

Proof We remark that $ij \in g = g(\hat{\ell})$ if and only if $\hat{\ell}_{ij} = \hat{\ell}_{ji} = 1$. The negation of the assertion stated in Lemma 1 applied to $\hat{\ell}_i j = 1$ and $\hat{\ell}_{ji} = 1$ independently now implies that $\hat{\ell}^{i\star}_{ji} = \hat{\ell}^{j\star}_{ij} = 1$. ■

We also require a partial characterisation of weakly monadically stable networks. This is stated in the following lemma.

Lemma 3 *Let the cost structure $c \gg 0$ be strictly positive. Then every weakly monadically stable network $g \in \mathbb{G}^N$ in the consent model with two-sided link formation costs (A^a, π^a) is link deletion proof for the network payoff function φ^a.*

Proof Suppose that $g \in \mathbb{G}^N$ is weakly monadic in the consent model with two-sided link formation costs (A^a, π^a). Then there exists some communication profile $\hat{\ell} \in A^a$ such that $g = g(\hat{\ell})$ and for every player $i \in N$: $\hat{\ell}_i \in A^a_i$ is a best response to $\hat{\ell}^{i\star}_{-i}$ for the game-theoretic payoff function π^a.

Suppose now that g is not link deletion proof for φ^a. Then there exists some $i \in N$ with $ij \in g$ for some $j \neq i$ and $\varphi^a(g - ij) > \varphi^a_i(g)$, implying that $\varphi_i(g - ij) + c_{ij} > \varphi_i(g)$. By definition, $\hat{\ell}^{j\star}_{ij} = 0$. Hence, from Lemma 1, $\ell_{ji} = 0$ is the unique best response to $\hat{\ell}^{j\star}$ for player j. Since $ij \in g$ by assumption it has to hold that $\hat{\ell}_{ji} = 1$. This contradicts the hypothesis that $\hat{\ell}_j$ is a best response to $\hat{\ell}^{j\star}_{-j}$.

This contradiction indeed shows that g has to be link deletion proof relative to φ^a. ■

The proof of Theorem 5.9 now proceeds as follows.

First we show that strict pairwise stability for φ^a implies monadic stability in (A^a, π^a) under the hypothesis that $c \gg 0$.

Let $g \in \mathbb{G}^N$ be a network that is strictly pairwise stable with regard to the network payoff function φ^a as given in the assertion. Then g is strong link deletion proof and satisfies the property that

$$ij \notin g \text{ implies that } \varphi^a_i(g + ij) < \varphi^a_i(g) \text{ as well as } \varphi^a_j(g + ij) < \varphi^a_j(g).$$

Hence, this can be rewritten as

$$ij \notin g \text{ implies } \varphi_i(g + ij) - c_{ij} < \varphi_i(g) \text{ as well as } \varphi_j(g + ij) - c_{ji} < \varphi_j(g). \tag{33}$$

With g we define for all $i \in N$:

$$\hat{\ell}_{ij} = 1 \text{ if } ij \in g$$
$$\hat{\ell}_{ij} = 0 \text{ if } ij \notin g.$$

Hence, $g(\hat{\ell}) = g$ and $\hat{\ell}$ is non-superfluous. We now investigate whether the given communication profile $\hat{\ell}$ is indeed a best response to the monadic belief system $\hat{\ell}^{i\star}$ for all $i \in N$ as required by the definition of weak monadic stability.

Case A: $ij \notin g$.

From (33) it follows immediately that $\hat{\ell}^{i\star}_{ji} = \hat{\ell}^{j\star}_{ij} = 0$. From the hypothesis that $c_{ij} > 0$ and $c_{ji} > 0$ and the definition of monadic belief systems, it follows with Lemma 1 that $\hat{\ell}_{ij} = 0$ is the unique best response to $\hat{\ell}^{i\star}_{-i}$ and that $\hat{\ell}_{ji} = 0$ is the unique best response to $\hat{\ell}^{j\star}_{-j}$.

Hence, for Case A the communication strategy $\hat{\ell}$ satisfies the condition of weak monadic stability.

Case B: $ij \in g$.
In this case $\hat{\ell}_{ij} = \hat{\ell}_{ji} = 1$. Link deletion proofness of g now implies that $\hat{\ell}^{i\star}_{ji} = 1$ or else (33) is contradicted.

Cases A and B now imply that

$$ij \in g \text{ if and only if } \hat{\ell}^{i\star}_{ji} = \hat{\ell}^{j\star}_{ij} = 1. \tag{34}$$

Applying strong link deletion proofness and the insight for Case A leads us to the conclusion that $\hat{\ell}_i$ is indeed the unique best response to $\hat{\ell}^{i\star}_{-i}$. This in turn implies that $\hat{\ell}$ supports g as a weakly monadically stable network.

Finally, it is immediately clear from (34) and the definition of $\hat{\ell}$ that for all $i, j \in N$: $\hat{\ell}^{i\star}_{ji} = \hat{\ell}_{ij}$, implying that the monadic beliefs are indeed confirmed.

Thus, we conclude that $\hat{\ell}$ supports g as a monadically stable network. This completes the proof of the first part of the assertion.
Second, we show that the monadic stability of a network for (A^a, π^a) implies strict pairwise stability for φ^a under the hypothesis that $c \gg 0$.

Let $g \in \mathbb{G}^N$ be monadically stable. Then there exists some action tuple $\hat{\ell} \in A^a$ such that $g = g(\hat{\ell})$ and for every player $i \in N$: $\hat{\ell}_i \in A^a_i$ is a best response to $\hat{\ell}^{i\star}_{-i}$ for the payoff function π^a. Furthermore, $\hat{\ell}^{i\star}_{-i} = \hat{\ell}_{-i}$.

From Lemma 3 we already know that g has to be link deletion proof for φ^a since g is weakly monadically stable. Hence, for every $ij \in g$ we have that $\varphi_i(g - ij) + c_{ij} \leqslant \varphi_i(g)$. Now through the definition of the monadic belief systems and the self-confirming condition of monadic stability we conclude that for every $ij \in g$:

$$\hat{\ell}_{ij} = \hat{\ell}^{j\star}_{ij} = \hat{\ell}_{ji} = \hat{\ell}^{i\star}_{ji} = 1. \tag{35}$$

Let $i \in N$ and $h \subset L_i(g)$. Now we define $\ell^h \in A^a_i$ by

$$\ell^h_{ij} = \begin{cases} \hat{\ell}_{ij} & \text{if } ij \notin h \\ 0 & \text{if } ij \in h. \end{cases}$$

Then $g\left(\ell^h, \hat{\ell}_{-i}\right) = g - h$. Since $\hat{\ell}_i$ is a best response to $\hat{\ell}^{i\star}_{-i} = \hat{\ell}_{-i}$ it has to hold that[20]

$$\pi^a_i\left(\ell^h, \hat{\ell}_{-i}\right) \leqslant \pi^a_i(\hat{\ell}).$$

Hence,

$$\varphi_i(g - h) + \sum_{ij \in h} c_{ij} \leqslant \varphi_i(g). \tag{36}$$

This in turn implies that $\varphi^a_i(g - h) \leqslant \varphi^a_i(g)$.

[20] Here we again apply the confirmation condition for monadic stability that is satisfied by $\hat{\ell}$.

Since, $i \in N$ and h are chosen arbitrarily, the network g has to be strong link deletion proof.

Next, let $ij \notin g$. Then $\hat{\ell}_{ij} = 0$ and/or $\hat{\ell}_{ji} = 0$. Suppose that $\hat{\ell}_{ji} = 0$. Then by the confirmation condition of monadic stability it follows that $\hat{\ell}^{i\star}_{ji} = \hat{\ell}_{ji} = 0$. Hence by Lemma 1, $\hat{\ell}_{ij} = 0$. Thus we conclude that for every $ij \notin g$:

$$\hat{\ell}_{ij} = \hat{\ell}^{j\star}_{ij} = \hat{\ell}_{ji} = \hat{\ell}^{i\star}_{ji} = 0. \tag{37}$$

This in turn implies through the definition of the monadic belief system that $\varphi_i(g + ij) - c_{ij} < \varphi_i(g)$ as well as $\varphi_j(g + ij) - c_{ji} < \varphi_j(g)$. Or $\varphi^a_i(g + ij) < \varphi^a_i(g)$ as well as $\varphi^a_j(g + ij) < \varphi^a_j(g)$. This shows the assertion that g is indeed strictly pairwise stable.

This completes the proof of Theorem 5.9.

References

1. Aumann, R.J. 1959. Acceptable points in general cooperative n-person games. In *Contributions to the theory of games*, vol. IV, ed. A.W. Tucker and R.D. Luce, 287–324. Princeton: Princeton University Press.
2. Aumann, R.J. 1974. Subjectivity and correlation in randomized strategies. *Journal of Mathematical Economics* 1: 67–96.
3. Aumann, R.J., and R.B. Myerson. 1988. Endogenous formation of links between coalitions and players: An application of the shapley value. In *The shapley value*, ed. A.E. Roth. Cambridge: Cambridge University Press.
4. Bala, V., and S. Goyal. 2000. A non-cooperative model of network formation. *Econometrica* 68: 1181–1230.
5. Barabási, A.-L. 2016. *Network science*. Cambridge: Cambridge University Press.
6. Bloch, F., and M.O. Jackson. 2006. Definitions of equilibrium in network formation games. *International Journal of Game Theory* 34: 305–318.
7. Bloch, F., and M.O. Jackson. 2007. The formation of networks with transfers among players. *Journal of Economic Theory* 133: 83–110.
8. Brueckner, J.K. 2006. Friendship networks. *Journal of Regional Science* 46: 847–865.
9. Chakrabarti, S., and R.P. Gilles. 2007. Network potentials. *Review of Economic Design* 11: 13–52 forthcoming.
10. Deroïan, F. 2003. Farsighted strategies in the formation of a communication network. *Economics Letters* 80: 343–349.
11. Dutta, B., S. Ghosal, and D. Ray. 2005. Farsighted network formation. *Journal of Economic Theory* 122 (2): 143–164.
12. Dutta, B., and S. Mutuswami. 1997. Stable networks. *Journal of Economic Theory* 76: 322–344.
13. Förster, M., A. Mauleon, and V. Vannetelbosch. 2016. Trust and manipulation in social networks. *Network Science* 4: 216–243.
14. Fudenberg, D., and D.K. Levine. 1993. Self-confirming equilibrium. *Econometrica* 61: 523–545.
15. Gilles, R.P., S. Chakrabarti, and S. Sarangi. 2006. Social network formation with consent: Nash equilibrium and pairwise refinements. Working paper, Department of Economics, Virginia Tech, Blacksburg, VA.
16. Gilles, R.P., S. Chakrabarti, S. Sarangi. 2011. Myopic and farsighted stability in network formation under consent. Working Paper, Queen's University Management School, Belfast, UK.

17. Gilles, R.P., S. Chakrabarti, and S. Sarangi. 2012. Nash equilibria of network formation games under consent. *Mathematical Social Sciences* 64 (2): 159–165.
18. Gilles, R.P., and S. Sarangi. 2010. Network formation under mutual consent and costly communication. *Mathematical Social Sciences* 60: 181–185.
19. Golub, B., and Y. Livne. 2010. Strategic random networks and tipping points in network formation. Working Paper, Stanford Graduate School of Business, Stanford, CA.
20. Goyal, S., and S. Joshi. 2006. Unequal connections. *International Journal of Game Theory* 34: 319–349.
21. Harari, Y.N. 2014. *Sapiens: A brief history of humankind*. London: Vintage Books.
22. Harrington, J. 2008. *Games, strategies and decision making*. New York: Worth Publishers.
23. Herings, P.J.J., A. Mauleon, and V. Vannetelbosch. 2009. Farsightedly stable networks. *Games and Economic Behavior* 67: 526–541.
24. Jackson, M.O. 2008. *Social and economic networks*. Princeton: Princeton University Press.
25. Jackson, M.O., and B.W. Rogers. 2005. The economics of small worlds. *Journal of the European Economic Association* 3: 617–627.
26. Jackson, M.O., and B.W. Rogers. 2007. Meeting strangers and friends of friends: How random are social networks? *American Economic Review* 97: 890–915.
27. Jackson, M.O., and A. van den Nouweland. 2005. Strongly stable networks. *Games and Economic Behavior* 51: 420–444; 71: 44–74.
28. Jackson, M.O., and A. Watts. 2002. The evolution of social and economic networks. *Journal of Economic Theory* 106: 265–295.
29. Jackson, M.O., and A. Wolinsky. 1996. A strategic model of social and economic networks. *Journal of Economic Theory* 71: 44–74.
30. Joshi, S., A.S. Mahmut, and S. Sarangi. 2020. Network formation with multigraphs and strategic complementarities. *Journal of Economic Theory*. https://doi.org/10.1016/j.jet.2020.105033.
31. Kirchsteiger, G., M. Mantovani, A. Mauleon, and V. Vannetelbosch. 2016. Limited farsightedness in network formation. *Journal of Economic Behavior & Organization* 128: 97–120.
32. Maschler, M., E. Solan, and S. Zamir. 2013. *Game theory*. Cambridge: Cambridge University Press.
33. Monderer, D., and L.S. Shapley. 1996. Potential games. *Games and Economic Behavior* 14: 124–143.
34. Myerson, R.B. 1977. Graphs and cooperation in games. *Mathematics of Operations Research* 2: 225–229.
35. Myerson, R.B. 1980. Conference structures and fair allocation rules. *International Journal of Game Theory* 9: 169–182.
36. Myerson, R.B. 1991. *Game theory: Analysis of conflict*. Cambridge: Harvard University Press.
37. Navarro, N. 2014. Expected fair allocation in farsighted network formation. *Social Choice and Welfare* 43: 287–308.
38. Newman, M.E.J. 2010. *Networks: An introduction*. Oxford: Oxford University Press.
39. Osborne, M.J. 2004. *An introduction to game theory*. New York: Oxford University Press.
40. Page, F.H., M.H. Wooders, and S. Kamat. 2005. Network and farsighted stability. *Journal of Economic Theory* 120: 257–269.
41. Seabright, P. 2010. *The company of strangers: A natural history of economic life*. Princeton: Princeton University Press. Revised and enlarged edn.
42. Song, Y., and M. van der Schaar. 2020. Dynamic network formation with foresighted agents. *International Journal of Game Theory*. https://doi.org/10.1007/s00182-020-00714-4.
43. van de Rijt, A., and V. Buskens. 2008. A nash equilibrium refinement for Myerson's network formation game. ISCORE paper 230, University of Utrecht, Utrecht, The Netherlands.
44. van den Nouweland, A. 1993. Games and graphs in economic situations. Ph.D. Thesis, Tilburg University, Tilburg, The Netherlands.
45. van den Nouweland, A. 2004. Models of network formation in cooperative games. In *Group Formation in Economics: Networks, Clubs, and Coalitions*, ed. G. Demange, M. Wooders, chap. 2. Cambridge: Cambridge University Press.
46. Wellman, B., P. Carrington, and A. Hall. 1988. Networks as personal communities. In *Social structures: A network approach*, ed. B. Wellman, and S. Berkowitz. Cambridge: Cambridge University Press.

Analysis of Biological Data by Graph Theory Approach Searching of Iron in Biological Cells

Mária Ždímalová, Martin Kopáni, Ondrej Sova, Helena Svobodová, and Anuprava Chatterjee

1 Introduction

Medical image analyses contain many things, like registrations, image-guided surgery, segmentation, and cell analyses. Image segmentation is the process of partitioning an image into a set of distinct clusters containing pixels with similar attributes. Segmentation can be performed manually by a human expert who simply examines the images and selects which one we consider as an object and which we consider as a background. This is called as semi-automatic segmentation. We consider many techniques during automatic methods. Both approaches, semi-automatic as well as automatic, have their advantages and disadvantages.

Semi-automatic segmentation methods investigate to define different region of interest types in the analyses of biological images. Some of these methods used a semi-automatic approach, which still needs some user interaction. We can mention graph cuts method, region growing methods, grab cuts methods, gradient flow active contour algorithms, intelligent scissors, level sets methods, methods based on fuzzy approach and techniques, hierarchical atlas registration and weighted schemes, and many others even their combination. Other methods were fully automatic and the

M. Ždímalová (✉) · O. Sova
Slovak University of Technolgy in Bratislava, Radlinského 11, 810 05 Bratislava, Slovak Republic
e-mail: zdimalova@math.sk

M. Kopáni
Faculty of Medicine, Institute of Medical Physics, Informatics and Telemedicine, Comenius University of Bratislava, Sasinkova 3, 812 72 Bratislava, Slovak Republic

H. Svobodová
Department of Simulation and Virtula Medial Education, Faculty of Medicine, Comenius University in Bratislava, Sasinkova 4, 812 72 Bratislava, Slovak Republic

A. Chatterjee
Department of Electrical Engineering, IIT Kharagpur, Kharagpur, West Bengal, India

S. Borkotokey et al. (eds.), *Game Theory and Networks*, Indian Statistical Institute Series,
https://doi.org/10.1007/978-981-16-4737-6_12

user only had a verification role. It is necessary to mention even the new methods of the last days—deep learning and its use in neural networks. In this contribution, we consider binary segmentation. Two types of regions are considered, object regions and background regions. The output of the segmentation is a binary image with extra information held representing "object" and "background" segments. We consider "objects" of all cells of interest and the background of the rest of the image. The main goal of the segmentation is to simplify and change the representation of an image in a more meaningful and easier way for the next analysis.

We deal with biological data in this paper. They are received from biological microscope and we aimed to analyze them and to get concrete results to medical requirements like to detect iron in the samples. Existence of the iron cells and their density in the data is linked with Alzheimer's problem. The disease appears in animals and humans. Because of the fact we cannot test real human data, the test appears on specially treated animals, where the brains coincide with the human brain. We observe the existence and increasing or decreasing of the existence of the iron in the brain of the mice during Alzheimer's problem which appears in animals as well as in humans. Nowadays, it is clear that iron play role in the etiology of Alzheimer's disease (Gong et al. 2019, [10]). There are more hypotheses of Alzheimer's origin. In the brains are cumulated tau proteins and amyloid plaques which lead to disruption of synaptic signal (Batista-Nascimento et al. 2012, [1]; Plascencia-Villa et al. 2016, [17]; Braidy et al. 2017, [3]). Iron is crucial because this metal bind to tau proteins and affects its phosphorylation and aggregation and also it binds to amyloid plaques during disease (Braidy et al. 2017, [3]; Daglas and Adlard 2018, [6]). An even higher amount of iron correlates with a higher incidence of amyloid beta (Telling et al. 2017, [23]) and all these actions can lead to oxidative stress, disruption of synaptic connection, and cell death (Zecca et al. 2004, [26]; Petillon et al. 2019, [16]). Many studies are aimed at changes in iron concentrations because changes arise before cognitive changes and iron appears as a good early diagnostic factor of Alzheimer's disease (Langkammer et al. 2010, [14]; Greenough et al. 2013, [12]; Ward et al. 2014, [24]; Belaidi and Bush 2016, [2]; Sokolov et al. 2017, [20]); therefore, a better post-staining analysis could improve the methodology of iron examination. In our study, we observed the existence and increasing or decreasing of the existence of the iron in the brain of the mice; during Alzheimer's issue, data analysis was done on photos from a light microscope.

The whole process with data analyses contains their pre-processing, Segmentation, and finally quantitative dealing analyses of segmented data.

The suggested method will localize regions of the detected area. Our aim is to give segmentation of iron and if it is possible also quantitative evaluation. Values getting by this approach we process and analyses by visual checking with the original picture or checking by quantitative numbers.

Currently, there appear many software even free on the Internet. We deal with that but we did not get the needed results. We need to mention Image J [18, 22, 28], Cell profile [19, 29], Meta Morph [30], and Illastic [31]. Because of the needs of the biologist for concrete types of results, we developed software on according and due to the requirements of bio-medical requirements.

2 Method Overview

The main object of this study is to find iron in biological samples for preparing data. The aim is to present a new filtering method and segmentation of biological data. We present a new suggested filter and its application as a pre-processing step. As follows, we deal with this pre-processed data and use segmentation techniques to find elements of irons in the samples.

Step 1: Upload data.
Step 2: Generate and apply the new filter method.
Step 3: Using of segmentation of data, using the semi-automatic technique of graph cuts.
Step 4: Post-analyses quantitative processing of data.

We describe obtained data, the steps of getting them, the quality of data, and characteristic signs which are important for measuring data. We study research scanned images of the cuts of tissue from the brain of mice. In the mice appeared genetic mutations leading to Alzheimer's issue. Scanned tissue comes from hippo-camp of mice connected with the memory and cognitive functions—like a possibility of learning. Scout mutation causes a decrease of iron in the scanned area, which does not appear under normal circumstances. For the analysis, we used photos of stained mice brains from light microscopy (microscope Zeiss Scope.A1 (Gottingen, Germany) with camera AxioCam MRc 5). The samples were collected from the well-known mouse model APP/PS1 with mutations that leads to Alzheimer's disease (Jankowsky et al. 2004, [13]; Webster et al. 2014, [24]). Brains were perfused and cut to 35 μm slices on a microtome (Leica SM 2000F, Wetzlar, Germany [24]). Then, histological staining was realized with Prussian blue (solution of potassium ferrocyanide) with DAB (3, 3'-Diaminobenzidine) (Falangola et al. 2005, [7]) and with congo red. The iron deposits were colored brown and amyloid plaques become red.

By scaling of the microscope is visible just the part of the cut, not the whole thickness, that is why every sample image was scanned more times, multiple times by different scalings of the microscope. This way we can get triples of the tissue samples. By observing the original data, we can see the appearance of small dark spots. These are marks of the existence and the presence of iron, which our method will find and measure.

We examine the data from a software perspective. Input data are the set of pixels. Every pixel keeps the information that carries only information and its color. This is exactly the beginning point in any method of image processing. We try to get information on how many pixels contain given tissue or keeps the marks about its presence, e.g., if the iron appears on the data with dark brown color, we define for every pixel how this pixel is similar alike to the dark brown. If we can detect which shade of dark brown color we should search, we can construct a filter that will show us quite exactly how much iron appears in the samples. This way we can get a series of images that will not catch the real color of the tissue, but it will give information about the occurrence of the given lattice. In the following, we describe the used filter. To get the exact value, we need to decide which pixels contain the mentioned tissue and which do not. At this point, we will use segmentation, which means the selection of background pixels (pixels without detected tissue) from object pixels (in our case pixels with contained tissue). The number of pixels is the resulting quality information that tells about the occurrence or the concentration of iron in the scanned area.

3 Fundamental Research

For such purposes mentioned in the paragraph above, we created our own software built-in "graph cutting" algorithms, for handling medical and biological data, which are in our case output images from the microscope. The advantage of this method is that it can provide global segmentation as well as local and we are also able to detect the edges, which represent the boundaries between cells. In our specific case, we need to provide the pre-processing of the images as well. We applied a newly created special filter for pre-processing of the first input data from the electronic microscope. The main benefit of this software is its complexity. It can do pre-processing of the images, segmentation of the image, which in our special case means finding the corresponding cells, and finally counting and categorization of the cells.

For purposes mentioned in the paragraph above we have created own software with built-in "graph cutting" algorithms, for handling biological data, which are in our case output images from a microscope. The advantage of this method is that it can provide global segmentation as well as the local one and we can detect the edges, which represent the boundaries between cells. In our specific case, we need to provide the pre-processing of the images as well. We have applied a newly created special filter for pre-processing of the first input data from the electronic microscope. The main benefit of this software is its complexity. It can do the pre-processing of the images, segmentation of the image, which in our case means finding the corresponding cells, and finally counting and categorization of the cells.

1. Data gathering from the microscope.
2. Pre-processing of the data.
3. Software initialization and specific input image loading.
4. Setting up the object and background pixels of the image.
5. Image segmentation process.
6. Output saving (image and numerical data).

The whole process consists of three main steps: Pre-processing, processing, and post-processing of data. Pre-processing is the whole first preparation data before segmentation. Our input data are medical images from a microscope, which are necessary to transform to a 24-bit map and afterward, correct contrast of input data image that resulting images will not be too dark, too light, see Fig. 1 and we can see well visible contours.

Under the processing of data, we understand handling data from a graph-theoretical approach. In this part are implemented and used all graphs algorithms used in the segmentation of the image. After pre-processing and correction soft input data in requested quality follows their processing. Input: We normalize the images by the histogram and filtration. The result of the segmentation (processing) is the image classified into two classes: object and background pixels. The fact and information, if the resulting pixel belongs after segmentation to the background of the object, is important by post analysis (post-processing).

The last step is devoted to the final quantitative analyses of the resulting data.

The aim of this work was a distinctive searching and counting of iron. The main aim was using a new suggested and constructed filter for the pre-processing of data. Consequently, on the prepared data, we use a graph cut modified algorithm and find iron by segmentation in the samples.

4 Pre-processing of Data—Filtering

Here, we describe the used filter and introduce our approach to create a special filter. The first step in the data analysis is preparing data usable for segmentation; see Fig. 1. We will transform the image, containing data of colors (standards are three values RGB) to an image that contains a concentration of the selected tissue (one value). One of the most important parts of image processing is using histogram and filtration.

Filtration based on histogram (black-and-white images)
The image histogram represents the intensity distribution in the digital image. Mathematically, it can be described as a function that assigns to every value of q from the set $0, \dots, Q-1$, where Q is the maximum possible intensity (in our case 255), the number of pixels having intensity q. The histogram provides important insights into the distribution of image intensity. It can determine whether the image is underexposed, overexposed, whether it uses a full range of intensities, and the like; see [21].

FSHS: The abbreviation stands for "full scale histogram stretch" and this operation is used to achieve the maximum pixel intensity range in black-and-white images.

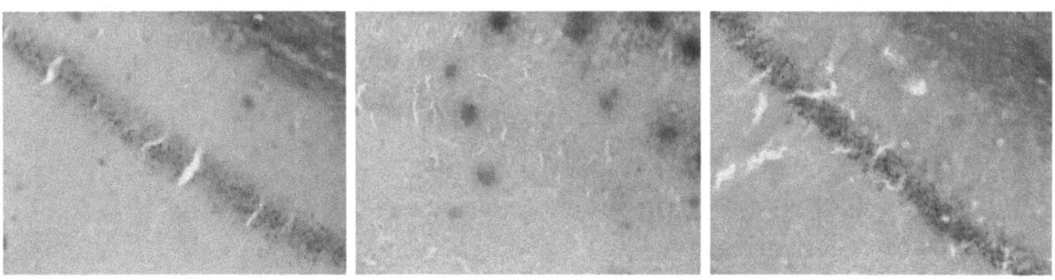

Fig. 1 Examples of origin data with iron

Adjusting the intensity of the original pixel (denoted by $p(x)$) to the new intensity, we can write

$$p_{corr}(x) = \frac{Q-1}{B-A}.(p(x) - A),$$

where Q is the maximum possible intensity (in our case 255), A, respectively, B is the minimum, respectively, maximum pixel intensities of the original image; see [21].

Filtration is, in general, separation of the useful information from not useful (if we consider background or noise). By handling with digital images, we consider reducing the noise or the background (non-useful information) and marking of selected objects (useful information).

Filtration based on RGB model (colored images)
According to [21], we define the RGB model as an additive colored model. They belong to the basic colors: red, green, and blue. Linear combination of these colors arise the whole scale, e.g., $(0, 0, 0)$ is black, and $(1, 1, 1)$ is white. Model RGB is used in image processing. In our case is the RGB information, which we want to catch, exactly a certain kind of color, connected with the concentration of the selected tissue.

By image processing of color, images are used on filters on this basic model RGB with the aim of processing of colors. In our case, the needed information is the information that we want to catch, certain kinds of colors connected with the concentration of the selected item. By image processing, we use filtering on the base RGB to modify colors. In general, we can define this kind of filtration as follows: for every pixel, we define this as a function of the color of the origin pixel, from which we can get and obtain new and modified coloring:

$$q_R(x) = f_R(p_R(x), p_G(x), p_B(x))$$

$$q_G(x) = f_R(p_R(x), p_G(x), p_B(x))$$

$$q_B(x) = f_B(p_R(x), p_G(x), p_B(x)),$$

where p_R, p_G, p_B and q_R, q_G, q_B are the only values of the original or the newly transported intensities in the RGB channels. In the easiest case, we consider a linear combination of the original values of colors, which we present by the matrices. The color of the pixel will be presented as a vector in three folders:

$$q^*(x) = Mp^*(x),$$

where $p^*(x)$ and $q^*(x)$ are the column vectors of the origin and the new color of the pixel, M is the matrix of the size 3×3, where the coefficients are the elements of the weighted filter. These coefficients show the measure of the new coefficient of colors versus original coefficients of colors (R, G, B).

For our purposes, we use a similar function, which transforms the value soft three chanels RGB into the new value-intensity of pixel:

$$q(x) = w_R.p_R(x) + w_G.p_G(x) + w_B.p_B(x).$$

Coefficients w_R, w_G, w_B are the values between 0 and 1, and they are given by the settings of the filter. These coefficients decide which one colored channel will be marked by the filter or suppressed by the filter. On application of the approach, the colored picture will be reduced into a white-and-black image. This way we reduce the amount of data to one-third. The goal of the filter is to assign each pixel a value between 0 and 1 (respectively, 0 and 255), depending on the RGB input values. In doing so, we require that the value of the output pixel 1, if its color perfectly matches the color of the object to be searched, and drops reasonably to 0 in the case of another coloring.

The filter sensitivity to the color ratio is ensured by assigning weights (marked as w_R, w_G, w_B) to the components R, G, B of the access pixel. The filter can be written as a function of the hp, color, R, G, B color of the input image (labeled p_R, p_G, p_B):

$$q(x) = w_R.p_R(x) + w_G.p_G(x) + w_B.p_B(x).$$

When processing diverse color data, a different color than the desired color can be affected by the filtering result. To suppress the effect of pixels other than of the tuned color, we introduce a penalty function to divide the filtering result, when

$$q(x) = \frac{w_R.p_R(x) + w_G.P_G(x) + w_B.p_B(x)}{d_R.p_R(x) + d_G.P_G(x) + d_B.p_B(x)}.$$

Penalization values d_R, d_G, d_B are values between 0 and 1 and determine the rate of suppression of a given color channel. When using this filter alone, the coefficients d_R, d_G, d_B must be set

$$d_R + d_G + d_B = 1$$

to apply to exceed the intensity range. In our implementation, this is handled by a specially modified FSHS method that scales any calculated values into the interval $< 0, 255 >$.

Now we show the filter setting for these samples. In this case of detecting irons in samples, one set of parameters is three frames. We compare two procedures for the processing of these images that differ in the order of operations applied:

The average-filter is a variant in which we first average the RGB values of three original frames, pixel by pixel. From these averages, we will compile one and then apply a filter to the image. The filter-average, on the other hand, consists of filtering on each from three shots, then averaged three filtered images.

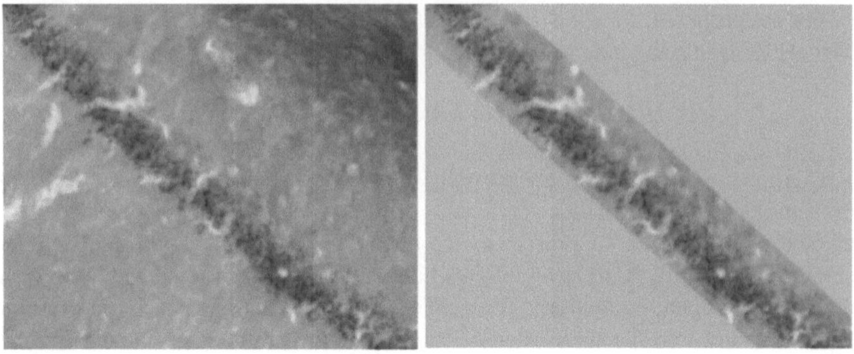

Fig. 2 Example of the cuts from images

The filter-average consists of filtering on each picture from three given shots. Consequently, we average three filtered images. This variant is much more computationally demanding.

Slide viewports—Cuts from the images: It was necessary to crop the image to remove areas that could skew the segmentation result (they contained spots or whole areas that are too similar to the searched object).

It was necessary to frame the image to remove areas that could distort the segmentation result (they contained spots or entire areas with color too similar to the searched object). In Fig. 2, the original image is on the left, and the viewport is on the right.

We created image slices in an external graphics program by replacing the deleted area with a color that is distinguishable from the area of interest (including the eyes of the software). In our case, this was done in pure green ($R = 0, G = 255, B = 0$), which the software can only recognize based on the maximum G value that none of the valid pixels of the image reaches. The pixels detected in this way are not taken into account even during the program run, for example, when initializing the network or correcting the filter results.

Filter application: The expected output of the filter is an image where the selected objects are distinguishable from the entire background and the intensity transitions at the edges of the cells should correspond as accurately as possible to the information in the images provided by the users.

The background itself should have the least variation in intensity. We can consider corrections. Depending on the filter parameters, a result that does not reach the maximum possible range of values and thus the maximum contrast may appear. Since we do not want to maintain the brightness of the image, on the contrary, the aim is to achieve the greatest possible contrast between the pixels of the detected objects and the pixels of the background, we can adjust the filtering result using the FSHS method. There was a problem in the detection of very small spots that had to be treated again.

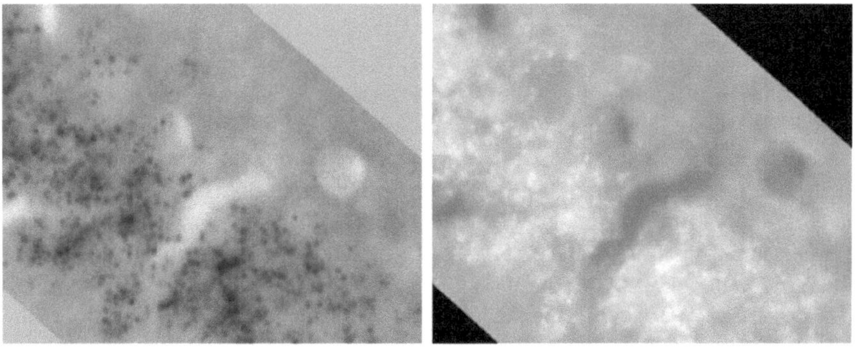

Fig. 3 Filtration example

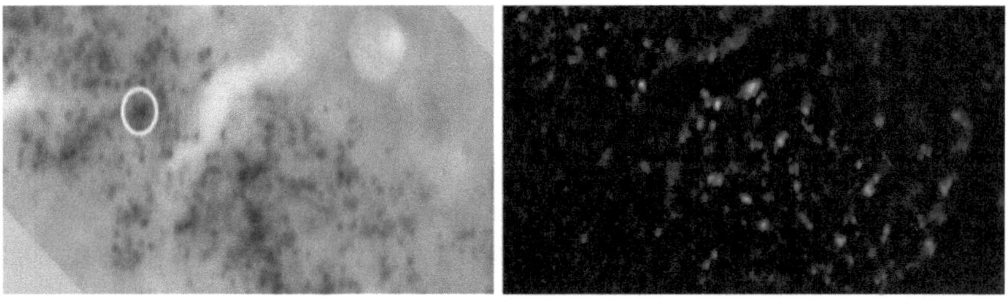

Fig. 4 Correction radius by subtracting the local average

The sample of the original image (in Fig. 3 on the left) and the filtering result (in Fig. 3 on the right) illustrates how large-scale spots on a larger scale conceal much finer local variations in values that should be evaluated as the presence of iron. (Visible to the original image with the naked eye.)

A significant portion of the small spots exhibited this condition, thereby severely distorting the segmentation result. We, therefore, decided to adjust the data by subtracting the local average. We determined the average of the surrounding values for each pixel and subtracted this from the pixel value. This correction is writeable as follows: $i_{corr}(x) = i(x) - 1/A(R). \sum i(y)$, where $i(x)$ is the intensity of the pixel x, $A(R)$ is the number of pixels of the averaging neighborhood, and R is the neighborhood radius, an optional parameter of this operation. We want it to be larger than the radius of the largest of the spots detected, but at the same time as small as possible so that the calculation is not unnecessarily prolonged.

By correcting with local averages, we achieved a much cleaner staining of the spots, thus ensuring consistent evaluation even in areas where the filter had previously evaluated them as a background; see example in Fig. 4.

5 Segmentation and Graph Cutting

Segmentation in image processing can be formulated in mathematics as a minimization problem. Segmentation can work as a powerful energy minimization tool producing a globally optimal solution. For segmentation, we use the mathematical method called "graph cutting". In the work, we focused primarily on Ford–Fulkerson and Edmonds–Karp algorithms. We process the 2D image, which we first abstract as a graph (in the means of graph theory—a part of mathematics) and then we try to find a maximum flow in it. After finding the maximum flow, we can segment the image. The graph cuts are used in medical and biological image segmentation following few dynamic algorithms, finding the local minimum of the energy. Compared to the threshold technique, this approach gives more realistic results.

The principle of Ford–Fulkerson and Edmonds–Karp, [8, 9] algorithms is based on increasing of the flow in the graph (net) through the augmenting paths. The algorithm progress while any augmenting path can be found. When there is no augmenting path available, the algorithm ends and the maximum flow is reached. The value of the maximum flow equals the sum of the capacities of the "minimal cut" edges. The minimal cut is the result of the graph cut algorithms (mentioned above) applied. The simplified explanation of finding the minimal cut is the process of pushing flow (imaginary units) from the source vertex named s to the tank vertex named t through the graph consisting of the vertices and edges while possible. Once the process is finished and there is no capacity of the edges to transport any other flow, the minimal cut can be found as the union of such edges. In the image segmentation process, the pixels of any 2D images can be abstracted into the graph vertices and the graph used in theoretical mathematics can be constructed. After that, the graph cut algorithms can be applied, the minimal cut can be found, and finally, the image can be represented by the objects and background.

6 Implementation in the Program and Evaluation of Capacities

Now we describe how we have implemented the graph theory approach in our program to obtain the segmentation. The program is written in the language C. For searching for the maximal flow of the network, we have used the Edmonds–Karp algorithm. We have modified some of the steps of this algorithm and we did the optimization of the fast running of the algorithm, as well. Mainly, we have decreased the number of iterations. If it was not possible to decrease the number of iterations, we have decreased the number of operations in these iterations to a minimum.

Capacity of the Edges We count the capacity for the corresponding links (edges). Those links, which connect exactly two neighboring pixels (vertices) p and q, we call N-links. Those links, which connect exactly one pixel with the source s and the sink t, we call T-links, see the Fig. 5, where the pixel is presented as a gray cube, N

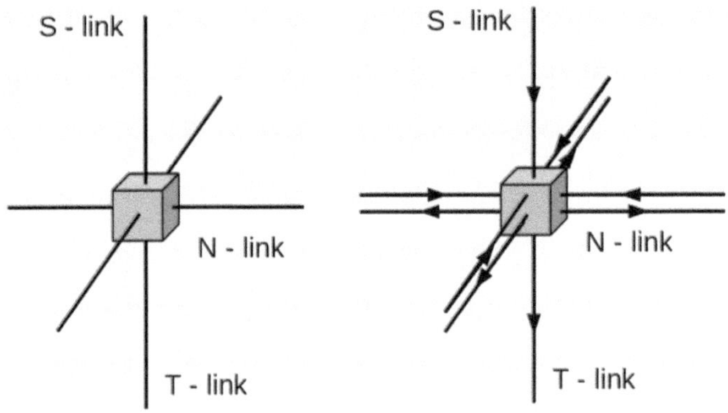

Fig. 5 One pixel and its links in the non-oriented as well as oriented segmentation network

links as horizontal links, and T-links as vertical lines. We use the following notation for counting the capacities:

P the set of all pixels,
(p, q) the edge connecting neighboring pixels p and q,
I_p the value of the intensity of the pixel p,
M maximal value of the intensity of the pixel (of the responsible figure),
D the difference between the maximal and minimal value of the intensity of the pixel (of the responsible figure),
O_{avr} average value of the intensity of object seed pixels,
B_{avr} average value of the intensity of the background seed pixels,
$S(p)$ capacity of the edge(link) connecting the sink (the vertex s) and corresponding pixel p,
$T(p)$ capacity of the edge (link) connecting output source (the vertex t) and concrete pixel (the vertex p),
$N(p, q)$ capacity of the edge (link) connecting neighbors pixels p a q,
λ weighing constant.

The weighing constant λ determines the result of segmentation. $N(p, q)$ expresses the relationship between intensities of p an q; $S(p)$ and $T(p)$ express the relationship between intensity values of pixels and the values O_{avr} and B_{avr}. More about connection between this variables and constants you can find in [4].

Linear Diffusion Coefficient Capacities of N-links and T-links are dependent on the intensities of the concrete pixel. Other values of capacities we count from the values of intensities of pixels as follows: Both N-line and T-line capacities depend on the intensity of the pixel. Therefore, the next capacity values are calculated from the pixel intensity values as follows [4, 15]:

Table 1 Capacities

Type	Edge	Capacity	
N-links	(p, q)	if $(p, q) \in P$	$N(p, q)$
T-links	(s, p)	if $p \in P \setminus \{o \cup B\}$	$\lambda S(p)$
		if $p \in O$	0
		if $p \in B$	0
	(p, t)	if $p \in P \setminus \{o \cup B\}$	$\lambda T(p)$
		if $p \in O$	0
		if $p \in B$	∞

$$
\begin{aligned}
N(p, q) &= D - |I_p - I_q| \\
S(p) &= M - |O_{avr} - I_p| \\
T(p) &= M - |B_{avr} - I_p|.
\end{aligned}
$$

It is precisely because of the character (definition) of the M and D constants that the capacities are non-negative. In extreme cases, some capacities may be zero. Taking into account all previous claims, we assign specific capacities to specific edges in the following way as shown in Table 1; see [4, 15].

Non-linear Diffusion Coefficient We suggest different and new approaches on how to give values to edges, as well. If we want to approach the assignment of N-line capacities in a way that takes greater account of the relative intensity of pixel intensities and penalizes their differences, it is necessary to choose a non-linear coefficient for calculating their capacity. The non-linear coefficient causes neighboring pixels with similar intensity values to have high-capacity edges (lines) and a certain drop in intensity of the neighboring pixels, and the edge-to-edge gain is almost zero. For the interpretation of the formulas, the following abbreviations are used:
s— the absolute value of intensity difference of two neighboring pixels,
σ—penalizing constant,
k—the penalty.

Thus, the value s is calculated as the absolute value of the difference of the two adjacent pixels in the picture, the penalizing constant is optional and affects the course of functions, especially how rapidly their first derivation changes, and the penalty k is given by the formula. We distinguish:

TV diffusion coefficient

$$d(s) = \frac{1}{s}, s \in N,$$

BFB diffusion coefficient

$$d(s) = \frac{1}{s^2}, s \in N,$$

Charbonnier's diffusion coefficient

$$d(s) = \frac{1}{\sqrt{(1 + \frac{s^2}{k^2})}}, s \in N,$$

Perona–Malik's diffusion coefficient

$$d(s) = \frac{1}{1 + (\frac{s}{k})^2}, s \in N,$$

$$d(s) = e^{-(\frac{s}{k})^2}, s \in N,$$

Weickert's diffusion coefficient

$$d(s) = \begin{Bmatrix} 1 & s = 0 \\ 1 - e^{\frac{-3.31488}{(s/K)^8}} & s > 0 \end{Bmatrix}, s \in \mathbb{N}.$$

Comparision of diffusion coefficient. The diagram in Fig. 6 shows the non-linear diffusion coefficients, depending on the intensity differences from $< 0, 255 >$ depended at the selected value:

(1) Linear coefficient—blue,

(2) TV coefficient—red,

(3) BFB coefficient—purple,

(4) Charbonnier coefficient—yellow,

(5) Perona–Malik coefficient—green, (-black),

(6) Weickert coefficient—orange.

We choose the Edmonds–Karp algorithm for the shortest growing augmenting paths. We implemented the algorithm with minor variations over the original one. From the programming point of view, the implementation can be divided into three main parts, which can also be further elaborated. They are as follows:

(a) marking procedure,

(b) path reconstruction,

(c) the distribution of vertices.

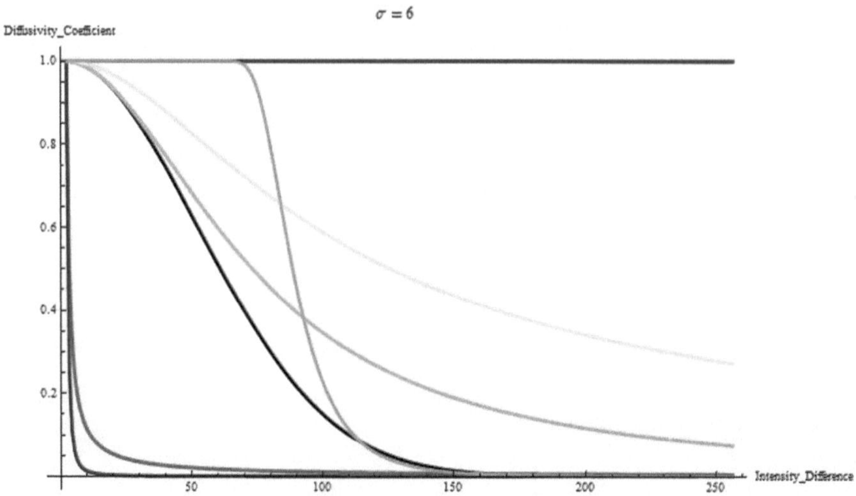

Fig. 6 Non-linear functions for evaluating the capacities

7 Results

The result was quantified by counting the pixels in the segmented region. This numerical result indicates the concentration of the desired substance, respectively. In the case of iron, it was necessary to frame the image and the result may have been affected by this cropping. When displaying the result, the program also specifies the number of valid pixels (the number of pixels in the image—the number of pixels removed) so that this cropping can be taken into account when evaluating the results. Since the segmentation result is relatively sensitive to network setup and, therefore, to the selection of representative points, we evaluate the results of O_{avr} and B_{avr}, which speak about these points. Based on the deviation in these two values, we can assess the objectivity of the segmentation result.

Comparison of approaches. We used two different approaches to averaging the three frames: Filter-Average and Average-Filter. An essential part of the evaluation is hence the comparison of values and deviations between these approaches. With a comparison of these results, we can sum up that both approaches gave the same results for detecting iron and quality of segmentation.

8 Discussion, Conclusion, and Remarks

Nowadays, for simpler real-time data, for more complex data, and global segmentation in tens of seconds and local segmentation of more complex data, it is also real time. Segmentation speed also depends on the quality of the input data, the

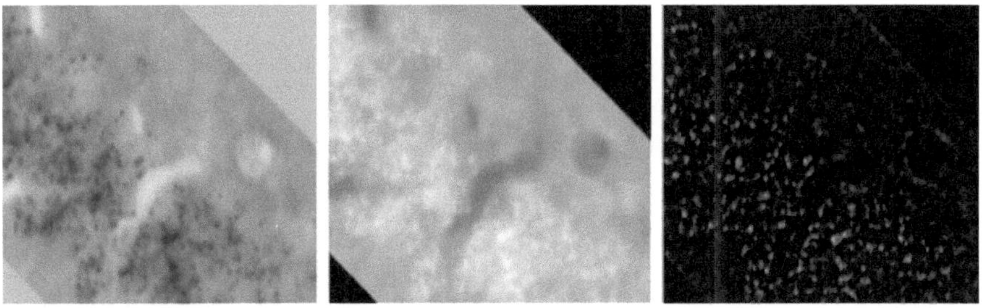

Fig. 7 Filter and correction sample

dimensions of the image, the selection, and the number of seed pixels. The program currently segments either the entire image (globally) or only part of it (locally), depending on what option the user chooses. In the near future, it will also be possible to segment the color images and we would also like to program some other parts so that it is understandable and clear to the normal user. Of course, further optimization of the program is also one of the goals.

We look at three images that capture the same area. In the images, a band is visible to the naked eye with the appearance of tiny dark spots. These are an indication of the presence of iron, which our method will measure. We processed triples of images, Fig. 7, with each of the triples being cut with the subject iron-containing band to eliminate disturbing data in other parts of the image (non-iron-related brown spots). For this case, the filter was set to highlight dark brown spots indicating the presence of iron, and we corrected the result by correcting the local average.

In Fig. 7, we see from left to right the original frame, the filtered frame, and the result after correction.

The filtration phase was carried out described above by the dual Filter-Diameter and Diameter-Filter method.

Segmentation was performed in six trials for each averaged image and for both averaging methods, recording the number of pixels selected and the $Oavg$, $Bavg$ values of the segmentation network, which we use as a control to evaluate the results.

There is a preview of the segmentation result on the image detail in Fig. 8: from left to right, the original image, the corrected filtered image, and the segmentation result are shown. The segmented pixels are marked in red. From the control values, it can be concluded that the two methods of filtering the triple frames give equivalently good results, and the deviations between them are derived from the segmentation method.

We succeeded in constructing a method by which we detected the presence of the substances sought and quantified their concentration, including the time course for iron. Using filtering and segmentation, we created segmented versions of the processed images that display the detected substance consistently and according to expected results. We detected variations in segmentation results from the recorded data and control values and suggested ways to obtain objective results from them. We have created and used a new filter in the pre-processing step to prepare data before

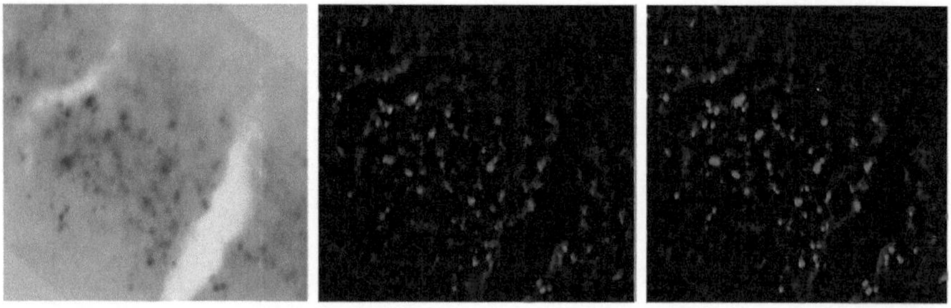

Fig. 8 Example of segmentation

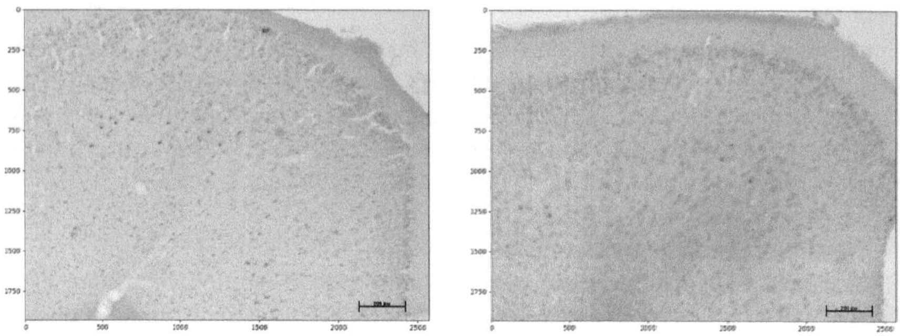

Fig. 9 Example of segmentation of data using Open CV

segmentation. This was done with the purpose to get better and more objective segmentation.

The results confirmed the existence of higher iron concentration in the sensed area of mouse brain tissue observed during Alzheimer's disease and quantified the difference from the control sample, i.e., healthy mouse.

Deficiencies identified and possible extensions: The implemented method for input data was tailor-made. By changing the filter settings, it can be used to process any medical data to detect and measure the presence of substances based on color. The implemented version of the program is fixed for processing images in the described two cases. Its drawbacks include the need for manual entry for each frame (representative object and background points must be selected), on which the segmentation result directly depends. The resulting deviations were solved by statistically processing several measurements to optimize the results.

In particular, we perceive the possible extensions of the implemented program to be greater automation of the process by the better interconnection of individual parts of the method. After a more in-depth analysis of the filter's behavior, it would be possible to develop (at least in part) the automatic selection of representative points, which would significantly speed up, simplify the work with the images, and reduce the deviations caused by this selection.

We mention another method using open source for some kind of special data for searching of iron: OPEN CV [27]. At least we can discuss remarks and observations: We are using adaptive thresholding here, see Fig. 9, and the threshold value is the weighted sum of neighborhood values where weights are a gaussian window. Then we obtained the black dots using connected components analysis and finally filtered them by size. At first, we were using the Otsu binarization, but the background was messing up the threshold calculation that Otsu did. Then we tried this Adaptive thresholding method, but a lot of false positives were coming. We figured out that these false positives were just portions of the image that were only mildly darker than their neighborhood. So, we increased the threshold while converting to the binary image as, what was working in our data:

$binary\_img = cv2.adaptiveThreshold\ (gray\_img, 255, cv2.ADAPTIVE\_$
$THRESH\_GAUSSIAN\_C,$
$cv2.THRESH_B INARY_I NV, 131, 20),$
instead of
$binary\_img = cv2.adaptiveThreshold(gray\_img, 255, cv2.ADAPTIVE\_$
$THRESH\_GAUSSIAN\_C,$
$cv2.THRESH\_BINARY\_INV, 131, 10).$

As a note, we consider the quality of segmentation of iron is a bit less like in the case of graph cutting. We can consider in the future the improvement of this method. But still, we want to present this as another possibility for detecting iron.

We can conclude we were successful in creating a new filter and software and finding better methods in segmenting iron in biological samples.

8.1 Ethics

Samples, which we used in our experiment,t are animal samples. One of the coauthors performed histological staining when she was visiting a collaborating laboratory in Finland. All procedures involving the animals were performed in compliance with the Principles of Laboratory Animal Care issued by the Ethical Committee of the collaborating laboratory.

Acknowledgements The first author acknowledges the scientific grant VEGA 1/0006/19. This study was accomplished with the financial support of Slovak Research and Development Agency APVV-16-0039.

References

1. Batista-Nascimento, L., C. Pimentel, Menezes R. Andrade, and C. Rodrigues-Pousada. 2012. Iron and neurodegeneration: From cellular homeostasis to disease. *Oxidative Medicine and Cellular Longevity*. https://doi.org/10.1155/2012/128647.

2. Belaidi, A.A., and A.I. Bush. 2016. Iron neurochemistry in Alzheimer's disease and Parkinson's disease: Targets for therapeutics. *Journal of Neurochemistry* 139: 179–197. https://doi.org/10.1111/jnc.13425.
3. Braidy, N., A. Poljak, C. Marjo, et al. 2017. Identification of cerebral metal ion imbalance in the brain of aging Octodon degus. *Frontiers in Aging Neuroscience* 9: 1–16. https://doi.org/10.3389/fnagi.2017.00066.
4. Boykov, Y., and M.-P. Jolly. 2001. Interactive graph cuts for optimal boundary and region segmentation of objects in N-D images. In *Proceedings of "International conference on computer vision"*, Vancouer, Canada, vol. 1, 105–112. ISBN 0-7695-1143-0.
5. Clarke, F., and I. Ekeland. 1978. Solutions périodiques, du période donnée, des équations hamiltoniennes. *Note CRAS Paris* 287: 1013–1015.
6. Daglas, M., and P.A. Adlard. 2018. The involvement of iron in traumatic brain injury and neurodegenerative disease. *Frontiers in Neuroscience*. https://doi.org/10.3389/fnins.2018.00981.
7. Falangola M.F., S.P. Lee, R.A. Nixon, et al. 2005. Histological co-localization of iron in Aβ plaques of PS/APP transgenic mice. Neurochemical Research 30: 201–205. https://doi.org/10.1007/s11064-004-2442-x .
8. Ford, L.R., and D.R. Fulkerson. 1956. Maximal flow through a network. *Canadian Journal of Mathematics* 8: 399–404.
9. Ford, L.R., and D.R. Fulkerson. 1962. *Flows in networks*. Princeton: Princeton University Press.
10. Gong, N., R. Dibb, M. Bulk, et al. 2019. Imaging beta amyloid aggregation and iron accumulation in Alzheimer's disease using quantitative susceptibility mapping MRI. *Neuroimage* 191: 176–185. https://doi.org/10.1016/j.neuroimage.2019.02.019.
11. Goldberg, A.V., and R.E. Tarjan. 1988. A new approach to the maximum flow problem. *Journal of the Association for Computing Machinery* 35 (4): 921–940.
12. Greenough, M.A., J. Camakaris, and A.I. Bush. 2013. Metal dyshomeostasis and oxidative stress in Alzheimer's disease. *Neurochemistry International* 62: 540–555. https://doi.org/10.1016/j.neuint.2012.08.014.
13. Jankowsky, J.L., D.J. Fadale, J. Anderson, et al. 2004. Mutant presenilins specifically elevate the levels of the 42 residue beta-amyloid peptide in vivo: Evidence for augmentation of a 42-specific gamma secretase. *Human Molecular Genetics* 13: 159–170. https://doi.org/10.1093/hmg/ddh019.
14. Langkammer, C., N. Krebs, W. Goessler, et al. 2010. Quantitative MR imaging of brain iron. *A Postmortem Validation Study, Radiology* 257: 455–462. https://doi.org/10.1148/radiol.10100495.
15. Loucký, J., and T. Oberhuber. 2010. *Graph cuts in segmentation of a left ventricle from MRI data*. Prague: Czech Technical University in Prague.
16. Petillon, C., R. Hergesheimer, H. Puy, et al. 2019. The relevancy of data regarding the metabolism of iron to our understanding of deregulated mechanisms in ALS; hypotheses and pitfalls. *Frontiers in Neuroscience* 13: 1–10. https://doi.org/10.3389/fnins.2018.01031.
17. Plascencia-Villa, G., A. Ponce, J.F. Collingwood, et al. 2016. High-resolution analytical imaging and electron holography of magnetite particles in amyloid cores of Alzheimer's disease. *Scientific Reports* 6: 24873. https://doi.org/10.1038/srep24873.
18. Rueden, C.T., M.C. Johannes Schindelin, et al. 2017. ImageJ2: ImageJ for the next generation of scientific image data. *BMC Bioinformatics* 18: 529. PMID 29187165. https://doi.org/10.1186/s12859-017-1934-z(on Google Scholar).
19. Carpenter, A.E., T.R. Jones, M.R. Lamprecht, C.Clarke, I.H. Kang, O. Friman, D.A. Guertin, C.J. Han, R.A. Lindquist, J. Moffat, P. Golland, and D.M. Sabatini. 2006. Genome CellProfiler: Image analysis software for identifying and quantifying cell phenotypes, Published: 31 October 2006, Open Access, Springer.
20. Sokolov, A.V., I.V. Voynova, V.A. Kostevich, et al. 2017. Comparison of Interaction between Ceruloplasmin and Lactoferrin/Transferrin: To Bind or Not to Bind. *Biochemistry (Mosc)* 82: 1073–1078. https://doi.org/10.1134/S0006297917090115.
21. Sonka, M., V. Hlavac, R. Boyle, 1993. London: Chapman Hall, 555.

22. Schneider, C.A., W.S. Rasband, and K.W. Eliceiri. 2012. NIH Image to ImageJ: 25 years of image analysis. *Nature Methods* 9(7): 671–675, PMID 22930834 (on Google Scholar).
23. Telling, N.D., J. Everett, J.F. Collingwood, et al. 2017. Iron biochemistry is correlated with amyloid plaque morphology in an established mouse model of Alzheimer's disease. *Cell Chemical Biology* 24: 1205-1215.e3. https://doi.org/10.1016/j.chembiol.2017.07.014.
24. Ward, R.J., F.A. Zucca, J.H. Duyn, et al. 2014. The role of iron in brain ageing and neurodegenerative disorders. *The Lancet Neurology* 13: 10450–1060. https://doi.org/10.1016/S1474-4422(14)70117-6.
25. Webster, S.J., A.D. Bachstetter, P.T. Nelson, et al. 2014. Using mice to model Alzheimer's dementia: An overview of the clinical disease and the preclinical behavioral changes in 10 mouse models. *Frontiers in Genetics* 5: 1–23. https://doi.org/10.3389/fgene.2014.00088.
26. Zecca L., M.B.H. Youdim, P. Riederer, et al. 2004. Iron, brain ageing and neurodegenerative disorders. National Review.
27. https://opencv-python-tutroals.readthedocs.io/en/latest/py_tutorials/py_imgproc/py_thresholding/py_thresholding.html
28. https://imagej.net/Citing
29. https://cellprofiler.org/citations/
30. https://www.moleculardevices.com/products/cellular-imaging-systems/acquisition-and-analysis-software/metamorph-microscopy
31. https://www.ilastik.org/

How Do You Defend a Network?

Marcin Dziubiński and Sanjeev Goyal

1 Introduction

Our nation's critical infrastructure is crucial to the functioning of the American economy...(It) is increasingly connected and interdependent and protecting it and enhancing its resilience is an economic and national security imperative [15].

Infrastructure networks—highways, aviation, shipping, pipelines, train systems, and posts—are a vital part of the modern economy. These networks face a variety of threats ranging from natural disasters to human attacks. The latter may take a violent form (guerrilla attacks, attacks by an enemy country, and terrorism) or a nonviolent form (as in political protest that blocks transport services).[1] A network can be made robust to such threats through additional investments in equipment and in personnel. As networks are pervasive, the investments needed could be very large; this motivates the study of targeted defense. What are the "key" parts of the network that should be protected to ensure maximal functionality? As defense is often a choice made by individual actors, we also wish to understand the relation between network structure and decentralized incentives. This paper develops a model to study these questions.

Consider a given infrastructure network consisting of nodes and links. The defender chooses to protect "nodes" of the network against damage/attacks; protecting a node is costly. Protection includes investments in security personnel, in training, in equipment, and in cybersecurity. These protection measures typically take time to implement and so we focus on ex ante investments in protection. We

[1]For an introduction to network based conflict, see [3, 37]; for news coverage of the effects of natural disasters and human attacks on infrastructure networks, see [18, 28, 31, 32].

Reprinted from: Theoretical Economics, Volume 12, Issue 1, pp. 331–376, 2017.

M. Dziubiński (✉)
Institute of Informatics, Faculty of Mathematics, Informatics, and Mechanics, Warsaw University, Warsaw, Poland
e-mail: m.dziubinski@mimuw.edu.p

S. Goyal
Faculty of Economics and Christ's College, University of Cambridge, Cambridge, England
e-mail: sg472@cam.ac.uk

S. Borkotokey et al. (eds.), *Game Theory and Networks*, Indian Statistical Institute Series,
https://doi.org/10.1007/978-981-16-4737-6_13

suppose that a defended node is immune to attack whereas an undefended node is eliminated by attack (along with all its links). The initial network, the defense, and the attack together yield a set of surviving nodes and links—the residual network. The defender chooses a defense strategy that maximizes the value of the residual network, net of the costs of defense.

Our model covers two scenarios. The first is that of an intelligent adversary who seeks to damage components and disrupt the flows in the network. The second is that of a natural threat: facing such a threat, the defender focuses on the worst case scenario. In both cases, the defender looks for the "maximin" solution. For expositional simplicity, we use the language of an intelligent adversary throughout. We study a game between a defender and an adversary and analyze the subgame perfect equilibrium of this game.

We consider network payoff functions in which the value to the defender of a network is component additive, and the payoff from each component is increasing and convex in the size of the component.[2] The convexity of value in component size is key to the appeal of connectivity in networks.

We begin with a study of optimal defense. Proposition 2 characterizes optimal defense and attack. Optimal attack targets two types of nodes: those that fragment the network into distinct components (the separators) and those that simply reduce the size of components (the reducing attacks). As payoffs are convex in component size, separators are particularly attractive targets for attack (as their elimination disconnects components). Anticipating this attack, optimal defense targets nodes that block the separators and reduce attacks. A set of nodes that block a collection of separators is referred to as a transversal. We prove that optimal defense either targets a minimal transversal or protects all nodes. Figures 3 and 4 illustrate these concepts.[3]

This characterization result allows us to study the relation between networks and conflict more closely. We find that the size of defense and attack are both nonmonotonic in the cost of attack; even more surprisingly, the size of defense and the payoff of the defender may fall with the addition of links in the network (Proposition 3).

We then turn to the intensity of conflict: this is the sum of expenditures of defense and attack. For a given configuration of costs of defense and attack, we derive the minimal intensity of conflict and then describe the networks that sustain it (Proposition 4). We then demonstrate that network architecture can create very large variations in the intensity of conflict. A feature of minimal conflict is that there is a single active player. We next discuss circumstances under which both players devote resources to conflict in equilibrium.

An important insight of the analysis is the optimality of strategic exposure: the defender may find it optimal to leave unprotected a key node (the elimination of

[2] This specification is consistent with Metcalfe's law (network value is proportional to the square of the number of nodes) and Reed's law (network value is exponentially increasing in the number of nodes). It is also in line with the large theoretical literature on network externalities [19, 27] and network economics [6, 26]. One way to define network value is the number of pairs of nodes connected (directly or indirectly) in the network. This is a special case of our value function.

[3] Appendix C provides a detailed application of the concepts to well-known families of networks (trees, core–periphery, interlinked stars).

which disconnects the network) and instead to protect an alternative, larger, set of nodes. We refer to this as the *queen sacrifice*. This leads us to identify a class of networks—*windmill graphs*—that minimize conflict and are also attractive for the defender. Figure 7 presents these networks.

In many situations, security decisions are made at the local level, e.g., individual airports choose their own security checks. This motivates the study of decentralized security.[4] Individual nodes care about surviving an attack and about being part of a large connected network. Observe that to block a separator it is sufficient for one node in the separator to protect itself. So, in the game among the nodes, defense choices within a separator are strategic substitutes. But for the network to remain connected, all separators must be blocked. Therefore, a node will protect itself only if other separators are being blocked: thus, defense choices also exhibit strategic complementarity. Proposition 5 shows that decentralized security choices can be characterized in terms of separators and transversals of the network. Finally, we demonstrate that a combination of incentive and coordination issues may lead to very large costs of decentralization.

Our paper contributes to the economic study of networks. The research on networks has been concerned with the formation, structure, and functioning of social and economic networks [22, 25, 35] The problem of key players has traditionally been studied in terms of Bonacich centrality, betweenness, eigenvectors, and degree centrality; see, e.g., [6, 7, 11, 14, 17, 20, 21]. Our paper suggests that for the problem of attack and defense, the key players are nodes that lie in separators and transversals. These nodes are typically distinct from nodes that maximize familiar notions of centrality. Appendix B discusses this distinction in detail. Thus, the principal contribution of our paper is to introduce two classical concepts from graph theory into economics and show how they address a problem of practical importance.

Individual defense is a public good, and so this conceptual contribution is also relevant for the study of games on networks more generally. Bramoullé and Kranton [9] draw attention to maximal independent sets. By contrast, our work brings out the role of minimal transversal of the separators. These sets are generally different from maximal independent sets.[5]

Our paper also contributes to the literature on network defense; see, e.g., [1, 5, 8, 12, 16, 23, 29]. To the best of our knowledge, our results on the role of separators and transversals in network conflict are novel, relative to the existing body of work. In particular, we note that the earlier work by [16, 23] focuses on optimal design and defense. In these papers, the optimal network takes on a very simple form—it is a star—and so the optimal defense takes on a correspondingly simple structure: protect the central hub node. By contrast, in the present paper the network is exogenous and arbitrary: this is a much broader problem and requires new conceptual tools.

---

[4] For an early contribution on interdependent security, see [30].

[5] For example, in a core–periphery network, all the core nodes are essential separators, while the maximal independent set can include at most one core node and must include peripheral nodes. See Appendix C for details on this.

We note that the problem of network defense has traditionally been studied in operations research, electrical engineering, and computer science; see, e.g., [2, 4, 24, 33]. In an early paper, Cunningham [13] looks at the problem of network design and defense with conflict on links. Relative to this literature, the novelty of our paper lies in the study of intensity of conflict and the externalities that arise in decentralized defense.

The rest of the paper is organized as follows. Section 2 presents the model of defense and attack. Section 3 introduces the main concepts and provides a characterization of equilibrium defense and attack. It also contains the study of comparative statics, active conflict, and conflict intensity. Section 4 takes up the case of decentralized defense. Section 5 concludes. All proofs are presented in Appendix A. Appendix B analyzes the relation between key nodes to attack and defend and other notions of centrality. Appendix C illustrates the notions of separators and transversals in well-known families of networks such as core–periphery networks, trees, interlinked stars, and bipartite graphs. In Appendix D, we discuss the role of sequentiality of moves and perfect defense in the results obtained in the paper.

2 The Model

We start with a given network and consider a two-player sequential move game with a defender and an adversary. In the first stage, the defender chooses an allocation of defense resources. In the second stage, given a defended network, the adversary chooses the nodes to attack. Successfully attacked nodes (and their links) are removed from the network, yielding a residual network. The goal of the defender is to maximize the value of the residual network, while the goal of the adversary is to minimize this value.[6]

Let $N = \{1, \ldots, n\}$, with $n \geq 3$, be a finite set of nodes. A link is a two element subset of N. The set of all possible links over $P \subseteq N$ is $g^P = \{ij : i, j \in P, i \neq j\}$ (where ij is an abbreviation for $\{i, j\}$. A *network* is set of links. Given set of nodes $P \subseteq N$, $\mathcal{G}(P) = 2^{g^P}$ is the set of all networks over P. The set $\mathcal{G} = \bigcup_{P \subseteq N} \mathcal{G}(P)$ is the set of all networks that can be formed over any subset of notes from N. Every network $g \in \mathcal{G}$ has a *value* $\Phi(g)$ associated with it : $\Phi : \mathcal{G} \to \mathbb{R}$ is called a *value function*.

The set of nodes $X \subseteq N$ chosen by adversary is called an attack. The set $X = \Phi$ is called the *empty attack*. A *defense* is set of nodes $\Delta \subseteq N$; node $i \in N$ is defended under Δ if and only if $i \in \Delta$. We assume that the defense is perfect a protected node cannot be removed by an attack, while any attacked unprotected node is removed with certainty. Given defense Δ and attack X, set $Y = X \setminus \Delta$ will be removed from the network. Removing a set of nodes $Y \subseteq N$ from a network creates a *residual network* $g - Y = \{ij \in g : i, j \in N \setminus Y\}$.

[6] The sequential move game formulation appears to be appropriate for the large-scale and time-consuming protection investments discussed in the Introduction. This two-stage model with observability of first-stage actions is consistent with the approach in the large literature on security and networks; see, e.g., [2, 34].

Defense resources are costly: the cost of defending a node is $c_D > 0$. Given network g, the defender's payoff from strategy $\Delta \subseteq N$, when faced with the adversary's strategy $X \subseteq N$, is

$$\Pi^D(\Delta, X; g, c_D) = \Phi(g - (X \setminus \Delta)) - c_D|\Delta|$$

Attack resources are costly: the cost of attacking a node is given by $c_A > 0$. Given defended network (g, Δ), the payoff to the adversary from strategy $X \subseteq N$ is

$$\Pi^A(\Delta, X; g, c_A) = -\Phi(g - (X \setminus \Delta)) - c_A|X| \quad (1)$$

We study the (subgame perfect) equilibria of this game.

Two nodes i and j are connected in network g if there is a sequence of nodes $i_o, \ldots, i_m$ such that $i = i_0$, $j = j_m$, and for all $0 < k \leq m$, $i_{k-1}i_k \in g$. A component of network g is a maximal and nonempty set of nodes $C \subseteq N$ such that any two distinct nodes $i, j \in C$ are connected in g. The set of components of g is denoted by $\mathcal{C}(g)$.

We assume that Φ is component additive. Given network g,

$$\Phi(g) = \sum_{C \in \mathcal{C}(g)} f(|C|)$$

where f satisfies the following assumption:

Assumption 1 *We have that* $f : \mathbb{R}_+ \rightarrow R_+$ is strictly increasing, strictly convex, and $f(0) = 0$.

2.1 Remarks on Model

We have assumed sequential moves; this is mainly for exposition. It is possible to show that our main results on characterization of conflict in terms of certain properties of the graph carries over with simultaneous moves. Perfect defense is a more substantial assumption. Smoother models of conflict such as the Tullock contest function would lead to modifications in parts of the main characterization results below. Appendix D discusses these points in greater detail. Finally, we have assumed that payoffs depend only on the sizes of the networks (or their components): so we abstract from other topological details of the network. This simplification allows us to make progress and should be seen as a first step in the study of network defense.

Since the game is finite and sequential, standard results guarantee the existence of (subgame perfect) equilibria. These equilibria are usually not unique, but generically, equilibrium outcomes are equivalent with respect to player's payoffs, sizes of defense and attack, and the value of residual network. This is the content of the following result.

Proposition 1 *For any network g and costs c_D and c_A, there exists a subgame perfect equilibrium. For generic values of c_A and c_D and generic f, the equilibrium attack and defense size and the payoffs of the players are unique.*

3 The Analysis

This section develops our main results for the two-person game between the defender and the adversary. Optimal attack focuses on sets of nodes that fragment the network (the separators), while optimal defense targets sets of nodes that block these separators (the transversals). The interest then moves on to the relation between network architecture and the intensity of conflict (the sum of resources allocated to attack and defense) and the prospects of active conflict (when the adversary eliminates some nodes while the defender protects others).

We begin with a study of a simple example that helps illustrate a number of interesting phenomena.

Example 1 (*Defense and attack on the star*) Consider the star network with $n = 4$ and $\{a\}$ as the central node (as in Fig. 1). The value function is $f(x) = x^2$.

As is standard, we solve the game by working backward. For every defended network (g, Δ) we characterize the optimal response of the adversary. We then compare the payoffs to the defender from different profiles, (g, Δ), and compute the optimal defense strategy. Equilibrium outcomes are summarized in Fig. 2. A number of points are worth noting.

(i) Observe that removing node a disconnects the network; this node is a separator. Moreover, there is a threshold level of cost of attack such that the adversary either attacks a or does not attack at all when $c_A > 7$. Protecting this node is also central to network defense.
(ii) The intensity of conflict exhibits rich patterns: when the cost of attack is very large there is no threat to the network and no need for defense. If the cost of attack is small, the intensity of conflict hinges on the level of defense costs. When they are low, all nodes are protected and there is no attack (the costs of conflict are nc_D); if they are high, then there is no defense but all nodes are eliminated (the costs of conflict are nc_A). For intermediate cost of attack and defense, both defense and attack are seen in equilibrium.
(iii) The size of defense may be nonmonotonic in the cost of attack. Fix the cost of defense at $c_D = 3.5$. At a low cost of attack ($c_A < 1$) the defender protects all nodes, in the range $c_A \in (1, 5)$ he protects 0 nodes, in the range $c_A \in (5, 13)$ he protects a, and then in the range $c_A > 13$, he stops all protection activity. Similarly, the size of the attack strategy may be nonmonotonic in the cost of attack.

The starting point of the general analysis is the nature of optimal attack. Given the convexity in the value function of networks, disconnecting a network is espe-

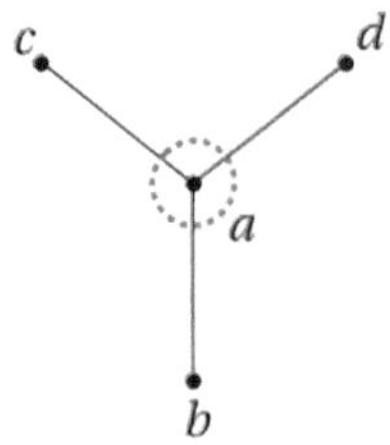

Fig. 1 Star network (n = 4)

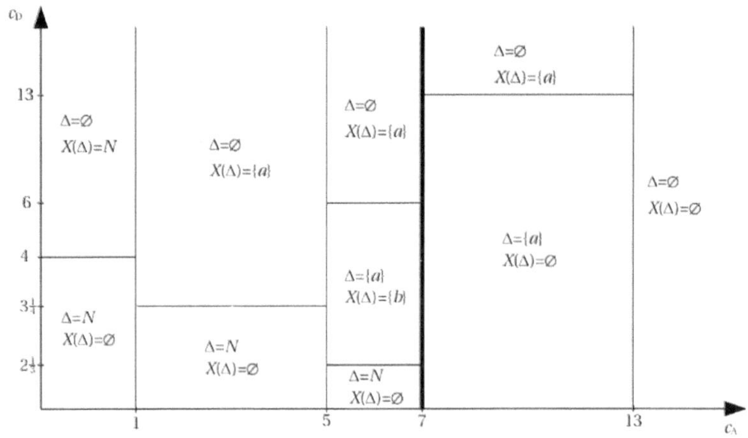

Fig. 2 Equilibrium outcomes: star network (n = 4) and $f(x) = x^2$

Fig. 3 Essential separators

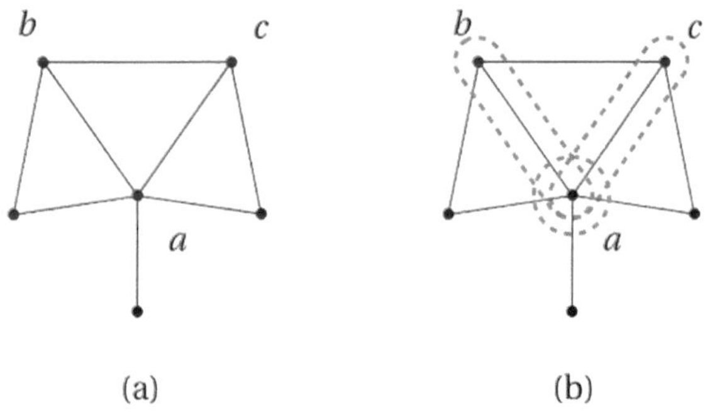

cially damaging. A set $X \subseteq N$ is a separator if $|\mathcal{C}(g)| < |\mathcal{C}(g - X)|$. In other words, a separator is a set of nodes the removal of which strictly increases the number of components in the network. A network will normally possess multiple separators and the adversary should target the most effective ones. A separator $S \subseteq N$ is essential for network $g \in G(N)$, if for every separator $S' \subseteq S$, $|\mathcal{C}(g - S)| > |\mathcal{C}(g - S')|$, i.e., a strict subset of eliminated nodes would lead to a strictly smaller number of components. The set of all essential separators of a network g is denoted by $\mathcal{E}(g)$. Figure 3 illustrates essential separators in an example.

The second element is the level of costs. As illustrated by Example 1, the network defense problem can be divided into two parts, depending on the cost of attack. Given $x \in N$, $\Delta f(x) = f(x+1) - f(x)$ is the marginal increase in the value of a component of size x when a single node is added to it. Under Assumption 1, $\Delta f(x)$ is strictly increasing. It is useful to separate two levels of costs: one, high costs with $c_A > \Delta f(n-1)$, and two, low costs with $c_A < \Delta f(n-1)$.

We start with the case of high cost as it brings out some of the main general insights in a straightforward way. Facing a high cost, the adversary must disconnect the network, i.e., choose a separator or not attack the network at all. Clearly, the adversary would never use an essential separator that yields a lower payoff than the empty attack. Given the cost of attack c_A and network g, the set of individually rational separators is $\epsilon(g, c_A) = X \in \mathcal{E}(g) : \Phi(g) - \Phi(g - X) \geq c_A|X|$.

When the cost of attack is low, it may be profitable for the adversary to use attacks that merely remove nodes from the network, without disconnecting it. A set $R \subseteq N$ is a reducing attack for a network g if there is no $X \subseteq R$ such that X is a separator for g. The set of all reducing attacks for a given network $g$is denoted by $\mathcal{R}(g)$.

The following lemma characterizes all the possible attacks of the adversary in terms of essential separators and reducing attacks. In addition, it provides a characterization of the attacks that are best responses in the adversary's sub game.

Lemma 1 *Fix a connected network g. Let $\Delta \subseteq N$ be a defense selected by the defender in the first stage. Any attack $X \subseteq N$ can be decomposed into two disjoint sets: a set E and a reminder set R such that the following statements hold:*

(i) The set E is either empty or $E \in \mathcal{E}(g)$.
(ii) The set R is a reducing attack for $g - E$.

Moreover, if X is a best response to Δ, then E is either empty or $E \in \mathcal{E}(g \in c_A)$.

The first part of the lemma says that any attack of the adversary can be seen as consisting of two phases. In one of the phases, the adversary fragments the network by removing a minimal set of nodes needed to obtain the desired components after the attack. This set is an essential separator of the network. In the other phase, the adversary reduces the size of the components (but without disconnecting any of them). Thus, the notion of essential separator captures exactly the attacks that serve the function of fragmenting the network. The characterization of attacks obtained in the first part of the lemma is useful in understanding the best responses of the adversary. If X is a best response to some strategy of the defender, then applying an essential separator phase of X after the reducing attack phase is applied must be worthwhile. But then, by convexity of f, it must be worthwhile even more to apply the essential separator phase before the reducing attack phase. Therefore, the essential separator phase must be individually rational.

We now turn to the equilibrium strategies of the defender. Again, it is instructive to start with the setting where the cost of attack is high. An optimal strategy of the defender should block a subset of individually rational essential separators in the most economical way. Given a family of sets of nodes, $\mathcal{H}$, and a set of nodes, M, $\mathcal{D}(M, \mathcal{H}) = \{X \in \mathcal{H} : X \cap M \neq \phi\}$ are the sets in $\mathcal{H}$ that are blocked (or *covered*) by M. The set M is called a transversal of $\mathcal{H}$ if $\mathcal{D}(M, \mathcal{H}) = \mathcal{H}$. The set of all transversals of $\mathcal{H}$ is denoted by $\mathcal{T}(\mathcal{H})$. Elements of $\mathcal{T}(\mathcal{H})$ that are minimal with respect to inclusion are called minimal transversals of $\mathcal{H}$. Elements of $\mathcal{T}(\mathcal{H})$ with the smallest size are called minimum transversals of $\mathcal{H}$. Let $\tau(\mathcal{H})$ denote the transversal number of $\mathcal{H}$, i.e., the size of a minimum transversal of $\mathcal{H}$. Given a family of sets $\mathcal{F} \in \mathcal{H}$, the set M is called a transversal of $\mathcal{F}$ in $\mathcal{H}$ if $\mathcal{D}(M, \mathcal{H}) = \mathcal{F}$. The set of all transversals of $\mathcal{F}$ in $\mathcal{H}$ is denoted by $\mathcal{T}(\mathcal{F}|\mathcal{H})$. Elements of $\mathcal{T}(\mathcal{F}|\mathcal{H})$ with the smallest size are called minimum transversals of $\mathcal{F}$ in $\mathcal{H}$. Let $\tau(\mathcal{F}|\mathcal{H})$ denote the transversal number of F in H, i.e., the size of a minimum transversal of $\mathcal{F}$ in $\mathcal{H}$. Notice that $\tau(\mathcal{F}|\mathcal{H}) \geq \tau(\mathcal{F})$. In other words, avoiding blocking some of the potential attacks of the adversary, and hence strategically exposing some parts of the network, may entail an additional cost. As we show below, strategic exposure may be a part of a rational defense strategy.

Let g be the network in Fig. 3. Let $\mathcal{H} = \mathcal{E}(g) = \{\{a\}, \{a, b\}, \{a, c\}\}$ be the set of all essential separators of g and let $F = \{\{a, b\}, \{a, c\}\}$ be its subset. Figure 4 illustrates the unique minimum (and, at the same time, minimal) transversal of $\mathcal{H}$, $\{a\}$. The unique minimum transversal of $\mathcal{F}$ in $\mathcal{H}$ is $\{b, c\}$. Thus, the most economic way to block exactly the separators from $\mathcal{F}$ out of all the separators from $\mathcal{H}$ is by blocking nodes b and c.

We provide more examples and a discussion of essential separators and their transversals in some well-known families of networks (trees, core–periphery, interlinked stars) in Appendix C.

We are now ready to state our first main result on optimal defense and attack.

Proposition 2 *Consider a connected network $g \in \mathcal{G}(N)$. Let (Δ^*, X^*) be an equilibrium.*

(i) If $c_A < \Delta f(n-1)$, then the following statements hold:

- *The variable $\Delta^* = N$ or Δ^* is a minimal transversal of $\mathcal{D}(\Delta^*, \mathcal{E}(g, c_A))$.*
- *We have $X^*(\Delta) = E \cup R$, where $E \in \mathcal{E}(g, c_A)$ and $R \in \mathcal{R}(g - E)$, with $X^*(\Delta) \cap \Delta = \phi$.*

(ii) If $c_A > \Delta f(n-1)$, then the following statements hold:

Fig. 4 Minimum transversal,{a}, of essential separators {{a},{a,b},{a,c}}

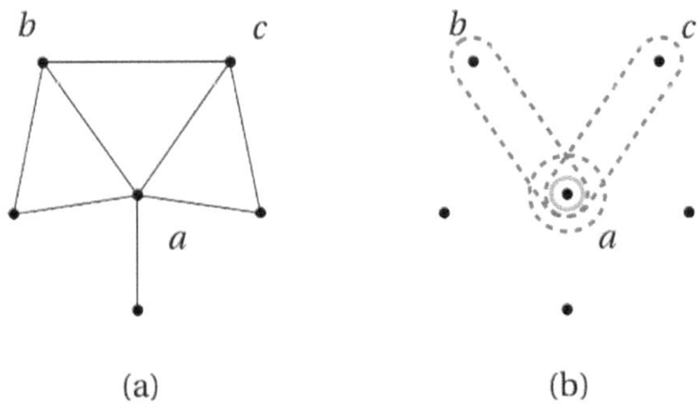

- *We have* $|\Delta^*| \leq \tau(\mathcal{E}(g, c_A))$ *and* Δ^* *is a minimum transversal of* $\mathcal{D}(\Delta^*, \mathcal{E}(g, c_A))$ *in* $\mathcal{E}(g, c_A)$.
- *We have* $X^*(\Delta) = \phi$ *if* $\Delta \in \mathcal{T}(\mathcal{E}(g, c_A))$*;* $X^*(\Delta) \in \mathcal{E}(g, c_A)$ *with* $X^*(\Delta) \cap \Delta = \phi$*, otherwise.*

The proposition brings out the economic tradeoffs in the network conflict. Essential separators—that are effective at fragmenting the network—are key to optimal attack and economical transversals that block these separators are key to optimal defense. Moreover, if the defender wishes to go beyond blocking the separator and protect nodes that merely expand the size of a component, then, due to convex character of network value function, it is optimal for him to protect all the nodes in the network.

More formally, optimal defense is defined in terms of the minimal transversal of the appropriate set of essential separators or defense must cover all nodes. If the cost of attack is such that elimination of single nodes is not worthwhile, optimal attack is bounded from above by the set of essential separators of the network. In this case, optimal defense can never exceed the size of the minimum transversal of the set of individually rational essential separators. If, alternatively, the cost of attack justifies the elimination of single nodes, optimal attack is constituted of nodes that comprise reducing attacks and essential separators. In this case, an interesting feature of optimal defense is that it may be larger than the smallest possible transversal (even when it does not cover all the nodes).

We now briefly describe the arguments underlying the proof. By Lemma 1, we know that any attack may be decomposed into two disjoint parts that comprise an essential separator and a reducing attack.

In the range of costs covered by part (ii), the adversary will not use reducing attacks. So, an optimal attack must be either empty or an individually rational essential separator. Next consider the optimal defense strategy, Δ^*. Clearly, Δ^* cannot be larger than the size of the minimum transversal of $\mathcal{E}(g, c_A)$, as that would be wasteful for the defender. If $|\Delta^*| = \tau(\mathcal{E}(g, c_A))$, then Δ^* must be a minimum transversal of $\mathcal{E}(g, c_A)$; choosing a defense other than a minimum transversal would simply lower payoffs. If $|\Delta^*| < \tau(\mathcal{E}(g, c_A))$, then Δ^* is a minimum transversal of $\mathcal{D}(\Delta^*, \mathcal{E}(g, c_A))$ in $\mathcal{E}(g, c_A)$.

We turn next to part (i) of Proposition 1. The proof proceeds by showing that a defense that exceeds a minimal transversal (of covered essential separators) must include some node that is being protected purely to prevent it from removal. Hence, the role of such a defense is to ensure the size of the component. This must mean that, in the absence of defense, the node would be eliminated in the subsequent optimal attack. We then exploit the convexity of f and the linearity of costs of defense and attack to establish that the adversary must find it optimal to eliminate all other unprotected nodes in the surviving component. Extrapolating from this, we establish that this must apply to all essential separators and then, by convexity, to single nodes in those components as well. In other words, if the defender finds it optimal to go beyond a minimal transversal of blocked essential separators, then he must protect all nodes.

We now consider the general comparative statics with respect to the costs and the network. It is worth noting some patterns in Example 1 above. Figure 2 suggests that defense size is falling in defense costs and is nonmonotonic in attack costs. The attack size is nonmonotonic in both attack cost and defense cost. These patterns are true more generally. They have payoff implications. The following result summarizes our analysis.

Proposition 3 *The equilibrium comparative statics are as follows.*

(i) The size of defense and the defender's payoff are both decreasing in the cost of defense. The defender's payoff increases in the cost of attack. However, depending on the costs and the network, the size of defense may increase or decrease when the cost of attack increases.
(ii) Depending on the costs and the network, the size of attack and adversary's payoff may increase or decrease when the cost of attack increases. The adversary's payoff increases in the cost of defense. However, depending on the costs and the network, the size of attack may increase or decrease when the cost of defense increases.
(iii) Depending on the costs and the network, adding links may increase or decrease the size of the optimal defense as well as the defender's payoff.

We note that the effect of defense cost on the size of attack may be nonmonotonic. This is because with a higher cost of defense, the defender may uncover some essential separators to which the adversary could switch. Their size might be smaller or larger than the size of separators chosen by the adversary under the lower cost of defense. As an example, consider the network g in Fig. 5 and suppose that $f(x) = x^2$, $c_A \in (31, 54)$, and $c_D \in (108, 121)$. Under these parameters, in every equilibrium the defender defends node a and the adversary responds with essential separator $\{b, c\}$. When the cost of defense rises to 122, equilibrium defense of the defender is ϕ to which the adversary responds with essential separator $\{a\}$. Alternatively, Example 1 illustrates that the size of attack might rise when the cost of defense is rising (cf. the case of $c_A \in (7, 13)$ in Fig. 2). Despite this nonmonotonic behavior of equilibrium attack size, the payoff to the adversary increases when the cost of defense rises. A similar observation also holds for the effect of attack cost on defense size and on payoffs.

An increase in attack cost has nonmonotonic effects on attack size and the adversary's payoff. This is illustrated by Example 1, e.g., when the cost of defense is in

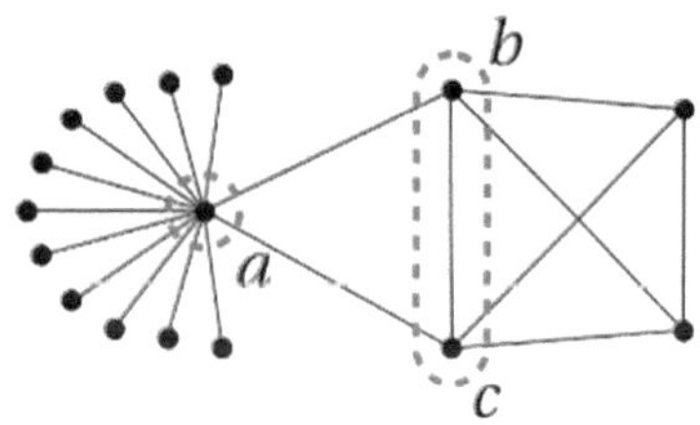

Fig. 5 Network where a rise in the cost of defense reduces the size of attack

the range (3.25, 4). The reason for these nonmonotonicities is as follows. When the cost of attack rises, some of the attacks stop being individually rational. This creates an opportunity for the defender to reduce defense, possibly at the expense of some value of the network. This, in turn, allows the adversary to execute attacks that were blocked when the cost of attack was lower. In the example, when $c_A \in (0, 1)$, it is individually rational for the adversary to remove any unprotected node. Therefore, with $c_D \in (3.25, 4)$, the defender defends all the nodes. When $c_A \in (1, 5)$, it is not individually rational for the adversary to remove single unprotected nodes. With the costs of defense in (3.25, 4), the defender prefers to leave the network undefended and loose the central node, saving on the cost of defense and loosing some value of the network. Such an attack is better for the adversary than not removing any node. The size of attack rises from 0 to 1 and the payoff of the adversary rises from -16 to $-3 - c_A \in (-9, -4)$. When $c_A > 7$, the size of attack falls back to 0 and the payoff to the adversary falls back to -16.

Finally, consider the effects of adding links. A first conjecture would be that adding links should always be good for the defender, as it creates more routes for connection and this should make the network easier to defend. The next example shows that this intuition is false: a denser network may induce a larger optimal defense with lower defender payoffs!

Example 2 (*Adding links may increase defense size and lower defender payoffs*) We consider the network given in Fig. 6. Suppose that payoff from a component of size x is $f(x) = x^2$.

Assume the cost of attack $c_A \in (23, 31)$ and the cost of defense $c_D \in (43, 85)$. The unique equilibrium outcome is $\Delta^\star = \{c\}$, $X^\star = \{d\}$. The equilibrium payoff to the defender is $101 - c_D$.

Now consider a network $g' = g \cup \{ef\}$, with a link added between the nodes e and f. With this additional link, the separator $\{d\}$ is replaced by separator $\{d, e\}$. Suppose that the cost of defense is $c_D \in (43, 62)$. Observe that with defense $\Delta^\star = \{c\}$, there exists an attack d, e that is optimal for the adversary and yields only $82 - c_D$ to the defender. Thus, the addition of a link, and retaining the same defense, may actually lower the defender's payoffs.

In the new network g', the unique equilibrium outcome is $\Delta^\star = \{d, e\}$ and $X^\star = \phi$. The equilibrium payoff to the defender is $144 - 2c_D < 101 - c_D$. So, *the optimal defense size increases and the defender's payoff falls as the network becomes denser*.

Alternatively, it is clear that as we keep adding links and arrive at the complete network, the optimal attack is empty (as $c_A > 23$) and so optimal defense is also the

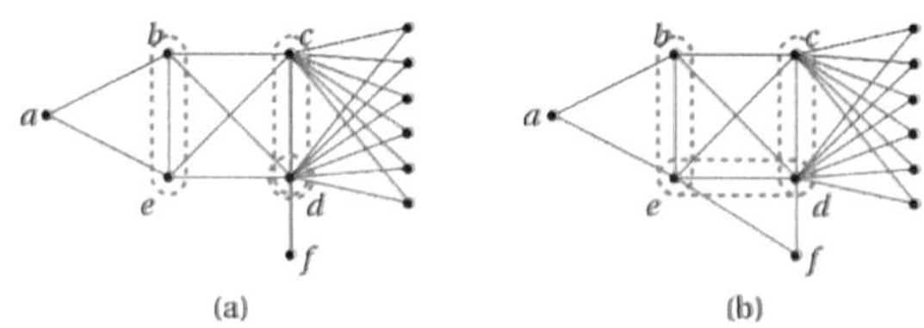

Fig. 6 Example 2. **a** Original network. **b** Network with added link

empty set. The defender's payoff is 144, which is the maximal attainable. Thus, the effects of adding links are nonmonotonic. ◇

This nonmonotonicity is not an artefact of the specifics of the network and the costs of attack and defense. It reflects a general feature of conflict in networks. To see this consider the case of the complete network. The first thought would be that a network that contains the most connections is the hardest to disrupt and always leads to the best outcomes for the defender. This is not true. The following example clarifies this point.

Example 3 (*Complete network vs. core–periphery network*) Suppose that n is large and that the cost of attack satisfies

$$f(n-2) - f(n-3) < c_A < f(n-1) - f(n-2)$$

With this cost of attack, the adversary removes two nodes from the complete network over n nodes, one node from the complete network containing $n-1$ nodes, and does not remove any nodes from the complete network containing $n-2$ or less nodes. Finally, suppose that the cost of defense satisfies

$$\frac{f(n) - f(n-2) - f(1)}{n} < c_D < \frac{f(n) - f(n-2)}{n}$$

With this cost of defense the defender protects all the nodes in a complete network with n nodes, because $f(n) - nc_D > f(n-2)$ (and we know that in a complete network the defender either protects all or no nodes, in equilibrium).

Now consider a network with $n-1$ nodes in a clique with one node linked to a single element of the core (let us call it i). This is a type of core–periphery network. If such a network is not protected, the adversary will remove node i only, disconnecting the network into a clique of size $n-2$ and a single isolated node. Now, we know that the defender is either inactive, protects i, or protects all the nodes in equilibrium. With the above cost of defense, the defender is inactive. First, note that $f(n) - nc_D < f(n-2) + f(1)$, so protecting everything is worse than being inactive. It can be checked that protecting i is worse, because in response the adversary would remove two nodes from the core of the network.

Thus, in the core–periphery network the equilibrium payoff to the defender is

$$f(n-2) + f(1) > f(n) - nc_D$$

So it is better than the complete network. ◇

This example illustrates the attractiveness of the queen sacrifice strategy: it is better to leave i unprotected because there is greater loss in value if it is protected! The idea of queen sacrifice and the suboptimality of the complete network will resurface in other contexts below.

3.1 Networks and Conflict

This section examines the relation between the network architecture and the nature of conflict more closely. We define the *intensity of conflict* as the sum of expenditures of defense and attack. Our analysis shows that for given costs of conflict, differences in network structure can lead to very large differences in conflict.

Proposition 1 tells us that the size of equilibrium attack and defense are generically unique. We start by defining the minimum intensity of conflict for given costs of attack and defense. Define minimal costs of conflict for given costs and f as

$$\mathrm{CC}(c_A, c_D, f) = \min_{g \in \mathcal{G}(N)} c_D|\Delta^\star(g, c_A, c_D, f)| + c_A|X^\star(g, c_A, c_D, f)|$$

Example 1 illustrates some of the forces at work. Observe that when the cost of attack is very large, $c_A > 13 = f(n) - (n-1)f(1)$, no attack is profitable, and, anticipating this, the defender abstains from defense. The intensity of conflict is 0. This lack of conflict for large costs of attack is independent of the architecture of the network.

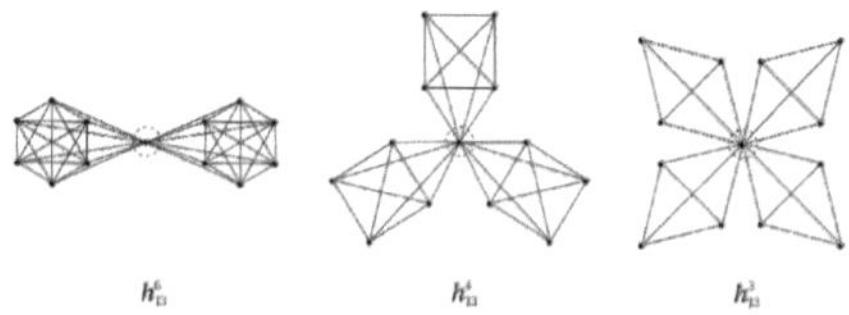

Fig. 7 Windmill graphs $(h_n^m) : n = 13, m = 6, 4, 3$

Turning to the lower cost of attack, an inspection of Fig. 1 in Example 1 tells us that the intensity of conflict also depends on the cost of defense. It will be useful to define a special class of networks, windmill graphs. These graphs are denoted by h_n^m, where $n \geq 2$ and $m \in \{1, \ldots, n-1\}$. There is one critical node that, when removed, disconnects the network. The remaining nodes are partitioned into cliques of size m and, possibly, one clique of smaller size (this implies that there are $(n-1)/m$ such cliques). Every member of a clique is connected to the critical node. We now define a key cost threshold for defense that equates the payoff from full defense with the payoff from an unprotected h_n^m network:

$$c(m.n) = \frac{f(n) - \lfloor \frac{n-1}{m} \rfloor f(m) - f((n-1)\mathrm{mod}\, m)}{n}$$

Figure 7 illustrates windmill graphs.

We are now ready to prove a general characterization of minimal conflict levels.

Proposition 4 *(i) If $c_A > f(n) - (n-1)f(1)$, then $CC(c_A, c_D, f) = 0$. It is attained on any connected network.*

(ii) If $c_A \in (f(n-1), f(n) - (n-1)f(1))$, then $CC(c_A, c_D, f) = 0$. It is attained on any connected network g with $\mathcal{E}(g, c_A) = 0$.

(iii) If $c_A \in (\Delta f(m-1), \Delta f(m))$ *with* $m \in \{1, \ldots, n-1\}$*, one of the following statements holds:*

- *(iv) If* $c_D > c(m, n)$*, then* $CC(c_A, c_D, f) = c_A$*. It is attained on a windmill network,* h_n^m*.*
- *(v) If* $c_D < c(m, n)$ *with* $m \in \{1, \ldots, n-1\}$*, then* $CC(c_A, c_D, f) = nc_D$*. It is attained on any connected network.*

In case (ii), when the cost of attack is high, $c_A > \Delta f(n-1)$, the minimal costs of conflict are 0, as it is not profitable for the adversary to attack any network with $\mathcal{E}(g, c_A) = \phi$. Such networks include the complete network, as well as networks that are robust to node removal in the sense that they require a large number of nodes to be removed to get disconnected. More generally, for any integer $t \geq 1$, a network is *t-connected* if it can be disconnected by removing t nodes and cannot be disconnected by removing less than t nodes. Any *t-connected* network with $t \geq (f(n) - nf(1)/(c_A - f(1))$ has empty $\mathcal{E}(g, c_A)$. Menger (1927) provides a characterization of such networks: a network is at least *t-connected* if and only if any two nodes that are not neighbors are connected through at least t node independent paths.[7] Thus, such networks have many redundant connections between nodes.

The last case, with lower attack costs $c_A < \Delta f(n-1)$, is much richer. Suppose that $c_A \in (\Delta f(m-1), \Delta f(m))$, where $m \in \{1, \ldots, n-1\}$. Now it is profitable to the adversary to attack any undefended node in a component of size greater than m. *Hence, the lower bound on costs of conflict is* $\min(c_A, nc_D)$. If the cost of defense is sufficiently low, $c_D < c(m, n)$, then complete defense is better than any other defense and the minimal costs of conflict are nc_D . If $c_D > c(m, n)$, then complete defense has higher costs as compared to the outcome with no defense and one attacked node. This leads to total costs of conflict of c_A. To sustain an equilibrium with such costs of conflict, we need a network that has a separator of size 1 and that all components in the residual network have size at most m. The windmill graph possesses exactly this characteristic. This motivates the windmill network: for $m \in \{1, \ldots, n-2\}$, the windmill network h_n^m has such an equilibrium and yields the minimal costs of conflict, c_A.

We now turn to the role of networks in shaping the intensity of conflict. Proposition 4 tells us that network architecture matters only if the costs are as in cases (ii) or (iii).

Consider case (ii). Proposition 1 tells us that $CC(c_A, c_D, f) = 0$ in this range. To see the impact of network architecture, consider a star network. If $cD < f(n) - (n-1)f(1)$, then in equilibrium the defender protects the center of the star and the costs of conflict are c_D. Alternatively, if $c_D > f(n) - (n-1)f(1)$, then in equilibrium the defender chooses the empty defense, the adversary attacks the center of the star, and the costs of conflict are c_A. So, when the costs of attack and defense reach their upper bound, the difference in the costs of conflict between the star network and the

[7] Two paths are node independent if the only nodes they have in common are the starting and the ending nodes.

minimal attainable is $f(n) - (n-1)f(1)$. It is easy to see that this can grow without bound as n gets large.

Next consider case (iii), with $m \in \{1, \ldots, n-2\}$. Proposition 4 tells us that the minimum conflict, attained on network h_n^m (for example) is c_A. Suppose $c_D \in (c(m,n), (f(n) - f(m))/n)$ and consider a complete network. The unique equilibrium outcome is full protection and so the costs of conflict are nc_D. When the cost of defense reaches its upper bound and the cost of attack reaches its lower bound, the difference in costs between this minimum and the complete network reaches $f(n) + f(m-1) - 2f(m)$, which is maximal, $f(n) - 2f(1)$, for $m = 1$. Again, the network architecture can have very large effects on the intensity of conflict.

Active conflict In Proposition 4, minimal conflict is associated with a single active player. An inspection of Fig. 3, in Example 1 above, shows us that both players can be active in equilibrium. This motivates the study of circumstances under which we should expect to see active conflict. Example 1 draws attention to the role of costs: neither the attack nor the defense costs can be too high. Here we briefly discuss the role of the network architecture and the network value function.

We start with an observation that draws upon Proposition 2: for active conflict to arise there must exist an individually rational essential separator. If such a separator does not exist, then convexity of function f together with linearity of costs implies that either none or all nodes are defended. In particular, if g is a complete network, then for all costs and all functions f (satisfying our assumptions), there is no equilibrium with active conflict.

Are there any other (connected) networks with the same property as complete networks? If the marginal value of f is growing sufficiently fast, then no active conflict is possible. Let f satisfy the property, for $x \geq 0$,

$$\Delta f(x) > x f(x) \tag{2}$$

where $\Delta f(x) = f(x+1) - f(x)$.

The property is satisfied by functions $f(x) = (x+1)! - 1$ and $(x+1)^x - 1$, for example. Marginal value in these functions grows so rapidly that adding a single node to a component of size m increases its value more than m times. In effect, the returns from protecting $m < n$ nodes are smaller than average returns from protecting additional $m - n$ nodes. Thus, if the defender prefers protecting the first m nodes to no protection, he is even more willing to protect the whole network. Formally, let

$$\Phi^\star(m; g, c_A) = \max_{\Delta \subseteq N, |\Delta| \leq m} \; \min_{X \in \mathrm{BR}(\Delta; g, c_A)} \Phi(g - X(\Delta) \setminus \Delta)$$

be a function that gives the maximum value of the residual network that can be attained from network g when up to m units of defense are used and the cost of attack is cA ($\mathrm{BR}(\Delta; g, c_A)$ denotes the set of best responses of the adversary to Δ, given g and c_A). Suppose that there is an equilibrium, $(\Delta^\star, X^\star)$, featuring active conflict. Let $|\Delta^\star| = m$. Since there is active conflict, so $1 \leq m \leq n-1$ and $|X^\star(\Delta^\star)| \geq 1$. Since $\Delta^\star$ is better than ϕ, so $c_D \leq (\Phi^\star(m; g, c_A) - \Phi^\star(0; g, c_A))/m \leq f(n-1)$.

Alternatively, since $\Delta^\star$ is better than N, so $c_D \geq (f(n) - \Phi^\star(m; g, c_A))/(n-m) \geq (f(n) - f(n-1))/(n-1)$. Combining both the inequalities we get $f(n) \leq nf(n-1)$, which contradicts (2).

4 Decentralized Defense

In many applications, security decisions are made at the individual node level. This section studies decentralized security choices in a network that is under attack. We begin by showing that the equilibrium choices of the nodes and the adversary can be characterized in terms of transversals and separators of the underlying network. We then show that the welfare gap between decentralized equilibrium and first best outcomes is unbounded: interestingly, individual choice may lead to too little and to too much protection, relative to the choice of a single (centralized) defender.

We consider a two-stage game. In the first stage, each of the nodes in the network decides whether to protect itself or to stay unprotected. These choices are observed by the adversary who then chooses the nodes to attack.

Let $N = \{1.2, \dots, n\}$, where $n \geq 3$ is the set of players, and let $S_i = \{0, 1\}$ denote the strategy set of node $i \in N$. Here $s_i = 1$ means that the node chooses to defend itself and $s_i = 0$ refers to the case of no defense. These choices are made simultaneously. There is a one-to-one correspondence between a strategy profile of the nodes, $s \in \{0, 1\}^N$, and the resulting set of defended nodes $\Delta \subseteq N$. So we will use Δ to refer to the strategy profile of the nodes in the first stage.

In the second stage the adversary observes the defended network (g, Δ) and chooses an attack $X \subseteq N$, which leads to a residual network $g - (X \setminus \Delta)$. The payoff to the adversary remains as in the case of the centralized defense and is defined in (1). The payoff to a node depends on whether the node is removed by the attack. A removed node receives payoff 0. Each of the surviving nodes receives an equal share of the value of its component in the residual network,

$$\Pi^i(\Delta, X; g, c_D) = \begin{cases} 0 & \text{if } i \in X \setminus \Delta \\ \frac{f(|C(i)|)}{|C(i)|} - s_i c_D & \text{otherwise,} \end{cases}$$

where $C(i)$ is the component in the residual network $g - (X \setminus \Delta)$ that contains i.

This completes the description of the *decentralized defense game*. We study the subgame perfect equilibria of this game, restricting attention to those without active conflict.

Let us solve the game starting from the second stage. As in the two-player game, the adversary chooses either the empty attack or an attack that is a combination of an essential separator and a reducing attack. If the cost of attack is low and there is no active conflict, then either the adversary removes all the nodes or all nodes are protected. In any other outcome the adversary must remove at least one node. If the cost of attack is high and there is no active conflict, then either none of the nodes protects or, anticipating the strategy of the adversary, the nodes choose a defense configuration that blocks all the individually rational essential separators. Therefore, in equilibrium, they must choose a minimal transversal of $\mathcal{E}(g, c_A)$. We build on these observations to provide the following characterization of equilibria with no active conflict in the decentralized defense game.[8]

Proposition 5 *Consider a connected network $g \in G(N)$. Let $\Delta^\star$ be the equilibrium defense.*

(i) If $c_D > f(n)/n$, then $\Delta^\star = \phi$ is the unique equilibrium defense.
(ii) If $c_D \leq f(n)/n$, one of the following statements holds.

(a) If $c_A < f(n) - f(n-1)$, then $\Delta^\star = N$ is an equilibrium defense.
(b) If $c_A > f(n) - f(n-1)$, then any minimal transversal of $\mathcal{E}(g, c_A)$ is an equilibrium defense.

The equilibrium strategy of the adversary is as in Proposition 2.

We now turn to discussing inefficiencies that may arise due to decentralized protection, as well as their sources. We compare the aggregate welfare of the nodes in the equilibrium of the two-player game with the aggregate welfare in the decentralized defense game. Let $\Pi^{D^\star}(g, c_A, c_D)$ denote the equilibrium payoff in the two-player game on network g with cost of defense c_D and cost of attack c_A. Aggregate welfare in the two-player game, starting from network g, and costs c_A and c_D, are defined as

$$W^F(g, c_A, c_D) = \Pi^{D^\star}(g, c_A, c_D)$$

Aggregate welfare under defense profile Δ and attack X, of the $n+1$-player game starting from network g, and given cost of defense c_D, is defined as

$$W^D(\Delta, X; g, c_D) = \sum_{i \in N} \Pi^i(\Delta, X; g, c_D)$$

Following Koutsoupias and Papadimitriou (1999), we study the cost of decentralization in terms of the price of anarchy (PoA): the ratio of welfare in the two-player game to the welfare in the worst equilibrium of the decentralized defense game. Let $E(g, c_A, c_D)$ denote the set of equilibria of the $n+1$-player game on network g with cost of attack c_A and cost of defense c_D. Let

$$\text{PoA} = \max_{g, c_A, c_D} \left(\frac{W^F(g, c_A, c_D)}{\min_{(\Delta, X) \in E(g, c_A, c_D)} W^D(\Delta, X(\Delta); g, c_D)} \right)$$

[8] We concentrate on equilibria with no active conflict, because, on one hand, it allows for providing a clean characterization and, on the other hand, it provides a sufficiently rich platform for discussing the sources of inefficiencies when defense decisions are decentralized. All other equilibria in decentralized defense game could be characterized in the same spirit as the characterization provided in Proposition 2 for the centralized defense game.

Our analysis highlights externalities and points to sources of inefficiency in decentralized defense. The first source is the familiar one of positive externalities: an individual's protection decision creates benefits for other nodes, which she does not take into account. Consider a star network and suppose that cost of attack is high, $c_A > f(n) - f(n-1)$, and $c_D \in (f(n)/n, f(n))$. In the equilibrium of the two-player game, the aggregate welfare is $f(n) - c_D$. However, in the equilibrium of the decentralized game, the central player does not find it profitable to defend itself, as $c_D > f(n)/n$. So aggregate welfare in equilibrium of the $n + 1$-player game is 0. The ratio of the two is unbounded for $c_D \in (f(n)/n, f(n))$.

Protection choices exhibit a threshold property: for a node to find it profitable to protect it is necessary that other nodes belonging to the same minimal transversal protect. Thus, protection decisions are strategic complements. This can generate coordination failures, resulting in large welfare losses. To see this, consider a tree with two hubs each of which is linked to $(n-2)/2$ distinct nodes. Suppose that

$$f(n) - f(n-1) < c_A < f\left(\frac{n}{2}\right) - \frac{(n-2)f(1)}{2}$$

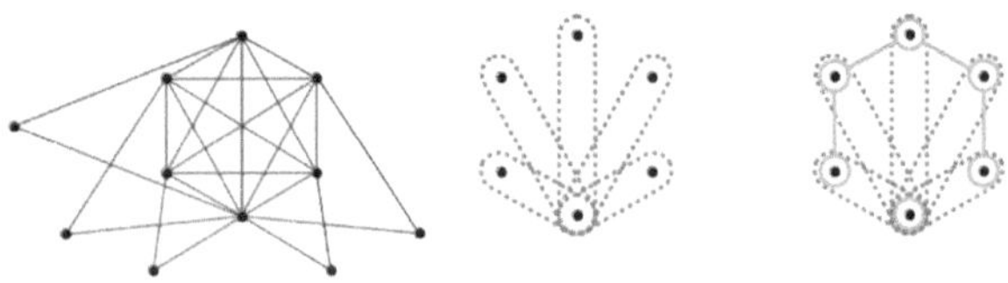

Fig. 8 Network with essential separators of size 2 having two minimal transversals: one of size 1 and one of size 5

so the adversary will only attack hub nodes. If $2f(n/2)/n < c_D < f(n)/n$, then the first best outcome is to defend the two hubs. One hub protecting itself gives incentives to the other hub to protect: two protected hubs is an equilibrium outcome. However, a hub node does not have unilateral incentives to protect: zero protection is also an equilibrium outcome. In this equilibrium the aggregate payoffs are $(n-2)f(1)$ as compared to first best outcome of $f(n) - 2c_D$. The cost of decentralization can be unbounded.

Third, at the local level, the game is clearly one of strategic substitutes. A node in a separator has incentives to protect only if no other node in the separator protects itself. Like public good games on networks (cf. [9]), the network protection game therefore displays multiple equilibria. This can generate very large efficiency losses. As an example consider network g depicted in Fig. 8.

Suppose that $f(x) = x^2$, $c_A \in (21, 28)$, and $c_D < 11$. Since the cost of attack is high, the adversary will not remove a node without disconnecting the network. The set of individually rational essential separators is a combination of sets as depicted in Fig. 8. Notice that the minimum transversal of $\mathcal{E}(g, c_A)$ is the node belonging to each of the separators, while the largest minimal transversal consists of one distinct node from each of the two element separators. Hence, the modified PoA in this case is $|\mathcal{E}(g, c_A)|$ and as the example in Fig. 8 suggests, it is possible to have a graph g such that $|\mathcal{E}(g, c_A)| \geq (n-1)/2$. Again, the cost of decentralization is unbounded.

The idea that personal security exhibits positive externalities is well known in the economic epidemiology literature (and has been noted in the recent research in this area; see, e.g., [1, 10, 36]. Moreover, in the standard disease setting security choices are strategic substitutes. Our model departs from this standard setting in two important ways: one, we have an intelligent adversary, and two, agents in our model care about the size of the component (not just about survival). This means that security choices exhibit features of both complements and substitutes. In addition due to the role of size effects, security choices can exhibit large coordination failures. These features of the model distinguish it from the existing literature and call for new methods of analysis and yield fresh insights.

5 Concluding Remarks

Infrastructure networks are a key feature of an economy. These networks face a variety of threats ranging from natural disasters to intelligent attacks. This paper develops a strategic model of defense and attack in networks.

We provide a characterization of equilibrium attack and defense in terms of two classical concepts in graph theory: separators and transversals. We show that the intensity of conflict (the resources spent on attack and defense) and the possibility of active conflict (when both adversary and defender target nodes for action) are both intimately related to the architecture of the network. Finally, we show that the welfare costs of decentralized defense can be very large.

We have assumed that the defender moves first and is followed by the attacker, and that the defense is perfect: it would be more natural to allow for outcomes of conflict to vary with resources of attack and defense allocated to a node. Appendix D presents a preliminary analysis of models where we relax these assumptions. A general analysis remains an important problem for future research.

Finally, we have assumed that payoffs depend only on the sizes of the networks (or their components). In future work, it would be important to study a model where payoffs depend on the details of the architecture of the components.

6 Appendix A: Proofs

We start with proving Proposition 1 that states generic equivalence of equilibrium outcomes of the defender adversary game in terms of payoffs, size of defense, and size of attack. We start with the following auxiliary lemmata.

Lemma 2 *Let g be a network over set of nodes N and let $\Delta \subseteq N$ be a set of defended nodes. Generically, for any best responses $X^\star$ and $X^{\star\star}$ to defense Δ, $\Phi(g - X^\star) = \Phi(g - X^{\star\star})$ and $|X^\star| = |X^{\star\star}|$.*

Proof Let g be a network and let Δ be a defense, as stated in the lemma. Let $X^\star$ and $X^{\star\star}$ be best responses to (g, Δ). Then we have

$$-\Phi(g - X^\star) - |X^\star|c_D = -\Phi(g - X^{\star\star}) - |X^{\star\star}|c_D$$

If $|X^\star| = |X^{\star\star}|$, then it follows that $\Phi(g - X^\star) = \Phi(g - X^{\star\star})$ and we are done. Otherwise, the equality is equivalent to

$$c_D = \frac{\Phi(g - X^\star) - \Phi(g - X^{\star\star})}{|X^{\star\star}| - |X^\star|}$$

The set of values on the right-hand side of the equality is finite (there are at most $2^{n+1} - 1$ values there). Hence, the equality can be satisfied for a finite number of values of $c_D \in \mathbb{R}_{++}$. This completes the proof. □

Lemma 3 *Let g be a network over the set of nodes N. Generically, for any two equilibria $(\Delta^\star, X^\star)$ and $(\Delta^{\star\star}, X^{\star\star})$, $\Phi(g - X^\star(\Delta^\star)) = \Phi(g - X^{\star\star}(\Delta^{\star\star}))$ and $|\Delta^\star| = |\Delta^{\star\star}|$.*

Proof Let $g, \Delta^\star, \Delta^{\star\star}.X^\star, X^{\star\star}$be as stated in the lemma. Since $\Delta^\star$ is a best response to $X^\star$, so

$$\Phi(g - X^\star(\Delta^\star)) - |\Delta^\star|c_D \geq \Phi(g - X^\star(\Delta^{\star\star})) - |\Delta^{\star\star}|c_D \tag{3}$$

and since $\Delta^{\star\star}$ is a best response to $X^{\star\star}$, so

$$\Phi(g - X^{\star\star}(\Delta^{\star\star})) - |\Delta^{\star\star}|c_D \geq \Phi(g - X^{\star\star}(\Delta^\star)) - |\Delta^\star|c_D \tag{4}$$

By Lemma 2, generically, $\Phi(g - X^{\star\star}(\Delta^\star)) = \Phi(g - X^\star(\Delta^\star))$ (as both $X^{\star\star}(\Delta^\star)$ and $X^\star(\Delta^\star)$ are best responses to $\Delta^\star$). This together with (3) and (4) implies

$$\Phi(g - X^{\star\star}(\Delta^{\star\star})) - |\Delta^{\star\star}|c_D \geq \Phi(g - X^\star(\Delta^\star)) - |\Delta^\star|c_D$$

Similarly, by Lemma 2, generically, $\Phi(g - X^\star(\Delta^\star)) = \Phi(g - X^{\star\star}(\Delta^\star))$. This together with (3) and (4) implies

$$\Phi(g - X^{\star\star}(\Delta^{\star\star})) - |\Delta^{\star\star}|cD = \Phi(g - X^\star(\Delta^\star)) - |\Delta^\star|c_D \tag{5}$$

If $|\Delta^\star| = |\Delta^{\star\star}|$, then $\Phi(g - X^\star(\Delta^\star)) = \Phi(g - X^{\star\star}(\Delta^{\star\star}))$ and we are done. Otherwise, (5) can be rewritten as

$$c_D = \frac{\Phi(g - X^\star(\Delta^\star)) - \Phi(g - X^{\star\star}(\Delta^{\star\star}))}{|\Delta^\star| - |\Delta^{\star\star}|}$$

Since the number of values on the right-hand side is finite, for almost every value of $c_D \in \mathbb{R}_{++}$ this equality is not satisfied. Hence, generically, $|\Delta^{\star}| = |\Delta^{\star\star}|$ and $\Phi(g - X^{\star}(\Delta^{\star})) = \Phi(g - X^{\star\star}(\Delta^{\star\star}))$. □

Lemma 4 *Let g be a network over set of nodes N and let $X, Y \subseteq N$ be two attacks such that $|X| \neq |Y|$. Generically, $\Phi(g - X) \neq \Phi(g - Y)$.*

Proof Let g, X, and Y be as stated in the lemma. Suppose that $\Phi(g - X) = \Phi(g - Y)$. This equality can be rewritten as

$$\sum_{C \in \mathcal{C}(g-X)} f(|C|) = \sum_{C \in \mathcal{C}(g-Y)} f(|C|)$$

Since $X \neq Y$ so there exists $s > 0$ such that $g - X$ has a component of size s and $g - Y$ has not or $g - Y$ has a component of such a size and $g - X$ has not. Suppose that $\Phi(g - X) = \Phi(g - Y)$. Hence, the equality above reduces to

$$f(s_1) + \cdots + f(s_p) = f(z_1) + \cdots + f(z_q) \tag{6}$$

where $s_1, lcdots, s_p$ and $z_1, \ldots, z_q$ are sizes of components such that $\{s_1, \ldots, s_p\} \cap \{z_1, \ldots, z_q\} = \phi$. Equation (6) puts very strict constraints on function f and perturbing it slightly (within the set of functions satisfying Assumption 1) destroys the equality. Thus, $\Phi(g - X) \neq \Phi(g - Y)$ for $|X| \neq |Y|$ is a nongeneric property of f. □

With Lemmas 2, 3, and 4 in hand, we are ready to prove Proposition 1.

Proof of Proposition 1 Generic equivalence of defense size and of payoff to the defender follow directly from Lemma 3. Consider equivalence of attack size and of payoff to the adversary. By Lemmata 3 and 4, generically $\Phi(g - X^{\star}(\Delta^{\star})) = \Phi(g - X^{\star\star}(\Delta^{\star\star}))$ and $|X^{\star}(\Delta^{\star})| = |X^{\star\star}(\Delta^{\star\star})|$. Thus, the points follow as well. □

Proofs of Lemma 1 and Proposition 2 exploit some properties of graphs. The first step is to establish these properties. Lemma 5 characterizes the essential separators as those separators that are "thin": every node of such separators is a neighbor of at least two components of the residual network. Given a set of nodes $X \subseteq N$ and a network g over N, $\partial_g(X) = \{k \in N \setminus X :$ there is $j \in X$ such that $jk \in g\}$ is the *neighborhood of X in g*. If X is a singleton, that is, $X = \{j\}$, then we will write $\partial_g(j)$ instead of $\partial_g(\{j\})$ ($\partial_g(j)$ is the set of neighbors of j in g). We will drop the subscript g in the notation if network g is clear from the context.

Lemma 5 *Let $g \in G(N)$ be a network over a set of nodes N. A set $X \subseteq N$ is an essential separator if and only if $X \neq \phi$ and, for every $i \in X$, there exist two distinct components $C_1, C_2 \in \mathcal{C}(g - X)$, $C_1 \neq C_2$, such that $\partial_{g-X}(i) \cap C_1 \neq \phi$ and $\partial_{g-X}(i) \cap C_2 \neq \phi$.*

Proof Let $g \in \mathcal{G}(N)$ be a network over a set of nodes N and let $X \subseteq N$.
The necessary part. Assume that X is an essential separator. Since X is a separator, so $X \neq \phi$. Assume, to the contrary, that there exists $i \in X$ such that there is at most one component $C \in \mathcal{C}(g - X)$ such that $\partial_{g-X}(i) \cap C \neq \phi$. Suppose first there is no such component. Then the attack $X' = X \setminus \{i\}$ results in the set of components $\mathcal{C}(g - X') = \mathcal{C}(g - X) \cup \{\{i\}\}$, larger than $\mathcal{C}(g - X)$, which contradicts the assumption that X is essential. Second, suppose that there is exactly one component $C \in \mathcal{C}(g - X)$ such that $\partial_{g-X}(i) \cap C \neq \phi$. Taking attack X', as before, leads to a residual network with set of components $\mathcal{C}(g - X') = (\mathcal{C}(g - X) \setminus \{C\}) \cup \{C \cup \{i\}\}$, which has the same cardinality as $\mathcal{C}(g - X)$. Therefore, X is not essential, a contradiction.

Sufficiency part Assume that $X \neq \phi$, and for every $i \in X$, there exist two distinct components $C_1, C_2 \in \mathcal{C}(g - X)$ such that $C_1 \cap \partial_{g-X}(i) \neq \phi$ and $C_2 \cap \partial_{g-X}(i) \neq \phi$. Then there exist two nodes, $j_1 \in C_1 \cap \partial_{g-X}(i)$ and $j_2 \in C_2 \cap \partial_{g-X}(i)$, that are connected in g and not connected in $g - X$. Hence, X is a separator and we have to show that it is essential. Suppose $X' \subsetneq X$, so there is some i such that $i \in X$ but $i \notin X'$. Given the definition of $i \in X$ it follows that $|C(X')| \leq |C(X)| - 1$. Since X' was arbitrary, the claim is established. □

We now develop a characterization of optimal attack strategies in terms of essential (individually rational) separators and reducing attacks.

Proof of Lemma 1 The proof of the first part is by induction on the number of nodes in X that violate the condition from Lemma 5. For the induction basis consider the set of all $X \subseteq N$ for which there are no nodes that violate the condition. Then, by Lemma 5, X is essential and so the remainder is ϕ and $E = X$ (in particular, it may be that $E = X = \phi$). The claim holds.

For the induction step, take any $X \subseteq N$ for which there are exactly m nodes that violate the condition from Lemma 5. Suppose that the claim holds for any $Y \subseteq N$ for which there are $l < m$ nodes that violate the condition. Let $i \in X$ be a node that violates the condition and let $Y = X \setminus \{i\}$. Since the condition is violated for $i \in X$, so $g - Y$ either contains one component more than $g - X$ (namely, component $\{i\}$) or it has the same number of components with one component C in $g - X$ replaced with $C \cup \{i\}$ in $g - Y$. Hence, the condition is violated for $l < m$ nodes from Y in $g - Y$. Thus, by the induction hypothesis, Y can be decomposed into two disjoint sets E and R as claimed. Since, as we argued above, adding i to Y does not increase the number of components in the residual network, so $R \cup \{i\}$ does not contain a separator of $g - E$ and so the decomposition of X into E and $R \cup \{i\}$ satisfies the conditions from the claim. Thus, points (i) and (ii) are shown.

Now we show that if g is connected and X is a best response to some defense $\Delta \subseteq N$, then either $E = \phi$ or $E \in \mathcal{E}(g, c_A)$.

We show first, for any attack X and any decomposition of X into two disjoint sets E and R satisfying points (i) and (ii), that

$$\Phi(g - E) - \Phi(g - X) \leq \Phi(g) - \Phi(g - R) \tag{7}$$

We use induction on R. For the induction basis, let $R = \phi$. Then (7) trivially holds. For the induction step, suppose that (7) holds for any $T \subsetneq R$. Take any $i \in R$, and let $T = R \setminus \{i\}$ and $Y = X \setminus \{i\}$. Let $C \in \mathcal{C}(g - Y)$ be the component with $i \in C$. Since R does not contain an essential separator of $g - X$ so $\mathcal{C}(g - X)$ and $\mathcal{C}(g - Y)$ differ at component C only: either $C \setminus \{i\} \in \mathcal{C}(g - X)$ or $C \setminus \{i\} = \phi$. Hence

$$\Phi(g - X) = \Phi(g - Y) - \big(f(|C|) - f(|C| - 1)\big) \tag{8}$$

Now let $C' \in \mathcal{C}(g - T)$ be the component with $i \in C'$. Applying attack $\{i\}$ to $g - T$ replaces C' with components $C'_1, \ldots, C'_m$ such that $\cup_{i=1}^{m} C'_i = C' \setminus \{i\}$. Hence

$$\begin{aligned}\Phi(g - R) &= \Phi(g - T) - \Big(f(|C'|) - \sum_{i=1}^{m} f(|C'_i|)\Big) \qquad (9)\\ &\leq \Phi(g - T) - \big(f(|C'|) - f(|C'| - 1)\big)\end{aligned}$$

(by the fact that f is strictly convex). By the induction hypothesis,

$$\Phi(g - E) - \Phi(g - Y) + (f(|C|) - f(|C| - 1)) \leq \Phi(g) - \Phi(g - T) + (f(|C|) - f(|C| - 1))$$

and, by the fact that $C \subseteq C'$ and by convexity of f,

$$\Phi(g - E) - \Phi(g - Y) + (f(|C|) - f(|C| - 1)) \leq \Phi(g) - \Phi(g - T) + (f(|C'|) - f(|C'| - 1))$$

Thus, by (8) and (9),

$$\Phi(g - E) - \Phi(g - X) \leq \Phi(g) - \Phi(g - R)$$

This shows the induction step. Hence, we have shown (7).

Now, let $\Delta \subseteq N$ be a defense chosen in the first stage and suppose that X is a best response to Δ. Whereas X is a better response to Δ than R, so

$$-\Phi(g - X) - c_A|X| \geq -\Phi(g - R) - c_A|R|$$

and, consequently,

$$\Phi(g - R) \geq \Phi(g - X) + c_A(|X| - |R|) = \Phi(g - X) + c_A|E|$$

From (7), we have

$$\Phi(g - X) \geq \Phi(g - E) + \Phi(g - R) - \Phi(g)$$

Putting the last two inequalities together, we arrive at

$$\Phi(g - R) \geq \Phi(g - E) + \Phi(g - R) - \Phi(g) + c_A|E|$$

Simplifying this yields

$$-\Phi(g - E) - c_A|E| \geq -\Phi(g)$$

In other words, $E \in \mathcal{E}(g, c_A)$ □

The proof of part (ii) of Proposition 2 now follows from the lemmata above and the arguments in the main text. We turn next to proving part (i) of Proposition 2.

To simplify some parts of the argument, we will make a tie-breaking assumption on the behavior of the adversary. It says that if two strategies yield equal payoffs to the adversary, then he will choose the strategy that yields a lower payoff to the defender.

Assumption 2 Given a network g and defense Δ, if two strategies $X \subseteq N$ and $X' \subseteq N$ yield the same payoff to the adversary, then he chooses the strategy that results in a residual network of lower value.

The first step here is to state and prove the following lemma.

Lemma 6 *Let $g \in \mathcal{G}(N)$ be a connected network over N, and let c_D and c_A be the costs of defense and attack, respectively. Suppose that $\Delta \subseteq N$ is an equilibrium defense and $X \subseteq N$ is a best response to it. Suppose that there exists $i \in \Delta$ such that $D(\Delta, \mathcal{E}(g, c_A)) = \mathcal{D}(\Delta \setminus \{i\}, \mathcal{E}(g, c_A))$. Let $X' \subseteq N$ be a best response to $\Delta' = \Delta \setminus \{i\}$.*

Then there exists a component $C \in \mathcal{C}(g - X)$ such that $C \subseteq \Delta$ and either $C = \{i\}$ or $C \setminus \{i\} \in \mathcal{C}(g - X')$. Moreover,

$$\Pi^D(\Delta, X; g) = \Pi^D(\Delta', X'; g) + f(|C|) - f(|C| - 1) - c_D \tag{10}$$

and

$$c_A \leq f(|C|) - f(|C| - 1) \tag{11}$$

Proof Let $\Delta \subseteq N$ be a defense, $i \in \Delta$ and $\Delta' = \Delta \setminus \{i\}$. Let X be a best response to Δ and let X' be a best response to Δ'.

Since X is a best response to Δ, so $X \cap \Delta = \phi$ and $\Phi(g - (X \setminus \Delta)) = \Phi(g - X)$, and analogously with X' and Δ'. We prove the lemma in the seven steps below.

(i) We have $\Phi(g - X) > \Phi(g - X')$. Since Δ is an equilibrium strategy of the defender, so $\Pi^D(\Delta, X; g) \geq \Pi^D(\Delta', X'; g)$, that is, $\Phi(g - X) - c_D|\Delta| \geq \Phi(g - X') - c_D(|\Delta| - 1)$. Hence, $\Phi(g - X) > \Phi(g - X')$.

(ii) We have $i \in X'$. Assume, to the contrary, that $i \notin X'$. Then $X' \cap \Delta = X' \cap \Delta' = \phi$. Similarly, since $X \cap \Delta = \phi$, so $X \cap \Delta' = \phi$. Hence, $\Pi^A(\Delta', X'; g) = \Pi^A(\Delta, X'; g)$ and $\Pi^A(\Delta', X; g) = \Pi^A(\Delta, X; g)$. By the fact that $\Pi^A(\Delta', X'; g) \geq \Pi^A(\Delta', X; g)$, as X' is a best response to Δ', this yields $\Pi^A(\Delta, X'; g) \geq \Pi^A(\Delta, X; g)$. Additionally, by point (i), $\Phi(g - X) > \Phi(g - X')$, so X'

results in a residual network of lower value than in the case of X. Hence, by the tie-breaking Assumption 2, X' is an equilibrium response to Δ, a contradiction. Thus, it must be that $i \in X'$.

Take any decomposition $E \cup R$ of Y, as described in Lemma 1. It cannot be that $i \in E$, as otherwise we would have $E \in \mathcal{D}(\Delta, \mathcal{E}(g, c_A))$, while $E \notin \mathcal{D}(\Delta', \mathcal{E}(g, c_A))$, as $Y \cap \Delta' = \phi$, and we would have a contradiction with the assumption that $\mathcal{D}(\Delta, \mathcal{E}(g, cA)) = \mathcal{D}(\Delta', \mathcal{E}(g, cA))$. Hence, $i \in R$ and there exists a component $C \in \mathcal{C}(g - E)$ such that $i \in C$. Let $C = \widetilde{C} \setminus R$ be what remains of C after the remainder R of Y is applied to $g - E$. Therefore, either $C = \phi$ (i.e., it is completely removed by R) or $C \in \mathcal{C}(g - Y)$ (i.e., it is a component in $g - Y$). Suppose that $C = \phi$, that is, $C \subseteq R$. Then $\partial_{g-E}(i) \subseteq R$ and $\partial_g(i) \subseteq E \cup R = Y$. Since $i \in Y$, so $\{i\} \cup \partial_g(i) \subseteq Y$. Suppose now that C is a component in $\mathcal{C}(g - Y)$. We will show that $i \in \partial_{g-E}(C)$. Assume the opposite. Then $\partial_{g-E}(C)$ must be a separator in $g - E$, as it separates C from a component containing i. But then $\partial_{g-E}(C)$ contains an essential separator for $g - E$. Since $\partial_{g-E}(C) \subseteq R$, this contradicts the assumption that R is a remainder and does not contain any essential separators of $g - E$. Hence, it must be that $i \in \partial_{g-E}(C)$ and, consequently, $i \in \partial_g(C)$.

(iii) For all $C' \in \mathcal{C}(g - X')$ with $i \in \partial_g(C')$, $C' \subseteq \Delta$. Assume the opposite. Then there exists $C' \in \mathcal{C}(g - X')$ with $i \in \partial_g(C')$ (and consequently $i \notin C\prime$) such that $i' \in C' \setminus \Delta$. Consider a strategy $X'' = (X' \setminus \{i\}) \cup \{i'\}$. Since $X \cap \Delta = \phi$ and $i' \notin \Delta, soX'' \cap \Delta' = \phi$. Notice that $\Phi(g - X'') \leq \Phi(g - X')$, as both the residual networks agree at all the components apart from what remains of $C' \cup \{i\}$ after i' is removed (at the least it is one component of the same size as C'). Since $|X'| = |X''|$ so $\Pi^A(\Delta', X''; g) \geq \Pi^A(\Delta', X'; g)$ and so X'' is a best response to Δ'. But then we get a contradiction with point (ii), as $i \notin X''$. Hence, it must be that $C' \subseteq \Delta$.

(iv) There exists $C' \in \mathcal{C}(g - X') \cup \{\phi\}$ such that $C = C' \cup \{i\} \in \mathcal{C}(g - X)$ and $C \subseteq \Delta$. Let $C' = \phi$ if $\{i\} \cup \partial_g(i) \subseteq X'$ or let C' be the unique $C' \in \mathcal{C}(g - X')$ with $i \in \partial_g(C')$, otherwise. By point (iii) such C' exists. By point (iv) and by the fact that $i \in \Delta$, $C \subseteq \Delta$. Thus, there exists a component $C'' \in \mathcal{C}(g - X)$ such that $C \subseteq C''$. Suppose that $C \subsetneq C''$. We will show that in this case $X \cup \{i\}$ is a better response to Δ' than X', a contradiction.

Notice that since $X \cap \Delta = \phi$ and $\Delta' = \Delta \setminus \{i\}$ so $(X \cup \{i\}) \cap \Delta' = \phi$. By point (iii) either $\{i\} \cup \partial_g(i) \subseteq X \cup \{i\}$ or there exists exactly one component $C''' \in \mathcal{C}(g - (X \cup \{i\}))$ such that $i \in \partial_g(C''')$. Hence, $C'' = C''' \cup \{i\}$ and C'' must be unique in $\mathcal{C}(g - X)$ with $i \in \partial_g(C'')$. The residual network $g - (X \cup \{i\})$ differs from $g - X$ at one component only: instead of $C\prime\prime$ it has $C'' \setminus \{i\}$. Thus, the value of residual network $g - (X \cup \{i\})$ is

$$g - (X \cup \{i\}) = \Phi(g - X) - f(|C''|) - f(|C''| - 1) \tag{12}$$

Similarly, since either $C' = \phi$ or $i \in \partial_g(C')$, so the residual network when using $X' \setminus \{i\}$ against Δ', $g - (X' \setminus \{i\})$, differs from $g - X'$ by one component: it has C instead of C'. Additionally, since $\Delta = \Delta' \cup \{i\}$ and $X' \cap \Delta' = \phi$ so

$X' \setminus \{i\} = X' \setminus \Delta$. Thus, the value of the residual network $g - (X' \setminus \{i\})$ can be written as

$$\Phi(g - (X' \setminus \{i\}) = \Phi(g - X') + f(|C|) - f(|C| - 1) \tag{13}$$

Since X is a best response to Δ, it is not worse than $X' \setminus \{i\}$. Hence

$$-\Phi(g - X) - c_A|X| \geq -\Phi(g - (X' \setminus \{i\}) - c_A(|X'| - 1)$$

This, together with (13), implies

$$\Phi(g - X) \leq \Phi(g - X') + f(|C|) - f(|C| - 1) - c_A(|X| - |X'| + 1) \tag{14}$$

Similarly, since X' is a best response to Δ', it is not worse than $X \cup \{i\}$. Hence

$$-\Phi(g - X') - c_A|X'| \geq -\Phi(g - (X \cup \{i\}) - c_A(|X| + 1)$$

This, together with (12), implies

$$\Phi(g - X) \geq \Phi(g - X') + f(|C''|) - f(|C''| - 1) - c_A(|X| - |X'| + 1) \tag{15}$$

From (14) and (15) we get

$$f(|C''|) - f(|C''| - 1) - \big(f(|C|) - f(|C| - 1)\big) \leq c_A(|X| + 1) - c_A|X| - \big(c_A|X'| - c_A(|X'| - 1\big) = 0$$

If $C \subsetneq C''$, then $|C| < |C''|$, and, by strict convexity of f, the left-hand side is greater than 0, a contradiction. Thus, it must be that $C'' = C$.

(v) We have $\Pi^D(\Delta, X; g) = \Pi^D(\Delta', X'; g) + f(|C|) - f(|C| - 1) - c_D$. Since X is a best response to Δ, it is not worse than $X' \setminus \{i\}$. Hence

$$-\Phi(g - X) - c_A|X| \geq -\Phi\big(g - (X' \setminus \{i\})\big) - c_A(|X'| - 1)$$

Adding $f(|C|) - f(|C| - 1)$ to both sides we get

$$\begin{aligned} &-\big(\Phi(g - X) - \big(f(|C|) - f(|C| - 1)\big)\big) - c_A|X| \\ &\quad \geq \big(\Phi(g - X' \setminus \{i\}) - \big(f(|C|) \quad - f(|C| - 1)\big)\big) - c_A(|X'| - 1) \end{aligned} \tag{16}$$

As we observed in the proof of point (v) ((12) and (13) and the fact that $C'' = C$),

$$\Phi\big(g - (X \cup \{i\})\big) = \Phi(g - X) - \big(f(|C|) - f(|C| - 1)\big) \tag{17}$$

$$\Phi(g - X') = \Phi\big(g - (X' \setminus \{i\})\big) - \big(f(|C|) - f(|C| - 1)\big) \tag{18}$$

Hence, from (16), we get

$$-\Phi\big(g - (X \cup \{i\})\big) - c_A(|X| + 1) \geq -\Phi(g - X') - c_A|X'|$$

Alternatively, since X' is a best response to Δ', so

$$-\Phi\big(g - (X \cup \{i\})\big) - c_A(|X| + 1) \leq -\Phi(g - X') - c_A|X'|$$

Combining these two inequalities we get

$$-\Phi\big(g - (X \cup \{i\})\big) - c_A(|X| + 1) = -\Phi(g - X') - c_A|X'| \tag{19}$$

Since X' is the equilibrium response to Δ', by tie-breaking Assumption 2,

$$\Phi(g - X') \leq \Phi\big(g - (X \cup \{i\})\big)$$

Additionally this, together with (17) and (18), implies

$$\Phi\big(g - (X' \setminus \{i\})\big) \leq \Phi(g - X) \tag{20}$$

From (19), (17), and (18) we get

$$-\Phi(g - X) - c_A|X| = -\Phi\big(g - (X' \setminus \{i\})\big) - c_A(|X'| - 1)$$

Again, since X is the equilibrium response to Δ, by tie-breaking Assumption 2,

$$\Phi(g - X) \leq \Phi\big(g - (X' \setminus \{i\})\big)$$

and, by (17), (18), and (20),

$$\Phi(g - X) = \Phi\big(g - (X' \setminus \{i\})\big)$$

$$\Phi\big(g - (X \cup \{i\})\big) = \Phi(g - X')$$

Thus, both X and $X' \setminus \{i\}$ are best responses to Δ and both X' and $X \cup \{i\}$ are best responses to Δ'. This, together with (18), implies

$$\Pi^D(\Delta, X; g) = \Pi^D(\Delta', X'; g) + f(|C|) - f(|C| - 1) - c_D$$

(vi) We have $c_A \leq f(|C|) - f(|C| - 1)$. Since X' is a better response to Δ' than $X' \setminus \{i\}$, so

$$-\Phi(g - X') - c_A|X'| \geq -\Phi\big(g - (X' \setminus \{i\})\big) - c_A(|X'| - 1)$$

and, consequently,

$$c_A \leq \Phi\big(g - (X' \setminus \{i\})\big) - \Phi(g - X')$$

By (17),

$$c_A \leq f(|C|) - f(|C| - 1)$$

□

Proof Proof of part (I) of Proposition 2 Characterization of the optimal strategies of the adversary follows directly from Lemma 1. Thus, in what follows we concentrate on the equilibrium defense.

Let Δ be an equilibrium defense. We will show first that if $\Delta \subsetneq N$, then Δ must be a minimal transversal of $\mathcal{D}(\Delta, \mathcal{E}(g, c_A))$.

Assume the opposite. Then there exists $i \in \Delta$ such that $\mathcal{D}(\Delta \setminus \{i\}, \mathcal{E}(g, c_A)) = \mathcal{D}(\Delta, \mathcal{E}(g, c_A))$. Let X be the equilibrium response to Δ and let X' be the equilibrium response to $\Delta' = \Delta' \setminus \{i\}$. Clearly $X \cap \Delta = \phi$ and $X' \cap \Delta' = \phi$.

Recall that $\mathcal{C}(g - X')$ is the set of components in the residual network when the strategies Δ' and X' are used by the players, and $\mathcal{C}(g - X)$ is the set of components in the residual network when Δ and X are used. By the assumption that $\Delta \subsetneq N$, both these sets are nonempty. We will show that either Δ' or Δ'' (described below) is a better strategy for the defender than Δ, which will contradict the assumption that Δ is an equilibrium strategy.

Let $C \in \mathcal{C}(g - X)$ be a component such that $C \subseteq \Delta$ and either $C = \{i\}$ or $C \setminus \{i\} \in \mathcal{C}(g - X')$. By Lemma 6 such C exists.

Since for all $j \in \partial_g(C)$, $\mathcal{D}(\Delta, \mathcal{E}(g, c_A)) \subsetneq \mathcal{D}(\Delta \cup \{j\}, \mathcal{E}(g, c_A))$, any such j belongs to an essential separator not covered by Δ. Take any $j \in \partial_g(C)$ and let $\{C_1, \ldots, C_m\} \subseteq \mathcal{C}(g - X)$ be all the components in $g - X$ such that $j \in \partial_g(C_l)$ for all $l \in \{1, \ldots, m\}$ (assume, without loss of generality, that $C_1 = C$; notice that in particular it may be that $m = 1$ and the argument below works for that case as well). Consider defenses $\Delta' = \Delta \setminus \{i\}$ and $\Delta'' = \Delta \cup \{j\} v \cup_{l=2}^{m} C_l$. We will show that either Δ' or Δ'' is a better strategy for the defender than Δ.

Let X'' be the equilibrium response of the adversary to Δ'' and let $C'' = \{j\} \cup_{l=1}^{m} C_l$. We show first that $C'' \subset \mathcal{C}(g - X'')$. Since Δ'' protects C'', there is component $C''' \in \mathcal{C}(g - X'')$ such that $C'' \subseteq C'''$. Suppose that $C'' \subsetneq C'''$. Then there exists $v \in C'''$ such that $v \notin C''$. We will show that $v \notin \Delta''$. If $v \in \partial_g(C_l)$ for some $l \in \{1, \ldots, m\}$, then it cannot be that $v \in \Delta$ (because these components are separated by X used as an equilibrium response to Δ). Thus, the only possibility is that $v \in \partial_g(\{j\})$. But then v would be one of the components C_l created by applying X to g and, consequently, it would be $v \in C''$, a contradiction with the assumption that $v \notin C''$. Since $v \notin \Delta$ and $v \notin C''$, then $v \notin \Delta''$. Now consider a response $X'' \cup \{v\}$ to Δ''. At the very least it removes a node from component C''' (it may additionally disconnect the component). Hence, $\Phi\big(g -$

$(X'' \cup \{v\})\big) \leq \Phi(g - X'') - f(|C'''|) + f(|C'''| - 1)$. Alternatively, by Lemma 6, (11), $c_A \leq f(|C|) - f(|C| - 1) < f(|C'''| - f(|C'''| - 1))$ (by convexity of f and $|C'''| \leq |C| + 1$). Thus, it follows that

$$-\Phi\big(g - (X'' \cup \{v\})\big) - c_A(|X''| + 1) > -\Phi(g - X'') - c_A|X''|,$$

which contradicts the assumption that X'' is a best response to Δ''. Therefore, it must be $C''' = C''$

As we have shown above, $C'' = \{j\}\cup_{l=1}^{m} \in \mathcal{C}(g - X)$. After attack $X \cup \{j\}$ is applied to g, component C'' is replaced with components $C_1, \ldots, C_m$. Hence

$$\Phi(g - X'') = \Phi\big(g - (X'' \cup \{j\})\big) + f\Big(1 + \sum_{l=1}^{m} |C_l|\Big) - \Big(\sum_{l=1}^{m} f(|C_l|)\Big) \tag{21}$$

Alternatively, since C is a component in $g - X$, every node in $\partial_g(C)$ is removed by X. Thus, when nodes in $\Delta \cup \{j\} \cup_{l=2}^{m} C_l$ are defended, the residual network $g - (X \setminus \{j\})$ differs from $g - X$ by having component C'' instead of components $C_1, \ldots, C_m$. Hence

$$\big(g - (X \setminus \{j\})\big) = \Phi\big(g - (X \setminus \{j\})\big) + f\Big(1 + \sum_{l=1}^{m} |C_l|\Big) - \Big(\sum_{l=1}^{m} f(|C_l|)\Big) \tag{22}$$

Since X'' is a better response to Δ'' than $X \setminus \{j\}$,

$$-\ \Phi(g - X'') - c_A|X''| \geq -\Phi\big(g - (X \setminus \{j\})\big) - c_A(|X| - 1) \tag{23}$$

and $\Phi(g - X'') \leq \Phi\big(g - (X \setminus \{j\})\big)$ in the case of equality (notice that $(X \setminus \{j\}) \cap \Delta'' = \phi$ as $X \cap C_l = \phi$ for all $l \in \{1, \ldots, m\}$ and $X \cap \Delta = \phi$).

Equations (21), (22), and (23) imply

$$-\Phi\big(g - (X \setminus \{j\})\big) - c_A|X''| \geq -\Phi(g - X) - c_A(|X| - 1)$$

Subtracting c A from both sides we get

$$-\ \Phi\big(g - X'' \cup \{j\}\big) - c_A(|X''| + 1) \geq -\Phi(g - X) - c_A|X| \tag{24}$$

Alternatively, since X is a best response to Δ than is $X'' \cup \{j\}$, we have

$$-\ \Phi(g - X) - c_A|X| \geq -\Phi\big(g - (X'' \cup \{j\})\big) - c_A(|X''| + 1) \tag{25}$$

and $\Phi(g - X) \leq \Phi\big(g - (X \setminus \{j\})\big)$, in the case of equality.

By (24) and (25), $X'' \cup \{j\}$ is a best response to Δ as well, and since X is an equilibrium response to Δ, it must be that $\Phi(g - X) \leq \Phi\big(g - (X'' \cup \{j\})\big)$. Combining this with (21) we get

$$\Phi(g - X'') \geq \Phi(g - X) + f\Big(1 + \sum_{l=1}^{m} |C_l|\Big) - \Big(\sum_{l=1}^{m} f(|C_l|)\Big) \tag{26}$$

and from (10) and (26) it follows that

$$\Pi^D(\Delta, X; g) = \Pi^D(\Delta', X'; g) + f(|C|) - f(|C| - 1) - c_D$$

$$\Pi^D(\Delta'', X''; g) \geq \Pi^D(\Delta, X; g) + f\Big(1 + \sum_{l=1}^{m} |C_l|\Big) - \Big(\sum_{l=1}^{m} f(|C_l|)\Big) - c_D$$

Since Δ is a better strategy than Δ', then $f(|C|) - f(|C| - 1) \geq c_D$. Alternatively since Δ is a better strategy than Δ'', then $c_D \geq f(1 + \sum_{l=1}^{m} |C_l|) - (\sum_{l=1}^{m} f(|C_l|))$. Hence, $f(|C|) - f(|C| - 1) \geq f(1 + \sum_{l=1}^{m} |C_l|) - (\sum_{l=1}^{m} f(|C_l|))$, which contradicts the convexity of f.

Thus, we have shown that $\Delta \subsetneq N$, and then Δ must be a minimal transversal of $\mathcal{D}(\Delta, \mathcal{E}(g, c_A))$. □

Proof Proof of proposition 3: The nonmonotonicities have been established in the text. Here we establish monotonicity of the defender's payoff in cost of attack and monotonicity of the adversary's payoff in cost of defense. We start with monotonicity of payoff to the defender in cost of attack. The argument here is straightforward in the generic case, where equilibrium payoffs are unique: suppose (Δ^*, X^*) is an equilibrium with network g and costs (c_A, c_D). Let $c_A^{'} > c_A$. If the defender retains defense strategy Δ^*, it must be the case that the attack strategy will be weakly smaller under high cost $c_A{}'$. This in turn implies that the defender's payoff must be weakly larger if he maintains the original strategy Δ^*. So, in equilibrium under $(c_a^{'}, c_D)$, he must also do better. However, the monotonicity holds for any values of the parameters. The problem here is the nonuniqueness of equilibrium payoffs. However, this is not a concern, because if this was the case, the more costly attacks would cease being equally good for the adversary as the less costly ones. The precise argument is as follows. Let c_A and $c_A^{'}$ be the costs of attack such that $c_A^{'} > c_A$. Let (Δ^*, X^*) be an equilibrium under c_A and let (Δ^{**}, X^{**}) be an equilibrium under $c_A^{'}$. Since $X^*(\Delta^*)$ is a best response to Δ^* under c_A, it is not worse than $X^{**}(\Delta^*)$; hence

$$-\Phi(g - X^*(\Delta^*)) - c_A|X^*(\Delta^*)| \geq -\Phi(g - X^{**}(\Delta^*)) - c_A|X^{**}(\Delta^*)|$$

Which yields

$$\Phi(g - X^*(\Delta^*)) - \Phi(g - X^{**}(\Delta^*)) \leq c_A\left(|X^{**}(\Delta^*)| - |X^*(\Delta^*)|\right) \tag{27}$$

Similarly,Since $X^{**}(\Delta^*)$ is a best response to Δ^* under $c_A^{'}$, it is not worse than $X^*(\Delta^*)$. This yields

$$\Phi(g - X^*(\Delta^*)) - \Phi(g - X^{**}(\Delta^*)) \geq c_A^{'}\left(|X^{**}(\Delta^*)| - |X^*(\Delta^*)|\right) \tag{28}$$

Equations (27) and (28) imply $c_A\,(|X^{**}(\Delta^*)| - |X^*(\Delta^*)|) \geq c'_A\,(|X^{**}(\Delta^*)| - |X^*(\Delta^*)|)$. By $c'_A > c_A$ it follows that

$$|X^{**}(\Delta^*)| \leq |X^*(\Delta^*) \tag{29}$$

Now assume to the contrary that

$$\Pi^D(\Delta^*, X^*(\Delta^*); g, c_D) > \Pi^D(\Delta^{**}, X^*(\Delta^{**}); g, c_D)$$

Since Δ^{**} is an equilibrium defence under c'_A,

$$\Pi^D(\Delta^{**}, X^{**}(\Delta^{**}); g, c_D) \geq \Pi^D(\Delta^*, X^{**}(\Delta^*); g, c_D)$$

The two equations above imply

$$\Pi^D(\Delta^*, X^*(\Delta^*); g, c_D) > \Pi^D(\Delta^*, X^{**}(\Delta^*); g, c_D)$$

that is,

$$\Phi(g - X^*(\Delta^*)) - c_D|\Delta^*| > \Phi(g - X^{**}(\Delta^*)) - c_D|\Delta^*|$$

and, consequently,

$$\Phi(g - X^*(\Delta^*)) - \Phi(g - X^{**}(\Delta^*)) > 0 \tag{30}$$

Equations (27) and (30) imply $c_A(|X^{**}(\Delta^*)| - |X^*(\Delta^*)|) > 0$. By $c_A > 0$, it follows that $|X^{**}(\Delta^*)| > |X^*(\Delta^*)|$, a contradiction with (29). Thus, it must be that $\Pi^D(\Delta^*, X^*(\Delta^*); g, c_D) \leq \Pi^D(\Delta^{**}, X^{**}(\Delta^{**}); g, c_D)$. Notice that this argument holds for any parameters of the model, not only in the generic case..

We now turn to the monotonicity of payoff to the adversary in cost of defense. Let c_D and c'_D be the costs of defense such that $c'_D > c_D$. Let (Δ^*, X^*) be an equilibrium under c_D and let (Δ^{**}, X^{**}) be an equilibrium under c'_D. Since $X^*(\Delta^*)$ is a best response to Δ^* and $X^{**}(\Delta^*)$ is a best response to Δ^* in the adversary's subgame,

$$-\Phi(g - X^*(\Delta^*)) - |X^*(\Delta^*)|c_A = -\Phi(g - X^{**}(\Delta^*)) - |X^{**}(\Delta^*)|c_A$$

Thus, another equilibrium under c_D is (Δ^*, X'), where X' equals X^* at all defense profiles but Δ^*, where it is equal to Δ^{**}. By Lemma 3, generically, $\Phi(g - X^*(\Delta^*)) = \Phi(g - X^{**}(\Delta^*))$. By analogous arguments, $\Phi(g - X^{**}(\Delta^{**})) = \Phi(g - X^*(\Delta^{**}))$. Since Δ^* is an equilibrium defense under c_D and Δ^{**} is an equilibrium defense under c'_D,

$$\Phi(g - X^*(\Delta^*)) - |\Delta^*|c_D \geq \Phi(g - X^*(\Delta^{**})) - |\Delta^{**}|c_D$$
$$\Phi(g - X^{**}(\Delta^{**})) - |\Delta^{**}|c'_D \geq \Phi(g - X^{**}(\Delta^*)) - |\Delta^*|c'_D$$

which can be rewritten as

$$\Phi(g - X^*(\Delta^{**})) - \Phi(g - X^*(\Delta^*)) \leq (|\Delta^{**}| - |\Delta^*|)c_D$$
$$\Phi(g - X^{**}(\Delta^{**})) - \Phi(g - X^{**}(\Delta^*)) \geq (|\Delta^{**}| - |\Delta^*|)c'_D$$

Since $c'_D > c_D$, these inequalities imply

$$\Phi(g - X^{**}(\Delta^{**})) - \Phi(g - X^{**}(\Delta^*)) > \Phi(g - X^*(\Delta^{**})) - \Phi(g - X^*(\Delta^*))$$

This, combined with $\Phi(g - X^*(\Delta^*)) = \Phi(g - X^{**}(\Delta^*))$ and $\Phi(g - X^{**}(\Delta^{**})) = \Phi(g - X^*(\Delta^{**}))$, leads to contradiction. Hence, it must be that the payoff to the adversary increases when the cost of defense increases. Notice that this argument holds for generic values of the parameters of the model. There are nongeneric examples where the payoff to the adversary decreases when the cost of defense increases.

Before proving Proposition 4, we need the following auxiliary lemma, stating a useful property of a convex function.

Lemma 7 *Let $f : R \longrightarrow R$ be a strictly convex and differentiable function. Then function*

$$g(x, y) = \frac{yf(x) - xf(y)}{x - y}$$

is strictly increasing in both arguments as long as $x > y$.

Proof To show the result we compute partial derivatives of h:

$$g_x(x, y) = \left(\frac{y}{x - y}\right)\left(f'(x) - \frac{f(x) - f(y)}{x - y}\right)$$
$$g_y(x, y) = \left(\frac{x}{x - y}\right)\left(\frac{f(x) - f(y)}{x - y} - f'(y)\right)$$

By strict convexity of f, $f'(y) < \left(\frac{f(x)-f(y)}{x-y}\right) < f'(x)$ as long as $x > y$; hence, $g_x, g_y > 0$ and g is strictly increasing in x and in y. This completes the proof. □

Now we are ready to prove Proposition 4.

Proof of Proposition 4 Point (i) follows directly and we omit the proof. For point (ii) observe, from Proposition 2, that with $c_A \in (\Delta f(n-1), f(n) - (n-1)f(1))$ and $g \in \mathcal{E}(g, c_A) = 0$, the optimal attack targets no nodes. So the optimal defense also consists of defending no nodes. Thus, the costs of conflict are 0.

For point (iii), assume that $c_A \in (\Delta f(m-1), \Delta f(m))$ with $m \in \{1, \ldots, n-1\}$. With such a cost of attack, on any connected network, the adversary best responds to any incomplete defense by removing at least one node. Therefore, the lower bound for the costs of conflict are $\min(c_A, nc_D)$ in this case.

Part 1. Suppose that $c_D > c(n, m)$. We show first that in every equilibrium on h_n^m the defender chooses the empty defense and the adversary responds to it with attack $\{1\}$(the separator of h_n^m). By Proposition 2, an equilibrium defense must be either empty, or complete, or equal to $\{1\}$. Moreover, the best response of the adversary to the empty defense either contains $\{1\}$, in which case the reducing attack part of it must be empty (because components of $h_n^m - \{1\}$ have sizes at most m), or does not contain $\{1\}$, in which case it must be a reducing attack leaving a residual network consisting of a single component of size m. It is easy to check that the former is the best response to the empty defense and the latter is the best response to defense $\{1\}$. Hence, empty defense is better than $\{1\}$. The payoff to the defender from using the empty defense is

$$\Pi^D(\phi, \{1\}; h_n^m, c_D) = \Phi(h_n^m - \{1\}) = \left\lfloor \frac{n-1}{m} \right\rfloor f(m) + f((n-1) \mod m)$$

With cost of defense $c_D > c(m, n)$, the payoff to the defender from the complete defense,

$$\Pi^D(N, \phi; h_n^m, c_D) = f(n) - nc_D,$$

is lower than the payoff from the empty defense. Hence, on the equilibrium path the defender chooses ϕ and the adversary responds with $\{1\}$.

Second, we show that for the ranges of costs in question, $nc_D > c_A$. Since $c_D > c(n, m)$,

$$nc_D > f(n) - \left\lfloor \frac{n-1}{m} \right\rfloor f(m) - f((n-1) \mod m)$$

The right-hand side of this inequality can be rewritten as

$$f(n) - \frac{n-1-(n-1) \mod m}{m} \cdot f(m) - f((n-1) \mod m)$$

Since f is strictly convex and $(n-1) \mod m < m$, then $((n-1) \mod m) f(m) > mf((n-1) \mod n)$. Therefore,

$$c_D > f(n) - \left(\frac{n-1}{m}\right) f(m)$$

The right-hand side can be rewritten as

$$\begin{aligned} f(n) - \left(\frac{n-1}{m}\right) f(m) &= \sum_{j=m}^{n-1} \Delta f(j) - (n-m-1)\left(\frac{f(m)}{m}\right) \\ &= \Delta f(m) + \sum_{j=m+1}^{n-1} \left(\Delta f(j) - \frac{f(m)}{m}\right) \end{aligned}$$

By convexity of f, for all $j > m$, $\Delta f(j) > f(m)/m$. Thus, $ncD > \Delta f(m)$ and, since $c_A \in (\Delta f(m-1), \delta f(m))$, $nc_D > cA$. Hence, the minimal costs of conflict are c_A.

Part 2. Suppose that $c_D < c(n, m)$. We will show that with such a cost of defense, in any equilibrium on a connected network the defender chooses the complete defense. Notice that with $c_D < c(n, m)$, on any connected network g, any defense Δ of size $|\Delta| \leq m$ is worse for the defender than the complete defense. This is because the residual network after the adversary best responding to Δ consists of components of sizes at most m and the upper bound on the value of such residual networks is $\lfloor (n-1)/m \rfloor f(m) + f((n-1) \mod m)$ (this upper bound is attained by h_n^m). With $c_D < c(n, m)$ the defender prefers complete defense to Δ.

Consider defense Δ of size $d = |\Delta|$ such that $m < d < n$. Let X be a best response to Δ. The payoff to the defender from Δ and X is

$$\begin{aligned}\Pi^D(\Delta, X; g, c_D) &= \Phi(g - X) - dc_D \\ &\geq f(d) + \left\lfloor \frac{n-d-1}{m} \right\rfloor + f((n-d-1) \mod m) - dc_D\end{aligned}$$

The upper bound on the value of the residual network above comes from the following observation. With $c_A \in (\Delta f(m-1), \Delta f(m))$, in any best response the adversary removes unprotected nodes from any component of size greater than m. Therefore, in the best case the adversary removes one node and the only component of size greater than m in the residual network is a fully protected component of size d (by convexity of f it is better to have one fully protected component of size d than several fully protected and smaller ones summing up to d). Thus, if

$$c_D < \frac{f(n) - f(d) - \left\lfloor \frac{n-d-1}{m} \right\rfloor - f((n-d-1) \mod m)}{n-d}$$

then the complete defense is better to Δ for the defender. We will show that $c(n, m)$ is lower than the right-hand side of the inequality above, which will imply that for the costs of defense under consideration, the complete defense is better for the defender.

The inequality

$$\begin{aligned}c(n, m) &= \frac{f(n) - \left\lfloor \frac{n-1}{m} \right\rfloor - f((n-1) \mod m)}{n} \\ &< \frac{f(n) - f(d) - \left\lfloor \frac{n-d-1}{m} \right\rfloor - f((n-d-1) \mod m)}{n-d}\end{aligned}$$

can be rewritten as

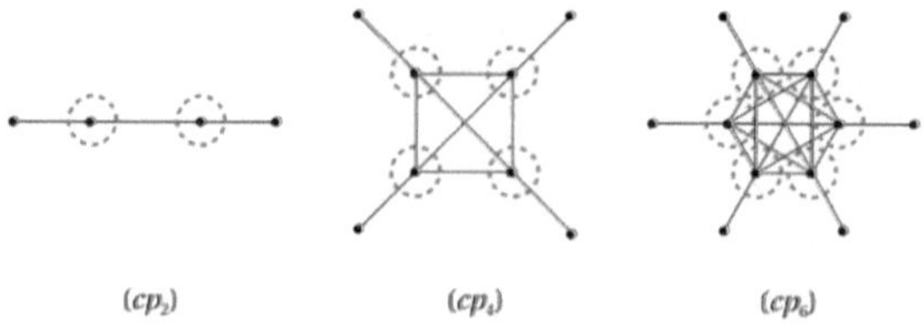

Fig. 9 Core–periphery networks cp_2, cp_4, and cp_6

$$df(n) - nf(d) - \Big(\frac{n-d}{m}\Big)(r_1 f(m) - mf(r_1)) + \Big(\frac{n}{m}\Big)(r_2 f(m) - mf(r_2)) > 0,$$

where $r_1 = (n-1) \mod m$ and $r_2 = (n-d-1) \mod m$.[9] Since $r_2 < m$ and f is convex, then $r_2 f(m) - mf(r_2) > 0$, and to show that the inequality above holds it suffices to show that

$$df(n) - nf(d) - \Big(\frac{n-d}{m}\Big)(r_1 f(m) - mf(r_1)) > 0 \tag{31}$$

Since $d < n$ and f is convex, $df(n) - nf(d) > 0$. Moreover, by Lemma 7,

$$\frac{df(n) - nf(d)}{n-d} > \frac{r_1 f(m) - mf(r_1)}{m - r_1}$$

(as $n > d > m > r_1$). Hence

$$\frac{df(n) - nf(d)}{n-d} - \Big(\frac{m-r_1}{m}\Big)\Big(\frac{r_1 f(m) - mf(r_1)}{m - r_1}\Big)$$

which implies (31), by multiplying both sides by $(n-d)$. Hence, any equilibrium defense is complete and the costs of conflict are nc_D. □

6.1 Examples of No Conflict Networks

Even if the marginals of f do not grow very rapidly, there may exist networks (other than complete network) that do not feature active conflict. Take $f(x) = x^2$, for example. Consider a family of core–periphery networks, $\{cp_k\}_{k \in \mathbb{N}}$. Given $k \in \mathbb{N}$, network cp_k has $2k$ nodes: a fully connected core of k nodes, and a periphery of k nodes. Each core node is connected to exactly one, unique, periphery node (cf. Fig. 9).

When the cost of attack is high, $c_A > 4m - 1$, then it is easy to verify that in equilibrium the defender will either defend all the core nodes or use an empty defense. When the cost of attack is low, $cA < 4m - 1$, then, again, there are two types of

[9] Recall that for integer x and y, $\lfloor x/y \rfloor = (x - x \mod y)/y$.

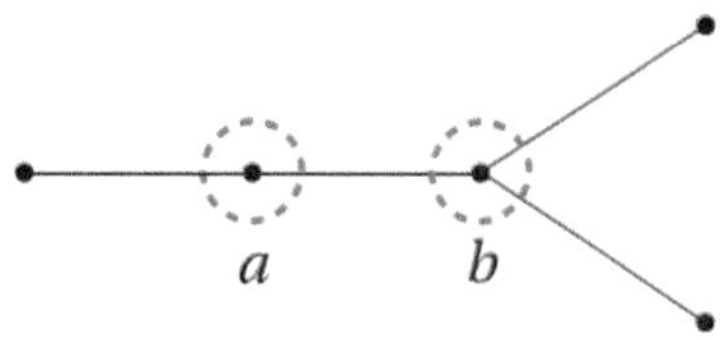

Fig. 10 Network that allows for active conflict (under $f(x) = x^2$)

equilibrium defense: either no node is defended or all nodes are defended. It is easy to verify that three types of defense would be candidates for equilibrium defense here: empty defense, complete defense, and defense with all core nodes protected. To rule out the last one, suppose that $2(2m - k) - 1 \leq c_A < 2(2m - k) + 1$, where $1 \leq k \leq m - 1$. Notice, in the example above, that if each core node was connected to a higher number of periphery nodes, active conflict would be possible (as illustrated by Example 1). With more periphery nodes per core node (and with suitable costs of defense and attack), protecting the separators may create enough value for such a defense to be attractive. Increasing the value of the residual network requires defending all the nodes, which is too high an investment and too low a gain to be profitable. This illustrates one reason for the possibility of active conflict in the model: blocking all the individually rational essential separators may secure a high value of the residual network at a relatively low cost, while increasing the value further may require a much higher cost.

To get more insight into why active conflict is possible, despite the convexity of f and the linearity of costs, consider the network in Fig. 10. Figure 11 illustrates function $\Phi^\star(m; g, c_A)$ under different ranges of costs of attack. The dotted line is an upper convex hull of that function. The optimal size of defense is at a point of that hull adjacent to a line with slope c_D. In the case of low cost of attack, if the convex hull contains any points of $\Phi^\star(m; g, cA)$ for $0 < m < n$, then active conflict is possible for some suitable range of costs of defense. In the case of high cost of attack, active conflict is possible if the convex hull contains any points of $\Phi^\star(m; g, cA)$ for $0 < m < \tau(\mathcal{E}(g, cA))$.

In Fig. 11, low cost of attack is $c_A < 9$ and $c_A > 9$ is high cost of attack. Active conflict is possible for $c_A \in (5, 9)$. When $c_A \in (5, 7)$ and $c_D \in (3.75, 4)$, then the unique equilibrium defense is $\Delta^\star = \{b\}$, and the best response to it in the adversary's subgame is $X^\star(\Delta^\star) = \{a\}$. When $c_A \in (7, 9)$ and $c_D \in (5, 9)$, then the unique equilibrium defense is $\Delta^\star = \{a, b\}$ and removing any unprotected node is a best response to it in the adversary's subgame. When $c_A \in (9, 15)$, $\tau(\mathcal{E}(g, c_A)) = 2$ and there is no equilibrium outcome with active conflict.

Proof Proof of Proposition 5: For point (i), suppose that $c_D > \frac{f(n)}{n}$. We will show that in this case the equilibrium defense $\Delta = \phi$. Assume, to the contrary, that $\Delta \neq \phi$ and let X be the equilibrium response to Δ. Pick any $i \in \Delta$ and let $C(i)$ be the component of i in the residual network $g - X$. The payoff to i is

$$\Pi^i(\Delta, X; g, c_d) = \frac{f(|C(i)|)}{|C(i)|} - c_D$$

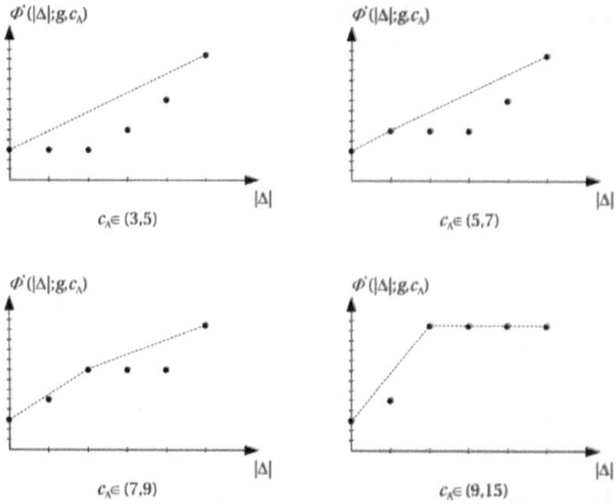

Fig. 11 Optimal defenses of different sizes for network in Fig. 10

By the fact that f is strictly increasing and strictly convex, $f(x)/x$ is increasing. Hence $\Pi^i(\Delta, X; g, c_d) \leq f(n)/n - c_D < 0$. Thus, i is better off by not protecting, a contradiction to the assumption that Δ is an equilibrium defense. Hence, it must be that $\Delta = \phi$.

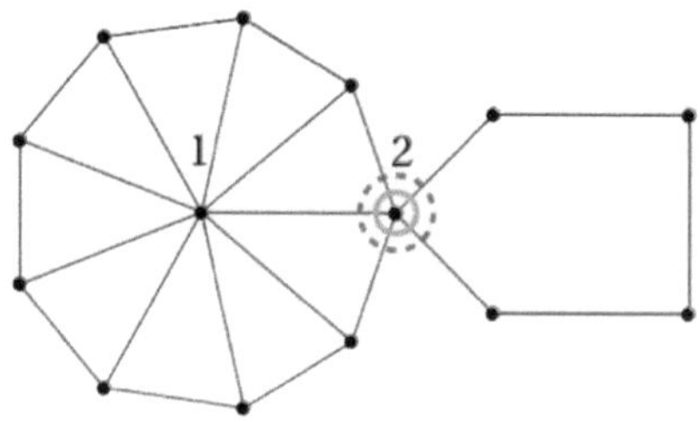

Fig. 12 Separators and other centrality measures

For point (ii), suppose that $c_D < \frac{f(n)}{n}$. Assume that $c_A < f(n) - f(n-1)$. We will show that $\Delta = N$ is an equilibrium defense. Assume otherwise. Then there exists $i \in \Delta$ that is better off by deviating and choosing no protection. Since $c_A < f(n) - f(n-1)$, the best response to $\Delta \setminus \{i\}$ is $X = \{i\}$, and so the deviating node gets removed, obtaining payoff 0 instead of $f(n)/n - cD \geq 0$. Hence, i is not better off by deviating and so $\Delta = N$ is an equilibrium defense. This proves point (a).

Assume that $c_A > f(n) - f(n-1)$. Let Δ be minimal transversal of $\mathcal{E}(g, c_A)$. We will show that Δ is an equilibrium defense. By Lemma 1, the best response to Δ is the empty attack $X = \phi$. Assume, to the contrary, that Δ is not an equilibrium defense. Then there exists $i \in \Delta$ that is better off by choosing no protection instead of protection. Since Δ is a minimal transversal, it must be that there exists an essential separator $E \in \mathcal{E}(g, c_A)$ such that $\Delta \setminus \{i\} \cap E = \phi$. Moreover, any such separator contains i. Since any such separator is better than the empty attack, the adversary responds to $\Delta \setminus \{i\}$ with one of these separators, removing i. But then i gets payoff 0

Fig. 13 **Table 1**. Centralities of nodes 1 and 2 in the network from Fig. 12

Centrality	Node 1	Node 2
Degree	9	5
Closeness	0.684	0.619
Betweenness	42.5	37.5
Eigenvector	0.5765	0.3036
Bonacich, high	532.2	281.18
Bonacich, medium	2.4311	1.8208
Bonacich, low	1.093	1.0519
Intercentrality, high	2940.9	2863.2
Intercentrality, medium	5.2438	3.1263
Intercentrality, low	1.1936	1.1061

instead of $f(n)/n - c_D \geq 0$. Hence, it is not better of by deviating, a contradiction. Therefore, Δ must be an equilibrium defense. This proves point (b).

Since the adversary's subgame remains as in the centralized defense game, an equilibrium response $X^{\star}$ is as described in Proposition 2. □

7 Appendix B: Key Players and Centrality

Essential separators and their transversals determine the key nodes in our study of attack and defense. These key groups of nodes give rise to new notions of centrality distinct from other notions such as closeness, betweenness, or eigenvector centralities. To see how these notions are different, consider the network in Fig. 12 (for simplicity the example is based on individual, rather than group, notions of centrality). Assume that the network value is based on function $f(x) = x^2$ and suppose that the cost of attack is $c_A \in (25, 89)$, so that the adversary attacks only the nodes that separate the network and so that removing node 2 is better than not attacking at all. Suppose also that $c_D \in (0, 89)$, so that defending node 2 constitutes an optimal defense as well. However, this node is less central than node 1 in the sense of degree, closeness, betweenness, eigenvector, Bonacich, and intercentrality measures.[10] The numerical values for these centralities are summarized in Fig. 13. For Bonacich centrality, we consider three values of the parameters: high ($\alpha = 0.237$), intermediate ($\alpha = 0.1$), and low ($\alpha = 0.01$).

[10] Following [7], we define for a parameter $\alpha \in \mathbb{R}$, $\mathbf{b}(g, \alpha) = \mathbf{M}(g, \alpha)\mathbf{1}$, where $\mathbf{M}(g, \alpha) = (\mathbf{I} - \alpha\mathbf{1G})^{-1}$, $\mathbf{I}$ is the identity matrix, and $\mathbf{G}$ is the adjacency matrix of the network. We require α to be relatively small so that $\mathbf{M}(g, \alpha)$ is well defined and nonnegative. The intercentrality measure we consider, also defined in that paper, is $c_i(g, \alpha) = b_i(g, \alpha)^2/M_{ii}(g, \alpha)$. We define closeness as $cl_i(g) = (n-1)/\sum_{j \neq i} d(i, j; g)$, where $d(i, j; g)$ is the length of the shortest path between i and j in g.

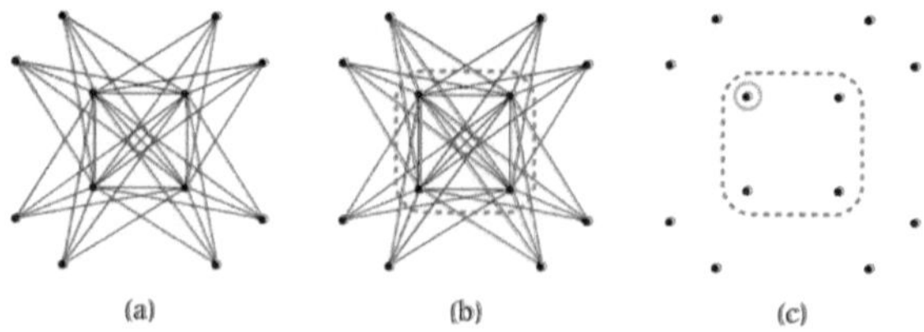

Fig. 14 Separators and transversals in interlinked stars (n = 12)

8 Appendix C: Separators and Transversals in Families of Networks

8.1 *Interlinked Stars*

Interlinked stars are networks with two disjoint nonempty sets of nodes: the set of *centers* C and the set of *periphery nodes* P. The centers are fully connected, forming a clique. Each of the periphery nodes is connected to all the centers. Interlinked stars have one essential separator: the set of all the centers, $\mathcal{E}(g) = \{C\}$. All minimal transversals of $\mathcal{E}(g)$ are singleton sets consisting of one central node. The essential separator and a minimal transversal for an interlinked star are illustrated in Fig. 14.

8.2 *Complete Bipartite Networks*

In a complete bipartite network the set of nodes, N, can be partitioned into two disjoint sets, N_1 and N_2, $N_1 \cap N_2 = \phi$, such that the set of links is the set of all possible links connecting nodes from N_1 and nodes from N_2. There are two essential separators in these networks, $\mathcal{E}(g) = N_1, N_2$. Every transversal consists of one node from N_1 and one node from N_2. Minimal essential separators and transversals for complete bipartite networks are illustrated in Fig. 15.

8.3 *Trees*

In any tree network, every nonempty set of internal nodes (nodes that are not leaves) constitutes a separator. Essential separators are sets of internal nodes such that no two of them are neighbors. Transversals of essential separators are subsets of internal nodes. In particular, there is a unique transversal of the set of all essential separators: the set of all internal nodes. Minimal essential separators and transversal for tree networks are illustrated in Fig. 16.

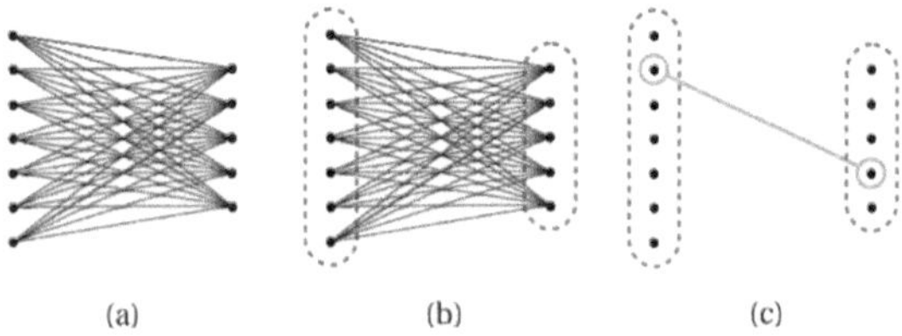

Fig. 15 Separators and transversals in complete bipartite networks (n = 12)

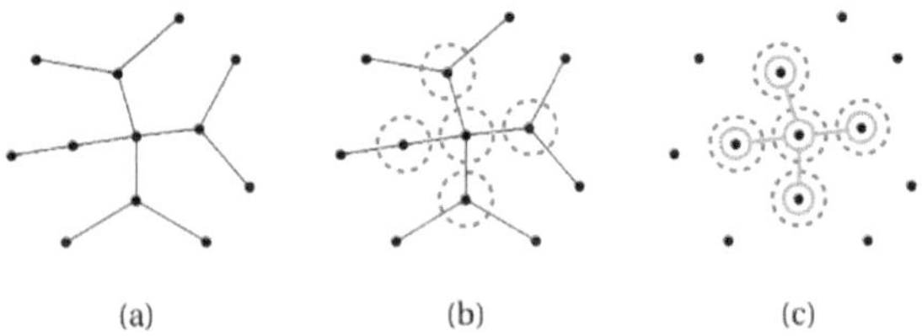

Fig. 16 Separators and transversals in trees (n = 12)

8.4 *Core–Periphery Networks*

Nodes are divided in two disjoint sets: the core and the periphery. Each node of the periphery is connected to exactly one node of the core, while the nodes of the core are connected with periphery nodes and the core constitutes a clique. Essential separators are subsets of the core. There is a unique transversal: the set of all core nodes. Minimal essential separators and transversals for core–periphery networks are illustrated in Fig. 17.

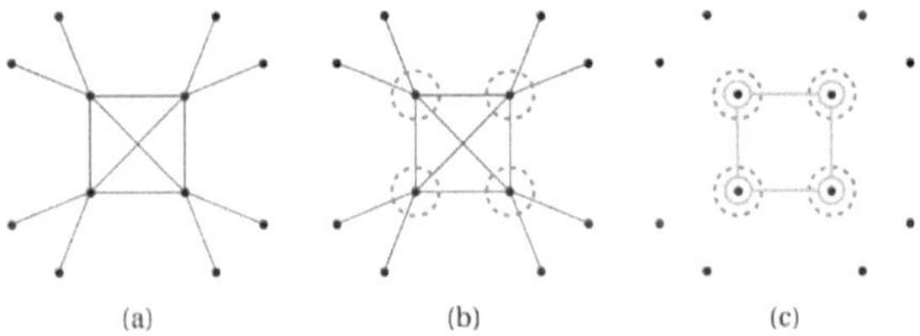

Fig. 17 Separators and transversals in core–periphery networks (n = 12)

9 Appendix D: Order of Moves and Nature of Conflict

This section explores the role of sequential choice and perfect defense.

9.1 *Simultaneous Moves*

Consider a variant of the model studied in the paper where the players make their choice simultaneously. In this case the set of strategies of the defender remains unchanged. A pure strategy of the adversary is now a set of nodes, $X \subseteq N$, chosen to attack. It is important to note that the timing of moves does not affect Lemma 1, which remains unchanged. Suppose that the cost of attack is high. Any strategy, X, in the support of the equilibrium strategy of the adversary must be an individually rational essential separator, i.e., $X \in \mathcal{E}(g, c_A)$. Similarly, any strategy, Δ, in the support of the equilibrium strategy of the defender must be a minimum transversal of the set of essential separators it blocks, $\mathcal{D}(\Delta, \mathcal{E}(g, c_A))$, in $\mathcal{E}(g, c_A)$.

The second observation is that, depending on the network, the players may use pure or mixed strategies in equilibrium. This is a departure from our existing results, where equilibrium always exists in pure strategies. But note that the use of mixed strategies is sensitive to the network. In particular, if the network is such that one unit of defense is sufficient to block all the individually rational essential separators of the adversary, then in equilibrium both players use pure strategies and equilibrium outcomes are the same as in the sequential model studied in the paper. When $\tau(\mathcal{E}(g, c_A)) > 1$, the defender may choose to block more individually rational essential separators by mixing across several transversals.

9.2 The Model of Conflict

We have assumed perfect defense. A more natural way to proceed would be to suppose that the number of resources assigned by each player to a node determines the probability of winning/losing the node. Following Tullock (1980), suppose that the probability of successfully attacking the node is given by a contest success function (CSF)

$$\pi(a, d) = \begin{cases} 0 & \text{if } a = 0 \\ \frac{d^{\gamma}}{a^{\gamma}+d^{\gamma}} & \text{otherwise,} \end{cases}$$

where $\gamma \in R_+$, and a and d are resources assigned by the adversary and defender, respectively. The probability of successfully defending the node is $\pi(d, a) = 1 - \pi(a, d)$.[11]

A strategy of the defender is a vector $\mathbf{d} \in \mathbb{N}^N$ such that d_i is the number of defense resources assigned to node i. A strategy of the adversary is a function $X : \mathbb{N}^N$ such that, given vector of defense allocation $\mathbf{d}$, it maps to a vector of attack allocation $\mathbf{a} = X(\mathbf{d})$ such that a_i is the number of attack resources assigned to node i. We will call the set of nodes that receive a positive number of defense resources the defended nodes and the set of nodes that receive a positive number of attack resources the attacked nodes. Given defense and attack allocations, $(\mathbf{d}, \mathbf{a})$, the probability that set $M \subseteq N$ of nodes is won by the adversary and removed from g is

[11] The perfect defense model studied in the paper can be seen as a limiting case of the general contest model: the probability of successful attack is given by $\alpha a^{\gamma}/(\delta d^{\gamma} + \alpha a^{\gamma})$ with $\alpha = 1$ and $\delta \longrightarrow +\infty$.

$$w(M|\mathbf{a}, \mathbf{d}) = \Pi_{j \in M} \pi(a_i, d_i)$$

The expected payoffs to the defender and the adversary from defense and attack allocations, $(\mathbf{a}, \mathbf{d})$, are

$$\Pi^A(\mathbf{a}, \mathbf{d}|g, c_A) = -\sum_{M \subseteq N} w(M|\mathbf{a}, \mathbf{d})(1 - w(N \setminus M|\mathbf{a}, \mathbf{d}))\Phi(g - M) - c_A \sum_{j \in N} a_i$$

$$\Pi^D(\mathbf{a}, \mathbf{d}|g, c_A) = \sum_{M \subseteq N} w(M|\mathbf{a}, \mathbf{d})(1 - w(N \setminus M|\mathbf{a}, \mathbf{d}))\Phi(g - M) - c_D \sum_{j \in N} d_i$$

Lemma 1 still obtains. The set of attacked nodes can be decomposed into an essential separator and a reducing attack. In what follows we restrict attention to high costs of attack and we focus on the benchmark model of linear contests: $\gamma = 1$. The main point we wish to make is that with Tullock contests, optimal defense will extend beyond minimal transversals and may cover multiple nodes in the same separator.

Consider an interlinked star with two core nodes: 1, 2, and $n-2$ periphery nodes ($n \geq 4$). Suppose that the cost of attack is high, $c_A > \Delta f(n-1)$. The unique essential separator of g is the set of core nodes, $\{1, 2\}$. Let a_1, a_2 be the amount of resources assigned by the adversary to the two core nodes and let d_1, d_2 be the defense resources assigned by the defender to the two core nodes. Expected payoff to the adversary from assignment (a_1, a_2, d_1, d_2) is

$$\begin{aligned}\Pi^A(\mathbf{d}, \mathbf{a}|g, c_A) &= -\pi(a_1, d_2)\pi(a_2, d_2)(n-2)f(1)\\ &\quad - (\pi(a_1, d_1) + \pi(d_2, a_2) - 2\pi(d_1, a_1)\pi(a_2, d_2))f(n-1)\\ &\quad - (1 - \pi(a_1, d_1) - \pi(d_2, a_2) + \pi(d_1, a_1)\pi(a_2, d_2))f(n)\\ &\quad - c_A(a1 + a2)\\ &= -f(n) + \pi(a_1, d_1)\pi(a_2, d_2)V_1(n)\\ &\quad + (\pi(a_1, d_1) + \pi(a_2, d_2) - \pi(a_1, d_1)\pi(a_2, d_2))V_2(n)\\ &\quad - c_A(a_1 + a_2),\end{aligned}$$

where $V_1(n) = f(n-1) - (n-2)f(1)$ and $V_2(n) = f(n) - f(n-1)$. Notice that $V_2(n)$ is the gain from removing the first node of the core, and $V_1(n)$ is the gain from removing the second node of the core. Since the cost of attack is high, $V_2(n) < c_A$. Hence, if $V_1(n) \leq V_2(n)$, then it is not profitable for the adversary to attack, and both players assign no resources to the nodes in equilibrium. Consider now the more interesting case where $V_1(n) > V_2(n)$.

The expected payoff to the defender is

$$\begin{aligned}\Pi^D(\mathbf{d}, \mathbf{a}|g, c_A) = f(n) - \pi(a_1, d_1)\pi(a_2, d_2)V_1(n) \\ - (\pi(a_1, d_1) + \pi(a_2, d_2) - \pi(a_1, d_1)\pi(a_2, d_2))V_2(n) \\ - c_D(d_1 + d_2)\end{aligned}$$

The defender chooses (d_1, d_2) to maximize his expected payoff subject to the constraints that $d_1, d_2 \geq 0$ and that the adversary chooses (a_1, a_2) to maximize his expected payoff subject to $a_1, a_2 \geq 0$.

It is simpler to begin with the case where the defender is given $2d \geq 0$ defense resources and the adversary is given $2a \geq 0$ attack resources. This turns the optimization problem above into a zero-sum bilevel optimization problem, where the defender chooses an allocation of $2d$ to maximize

$$\pi(a_1, d_1)\pi(a_2, d_2)V_1(n) + (\pi(a_1, d_1) + \pi(a_2, d_2) - \pi(a_1, d_1)\pi(a_2, d_2))V_2(n)$$

It is possible to show that the partition (d, d) is a maximizer of both $\pi(a_1, d_1)$ $\pi(a_2, d_2)V_1(n)$ and $(\pi(a_1, d_1) + \pi(a_2, d_2) - \pi(a_1, d_1)\pi(a_2, d_2))V_2(n)$, and hence of the whole expression above. In response, the adversary chooses the partition (a, a). Thus, (d, d) and (a, a) are the equilibrium defense and the attack strategies as well.

When both players distribute their resources evenly, the payoff to the adversary is

$$\Pi^A(\mathbf{d}, \mathbf{a}|g, c_A) = -f(n) + \pi(a, d)^2 V_1(n) + (2\pi(a, d) - \pi(a, d)^2)V_2(n) - 2c_A a$$

If $d \geq V_2(n)/c_A$, it is not profitable for the adversary to attack. Thus, with sufficiently low ratio c_D/c_A, the defender distributes his resources evenly and the adversary does not attack. Otherwise, both players compete, choosing optimal levels of attack and defense resources and distributing them evenly.

References

1. Acemoglu, D., A. Malekian, and A. Ozdaglar. 2016. Network security and contagion. *Journal of Economic Theory* 166: 536–585. https://doi.org/10.1016/j.jet.2016.09.009. https://www.sciencedirect.com/science/article/pii/S0022053116300837
2. Alpcan, T., and T. Başar. 2011. Network security. Cambridge: Cambridge University Press. A decision and game-theoretic approach.
3. Arquilla, J., and D. Ronfeldt. 2001. *Networks and Netwars: The Future of Terror, Crime, and Militancy*. Santa Monica, CA: RAND Corporation. https://doi.org/10.7249/MR1382.
4. Aspnes, J., K. Chang, and A. Yampolskiy. 2006. Inoculation strategies for victims of viruses and the sum-of-squares partition problem. *Journal of Computer and System Sciences* 72 (6): 1077–1093. https://doi.org/10.1016/j.jcss.2006.02.003.
5. Baccara, M., and H. Bar-Isaac. 2008. How to organize crime. *The Review of Economic Studies* 75 (4): 1039–1067.

6. Bala, V., and S. Goyal. 2000. A noncooperative model of network formation. *Econometrica* 68 (5): 1181–1229. https://doi.org/10.1111/1468-0262.00155.
7. Ballester, C., A. Calvó-Armengol, and Y. Zenou. 2006. Who's who in networks. Wanted: the key player. *Econometrica* 74 (5): 1403–1417. https://doi.org/10.1111/j.1468-0262.2006.00709.x.
8. Bier, V., S. Oliveros, and L. Samuelson. 2007. Choosing what to protect: Strategic defensive allocation against an unknown attacker. *Journal of Public Economic Theory* 9 (4): 563–587. https://doi.org/10.1111/j.1467-9779.2007.00320.x.
9. Bramoullé, Y., and R. Kranton. 2007. Public goods in networks. *Journal of Economic Theory* 135 (1): 478–494. https://doi.org/10.1016/j.jet.2006.06.006.
10. Cerdeiro, D., M. Dziubiński, and S. Goyal. 2014. Individual security and network design. In *Proceedings of the Fifteenth ACM Conference on Economics and Computation, EC '14*, 205–206. New York, NY, USA: Association for Computing Machinery. https://doi.org/10.1145/2600057.2602894.
11. Choi, S., A. Galeotti, and S. Goyal. 2017. Trading in networks: theory and experiments. *Journal of the European Economic Association* 15 (4): 784–817. https://doi.org/10.1093/jeea/jvw016.
12. Clark, D.J., and K.A. Konrad. 2007. Asymmetric conflict: weakest link against best shot. *Journal of Conflict Resolution* 51 (3): 457–469. https://doi.org/10.1177/0022002707300320.
13. Cunningham, W.H. 1985. Optimal attack and reinforcement of a network. *Journal of the Association for Computing Machinery* 32 (3): 549–561. https://doi.org/10.1145/3828.3829.
14. DeMarzo, P.M., D. Vayanos, and J. Zwiebel. 2003. Persuasion bias, social influence, and unidimensional opinions*. *The Quarterly Journal of Economics* 118 (3): 909–968. https://doi.org/10.1162/00335530360698469.
15. Department of Homeland Security. 2012. Office of infrastructure protection strategic plan: 2012–2016. Technical report
16. Dziubiński, M., and S. Goyal. 2013. Network design and defence. *Games and Economic Behavior* 79: 30–43. https://doi.org/10.1016/j.geb.2012.12.007.
17. Elliott, M., and B. Golub. 2019. A network approach to public goods. *Journal of Political Economy* 127 (2): 730–776. https://doi.org/10.1086/701032.
18. H.O. Eun. 2010. mpact analysis of natural disasters on critical infrastructure, associated industries, and communities. Ph.D. thesis, West Lafayette
19. Farrell, J., and G. Saloner. 1986. Installed base and compatibility: Innovation, product preannouncements, and predation. *The American Economic Review* 76 (5): 940–955.
20. Galeotti, A., S. Goyal, M.O. Jackson, F. Vega-Redondo, and L. Yariv. 2010. Network games. *The Review of Economic Studies* 77 (1): 218–244. https://doi.org/10.1111/j.1467-937X.2009.00570.x.
21. Golub, B., and M.O. Jackson. 2010. Naïve learning in social networks and the wisdom of crowds. *American Economic Journal: Microeconomics* 2 (1): 112–49.
22. Goyal, S. 2007. *Connections: An introduction to the economics of networks*. Princeton, NJ: Princeton University Press.
23. Goyal, S., and A. Vigier. 2014. Attack, defence, and contagion in networks. *The Review of Economic Studies* 81 (4): 1518–1542. https://doi.org/10.1093/restud/rdu013.
24. Grötschel, M., C.L. Monma, and M. Stoer. 1995. Design of survivable networks. In *Network models*, vol. 7. Handbooks in Operations Research and Management Science, 617–672. Amsterdam: North-Holland. https://doi.org/10.1016/S0927-0507(05)80127-6.
25. Jackson, M.O. 2008. *Social and economic networks*. Princeton, NJ: Princeton University Press.
26. Jackson, M.O., and A. Wolinsky. 1996. A strategic model of social and economic networks. *The Journal of Economic Theory* 71 (1): 44–74. https://doi.org/10.1006/jeth.1996.0108.
27. Katz, M.L., Shapiro, C.: Network externalities, competition, and compatibility. The American Economic Review **75**(3), 424–440 (1985). http://www.jstor.org/stable/1814809.
28. K.L. Kliesen. 1994. The economics of natural disaster. The Regional Economist.
29. Kovenock, D., and B. Roberson. 2012. Conflicts with multiple battlefields. In *The Oxford Handbook of the Economics of Peace and Conflict*, ed. M.R. Garfinkel and S. Skaperdas. Oxford: Oxford University Press.

30. Kunreuther, H., and G. Heal. 2003. Interdependent security. *Journal of Risk and Uncertainty* 26 (2/3): 231–249. http://www.jstor.org/stable/41755017.
31. Luft, G. 2005. Pipeline sabotage is terrorist's weapon of choice. *Pipeline and Gas Journal* 232: 42–44.
32. M. Pandey. 2011. Political agitations affect railway service. India today. https://www.indiatoday.in/india/north/story/political-agitations-in-the-country-affect-railway-service-131001-2011-03-26.
33. Smith, J.C. 2008. Preface [Special issue on games, interdiction, and human interaction problems on networks]. *Networks* 52 (3): 109–110. https://doi.org/10.1002/net.20234.
34. Tambe, M. 2012. *Security and game theory: Algorithms, deployed systems, lessons learned*. Cambridge: Cambridge University Press.
35. Vega-Redondo, F. 2007. Complex social networks. In *Econometric Society Monographs*, vol. 44. Cambridge: Cambridge University Press. https://doi.org/10.1017/CBO9780511804052.
36. Zawadowski, A. 2013. Entangled financial systems. *The Review of Financial Studies* 26 (5): 1291–1323. https://doi.org/10.1093/rfs/hht008.
37. Zhu, S., and D. Levinson. 2011. Disruptions in transport networks: A review. In *Network Reliability in Practice*, ed. M.H.X.L. David Levinson and M. Bell, 5–20. New York: Springer.

Macroeconomic and Financial Networks: Review of Some Recent Developments in Parametric and Non-parametric Approaches

Anindya S. Chakrabarti, Anirban Chakraborti, and Suryansh Upmanyu

1 Introduction

Networks are ubiquitous in the natural and social worlds. In social lives, networks of peers (friends, families, colleagues, and so on) influence the decisions that people make and are simultaneously impacted by the decisions made by people. As the world becomes more connected in the digital age, it becomes more and more likely that any decision-making entity will be impacted by the network it belongs to, and it has to gauge the impact of its decision on its neighbors in the network. Evidently, financial markets have become increasingly more complex and entangled with time. Economies have become more interdependent, both within and across countries, due to natural growth processes and globalization. Therefore, an interesting question to ask is "While it is correct that the nature of linkage across economic entities are granular, does it really impact the economic behavior in a substantial manner?" In other words, while the network description of an economy might be more realistic than the earlier homogeneous and representative single entity paradigm, does it provide any new insights into the working of the economy? In this review article, we would argue that indeed the network view goes beyond descriptive accuracy and provides a more complete and useful view of the economic mechanisms.

A. S. Chakrabarti (✉)
Indian Institute of Management, Ahmedabad, Gujarat, India
e-mail: anindyac@iima.ac.in

A. Chakraborti
School of Engineering and Technology, BML Munjal University, Gurugram, Haryana 122413, India
e-mail: anirban@jnu.ac.in

S. Upmanyu
School of Computational and Integrative Sciences, Jawaharlal Nehru University, New Delhi, India

S. Borkotokey et al. (eds.), *Game Theory and Networks*, Indian Statistical Institute Series,
https://doi.org/10.1007/978-981-16-4737-6_14

At the outset, we would like to mention that there are already some very well-written books and reviews on networks in economics. For example, [87, 97] provide applications of networks in the context of microeconomics; [1] reviewed macroeconomics and financial implications of networks. Textbooks on networks are abounded —for different treatments in the domains of economics and finance, interested readers can refer to [66, 86, 96] for a microeconomics-oriented description; [63] considers a more financial econometrics-oriented viewpoint. This list is merely indicative and by no means exhaustive. Though there has been substantial developments in different facets of economic networks, we could not find a single reference that brings both microeconomic and macroeconomic (and financial) networks in one place; this provided us the motivation for writing this review. However, this review should be seen more as a compendium than a stand-alone complete reference. We have also included in this review, some recent developments in the statistical physics literature that found applications in high-dimensional financial data. These are mostly non-parametric in nature, as opposed to more standard parametric economic and finance models. In our view, such non-parametric approaches provide useful and complementary methods of analyzing the underlying network structures.

Network description of a system is more computationally intensive than a representative agent description, owing to the heterogeneity displayed by the constituent parts. Even a couple of decades back, the computational burden was too much to gain reasonable magnification of an economic system into the nodes and linkages between them. A tremendous improvement in computational power in the last few decades and the need to develop more realistic models have contributed to the current state-of-the art knowledge in networks. Across all the topics that we will discuss below, a common thread that ties the significant developments due to this approach is the explicit modeling and analysis of "externalities" or "spillovers". The outcome for a particular node may depend on multiple external factors other than its own decision. Using networks, we are able to analyze the effect of these external factors. It is important to note that the spillover effects might often be of second-order importance, whereas the aggregate dynamics of a system can potentially have first-order importance. As we will emphasize below, that in both macroeconomics and finance, the network architecture at the macro-level does influence the aggregate behavior of the economy as well.

The article is organized as follows: we start by discussing the recent literature on production networks in Sect. 2. Firms typically rely on other firms in the supply chain for inputs for the production process. These dependencies manifest themselves in the form of a production network, where firms across different industries are linked with each other (either directly or indirectly) as sellers or buyers of products that each firm produces. The study of production networks focuses on the role that these connections play in shock transmission across the network, i.e., it studies the impact of an exogenous shock to a particular firm on the rest of the firms in the network to which it is connected directly or indirectly. Next, we explore the connections between different countries in the form of international trade networks. With the advent of globalization, local industries have benefited due to the possibility of cheaper production technologies abroad. Similar to the field of production networks, international trade networks study linkages between economies through the sale and purchase of goods, although in different countries. An important question in this field studies the dynamics of link formations, i.e., how agents decide whether to form or remove a link with other agents. These decisions are taken in a cost-minimizing way and affect the efficiency of the equilibrium outcome and distribution of surplus among market participants.

Section 3 focuses on the propagation of risk through financial networks. Financial networks are formed when there is a transfer of funds (or assets) between agents, either due to a lack of funds for the borrower, or as a means of insurance against future uncertainty and risk. Ever since the 2007–08 financial crisis, there has been a rising interest in the study of the role of networks in transmitting shocks throughout the financial system. This strand of literature focuses on the reasons for the formation of different network structures, and the analysis of shock propagation through them. A complementary approach to study financial network focuses on inferring linkages based on time series properties of multiple financial assets.

In Sect. 4, we analyze social networks. Given the situation, a person may interact with others through different media, thus forming a social network. These kinds of networks can be seen all around us. We find information about our friends, and the friends of our friends, through online social media platforms like Facebook. Interaction with the people in our neighborhood leads to a transmission of information. We rely on our contacts, and online job portals, to search for new employment opportunities. All the above-mentioned situations explore the concept of different forms of social networks based on their use. As people become more connected with the rapid growth in technology, social networks emerge as powerful and useful tools, as a means of communication and information transmission. This section discusses the impacts of social networks on mechanisms including informal risk-sharing and information transmission across economic agents.

Next, we present some empirical work on networks in Sect. 5, where we discuss some recent developments in econometrics-based network approaches to networks, which are mostly parametric in nature. Finally, we end our discussions with non-parametric approaches to networks in Sect. 6.

2 Macroeconomic Networks

We discuss a benchmark model used in production networks, popularized by Ref. [3]. We then discuss a few extensions of this model, followed by the role of production networks in competition policy. For a discussion on recent empirical work on these topics, we refer the reader to Ref. [43].

2.1 Input–Output Networks

The baseline model is a variant of the model developed in Ref. [108]. Following Ref. [3], the model considers a static economy with n granular industries, each producing a distinct good. The production function for the ith industry is assumed to be a constant returns to scale Cobb–Douglas function:

$$y_i = z_i \tau_i h_i^{\alpha_i} \prod_{j=1}^{n} x_{ij}^{a_{ij}}, \tag{1}$$

where h_i denotes the amount of labor hired by industry i, x_{ij} is the quantity of good j used to produce good i, α_i gives the share of labor in industry i's production technology, $a_{ij} \geq 0$ is a measure of the importance of good j as an input for good i, z_i is a Hicks-neutral productivity shock, and τ_i is a normalization constant. Similarly, the economy consists of a representative household providing an inelastic supply of 1 unit of labor. The utility function of the representative household over the n goods produced by the industries is given by

$$u(c_1, \ldots, c_n) = \sum_{i=1}^{n} \beta_i log(c_i/\beta_i), \tag{2}$$

where c_i is the amount of good i consumed and β_i gives good i's share in the utility function of the household. In equilibrium, quantities, and prices are such that firms maximize their profits conditional on prices and wages, households maximize their utility, and all markets clear.

In this model, $\mathbf{A} = [a_{ij}]$ denotes the input–output matrix of this economy, where a_{ij} is defined above. The Domar Weight of an industry is given by the industry's sales as a fraction of GDP ($\lambda_i = p_i y_i / GDP$). Finally the Leontief Inverse of this economy is denoted by $\mathbf{L} = (\mathbf{I} - \mathbf{A})^{-1}$. Given that the spectral radius of $\mathbf{A}$ is less than 1, an element of $\mathbf{L}$ - l_{ij}—shows the importance of industry j both as a direct supplier and an indirect supplier (as a supplier to i's supplier, and so on) to i.

After solving for equilibrium, Ref. [3] derived the following 2 results:

Theorem 1 *The log output of industry i (where $i \in \{1, \ldots, n\}$) is given by*

$$log(y_i) = \sum_{j=1}^{n} l_{ij} \epsilon_j + \delta_i, \tag{3}$$

where $\epsilon_j = log(z_j)$ denotes a productivity shock to industry j, and δ_i is some constant which is independent of shocks.

and

Theorem 2 *The real value added aggregated across the industries is given by*

$$log(GDP) = \sum_{i=1}^{n} \lambda_i \epsilon_i, \tag{4}$$

where

$$\lambda_i = \sum_{j=1}^{n} \beta_j l_{ji}. \tag{5}$$

These two theorems imply that shocks can transmit through input–output linkages across different industries. Since the matrix in consideration here is the Leontief inverse (which is dependent on the input–output matrix), it implies that along with direct effects, even the indirect effects of shocks matter across the network.

Another observation is that any shock to industry i will be propagated downstream to those industries which require i's good as an input in their own production. This effect is further propagated throughout the network. Theorem 2 also implies that the Domar Weights are a sufficient statistic for measuring how idiosyncratic shocks to different industries affect aggregate output. The downstream propagation of shocks from an industry to its customers (direct and indirect) means that the economy is more sensitive to shocks impacting industries which are important input suppliers.

2.1.1 Demand-Side Shocks

To incorporate demand-side shocks, Ref. [1] introduces government purchases for a good i as g_i, which is exogenously given. A change in government spending in this model is similar to an exogenous shock to the demand for the goods of individual industries. In this model, after solving for equilibrium output of each industry, we see that the impact of a productivity shock has an upstream effect, i.e., a shock to firm i propagates through the network by affecting i's suppliers, and further their suppliers and so on. This happens because if industry i is affected by a positive demand shock, it would increase i's input demand. This would generate a rise in demand for the products of i's suppliers, and so on.

2.2 Extensions

2.2.1 Relaxing the Cobb–Douglas Assumption

The baseline model assumes that production technologies of firms take a Cobb–Douglas form. This implies that the realization of shocks does not affect an industry's expense on inputs as a fraction of its sales. Some papers, like Refs. [21, 42] focus on nested CES production functions. They find that—up to a first-order approximation—

when elasticities of substitution are different from 1, there are 2 propagation channels for a productivity shock.

- A shock to industry i affects other industries through a downstream propagation of the shock.
- Productivity shocks can also lead to a reallocation of resources across industries which is affected by the elasticities of substitution between inputs.

2.2.2 Hulten's Theorem

Theorem 2 stated that Domar Weights are a sufficient statistic for how shocks to industries affect aggregate output. Hulten's theorem Ref. [95] makes a more general statement: In any efficient economy, the impact of a productivity shock to industry i (denoted by z_i) on aggregate output is equal to i's Domar weight, up to a first-order approximation. This can be written as

$$\frac{dlog(GDP)}{dlog(z_i)} = \lambda_i. \tag{6}$$

An important consequence of Hulten's theorem is that the effect on the economy of an idiosyncratic shock only depends on its size, and not on its location in the production network. References [41, 73] use this theorem to analyze the macroeconomic implications of idiosyncratic shocks to a production network. Generally, positive shocks to industry i would impact aggregate output in two ways. First, it leads to an outward shift of the production possibility frontier of the economy. Second, it may reallocate resources across different industries. If the original allocation is efficient, the aggregate effect due to the latter channel is second order and can be ignored in a first-order approximation. This would then imply that if economies are inefficient, then Hulten's theorem may not hold as the reallocation effect may be significant. Since Hulten's theorem focusses only on first-order approximations, if the focus is on second-order effects in a general economy, then it is seen that these second-order effects can be significant, and they depend on the structure of the network [22].

2.2.3 Frictions and Market Imperfections

The baseline model described previously assumes perfect competition. Reference [32] study the impact of productivity shocks by introducing exogenous wedges between marginal revenue and marginal cost of firms through markups. Their finding is that resource misallocation and its resulting inefficiency depends on the distribution of the firms' markups. Distortions are also studied by Ref. [23] in the form of CES production functions with exogenous wedges. They show that the first-order impact of productivity shocks in the network can be decomposed into two terms: (a) a term that accounts for the shocks' pure technology effect and (b) another term that accounts for changes in the economy's allocative efficiency. They also show that if we relax the assumption of Cobb–Douglas technology, the second effect could be substantial.

Reference [89] relaxes the perfect competition assumption, introducing oligopolistic markets in a model of production networks. In this model, shocks affect both prices and markups as the competitiveness of firms of the same industry changes in the network. Here, changes in market concentration lead to changes in the demand of industries for intermediate inputs, leading to an upstream transmission of shocks that would be absent if the markups were exogenously given.

2.2.4 Endogenous Production Networks

In the discussion above, the production networks were taken as exogenously given. One extension of the baseline model explicitly models the link formation decisions of the nodes/agents, thus making the network endogenous. Reference [14] develop a model using preferential attachment as the basis of link-creation between firms. Preferential attachment means that new edges in a network are more likely to be formed with nodes which already have more edges. Reference [44] modify the friendship model of Ref. [98] to form an industry-level network formation model. In this type of model, existing links between firms are used to search for new links to provide inputs for production. This model shows that a higher proximity in the network raises the likelihood that a firm will adopt another firm's product to use as an input in its own production.

Link formation incentives are introduced by Ref. [118] into a dynamic network formation model in which the set of suppliers to a firm keeps evolving and firms have to optimally choose one input from this randomly evolving set. Reference [2] consider an alternative model where firms in each industry select their input suppliers as a subset of other industries in the economy, knowing that each input combination would lead to a different CRS production technology.

2.3 Business Cycle from I-O Networks

We now discuss whether idiosyncratic shocks can build up in aggregate in the economy through the network structure. According to Lucas, the standard deviation of aggregate fluctuations is proportional to $\frac{1}{\sqrt{n}}$. This means that as the number of industries increases, idiosyncratic shocks get dissipated across the network, having a negligible effect on the economy. But as shown in Ref. [3], this argument breaks down if sectoral Domar weights show significant heterogeneity. These observations are also discussed in Ref. [73] through his granularity hypothesis, which states that in the presence of significant heterogeneity at the micro level, the grains of economic activity (comprised of firms or disaggregated industries) can matter for the behavior of macroeconomic aggregates. Specifically, he shows that even if there is a high level of disaggregation, aggregate volatility could be much larger than what Lucas hypothesized if the Domar weights have a heavy-tailed distribution.

Reference [3] then goes on to discuss that if the industries acting as input suppliers are sufficiently heterogeneous, they can generate very high levels of aggregate volatility, contrary to Lucas's hypothesis. This effect depends on the Leontief inverse and Bonacich centrality of the nodes of the production network. If the Bonacich centrality is high, it means that an industry is an important supplier to other central industries. Therefore, a shock to such an industry might not die down and could lead to substantial aggregate fluctuations in the economy.

Economies with heterogeneous production networks can show significant comovement as well. Reference [43] show that even if two economies have identical Domar weights distributions, the economy which is more interconnected will have a higher average pairwise correlation of output and it will be less volatile. This happens because the interconnected economy will have industries which are more diversified with respect to upstream risk from suppliers to other industries in the economy.

2.4 Policy Impact on Production Network

We now discuss the importance of considering production networks for analysis in competition policy [88]. For this, we define market power and explain how it is measured by competition authorities. A firm's market power denotes its ability to increase prices above marginal costs to raise its profits. The amount by which price can be raised over the marginal cost is referred to as a markup. As markups can be difficult to measure (since there is no reliable way to measure marginal costs), competition authorities work with concentration ratios instead to infer firms' market power. A popular way to measure concentration ratios is given by the Herfindahl–Hirschmann Index (HHI). This index is given as the sum of firms' squared market shares. To calculate HHI, only the sales data for different firms is required. Since this is readily available, the index is easy to calculate.

HHI is a good indicator of market power as it is directly proportional to Lerner's index (which is the difference in price and marginal cost of a firm, divided by the price). A higher level of concentration (denoted by a high value of HHI) would then imply a high Lerner's index, which implies that firms have the ability to increase the markup by a large margin. This can only happen if competition in the market is low, leading to some firms having significantly high market power. One specific use of HHI by competition agencies is to assess the chances of a merger being anti-competitive. For this exercise, the difference between pre-merger HHI and post-merger HHI is analyzed. If there is a significant increase in HHI, it would imply that the merger is anti-competitive, substantially increasing the market power for the newly formed firm. As discussed by Refs. [71, 112, 130], in Cournot competition (and some other market forms), such horizontal mergers can harm consumers by leading to a rise in prices and reduction in output.

Competition authorities do not usually consider the impact of a merger outside of the market where the merger takes place. Such partial equilibrium analysis could lead to an underestimation of the anti-competitive behavior of firms. This is where network analysis can play a role. Theory suggests that if a merger takes place, it would result in an increase in prices downstream because of a rise in input costs. Simultaneously, the reduction in competition in the particular market would lead to a reduction in quantities (in Cournot competition), which would imply a decrease in demand for inputs from upstream markets.

2.5 *International Trade Networks*

The discussion till now focused only on the linkages between industries located within the same economy. But in this era of globalization, profit-maximizing firms looking for cheaper production technologies also have the option of forming connections with industries located abroad. This leads to the formation of international trade networks. Here we discuss the model developed in Ref. [114]. This model has formed an important base for subsequent research in the field of international trade networks (some of which are discussed in Ref. [30]).

Consider an economy where firms show heterogeneity in terms of productivity and quality. These firms can buy inputs from multiple suppliers and sell the produced good either to consumers, or to other firms (as inputs). The production function for a firm i is

$$y_i = \kappa z_i l_i^{\alpha} \left(\left(\sum_{k \in S_i} (\phi_{ki} \nu_{ki})^{(\sigma-1)/\sigma} \right)^{\frac{\sigma}{(\sigma-1)}} \right)^{1-\alpha}, \tag{7}$$

where y_i is output, z_i is productivity, l_i is labor, α is the labor share, $\kappa > 0$ is a constant, ν_{ki} is the amount of inputs bought from the kth firm, and S_i is the set of firms supplying to the ith firm. Here, $\sigma > 1$ denotes the elasticity of substitution across i's suppliers. ϕ_{ki} is a measure of shift in demand that captures the idea that firms use different production technologies, affecting their demand from a particular firm. The input price index for a particular good is given by

$$P_i^{1-\sigma} - \sum_{k \in S_i} (p_{ki}/\phi_{ki})^{1-\sigma}, \tag{8}$$

where p_{ki} denotes the price that the kth supplier charges the ith firm for its product. A firm's marginal cost of production is defined as $c_i = \frac{w^{\alpha} P_i^{1-\alpha}}{z_i}$ where w denotes the wage rate. The sales of a firm are given by

$$s_i = \sum_{j \in C_i} \left(\frac{\phi_{ij}}{p_{ij}} \right)^{\sigma-1} P_j^{\sigma-1} M_j + F_i, \tag{9}$$

where M_j denotes the amount of intermediate purchases of firm j and F_i gives the amount of sales pertaining to final demand.

This model gives us two important observations. First, the marginal cost of firm i increases as the marginal costs of its suppliers increase, through P_i. This implies that a change in firm productivity (z_k) will affect marginal costs of all firms located downstream for which k acts as a direct or indirect supplier. Similarly, the set of suppliers S_i will also impact firms' production costs. Production costs are affected by international trade costs because they impact the set of suppliers, and they affect the cost of procuring from a supplier k through the price p_{ki}. Second, it is observed that the ith firm's sales s_i depend on two important factors: the set of customers C_i, and the amount which is sold to each customer (which depends on the price p_{ij} and the effective demand of the customer $(\phi_{ij} P_j)^{\sigma-1} M_j$). A change in trade costs can lead to a change in both these factors. For example, a rise in tariffs can either reduce a customer's demand or remove it from the set of customers entirely.

2.6 Matching in Trade Networks

2.6.1 Bipartite Networks

One section of the literature on international trade networks models buyer–seller relationships using bipartite graphs. For example, Ref. [31] use bipartite networks where one group of firms act as buyers (the set S_i for such firms is empty) and the other group acts as suppliers (for whom set C_i is empty). This model assumes full information for all agents, and costly link formation. As with the general theory on networks, this model also suggests that the distribution of customers per firm can be well approximated using a Pareto distribution.

This model shows a unique feature of negative degree assortativity. Low productivity suppliers are more likely to connect with high productivity buyers as only these buyers are able to incur the relatively high cost of linking with a low productivity seller. Similarly, high productivity sellers are more likely to connect with low productivity buyers. This also implies that high productivity firms are also highly connected.

In this model, high relationship-specific costs can lead to a reduction in welfare in the economy. These costs dampen trade flows and therefore reduce consumers' income. This happens because higher relation-specific costs make link formation more expensive and result in fewer links between firms. Even though having more suppliers is beneficial for each firm, these high production costs prevent them from doing so. This could reduce welfare if the firm is not able to optimally specialize in production due to a lack of suppliers. The resulting higher production costs would lead to an increase in consumer prices and subsequently reduced real wages for consumers. The introduction of tariffs also leads to adverse impacts on the economy. Tariffs increase the costs of procuring inputs from foreign suppliers and could lead to the breaking of links because of this rise in costs. Reference [28] relaxes the full information assumption by modeling costly information acquisition for firms. Therefore, an exporter engages in both production of a good, and search for consumers to link with.

2.6.2 Networks and Outsourcing

Another section of the literature aims at modeling the full production network. Reference [69] model such a network to analyze the impact of outsourcing. Here, a firm's production technology is such that its labor input and other intermediate inputs are perfectly substitutable for producing a good. Sellers meet potential buyers at random and buyers then optimally choose whether or not to outsource production. The probability of outsourcing is higher if own labor is costly (high wages) or foreign firms have low costs (due to better technology).

This paper assumes that labor is heterogeneous and consists of production and non-production workers, where only production workers can be outsourced. One observation is that trade liberalization increases the likelihood of goods getting outsourced. This is because if trade costs are reduced, it increases the probability of finding a good match abroad. It is then theoretically possible that trade liberalization can reduce real wages for production workers and increase real wages for non-production workers. This would happen since non-production workers would benefit from cheaper goods, whereas goods produced by production workers are likely to be outsourced due to liberalization. This paper suggests that the skill premium in the economy could be affected by the network structure.

Reference [118] relaxes the assumption that labor and inputs are perfect substitutes. Here firms meet possible matches randomly and decide whether to form a link with some other firm or not. Therefore, firms may not always get to match with the lowest cost supplier in the market. In equilibrium, it is seen that the distribution of customer firms asymptotically tends toward a power law distribution.

2.7 *Dynamic Networks*

2.7.1 Full Information

Reference [107] considers a model similar to the benchmark model presented above but extends it from a static to a dynamic setting. Using a Poisson process, firms are selected at random to decide the possible linking or dropping of matches in each period. Firms have rational expectations about the future and establish a link only if the relationship is profitable in the future. Therefore, a match may happen even if no profits are realized in the current period. Calibration of this model leads to very different shock propagation patterns compared to the static setting.

2.7.2 Search Frictions

Reference [48] develops a model where firms search for potential customers. This allows for a geographical dimension in the network model which is missing from Ref. [107]. As in the friendship model developed in Ref. [98], existing links can lower the cost of searching for new links for a firm. This implies that information flows faster through the network channels already established. Chaney's model then predicts that superstar firms will emerge in the economy, where few firms with already high number of connections grow ever larger.

Reference [68] deviate from the existing literature by allowing both sellers and buyers to search in the market. This paper models trade between producers (acting as exporters) and retailers (acting as importers), therefore allowing for many-to-many matching. In this model, the chances of a firm forming a link with a retailer are affected by multiple factors such as search intensity and the existing links of a firm. The latter feature of the model leads to the generation of fat-tailed in-degree and out-degree distributions.

2.7.3 Learning in Trade Networks

In Ref. [67], since perfect information is not available, exporters and importers engage in costly searching to form new links. Also, the firm can learn from its interaction with other firms. When a firm forms a link, it receives an imperfect signal about the attractiveness of its product in the market. The firm updates its belief about the potential of profits in a Bayesian manner, adjusting its search intensity accordingly. This implies that firms learn about their attractiveness over time. Popular firms are more likely to search more intensively, while less-popular firms will also search less.

Another type of information friction is present when firms cannot observe the productivity of a potential partner perfectly. This is analyzed in Ref. [115] where importers have the ability to learn about the reliability of potential supplier firms. Suppliers can either shirk or comply. The importer cannot directly determine the type of the supplier as contract enforcement forces myopic firms to comply with an exogenous probability. In every period after the link is formed, the importer observes a noisy signal about whether the supplier exerts effort or not. If it does, then it increases the likelihood that the supplier is reliable. If the exporter shirks, then the relationship is terminated.

There are two important observations of this model. First, the volume of trade conducted by a buyer and seller pair increases with time, since expected costs decrease as the buyer becomes more assured about the reliability of the supplier. Therefore lower prices lead to more sales and rise in intermediate input demand. Second, the likelihood of the survival of a relationship improves with time since unreliable suppliers reveal themselves early in the relationship. An important observation of this model is that learning leads to significantly higher aggregate trade than a scenario where learning does not take place.

3 Systemic Risk and Contagion in Financial Systems

In this section, we will start with a discussion of the "robust yet fragile" property commonly shown by financial networks. This will be followed by a brief explanation of the different sources of systemic risk, followed by a description of some popular measures to quantify this risk. Finally, we will discuss the increasingly important role of macroprudential stress testing for the stability of the financial network system.

3.1 Robust-Yet-Fragile Properties

We consider a simple model developed in Ref. [76] where a financial network consists of n banks forming links randomly through unsecured lending and borrowing activities. In the network, every node represents a particular bank, and each edge in the network denotes the bilateral unsecured interbank exposures between two banks. This network consists of directed edges, signifying that both lending and borrowing activities take place between banks. In the model, a bank i has j_i interbank lending links and k_i interbank borrowing links. The connectivity between the banks is given by the average degree of the interbank network which is denoted by z.

The balance sheet of a typical bank in the network looks as follows. The total assets of a particular bank i are given by its unsecured interbank assets, IA_i, and illiquid external assets, EA_i. It is assumed that a bank's total amount of interbank assets are spread evenly across its lending links.

Every interbank asset of bank i would be a liability for some other bank j. Therefore, unsecured interbank liabilities of bank i, IL_i, will be endogenously determined within the network. Each bank also has other liabilities given by exogenous customer deposits, EL_i. For each bank i in the network, the solvency condition is given by

$$(1-\phi)IA_i + EA_i - IL_i - EL_i > 0, \tag{10}$$

where ϕ denotes the fraction of banks which have taken loans from bank i but have defaulted. There is an implicit zero recovery assumption, which implies that if a counterparty defaults, all the assets of bank i held by that counterparty are lost and bank i is unable to recover anything. The solvency condition above could be simplified as $\phi < K_i/IA_i$, where $K_i = IA_i + EA_i - IL_i - EL_i$ gives us the capital buffer of bank i.

Now assume that all banks are identical. This would imply that $j_i = k_i = z$ for all banks. If some counterparty to bank i goes into default, then $\phi = i/z$ as i's assets are uniformly distributed among its counterparties. Contagion would spread beyond the first bank if another neighboring bank exists for which $z < IA/K$. This model also highlights the situations when systemic default can take place. If capital ratios are low or unsecured interbank lending is high, then it is more likely that systemic default occurs. The above equation then suggests that there exists a tipping point in the network. If the above equation is satisfied and z is sufficiently large, then an individual bank's default could induce all the other banks in the network to subsequently default as well. On the other hand, if the above condition is violated, then it implies that the bank's default has no systemic implications.

Reference [75] simulate the model with the assumption that links in the network are distributed uniformly, where the probability that a link exists between two banks is independently given by the probability p (a Poisson network). Their aim is to analyze the impact of the failure of a bank on the whole network. They specifically study (i) the probability of contagion across the network and (ii) the proportion of the network which is impacted by contagion, given different values of z (which gives the average connectivity of the network). Simulation results show that an increase in connectivity z does not have a monotonic effect on the likelihood that system-wide contagion will occur in the interbank network, since benefits of sharing risk eventually dominate the cost of risk-spreading. But even though the probability of contagion reduces as z increases, its impact is felt throughout the network. Therefore, the system exhibits a robust-yet-fragile tendency.

3.2 Sources of Contagion

3.2.1 Default Contagion

The line of work focusing on default contagion was first explored by Refs. [10, 70, 145]. Default contagion is described as follows. An exogenous shock to bank i's asset value could reduce its net worth and reduce its ability to repay its lenders. If the shock is large enough, it could lead to a default by bank i. If the loss due to bank i's default is large enough, it could lead to bank i's lenders defaulting as well, and so on. Recent work was accelerated after the emergence of the 2008 financial crisis [4, 74, 75, 111].

3.2.2 Distress Contagion

One stream of work explores distress contagion, where financial distress can spread even though an actual default by a borrower may not take place [24, 137]. This could happen if the market value of bank i declines due to a reduction in its net worth, even though it remains solvent. Even if a default does not happen, this could lead to a loss for bank j if the value of i's obligation to j is "marked to market".

3.2.3 Common Asset Contagion

Another line of work explores common asset contagion and fire sales. Banks can be connected indirectly if they have investments in the same assets. If a shock leads to change in asset prices, a bank might sell a significant amount of this asset so that its price falls significantly. If this asset was held by other banks too, they would be affected by the secondary shock (due to the sale of asset) as well, causing them to sell the asset. This would trigger a devaluation spiral [40, 59, 100].

3.2.4 Funding Liquidity Contagion

Contagion could also spread from the liability side. Institutions may be affected adversely if creditors start hoarding liquidity [6, 72, 74, 77]. This could lead to a funding run if a liquidity shock occurs unexpectedly, as in Ref. [58].

3.3 Systemic Risk: Measurements and Impact

3.3.1 MES and SES

Suppose there are N financial firms in the economy. Let r_{it} denote the return on firm i's equity in time period t. Therefore, we can calculate market return or index return as the weighted average of the asset returns across all the individual firms, $r_{mt} = \sum_{i=1}^{N} w_{it} r_{it}$. Here, the weight w_{it} assigned to each return series signifies the value of relative market capitalization of the ith firm. MES is calculated as each individual firm's marginal contribution to systemic risk, which in turn is evaluated by the system's expected shortfall, ES. This measure was introduced in Ref. [7].

Given the available information till time $t-1$, the ES in time period t is calculated as

$$ES_{mt}(C) = E_{t-1}(r_{mt}|r_{mt} < C) = \sum_{i=1}^{N} w_{it} E_{t-1}(r_{it}|r_{mt} < C), \tag{11}$$

where C is some threshold value (Ref. [7] takes $C = -VaR_\alpha$, where VaR_α is defined as the largest amount that an institution loses with confidence $1-\alpha$, that is, $P(r_{it} < -VaR_\alpha) = \alpha$). The MES is then given by calculating the partial derivative of the system's ES with respect to firm i's relative market capitalization in the economy [131]:

$$MES_{it}(C) = \frac{\partial ES_{mt}(C)}{w_{it}} = E_{t-1}(r_{it}|r_{mt} < C). \tag{12}$$

Intuitively, MES is a measure of the increase in risk due to an infinitesimal change in the relative market capitalization of the ith firm. The SES modifies this measure and signifies the level by which the equity of a bank can drop below a particular threshold (which is given as k, a fraction of the bank's assets) during a crisis, conditional on aggregate capital being less than k times the value of aggregate assets:

$$\frac{SES_{it}}{W_{it}} = kL_{it} - 1 - E_{t-1}\left(r_{it} | \sum_{i=1}^{N} W_{it} < k \sum_{i=1}^{N} A_{it}\right). \tag{13}$$

Here A_{it} gives the total assets of firm i in time t, W_{it} denotes the market value of firm i's equity, and L_{it} gives a measure of the firm's leverage, which is equal to A_{it}/W_{it}.

3.3.2 SRISK

$SRISK$ was introduced in Refs. [5, 39]. It extends the MES to allow for the consideration of a financial firm's size and liabilities. $SRISK$ is defined as a firm's expected shortfall in capital, when the entire system is affected by a crisis. When a firm has a larger capital shortfall, it has a higher likelihood of contributing to a financial crisis. Therefore, such a firm is systemically riskier. $SRISK$ is calculated as

$$SRISK_{it} = max\,[0, k(D_{it} + (1 - LRMES_{it})W_{it}) - (1 - LRMES_{it})W_{it}], \tag{14}$$

where k denotes the prudential capital ratio, $LRMES$ is defined as the long-run MES and the book value of aggregate liabilities is denoted by D_{it}. Intuitively, $LRMES$ provides us a measure of the expected future drop in a firm's equity value, conditional on the market falling below a specific threshold within a given time period (taken as 6 months here). Substituting $L_{it} = (D_{it} + W_{it})/W_{it}$, the above expression can be modified as:

$$SRISK_{it} = max\,[0, [kL_{it} - 1 + (1 - k)LRMES_{it}]W_{it}]. \tag{15}$$

3.3.3 CoVaR

$CoVaR$ is a systemic risk measure given by Ref. [8]. Let $CoVaR_{it}^{m|C(r_{it})}$ be a term related to the value at risk (VaR) of the realized market return, conditional on the observation of some event for firm i (denoted by $C(r_{it})$):

$$P(r_{mt} \leq CoVaR_{it}^{m|C(r_{it})} | C(r_{it}) = \alpha. \tag{16}$$

$CoVaR$ for the ith firm is then calculated as the difference of two terms: (i) the VaR of the entire system when the ith firm is in financial distress, and (ii) the VaR of the system when firm i is at the median state. Here, we can define distress in multiple ways depending on the definition of $C(r_{it})$. Reference [8] assumes that the loss is equal to its VaR and subsequently uses a quantile regression approach to analyze this situation:

$$CoVaR_{it}(\alpha) = CoVaR_{it}^{m|r_{it}=VaR_{it}(\alpha)} - CoVaR_{it}^{m|r_{it}=Median(r_{it})}. \tag{17}$$

3.3.4 Debtrank

Reference [25] developed a measure called DebtRank to find systemically important financial institutions. Debtrank is similar to PageRank by Google, and it is an eigenvector centrality measure which can be used to assess the influence of a bank on the interbank network as a whole. Suppose a bank i is connected to other highly connected banks in the network, then bank i would have a higher centrality. Therefore, a bank would have a higher DebtRank value when it is connected to other banks which have high values of DebtRank themselves.

3.4 Macroprudential Stress Testing

Central banks around the world regularly conduct stress tests aiming to measure the robustness of financial firms (e.g., banks) to adverse shocks. But it is also necessary to analyze the impact of network contagion as well in potentially amplifying systemic risk. As mentioned before, evidence suggests that most of the observed interbank networks show a core-periphery structure [51, 54, 106]. Such network structures showcase the robust-yet-fragile tendency described before.

Reference [70] describe a model which—under certain assumptions- proves the existence of a unique clearing vector after at least one bank in the network defaults. A particular assumption of the EisenbergNoe model is the absence of deadweight losses after a bank defaults. This leads to the clearing mechanism redistributing existing assets among the surviving banks with the aim of maximizing payments. Reference [127] relax this assumption, allowing for default costs after the failure of banks. Their model leads to multiple clearing vectors which includes a Pareto-dominant clearing vector. This clearing vector is found by allowing banks to fail one by one till there is only a single solvent bank remaining.

Reference [76] states that most of the models used for macroprudential stress testing mainly focus on post-default contagion. Recent developments in this literature show extensions where other sources of risk are also considered. For example, Ref. [99] study liquidity risk and contagion, focusing on the cash-flow constraint of banks. Reference [52] explore the theory of fire sales. In their model, portfolios are constrained by leverage or capital considerations, resulting in shocks to asset values leading to a rapid sale of the asset. The fire sale that follows leads to further deleveraging. Reference [53] attempt to analyze counterparty credit risk, liquidity hoarding, and fire sales in the same framework.

An important question for stress testing in the future could be to analyze the impact of contagion, not just on financial networks, but on the real economy as well. During the 2008 financial crisis, a debt overhang and reduction in credit supply led to both a rise in unemployment and a significant decrease in GDP growth rates. Reference [88] explore input–output networks as an alternative channel for shock propagation throughout the economy.

4 Social Networks

In this section, we first discuss the applications of social networks in labor markets, specifically focusing on job referrals by individuals for employment opportunities. We then explore different network models of information flows among people. The last part of this section focuses on the importance of social networks in the domain of informal risk sharing. For a detailed discussion on labor markets and social networks, see Ref. [140]. For information flows and risk sharing, see Ref. [37].

4.1 Labor Markets and Referrals

Job seekers frequently rely on their social networks to obtain information on possible employment prospects and recommend them for job opportunities, either formally or informally. Here, we discuss the role that social networks play for recommendations (or referrals) in the labor market. The literature presents various types of models to analyze referrals. The model of *asymmetric information* (analyzed by Refs. [45, 116]) argues that referrals reduce the information asymmetry between the candidate and employer about the candidate's quality. A person tends to have like-minded people in his network, so it is likely that high-quality workers will provide referrals for people who are themselves highly skilled. This would act as a signal for a prospective employer to gauge the quality of a candidate. Moreover, since the referrer's reputation is also at stake, he would only refer good quality candidates.

The model of *symmetric uncertainty* suggests that both the employer and job-candidate are uncertain about their match. Therefore, referrals can provide better information to both parties compared to other employment channels. This model is explored in Refs. [65, 78, 135]. According to this theory, employers would be more willing to provide referred hires with a higher wage subject on getting hired (since referrals would provide a better indication of the candidate's quality). Additionally, as match quality becomes apparent over time, referred hires would have lower separation rates than non-referred hires.

Another set of models discussed in Refs. [90, 101] focuses on the *moral hazard* aspect of referrals. The moral hazard interpretation explains that employers may not be able to monitor a newly hired worker properly. In this case, the referrer can act as a monitor, since the performance of the new worker affects his reputation as well. This allows the employer to motivate better performance from the worker. Empirical work on referrals studies the impact of this hiring channel on hiring probabilities, wages, and employee performance, compared to other hiring channels. References [93, 94] study the employee side of the market and find that hiring probabilities are higher when candidates use personal contacts rather than using other formal hiring channels. Similarly, using data obtained from employers' referral systems, Ref. [38] find that even though only 6 percent candidates use referrals, they make up 30 percent of all eventual hires in the dataset analyzed. References [26, 91, 92, 132] find that referrals lead to a greater likelihood of getting high wages compared to other hiring channels. References [38, 65, 92] state that candidates coming through the referral channel also tend to stay longer at their jobs. This would imply that referrals are associated with lower turnover rates. There is also some heterogeneity of referral effects. Referrals are more likely to be used by the younger demographic, ethnic and racial minorities, and individuals with lower socio-economic status. But this does not imply that the probability of getting hired for these groups is high as well. A study by Ref. [93] observes that conditional on usage of referrals, probability of getting hired is higher for whites than for blacks. Reference [27] find a similar result when comparing women to men.

The impact of business cycles on social networks is still largely unexplored. Local labor market conditions depend on business cycles, subsequently affecting the formation of social networks and their use in providing referrals. The change in composition of employed and unemployed contacts in an individual's network at different phases of a business cycle would impact the individual's decisions about the people he forms links with and the use of his network for different job prospects. Papers which have started addressing these questions include the works of Refs. [78, 80, 82, 105]. The study of social networks has been greatly hindered by a lack of good quality data. This has changed in recent years with the availability of social media and professional networking data. Recent studies like Refs. [15, 83] use data from Facebook to study the impact of social networks in decisions related to housing investment and employment prospects respectively. References [20, 110] utilize data obtained from online search portals to analyze employment prospects of workers. The use of referrals to improve a person's employment outcomes leads to a role for government policy as well. Even though referrals lead to many benefits, for both job candidates and employers, they also have some disadvantages. Referrals can lead to rising inequalities between different socio-economic groups as people tend to refer like-minded individuals. Reference [38] find significant evidence for assortative matching between referrers and referred individuals based on race, gender, and education in their dataset. This suggests a role for government intervention.

4.2 *Information Flows*

Information transmission is an important aspect of many programs, whether it is the marketing of a new product by a consumer brand, or a policy intervention by the local government. An important question in information transmission is the selection of groups (or specifically, individuals) to be targeted (or seeded) so that the program has maximum impact. Reference [29] observe that if peer farmers are provided incentives to transfer information regarding a new technology, the technology's adoption is 10–14 percent greater relative to a control group. Similarly, Ref. [12] provide evidence of higher adoption of a new product or trend if viral campaigns are used.

This implies that whether the use of networks can improve information transmission is important and has relevance in multiple areas. Networks can potentially provide huge benefits, but they incur a cost as well. Identifying whether networks do provide substantial advantage over traditional information transmission channels, and the subsequent identification of individuals (or seeds) can be an expensive and time-consuming process. Reference [18] show that a microfinance scheme's adoption depends heavily on the initial individuals chosen as seeds. Reference [9] find a similar result for the transfer of messages in the environment of an online social network.

We now consider different models which are used to study information transmission in networks. In a *viral process*, an informed individual (infected node) in a network transmits information to all the nodes it is connected to. This form of transmission is deterministic and irreversible and is called diffusion. This is a fairly simple model and thus, may not be of much use in explaining real-world phenomena. In *aggregation models*, the object of study is the change in intensity of beliefs as information spreads through a social network. The DeGroot model (used by Ref. [56]) is one such aggregation model. In this model, everyone receives signals at the initial stage. In subsequent stages, communication takes place among individuals and their beliefs are updated by averaging their and their neighbors' beliefs. This goes on until a steady state is reached where beliefs are not changed further and everyone reaches a consensus. One important fact about this model is that the consensus in the steady state is a weighted average of initial opinions, where the weights are given by the eigenvector centrality of each individual (Ref. [84]). In other words, a person is influential if he is connected to other influential individuals, and such a person would have a large impact on the final consensus which is reached.

The DeGroot model analyzes the speed of convergence toward a consensus in a given social network. Reference [85] discuss that networks showing homophily show a very slow rate of convergence. In such networks, people similar to each other reach a consensus within themselves first, and only then start moving toward a group consensus. References [17, 113] document this type of behavior when networks consist of different castes, religions and ethnicities. This model also provides an insight on good candidates to select as seeds. If a policymaker can persuade individuals to spread the correct information and he cares about the speed of transmission, then individuals with higher eigenvector centrality would be better candidates to act as seeds.

Reference [16] generalizes the DeGroot model to merge both diffusion and aggregation. Initially, the chosen seeds transmit information to inactive neighbors through the diffusion process. In each subsequent period, this process continues as active nodes spread information to their inactive neighbors. Simultaneously, the process of aggregation takes place in each period as individuals update their beliefs as in the DeGroot model. Therefore, this model leads to the existence of domains of influence, where an individual is influenced more by seeds which are closer to him (in a network sense).

Other models also discuss the role of strategic interactions in the process of information transmission. Reference [129] explores the presence of strategic complementarities in networks. These can accelerate diffusion. For example, a person who hears about Whatsapp for the first time is more likely to use it if he knows that his friends and relatives use it as well. Reference [79] study strategic substitutes, where adoption is less likely if more people in one's network are adopting.

4.3 Risk Sharing

In communities where formal insurance is not prevalent, the role of social networks is important to reduce risk through informal insurance channels. This is seen in multiple countries, as documented in References [19, 50, 128], among others.

Informal insurance allows risk-averse individuals with uncertain future incomes to opt for state-contingent monetary transfers which leads to a Pareto improvement. In the benchmark model [147], individuals are uncertain about their future incomes, there is perfect information about individuals' characteristics, and all agents can commit to contracts. In this case, the equilibrium consumption is distributed with the aim of maximizing expected utilitarian welfare along with Pareto weights to account for the heterogeneity among people. Therefore, this model fully insures agents against all diversifiable risks.

One extension of the benchmark model focuses on understanding how a given risk-sharing network is formed. If maintaining links between agents entails a social cost, then the efficient network is one which satisfies full risk sharing and in which every individual forms connections in a cost-minimizing way. It is possible that the equilibrium network is different from the efficient one. The equilibrium network is said to be stable [96] if no agent wants to deviate from the existing network by removing one of his links. Reference [36] give an alternate definition of stability by requiring that no pair of agents should profit by creating a link between themselves. Stable networks are usually smaller than the efficient network as individual agents do not consider the positive externality of a better diversification of risk in the network when they form links.

Another possible network structure is the bargaining model studied in Ref. [11]. In this model, agents with existing links can renegotiate between themselves by threatening to break the connection if their demands are not met. In case the connection is broken, it would have a significant adverse impact on the agent who is less well connected in the network, thus reducing his risk-sharing prospects. Therefore well-connected individuals are in a better position to negotiate and get a higher surplus. This implies that individuals tend to over-invest in the formation of links, to allow for better prospects of renegotiation with others. As costs of link formation increase, the star network is the only stable network structure left, where one central individual is connected to all other individuals, and no other link is present. Therefore high costs would imply high inequality in relationship patterns as well.

A different extension of the benchmark model relaxes the commitment assumption and looks at limited commitment risk sharing. Reference [33] study such a model, where the network is given exogenously. In this model the same network structure is used for risk sharing as well as the transmission of information about the deviation of agents. They find that the stability of the network first decreases, and then increases with the density of the network. In sparse networks, deviation is less likely as agents have few connections, so the breaking of a link leaves less opportunities for a person to undertake risk sharing. On the other hand, there are ample risk-sharing opportunities in dense networks, but information about deviations travels fast as well, leading to reduced incentives for an agent to deviate.

5 Econometric Modeling of Networks

In this section, we discuss the framework described in Ref. [63], a method developed by Diebold and Yilmaz in a series of articles to study the interdependence between multivariate time series.

5.1 Variance Decomposition and Connectedness Measures

The Diebold–Yilmaz (DY) approach tries to measure volatility spillovers in the economy given by the impact of an idiosyncratic shock to a firm on the rest of the firms. The main question that this framework seeks to answer is this: *How much of an entity i's future uncertainty at horizon H can be explained by shocks arising with entity j?* The foundation of this approach lies in the concept of Variance Decomposition. Given a Vector Autoregression (VAR) model, the variance decomposition matrix measures the proportion of forecast error variance explained by idiosyncratic shocks to other variables.

Formally, suppose we define an N-variable p lag VAR model as

$$x_t = \sum_{i=1}^{p} \phi_i x_{t-i} + \epsilon_t, \tag{18}$$

where $\epsilon_t \sim (0, \Sigma)$. The equivalent moving average representation of this model is given by $x_t = \sum_{i=0}^{\infty} A_i \epsilon_{t-i}$. It is assumed that the matrices satisfy the recursive relationship: $A_i = \phi_1 A_{i-1} + \phi_2 A_{i-2} + \cdots + \phi_p A_{i-p}$, where A_0 is an identity matrix.

In general, the shocks to entities in the economy can be correlated. To account for the correlation while calculating the Variance Decomposition matrix, the DY approach discusses two methods which allow us to work with correlated shocks. Reference [61] uses Cholesky Factorization to orthogonalize the shocks. One particular disadvantage of this approach is that it may give different results if the ordering of variables changes. To deal with this issue, later papers use the Generalized Variance Decomposition (GVD) framework for orthogonalization. This approach accounts for correlated shocks, assuming that the shocks follow a normal distribution. The GVD approach is described as follows: Suppose an element in the ith row and jth column of the Variance Decomposition matrix is given by $\theta_{ij}^{g}(H)$, where H specifies the horizon for which the forecast is made. Following the GVD approach, the Variance Decomposition matrix is given by

$$\theta_{ij}^{g}(H) = \frac{\sigma_{jj^{-1}} \sum_{h=0}^{H-1} (e_i' A_h \Sigma e_j)^2}{\sum_{h=0}^{H-1} (e_i' A_h \Sigma A_h' e_j)}, \; H = 1, 2, \ldots \tag{19}$$

where Σ denotes the covariance matrix of ϵ, the standard deviation of the disturbance in the jth equation is given by σ_{jj}, and e_i is a vector of zeros with a one in the ith entry.

Usually, $\sum_{j=1}^{N} \theta_{ij}^{g}(H) \neq 1$. So each entry is normalized by the row sum to determine pairwise directional connectedness from firm j to firm i:

$$\tilde{\theta}_{ij}^{g}(H) = \frac{\theta_{ij}^{g}(H)}{\sum_{j=1}^{N} \theta_{ij}^{g}(H)}. \tag{20}$$

Let $\tilde{\theta}_{ij}^{g}(H)$ be written as $C_{i \leftarrow j}^{H}$. Then we say that $C_{i \leftarrow j}^{H}$ gives us the *pairwise directional connectedness* from firm j to firm i. Additionally, the value $C_{ij}^{H} = C_{i \leftarrow j}^{H} - C_{j \leftarrow i}^{H}$ gives us the *net* pairwise directional connectedness between i and j. Henceforth, we call the Variance Decomposition matrix as the Connectedness matrix. The Connectedness matrix allows us to answer other relevant questions as well. Suppose we wanted to know the impact of exogenous shocks to other firms on firm i's forecast error variance. This can be calculated from the Connectedness matrix by adding all non-diagonal entries in the ith row of the matrix, which gives us the *Total directional connectedness* to firm i from all other firms j

$$C_{i \leftarrow \cdot}^{H} = \frac{\sum_{\substack{j=1 \\ j \neq i}}^{N} \tilde{\theta}_{ij}^{g}(H)}{\sum_{i,j=1}^{N} \tilde{\theta}_{ij}^{g}(H)} = \frac{\sum_{\substack{j=1 \\ j \neq i}}^{N} \tilde{\theta}_{ij}^{g}(H)}{N}. \tag{21}$$

On the other hand, if we wanted to calculate the impact of an exogenous shock on i to other firms in the economy, we can take the sum of all non-diagonal entries in the ith column of the Connectedness matrix. This gives us the *Total directional connectedness* from firm i to all other firms j:

$$C_{\cdot \leftarrow i}^{H} = \frac{\sum_{\substack{j=1 \\ j \neq i}}^{N} \tilde{\theta}_{ji}^{g}(H)}{\sum_{i,j=1}^{N} \tilde{\theta}_{ji}^{g}(H)} = \frac{\sum_{\substack{j=1 \\ j \neq i}}^{N} \tilde{\theta}_{ji}^{g}(H)}{N}. \tag{22}$$

As before, *net* Total directional connectedness is given by $C_{i}^{H} = C_{i \leftarrow \cdot}^{H} - C_{\cdot \leftarrow i}^{H}$. Finally, the *Total Connectedness* can be calculated as

$$C^{H} = \frac{\sum_{\substack{i,j=1 \\ i \neq j}}^{N} \tilde{\theta}_{ij}^{g}(H)}{\sum_{i,j=1}^{N} \tilde{\theta}_{ij}^{g}(H)} = \frac{\sum_{\substack{i,j=1 \\ i \neq j}}^{N} \tilde{\theta}_{ij}^{g}(H)}{N}. \tag{23}$$

Total Connectedness can be calculated as the ratio of the sum of the non-diagonal entries of the connectedness matrix to the sum of all the entries of the matrix. The Variance Decomposition matrix is a useful tool to analyze the impact of shocks on the entities in an economy. It allows us to measure not only the impact of an entity's own shock, but also the spillover from a shock affecting some other entity in the economy. Similarly, the Variance Decomposition matrix also helps us to estimate the transmission of shocks between a firm and the rest of the economy as whole. Moreover, it gives us the degree of connectedness in the economy, which can be very useful for policymakers.

5.1.1 Variance Decomposition Matrices as Networks

The Connectedness Matrix is a network adjacency matrix with some modifications. First, the elements of the Connectedness matrix are not restricted to 0 or 1, but instead can take any value between these 2 numbers. This implies that the links are *weighted*, i.e., they show the strength of the bonds between two entities. Secondly, the matrix is *directed*. Lastly, the entries of the Connectedness matrix are *dynamic*, so that they may change over time. The observation that the Connectedness matrix can be defined as a network means that the total directional connectedness measures are equivalent to node in-degree and out-degree. Similarly, total connectedness is given by the mean degree of the network.

5.2 *Empirical Results*

In this section we will discuss a few recent applications of the Diebold–Yilmaz approach. For each application, we will state the dataset used, explain the methodology and then discuss the important results.

5.2.1 Global Bank Networks

Reference [57] uses data for 96 banks across 29 countries provided by Thomson–Reuters for the period September 12, 2003–February 7, 2014. These banks are among the world's largest 150 banks (by assets) which were publicly traded during the given time period. These banks include all the "globally systemically important banks" which were publicly traded in this time period.

To measure volatility from returns, daily range-based realized volatility is calculated using the data on stock returns. Using the methodology introduced by Ref. [81], this is measured as

$$\hat{\sigma}_{it}^2 = 0.511(H_{it} - L_{it})^2 - 0.019[(C_{it} - O_{it})(H_{it} + L_{it} - 2O_{it}) - 2(H_{it} - O_{it})(L_{it} - O_{it})] - 0.383(C_{it} - O_{it})^2, \tag{24}$$

where H_{it}, L_{it}, O_{it} and C_{it} are the log values of daily high, low, opening, and closing prices for bank stock i on day t. The high dimensionality of global bank networks can lead to difficulties in estimating the connectedness in these networks. To mitigate this problem, the paper uses LASSO methods [138] which allows for both shrinkage and selection of variables, thus reducing the dimensionality of the problem. Formally, a penalized estimation problem is given as

$$\hat{\beta} = \underset{\beta}{\arg\min}\left[\sum_{t=1}^{T}\left(y_t - \sum_i \beta_i x_{it}\right)^2 + \lambda\sum_{i=1}^{K}|\beta_i|^q\right]. \tag{25}$$

This problem puts a penalty on the calculated β, depending on the value of q. For LASSO methods, $q = 1$, which leads to both selection and shrinkage of parameters. This paper uses a variant of the LASSO, called the adaptive elastic net [148]. This method has the "oracle property", which means that the generated model is consistent for the best KullbackLiebler approximation to the true data generating process. It is given as

$$\hat{\beta}_{AENet} = \underset{\beta}{\arg\min}\left[\sum_{t=1}^{T}\left(y_t - \sum_i \beta_i x_{it}\right)^2 + \lambda\sum_{i=1}^{K} w_i\left(\frac{1}{2}|\beta_i| + \frac{1}{2}\beta_i^2\right)\right], \tag{26}$$

where $w_i = 1/|\hat{\beta}_{i,OLS}|$ and λ is selected using tenfold cross validation. The weights allow the shrinking of the smallest OLS coefficients toward 0. The paper then proceeds as follows. The adaptive elastic net method is used to estimate the VAR model for log volatilities at horizon $H = 10$. This provides us with the Variance Decomposition matrix through which the different connectedness measures are calculated.

We first discuss the full sample static analysis. The main result is that the network graph shows strong clusters within and across countries. This is an important observation as it is not entirely obvious that location (rather than other factors like bank size) would be a dominant factor in the formation of the network. Another important observation from the Connectedness matrix is that North America and Europe are net transmitters of future volatility uncertainty. Moreover, looking at the country bank network (where each node denotes a particular country), it is seen that USA is highly connected, with links showing a strong connection from USA to Canada, Australia, and UK.

For dynamic analysis, the paper uses rolling estimation, analyzing the data for a 150-day window at a time. Analyzing the impact of the collapse of Lehman Brothers on the US banks, we see a sharp increase in the connectedness of these banks with others. This could explain the global spread of volatility, leading to a crisis worldwide. A similar observation is seen for the European Debt Crisis in 2011. Taking the static analysis network as a benchmark, the network for October 7, 2011 shows a marked difference. This network is much more tightly clustered, indicating a rise in volatility connectedness compared to the full sample benchmark.

The paper then discusses system-wide connectedness. First, we decompose the connectedness measures into its trend and cyclical variation. The cycles show a sharp increase in connectedness during the 2008 Recession and 2011 European Debt Crisis. The trend line first increases, hitting its peak during the Lehman bankruptcy, and then decreases, although at a slower rate. Alternatively, decomposing the connectedness measures into cross-country and within-country variation, it is observed that cross-country variation dominates the movements in system-wide connectedness. The authors also try to observe the relation between bank size and eigenvector centrality. Using a rank regression, it is seen that bank eigenvector centrality is highly correlated with bank size. But this relation weakens during the 2008 financial crisis and 2011 European debt crisis. This implies that during bad times, smaller banks can become central to the network, leading to idiosyncratic volatilities generating system-wide fluctuations.

5.2.2 Equity Volatility Network

Reference [64] uses stock return data for 35 major financial institutions for the time period January 2004–June 2014. 17 financial institutions are from USA and include 7 commercial banks, 2 investment banks, and 1 credit card company. The other institutions are those which were either acquired, went bankrupt, or were taken in government custody after the 2008 Financial Crisis. Commercial banks located in Europe form the rest of the sample. First, the full sample analysis shows the formation of 2 clusters based on whether the banks are located in USA or Europe. This means that location has an important role to play in volatility transmissions between banks. At the national level, the highest pairwise connectedness is observed between USA and UK, probably because these countries are home to the London Stock Exchange and the New York Stock Exchange. Comparing net connectedness measures, Belgium has the highest negative net connectedness, while France and USA have the highest positive net connectedness.

For dynamic analysis, a 200-day rolling sample was used for estimating connectedness measures. Plotting the values for total connectedness over the given time period, we see sharp increases in connectedness during the 2008 Financial Crisis and 2011 European Debt Crisis. The most important result of this paper is based on Total Directional Connectedness. Before Lehman Brothers collapsed, US financial institutions were net transmitters of volatility to European financial institutions. After a full-blown global crisis emerged due to the collapse of Lehman Brothers, net connectedness from USA to Europe declined. Finally, net connectedness from US to Europe went below zero as the European Debt Crisis intensified. The paper further analyses connectedness at the country and institution level, specifically comparing volatility transmissions to and from nodes at different periods of time.

5.2.3 Commodity Network

Reference [60] studies 19 different sub-indices provided by the Bloomberg Commodity Price Index. These sub-indices are for the following commodities: precious metals (silver, gold), livestock commodities (lean hogs, live cattle), energy (natural gas, heating oil, unleaded gasoline, crude oil), grains (soybeans, wheat, soybean oil, corn), industrial metals (nickel, copper, zinc, aluminum) and "softs" (sugar, cotton, coffee). Data is collected daily (except for weekends and holidays) from May 11, 2006 to January 25, 2016. Ordering the commodities from largest to smallest (first according to to-degrees and then according to from-degrees), it is observed that the rankings are almost similar, implying that the commodities transmitting significant volatility to others also receive significant volatility from others. The network itself shows low system-wide connectedness, but clusters are formed according to the traditional industry groupings. These clusters show high within-group connectedness. At the industry level, industrial metals, energy, and precious metals are close together, and energy has a high value of total directional connectedness to the other groupings in the network.

A dynamic analysis of connectedness measures is also conducted using a rolling window of 150 days. It is observed that commodity return volatilities have a lower connectedness compared to global stock market returns, global bank returns, and bond yield volatilities. There is a sharp rise in total connectedness during the 2008 Financial Crisis. This also led to a fall in commodity prices till 2009. Post—2009, connectedness dropped as markets recovered. It is also seen that system-wide connectedness was heavily affected by oil price volatility. This volatility was largely due to demand and supply shocks over the world. The paper also analyzes the total directional connectedness of each commodity over the specified time period. It is again confirmed that energy commodities (especially oil) are major transmitters of volatilities to others.

5.2.4 Global Business Cycle Network

Reference [62] studies business cycle connectedness using monthly seasonally—adjusted industrial production (IP) for G-7 countries excluding Canada. the time period for which data is considered is January 1958–December 2011. Firstly, the data is tested for any possible cointegration. Testing for unit roots using augmented Dickey–Fuller tests, there was no evidence against the unit root in any log IP series, and substantial evidence against the unit root in every differenced log IP series. To test for cointegration status, Johansen's maximum eigenvalue and trace tests are conducted. These show the possibility of at most one cointegrating relationship among the IP series. Therefore, a vector error-correction (VEC) model is used for approximation purposes.

Calculating the connectedness measures from the VEC model, the total connectedness is found to be 29.1 percent. Japan and USA are found to be the largest net transmitters of industrial production shocks. Similarly, Italy and Germany are the largest recipients of business cycle shocks. Using 5-year rolling windows to analyze dynamic connectedness, the paper shows that almost all US recessions were related to an increase in connectedness. The same is observed for recessions in Germany, France, Japan, and Italy ending in 1993–94. Late 1980s onwards, the rise of globalization led to an increase in connectivity among all countries. This also led to the upward movement of the band within which the connectedness index fluctuates. During this period, each subsequent cycle was longer and had a larger bandwidth than the previous one, showing that the business cycles have become more synchronized due to globalization. This behavior culminated with a sharp rise in connectedness during the 2008 Financial Crisis. Finally, it is shown that trade balance can be used to determine if a country is a net transmitter or receiver of business cycle volatility. If a country has a trade surplus, then it would have a tendency to be a net receiver of shocks. On the other hand, countries with trade deficits are more likely to be net transmitters.

5.3 Ripples on Financial Networks

Reference [102] analyzes the impact a volatility shock may have across a financial network by providing an algorithm characterizing ripples on financial networks. In the discussion below, we briefly discuss the algorithm. The paper uses data for the largest $N = 100$ stocks based on market capitalization at the New York Stock Exchange over the time period 2002–2017. The stocks were selected so that data was available for them throughout the given time period. The paper divides this period into four equal length intervals—2002–05, 2006–09, 2010–13, and 2014–17. During the first window, the US economy was experiencing a boom, the second window was marked by the 2008 financial crisis, the third period contains the phase of recovery after the crash, and the fourth period was marked by a period of relative stability.

The paper then constructs the return series using the first difference of log price series for each stock. To construct the conditional volatility series from returns, the paper uses the GARCH framework. The paper then aims to construct the maximally connected component of the network. Using the adaptive Lasso technique developed by Ref. [148], the maximally connected component is constructed by removing those stocks for which in-degree and out-degree is less than 10 percent. Hierarchical networks are then constructed using sample correlation matrices calculated using return and volatility series. Since correlations can be negative, distance matrices are calculated instead using the metric $d_{ij} = \sqrt{2(1 - \rho_{ij}}$ discussed by Ref. [109], where ρ denotes the correlation between firms i and j. Minimal Spanning Trees are filtered out from the network to provide maximum information from a minimal sized network. Using eigenvector centrality as an exogeneity criterion over the return correlation matrix, the stocks are ordered so that Cholesky decomposition can be used to derive orthogonalized impulse response functions from a VAR model for the volatility series. These impulse response functions are then used to study shocks across the network.

This approach is different from the Diebold–Yilmaz approach where they use Generalized Variance Decomposition (GVD) instead. Since GVD requires the assumption of normality, it could be a strong and incorrect assumption in many settings. Therefore, this paper gives an alternative approach to analyze spillovers.

6 Non-parametric Approaches

Finally, we briefly summarize findings from some recent works on non-parametric approaches to economic and financial networks that have mainly originated from the "econophysics" literature [35, 47, 109, 136].

6.1 Correlation-Based Networks

In order to gain insight about the co-movements among price returns in a stock market, correlation-based networks are constructed from the empirical correlation matrix. Such networks provide a visual representation of the co-movements as well as information about the underlying market dynamics [126]. By continuously monitoring the structure of the correlation-based network, one can find different patterns that appear time and again, and reveal the underlying trends in the system. Multiple methods have been proposed to construct networks [13, 142–144] from the empirical correlation matrices, such as the minimum spanning tree [34, 46, 103, 117, 119–122, 139], planar maximally filtered graph [141], threshold network [49], etc.

To initiate the discussion, below we describe the network construction algorithm and a standard filtering method. Readers are requested to consult [123] for a very nice detailed exposition of this methodology. Consider N number of daily return series being constructed from N asset prices for T days: $r_{it} = \log(p_{it}) - \log p_{i,t-1}$ for the ith series where $i = 1, \ldots, N$ and tth day where $t = 1, \ldots, T$. From these N number of return series, one can construct a correlation matrix of size $N \times N$ that we denote by $\Sigma_{N\times N}$ where σ_{ij} is the correlation coefficient between assets i and j. One can conduct an eigendecomposition of this typically large-dimensional matrix to analyze the corresponding eigenspectra:

$$\Sigma = \sum_{i=1}^{N} \lambda_i e_i e_i', \tag{27}$$

where λ_i denotes the ith eigenvalue and e_i is the corresponding eigenvector (e_i' is the transpose of e_i). Then the correlation matrix can be decomposed into three parts:

$$\Sigma = \Sigma^{market} + \Sigma^{group} + \Sigma^{random}, \tag{28}$$

where the *market* mode corresponds to the top eigenvalue, *group* mode corresponds to all deviating eigenvalues except the top one, and the *random* mode corresponds to the remaining eigenvalues. A natural question arises as to how to find the *deviating* eigenvalues? The method popularized by [126] is to apply Marcenko–Pastur distribution to decide the cut-off. Essentially, all eigenvalues above the cut-off given by Marchenko–Pastur distribution are statistically significant and can be taken as *deviating eigenvalues*. In practice, the top eigenvalue seems to capture the market dynamics quite well, and the ones in the *group* mode seems to represent sectoral dynamics [123]. However, some recent work shows that one needs to look further deep into the core-periphery structure of the implied networks for sectoral dynamics [104].

An important consideration in computational finance is the time period over which the computation of empirical cross-correlation matrix takes place. In general, incorrect choices of time periods could lead to non-stationarity issues or too much noise in the correlation matrices. Reference [124] applies random matrix theory in financial markets to address this problem. Random matrix theory is used to analyze the eigenvalues derived from random matrices, and had its original application in nuclear physics. Pharasi et al. [124] use the power mapping method where short epoch correlation matrices are subjected to non-linear distortions. Following the literature [126], this paper also conducts an eigenvalue decomposition of the empirical cross-correlation matrix. Resultant modes can be classified as the market mode, group mode, and random mode; the bulk of the eigenvalues constitute the random modes and is described by a Marcenko–Pastur distribution. In another paper, Pharasi et al. [125] utilized random matrix theory to find correlation patterns that may emerge during times of crisis vis-a-vis relatively stable periods. They attempted to categorize different "market states" and to find evidence for long-term precursors to the market crashes (see also, [104]).

While most research on network properties typically focus on individual networks in isolation from the rest of the world, there are many large-scale networks which show interdependence [55, 146]. In [146], the authors analyzed the foreign exchange and stock market networks for 48 countries based on complex Hilbert principal component analysis to quantify lead-lag relationships across the markets. They also constructed a coupled synchronization network to identify the formation of stable network communities.

6.2 Mesoscopic Networks

This paper [134] analyzes the economy at the mesoscopic (sectoral) level. An important finding of this paper is that the core of the return networks mainly consists of sectors of the economy which are relatively large. On the other hand, the periphery of such networks mostly consists of sectors which are relatively smaller in size. This observation hints at a connection between sector-level nominal return dynamics and the real size effect. Data for sectoral price indices collected for 65 sectors across

27 countries is analyzed over the time periods: Jan. 08–Dec. 09, Oct. 12–Sept. 13, and Oct. 14–Sept. 16. Data for the real variables (such as number of employees in each sector, revenue , and market capitalization) are available at the company level. Therefore, these are aggregated to get sectoral level data.

Return series is constructed using the first difference of log price series for each sector. The return series is first used to calculate the pairwise Pearson correlation coefficients which are then used to construct the distance matrix using the transformation $d_{ij} = \sqrt{2(1 - \rho_{ij})}$, where ρ denotes the correlation between firms i and j. Clustering algorithms of Multi-Dimensional Scaling (MDS) and Minimum Spanning Tree (MST) are used to study the network structure in all the countries considered in the sample. Both methods indicate that the network is in the form of a core-periphery structure.

This paper shows that the structure of the network derived from the return correlation matrix has a robust relationship with the measures of sectoral size. For this exercise, the eigenvector centrality is regressed on sized, where size is defined by either market capitalization, revenue, or employment, aggregated across all firms within a sector. On analyzing the results from twenty-seven countries, there is an indication that the variation in the dispersion of sectoral centralities in the sectoral return a correlation matrix can be explained by the dispersion in economic size.

6.3 *Multi-layered Economic and Financial Networks*

Reference [133] studies the empirical connections between financial networks and macroeconomic networks using the concept of multi-layered networks. This paper finds that the different network structures considered here take the form of a core-periphery structure, where the core consists of similar countries in each network. The paper also shows that if a country has high trade connectivity, it is more likely to have higher financial return correlations as well.

Moreover, the paper shows that the Economic Complexity Index is positively related to the equity markets. To reveal the dynamics and structure of the global market indices, the paper studies minimum spanning tree. It is observed that geographical proximity is an important factor in determining the correlation structure across different markets.

7 Concluding Remarks

In this article, we have reviewed a number of different approaches to describe, analyze, and study economic and financial networks. Future developments in the digital economy will usher in further interconnectedness in our existing economic system, leading to creative destructions and disruptions in the economic and financial networks along with new forms of networks being formed. Probably, theory of networks is going to take the center stage in economic analysis in such a world.

References

1. Acemoglu, D., U. Akcigit, and W. Kerr. 2016. Networks and the macroeconomy: An empirical exploration. *NBER Macroeconomics Annual* 30 (1): 273–335.
2. Acemoglu, D., and P.D. Azar. 2020. Endogenous production networks. *Econometrica* 88 (1): 33–82.
3. Acemoglu, D., V.M. Carvalho, A. Ozdaglar, and A. Tahbaz-Salehi. 2012. The network origins of aggregate fluctuations. *Econometrica* 80 (5): 1977–2016.
4. Acemoglu, D., A. Ozdaglar, and A. Tahbaz-Salehi. 2015. Systemic risk and stability in financial networks. *American Economic Review* 105 (2): 564–608.
5. Acharya, V., R. Engle, and M. Richardson. 2012. Capital shortfall: A new approach to ranking and regulating systemic risks. *American Economic Review* 102 (3): 59–64.
6. Acharya, V.V., and O. Merrouche. 2013. Precautionary hoarding of liquidity and interbank markets: Evidence from the subprime crisis. *Review of Finance* 17 (1): 107–160.
7. Acharya, V.V., L.H. Pedersen, T. Philippon, and M. Richardson. 2017. Measuring systemic risk. *The Review of Financial Studies* 30 (1): 2–47.
8. Adrian, T., and M.K. Brunnermeier. 2011. Covar. : Technical Report. National Bureau of Economic Research.
9. Alatas, V., A.G. Chandrasekhar, M. Mobius, B.A. Olken, and C. Paladines. 2019. When celebrities speak: A nationwide twitter experiment promoting vaccination in Indonesia. Technical Report. National Bureau of Economic Research.
10. Allen, F., and D. Gale. 2000. Financial contagion. *Journal of Political Economy* 108 (1): 1–33.
11. Ambrus, A., and M. Elliott. 2015. Investments in social ties, risk sharing and inequality. Economic Research Initiatives at Duke (ERID) Working Paper (179).
12. Aral, S., and D. Walker. 2011. Creating social contagion through viral product design: A randomized trial of peer influence in networks. *Management Science* 57 (9): 1623–1639.
13. Aste, T., T. Di Matteo, M. Tumminello, and R.N. Mantegna. 2005. Correlation filtering in financial time series. In *SPIE Noise and fluctuations in econophysics and finance*, vol. 5848, ed. D. Abbott, J.P. Bouchaud, X. Gabaix, and J.L. McCauley, 100–109. Bellingham: International Society for Optics and Photonics.
14. Atalay, E., A. Hortacsu, J. Roberts, and C. Syverson. 2011. Network structure of production. *Proceedings of the National Academy of Sciences* 108 (13): 5199–5202.
15. Bailey, M., R. Cao, T. Kuchler, and J. Stroebel. 2018. The economic effects of social networks: Evidence from the housing market. *Journal of Political Economy* 126 (6): 2224–2276.
16. Banerjee, A., E. Breza, A.G. Chandrasekhar, and M. Mobius. 2019. Naive learning with uninformed agents. Technical Report. National Bureau of Economic Research.
17. Banerjee, A., A.G. Chandrasekhar, E. Duflo, and M.O. Jackson. 2013. The diffusion of microfinance. *Science* 341 (6144): 1236498.
18. Banerjee, A., A.G. Chandrasekhar, E. Duflo, and M.O. Jackson. 2019. Using gossips to spread information: Theory and evidence from two randomized controlled trials. *The Review of Economic Studies* 86 (6): 2453–2490.
19. Banerjee, A.V., and E. Duflo. 2007. The economic lives of the poor. *Journal of Economic Perspectives* 21 (1): 141–168.
20. Banfi, S., S. Choi, and B. Villena-Roldán. 2019. Deconstructing job search behavior. Available at SSRN 3323545.
21. Baqaee, D.R., and E. Farhi. 2018. Macroeconomics with heterogeneous agents and input-output networks. Technical Report. National Bureau of Economic Research.

22. Baqaee, D.R., and E. Farhi. 2019. The macroeconomic impact of microeconomic shocks: Beyond hulten's theorem. *Econometrica* 87 (4): 1155–1203.
23. Baqaee, D.R., and E. Farhi. 2020. Productivity and misallocation in general equilibrium. *The Quarterly Journal of Economics* 135 (1): 105–163.
24. Battiston, S., D.D. Gatti, M. Gallegati, B. Greenwald, and J.E. Stiglitz. 2012. Liaisons dangereuses: Increasing connectivity, risk sharing, and systemic risk. *Journal of Economic Dynamics and Control* 36 (8): 1121–1141.
25. Battiston, S., M. Puliga, R. Kaushik, P. Tasca, and G. Caldarelli. 2012. Debtrank: Too central to fail? Financial networks, the fed and systemic risk. *Scientific Reports* 2: 541.
26. Bayer, P., S.L. Ross, and G. Topa. 2008. Place of work and place of residence: Informal hiring networks and labor market outcomes. *Journal of Political Economy* 116 (6): 1150–1196.
27. Beaman, L., N. Keleher, and J. Magruder. 2018. Do job networks disadvantage women? Evidence from a recruitment experiment in malawi. *Journal of Labor Economics* 36 (1): 121–157.
28. Benguria, F. 2015. The matching and sorting of exporting and importing firms: Theory and evidence. Available at SSRN 2638925.
29. BenYishay, A., and A.M. Mobarak. 2019. Social learning and incentives for experimentation and communication. *The Review of Economic Studies* 86 (3): 976–1009.
30. Bernard, A.B., and A. Moxnes. 2018. Networks and trade. *Annual Review of Economics* 10: 65–85.
31. Bernard, A.B., A. Moxnes, and K.H. Ulltveit-Moe. 2018. Two-sided heterogeneity and trade. *Review of Economics and Statistics* 100 (3): 424–439.
32. Bigio, S., and J. La'O. 2016. Distortions in production networks. Technical Report National Bureau of Economic Research.
33. Bloch, F., G. Genicot, and D. Ray. 2008. Informal insurance in social networks. *Journal of Economic Theory* 143 (1): 36–58.
34. Bonanno, G., G. Caldarelli, F. Lillo, and R.N. Mantegna. 2003. Topology of correlation-based minimal spanning trees in real and model markets. Physical Review E 68: 046130.
35. Bouchaud, J.P., and M. Potters. 2003. *Theory of financial risk and derivative pricing: From statistical physics to risk management*. Cambridge: Cambridge University Press.
36. Bramoullé, Y., and R. Kranton. 2007. Risk-sharing networks. *Journal of Economic Behavior & Organization* 64 (3–4): 275–294.
37. Breza, E., A. Chandrasekhar, B. Golub, and A. Parvathaneni. 2019. Networks in economic development. *Oxford Review of Economic Policy* 35 (4): 678–721.
38. Brown, M., E. Setren, and G. Topa. 2016. Do informal referrals lead to better matches? Evidence from a firm's employee referral system. *Journal of Labor Economics* 34 (1): 161–209.
39. Brownlees, C., and R.F. Engle. 2017. Srisk: A conditional capital shortfall measure of systemic risk. *The Review of Financial Studies* 30 (1): 48–79.
40. Caballero, R.J., and A. Simsek. 2013. Fire sales in a model of complexity. *The Journal of Finance* 68 (6): 2549–2587.
41. Carvalho, V., and X. Gabaix. 2013. The great diversification and its undoing. *American Economic Review* 103 (5): 1697–1727.
42. Carvalho, V.M., M. Nirei, Y. Saito, and A. Tahbaz-Salehi. 2016. Supply chain disruptions: Evidence from the great east japan earthquake. Columbia Business School Research Paper (17-5)
43. Carvalho, V.M., and A. Tahbaz-Salehi. 2019. Production networks: A primer. *Annual Review of Economics* 11: 635–663.
44. Carvalho, V.M., and N. Voigtländer. 2014. Input diffusion and the evolution of production networks. Technical Report. National Bureau of Economic Research.
45. Casella, A., and N. Hanaki. 2008. Information channels in labor markets: On the resilience of referral hiring. *Journal of Economic Behavior & Organization* 66 (3–4): 492–513.
46. Chakraborti, A. 2006. An outlook on correlations in stock prices. In *Econophysics of Stock and other Markets*, 13–23. Berlin: Springer.

47. Chakraborti, A., I. Muni Toke, M. Patriarca, and F. Abergel. 2011. Econophysics review: I. Empirical facts. *Quantitative Finance* 11 (7): 991–1012.
48. Chaney, T. 2014. The network structure of international trade. *American Economic Review* 104 (11): 3600–3634.
49. Chi, K.T., J. Liu, and F.C. Lau. 2010. A network perspective of the stock market. *Journal of Empirical Finance* 17 (4): 659–667.
50. Collins, D., J. Morduch, S. Rutherford, and O. Ruthven. 2009. Portfolios of the poor: How the world's poor live on $2 a day. Princeton: Princeton University Press.
51. Cont, R., A. Moussa, and et al. 2010. Network structure and systemic risk in banking systems. Edson Bastos e, Network Structure and Systemic Risk in Banking Systems (December 1, 2010).
52. Cont, R., and E. Schaanning. 2017. Fire sales, indirect contagion and systemic stress testing. Indirect Contagion and Systemic Stress Testing (June 13, 2017).
53. Covi, G., M. Montagna, and G. Torri. 2019. On the origins of systemic risk. Technical Report. European Central Bank Working Paper, forthcoming.
54. Craig, B., and G. Von Peter. 2014. Interbank tiering and money center banks. *Journal of Financial Intermediation* 23 (3): 322–347.
55. Curme, C., H.E. Stanley, and I. Vodenska. 2015. Coupled network approach to predictability of financial market returns and news sentiments. *International Journal of Theoretical and Applied Finance* 18 (07): 1550043.
56. DeMarzo, P.M., D. Vayanos, and J. Zwiebel. 2003. Persuasion bias, social influence, and unidimensional opinions. *The Quarterly journal of economics* 118 (3): 909–968.
57. Demirer, M., F.X. Diebold, L. Liu, and K. Yilmaz. 2018. Estimating global bank network connectedness. *Journal of Applied Econometrics* 33 (1): 1–15.
58. Diamond, D.W., and P.H. Dybvig. 1983. Bank runs, deposit insurance, and liquidity. *Journal of Political Economy* 91 (3): 401–419.
59. Diamond, D.W., and R.G. Rajan. 2011. Fear of fire sales, illiquidity seeking, and credit freezes. *The Quarterly Journal of Economics* 126 (2): 557–591.
60. Diebold, F.X., L. Liu, and K. Yilmaz. 2017. Commodity connectedness. Technical Report. National Bureau of Economic Research.
61. Diebold, F.X., and K. Yilmaz. 2009. Measuring financial asset return and volatility spillovers, with application to global equity markets. *The Economic Journal* 119 (534): 158–171.
62. Diebold, F.X., and K. Yilmaz. 2013. Measuring the dynamics of global business cycle connectedness.
63. Diebold, F.X., and K. Yılmaz. 2015. *Financial and macroeconomic connectedness: A network approach to measurement and monitoring*. USA: Oxford University Press.
64. Diebold, F.X., and K. Yilmaz. 2015. Trans-atlantic equity volatility connectedness: Us and European financial institutions, 2004–2014. *Journal of Financial Econometrics* 14 (1): 81–127.
65. Dustmann, C., A. Glitz, and U. Schönberg. 2009. Job search networks and ethnic segregation in the workplace. University College London, Working Paper.
66. Easley, D., J. Kleinberg, et al. 2010. *Networks, crowds, and markets*, vol. 8. Cambridge: Cambridge University Press.
67. Eaton, J., M. Eslava, C.J. Krizan, M. Kugler, and J. Tybout. 2014. A search and learning model of export dynamics. Unpublished manuscript.
68. Eaton, J., D. Jinkins, J. Tybout, and D. Xu. 2016. Two-sided search in international markets. In *2016 Annual Meeting of the Society for Economic Dynamics*.
69. Eaton, J., S. Kortum, F. Kramarz, and et al. 2015. Firm-to-firm trade: Imports, exports, and the labor market. Brown University, unpublished manuscript.
70. Eisenberg, L., and T.H. Noe. 2001. Systemic risk in financial systems. *Management Science* 47 (2): 236–249.
71. Farrell, J., and C. Shapiro. 1990. Horizontal mergers: An equilibrium analysis. *The American Economic Review* 107–126.

72. Fourel, V., J.C. Heam, D. Salakhova, and S. Tavolaro. 2013. Domino effects when banks hoard liquidity: The French network.
73. Gabaix, X. 2011. The granular origins of aggregate fluctuations. *Econometrica* 79 (3): 733–772.
74. Gai, P., A. Haldane, and S. Kapadia. 2011. Complexity, concentration and contagion. *Journal of Monetary Economics* 58 (5): 453–470.
75. Gai, P., and S. Kapadia. 2010. Contagion in financial networks. *Proceedings of the Royal Society A: Mathematical, Physical and Engineering Sciences* 466 (2120): 2401–2423.
76. Gai, P., and S. Kapadia. 2019. Networks and systemic risk in the financial system. *Oxford Review of Economic Policy* 35 (4): 586–613.
77. Galbiati, M., and K. Soramaki. 2010. Liquidity-saving mechanisms and bank behaviour.
78. Galenianos, M. 2012. Learning about match quality and the use of referrals. Available at SSRN 2053376.
79. Galeotti, A., B. Golub, and S. Goyal. 2017. Targeting interventions in networks. arXiv:1710.06026
80. Galeotti, A., and L.P. Merlino. 2014. Endogenous job contact networks. *International Economic Review* 55 (4): 1201–1226.
81. Garman, M.B., and M.J. Klass. 1980. On the estimation of security price volatilities from historical data. *Journal of business* 67–78.
82. Gavazza, A., S. Mongey, and G.L. Violante. 2018. Aggregate recruiting intensity. *American Economic Review* 108 (8): 2088–2127.
83. Gee, L.K., J. Jones, and M. Burke. 2017. Social networks and labor markets: How strong ties relate to job finding on facebook's social network. *Journal of Labor Economics* 35 (2): 485–518.
84. Golub, B., and M.O. Jackson. 2010. Naive learning in social networks and the wisdom of crowds. *American Economic Journal: Microeconomics* 2 (1): 112–49.
85. Golub, B., and M.O. Jackson. 2012. How homophily affects the speed of learning and best-response dynamics. *The Quarterly Journal of Economics* 127 (3): 1287–1338.
86. Goyal, S. 2012. *Connections: An introduction to the economics of networks*. Princeton: Princeton University Press.
87. Goyal, S. 2016. Networks and markets. Cambridge-INET Working Paper Series No: 2016/16. http://www.econ.cam.ac.uk/research-files/repec/cam/pdf/cwpe1652.pdf.
88. Grassi, B., and J. Sauvagnat. 2019. Production networks and economic policy. *Oxford Review of Economic Policy* 35 (4): 638–677.
89. Grassi, B., and et al. 2018. Io in io: Size, industrial organization, and the input-output network make a firm structurally important. Technical Report.
90. Heath, R. 2018. Why do firms hire using referrals? Evidence from bangladeshi garment factories. *Journal of Political Economy* 126 (4): 1691–1746.
91. Hellerstein, J.K., M.J. Kutzbach, and D. Neumark. 2014. Do labor market networks have an important spatial dimension? *Journal of Urban Economics* 79: 39–58.
92. Hellerstein, J.K., M.J. Kutzbach, and D. Neumark. 2015. Labor market networks and recovery from mass layoffs: Evidence from the great recession period. Technical Report. National Bureau of Economic Research.
93. Holzer, H.J. 1987. Informal job search and black youth unemployment. *The American Economic Review* 77 (3): 446–452.
94. Holzer, H.J. 1988. Search method use by unemployed youth. *Journal of Labor Economics* 6 (1): 1–20.
95. Hulten, C.R. 1978. Growth accounting with intermediate inputs. *The Review of Economic Studies* 45 (3): 511–518.
96. Jackson, M.O. 2010. *Social and economic networks*. Princeton: Princeton University Press.
97. Jackson, M.O. 2014. Networks in the understanding of economic behaviors. *Journal of Economic Perspectives* 28 (4): 3–22.
98. Jackson, M.O., and B.W. Rogers. 2007. Meeting strangers and friends of friends: How random are social networks? *American Economic Review* 97 (3): 890–915.

99. Kapadia, S., M. Drehmann, J. Elliott, and G. Sterne. 2012. Liquidity risk, cash flow constraints, and systemic feedbacks. In *Quantifying systemic risk*, 29–61. Chicago: University of Chicago Press.
100. Kiyotaki, N., and J. Moore. 2002. Balance-sheet contagion. *American Economic Review* 92 (2): 46–50.
101. Kugler, A.D. 2003. Employee referrals and efficiency wages. *Labour Economics* 10 (5): 531–556.
102. Kumar, S., A. Bansal, and A.S. Chakrabarti. 2019. Ripples on financial networks. IIMA Working Papers WP 2019-10-01, Indian Institute of Management Ahmedabad, Research and Publication Department. https://ideas.repec.org/p/iim/iimawp/14613.html.
103. Kumar, S., and N. Deo. 2012. Correlation and network analysis of global financial indices. *Physical Review E* 86 (2): 026101.
104. Kuyyamudi, C., A.S. Chakrabarti, and S. Sinha. 2019. Emergence of frustration signals systemic risk. *Physical Review E* 99 (5): 052306.
105. Kuzubas, T.U., and et al. 2009. Endogenous social networks in the labor market. Unpublished, Unpublished manuscript, University of Minnesota.
106. Langfield, S., Z. Liu, and T. Ota. 2014. Mapping the uk interbank system. *Journal of Banking & Finance* 45: 288–303.
107. Lim, K., and et al. 2017. Firm-to-firm trade in sticky production networks. In *2017 Meeting Papers*, vol. 280. Society for Economic Dynamics.
108. Long, J.B., Jr., and C.I. Plosser. 1983. Real business cycles. *Journal of Political Economy* 91 (1): 39–69.
109. Mantegna, R.N., and H.E. Stanley. 2007. *An introduction to econophysics: Correlations and complexity in finance*. Cambridge: Cambridge University Press.
110. Marinescu, I.E., and D. Skandalis. 2019. Unemployment insurance and job search behavior. Available at SSRN 3303367.
111. May, R.M., and N. Arinaminpathy. 2010. Systemic risk: The dynamics of model banking systems. *Journal of the Royal Society Interface* 7 (46): 823–838.
112. McAfee, R.P., and M.A. Williams. 1992. Horizontal mergers and antitrust policy. *The Journal of Industrial Economics* 181–187.
113. McPherson, J.M., and L. Smith-Lovin. 1987. Homophily in voluntary organizations: Status distance and the composition of face-to-face groups. *American Sociological Review* 370–379.
114. Melitz, M. 2003. The impact of trade on intra-industry reallocations and aggregate industry productivity. *Econometrica* 71 (6): 1695–1725. https://EconPapers.repec.org/RePEc:ecm:emetrp:v:71:y:2003:i:6:p:1695-1725.
115. Monarch, R., and T. Schmidt-Eisenlohr. 2017. Learning and the value of trade relationships. FRB International Finance Discussion Paper (1218).
116. Montgomery, J.D. 1991. Social networks and labor-market outcomes: Toward an economic analysis. *The American Economic Review* 81 (5): 1408–1418.
117. Nobi, A., S.E. Maeng, G.G. Ha, , J.W. Lee. 2013. Network topologies of financial market during the global financial crisis. arXiv:1307.6974.
118. Oberfield, E. 2018. A theory of input-output architecture. *Econometrica* 86 (2): 559–589.
119. Onnela, J.P., A. Chakraborti, K. Kaski, and J. Kertesz. 2003. Dynamic asset trees and black monday. *Physica A: Statistical Mechanics and its Applications* 324 (1): 247–252.
120. Onnela, J.P., A. Chakraborti, K. Kaski, J. Kertesz, and A. Kanto. 2003. Asset trees and asset graphs in financial markets. *Physica Scripta* 2003 (T106): 48.
121. Onnela, J.P., A. Chakraborti, K. Kaski, J. Kertesz, and A. Kanto. 2003. Dynamics of market correlations: Taxonomy and portfolio analysis. *Physical Review E* 68 (5): 056110.
122. Onnela, J.P., A. Chakraborti, K. Kaski, and J. Kertiész. 2002. Dynamic asset trees and portfolio analysis. *The European Physical Journal B-Condensed Matter and Complex Systems* 30 (3): 285–288.
123. Pan, R.K., and S. Sinha. 2007. Collective behavior of stock price movements in an emerging market. *Physical Review E* 76 (4): 046116.

124. Pharasi, H.K., K. Sharma, A. Chakraborti, and T.H. Seligman. 2019. Complex market dynamics in the light of random matrix theory. In *New perspectives and challenges in econophysics and sociophysics*, 13–34. Berlin: Springer.
125. Pharasi, H.K., K. Sharma, R. Chatterjee, A. Chakraborti, F. Leyvraz, and T.H. Seligman. 2018. Identifying long-term precursors of financial market crashes using correlation patterns. *New Journal of Physics* 20 (10): 103041.
126. Plerou, V., P. Gopikrishnan, B. Rosenow, L.N. Amaral, and H.E. Stanley. 2000. A random matrix theory approach to financial cross-correlations. *Physica A: Statistical Mechanics and its Applications* 287 (3): 374–382.
127. Rogers, L.C., and L.A. Veraart. 2013. Failure and rescue in an interbank network. *Management Science* 59 (4): 882–898.
128. Rosenzweig, M.R. 1988. Risk, implicit contracts and the family in rural areas of low-income countries. *The Economic Journal* 98 (393): 1148–1170.
129. Sadler, E. 2020. Diffusion games. *American Economic Review* 110 (1): 225–70.
130. Salant, S.W., S. Switzer, and R.J. Reynolds. 1983. Losses from horizontal merger: the effects of an exogenous change in industry structure on cournot-nash equilibrium. *The Quarterly Journal of Economics* 98 (2): 185–199.
131. Scaillet, O. 2004. Nonparametric estimation and sensitivity analysis of expected shortfall. *Mathematical Finance: An International Journal of Mathematics, Statistics and Financial Economics* 14 (1): 115–129.
132. Schmutte, I.M. 2015. Job referral networks and the determination of earnings in local labor markets. *Journal of Labor Economics* 33 (1): 1–32.
133. Sharma, K., A.S. Chakrabarti, and A. Chakraborti. 2019. Multi-layered network structure: Relationship between financial and macroeconomic dynamics. In *New perspectives and challenges in econophysics and sociophysics*, 117–131. Berlin: Springer.
134. Sharma, K., B. Gopalakrishnan, A.S. Chakrabarti, and A. Chakraborti. 2017. Financial fluctuations anchored to economic fundamentals: A mesoscopic network approach. *Scientific Reports* 7 (1): 1–11.
135. Simon, C.J., and J.T. Warner. 1992. Matchmaker, matchmaker: The effect of old boy networks on job match quality, earnings, and tenure. *Journal of Labor Economics* 10 (3): 306–330.
136. Sinha, S., A. Chatterjee, A. Chakraborti, and B.K. Chakrabarti. 2010. Econophysics: An introduction. New York: Wiley.
137. Tasca, P., and S. Battiston. 2016. Market procyclicality and systemic risk. *Quantitative Finance* 16 (8): 1219–1235.
138. Tibshirani, R. 1996. Regression shrinkage and selection via the lasso. *Journal of the Royal Statistical Society: Series B (Methodological)* 58 (1): 267–288.
139. Tilak, G., T. Széll, R. Chicheportiche, and A. Chakraborti. 2013. Study of statistical correlations in intraday and daily financial return time series. In *Econophysics of systemic risk and network dynamics*, 77–104. Berlin: Springer.
140. Topa, G. 2019. Social and spatial networks in labour markets. *Oxford Review of Economic Policy* 35 (4): 722–745.
141. Tumminello, M., T. Aste, T. Di Matteo, and R.N. Mantegna. 2005. A tool for filtering information in complex systems. *Proceedings of the National Academy of Sciences of the United States of America* 102 (30): 10421–10426.
142. Tumminello, M., C. Coronnello, F. Lillo, S. Micciche, and R.N. Mantegna. 2007. Spanning trees and bootstrap reliability estimation in correlation-based networks. *International Journal of Bifurcation and Chaos* 17 (07): 2319–2329.
143. Tumminello, M., T. Di Matteo, T. Aste, and R. Mantegna. 2007. Correlation based networks of equity returns sampled at different time horizons. *The European Physical Journal B* 55 (2): 209–217.
144. Tumminello, M., F. Lillo, and R.N. Mantegna. 2010. Correlation, hierarchies, and networks in financial markets. *Journal of Economic Behavior & Organization* 75 (1): 40–58.
145. Upper, C., and A. Worms. 2004. Estimating bilateral exposures in the german interbank market: Is there a danger of contagion? *European Economic Review* 48 (4): 827–849.

146. Vodenska, I., H. Aoyama, Y. Fujiwara, H. Iyetomi, and Y. Arai. 2016. Interdependencies and causalities in coupled financial networks. *PloS One* 11 (3).
147. Wilson, R. 1968. The theory of syndicates. *Econometrica: Journal of the Econometric Society* 119–132.
148. Zou, H., and H.H. Zhang. 2009. On the adaptive elastic-net with a diverging number of parameters. *Annals of Statistics* 37 (4): 1733.